BIG IDEAS
MATH®

Red Accelerated

A Common Core Curriculum

Ron Larson
Laurie Boswell

**BIG IDEAS
LEARNING®**

Erie, Pennsylvania
BigIdeasLearning.com

Big Ideas Learning, LLC
1762 Norcross Road
Erie, PA 16510-3838
USA

For product information and customer support, contact Big Ideas Learning
at **1-877-552-7766** or visit us at ***BigIdeasLearning.com***.

About the Cover
The cover images on the *Big Ideas Math* series illustrate the advancements in
aviation from the hot-air balloon to spacecraft. This progression symbolizes the
launch of a student's successful journey in mathematics. The sunrise in the
background is representative of the dawn of the Common Core era in math
education, while the cradle signifies the balanced instruction that is a pillar
of the *Big Ideas Math* series.

Printed in the U.S.A.

ISBN 13: 978-1-60840-505-3
ISBN 10: 1-60840-505-2

5 6 7 8 9 10 WEB 17 16 15 14

AUTHORS

Ron Larson is a professor of mathematics at Penn State Erie, The Behrend College, where he has taught since receiving his Ph.D. in mathematics from the University of Colorado. Dr. Larson is well known as the lead author of a comprehensive program for mathematics that spans middle school, high school, and college courses. His high school and Advanced Placement books are published by Holt McDougal. Ron's numerous professional activities keep him in constant touch with the needs of students, teachers, and supervisors. Ron and Laurie Boswell began writing together in 1992. Since that time, they have authored over two dozen textbooks. In their collaboration, Ron is primarily responsible for the pupil edition and Laurie is primarily responsible for the teaching edition of the text.

Laurie Boswell is the Head of School and a mathematics teacher at the Riverside School in Lyndonville, Vermont. Dr. Boswell received her Ed.D. from the University of Vermont in 2010. She is a recipient of the Presidential Award for Excellence in Mathematics Teaching. Laurie has taught math to students at all levels, elementary through college. In addition, Laurie was a Tandy Technology Scholar, and served on the NCTM Board of Directors from 2002 to 2005. She currently serves on the board of NCSM, and is a popular national speaker. Along with Ron, Laurie has co-authored numerous math programs.

ABOUT THE BOOK

Big Ideas Math Red Accelerated is the newest book in the *Big Ideas Math* series. The program uses the same research-based strategy of a balanced approach to instruction that made the *Big Ideas Math* series so successful. This approach opens doors to abstract thought, reasoning, and inquiry as students persevere to answer the Essential Questions that drive instruction. The foundation of the program is the Common Core Standards for Mathematical Content and Standards for Mathematical Practice. The *Big Ideas Math* series exposes students to highly motivating and relevant problems. Woven throughout the series are the depth and rigor students need to prepare for Calculus and other college-level courses. The *Big Ideas Math Red Accelerated*, along with the Algebra 1 book, completes the compacted pathway for middle school students.

Ron Larson

Laurie Boswell

TEACHER REVIEWERS

- Lisa Amspacher
  Milton Hershey School
  Hershey, PA

- Mary Ballerina
  Orange County Public Schools
  Orlando, FL

- Lisa Bubello
  School District of Palm
      Beach County
  Lake Worth, FL

- Sam Coffman
  North East School District
  North East, PA

- Kristen Karbon
  Troy School District
  Rochester Hills, MI

- Laurie Mallis
  Westglades Middle School
  Coral Springs, FL

- Dave Morris
  Union City Area
      School District
  Union City, PA

- Bonnie Pendergast
  Tolleson Union High
      School District
  Tolleson, AZ

- Valerie Sullivan
  Lamoille South
      Supervisory Union
  Morrisville, VT

- Becky Walker
  Appleton Area School District
  Appleton, WI

- Zena Wiltshire
  Dade County Public Schools
  Miami, FL

STUDENT REVIEWERS

- Mike Carter
- Matthew Cauley
- Amelia Davis
- Wisdom Dowds
- John Flatley
- Nick Ganger

- Hannah Iadeluca
- Paige Lavine
- Emma Louie
- David Nichols
- Mikala Parnell
- Jordan Pashupathi

- Stephen Piglowski
- Robby Quinn
- Michael Rawlings
- Garrett Sample
- Andrew Samuels
- Addie Sedelmyer
- Tyler Steffy
- Erin Taylor
- Reid Wilson

CONSULTANTS

● **Patsy Davis**
Educational Consultant
Knoxville, Tennessee

● **Bob Fulenwider**
Mathematics Consultant
Bakersfield, California

● **Linda Hall**
Mathematics Assessment Consultant
Norman, Oklahoma

● **Ryan Keating**
Special Education Advisor
Gilbert, Arizona

● **Michael McDowell**
Project-Based Instruction Specialist
Fairfax, California

● **Sean McKeighan**
Interdisciplinary Advisor
Norman, Oklahoma

● **Bonnie Spence**
Differentiated Instruction Consultant
Missoula, Montana

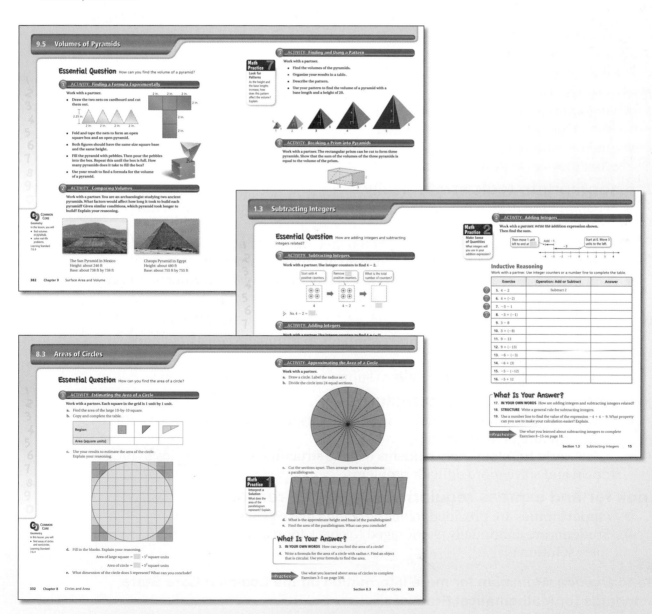

Common Core State Standards for Mathematical Practice

Make sense of problems and persevere in solving them.
- Multiple representations are presented to help students move from concrete to representative and into abstract thinking
- *Essential Questions* help students focus and analyze
- *In Your Own Words* provide opportunities for students to look for meaning and entry points to a problem

Reason abstractly and quantitatively.
- Visual problem solving models help students create a coherent representation of the problem
- Opportunities for students to decontextualize and contextualize problems are presented in every lesson

Construct viable arguments and critique the reasoning of others.
- *Error Analysis*; *Different Words, Same Question*; and *Which One Doesn't Belong* features provide students the opportunity to construct arguments and critique the reasoning of others
- *Inductive Reasoning* activities help students make conjectures and build a logical progression of statements to explore their conjecture

Model with mathematics.
- Real-life situations are translated into diagrams, tables, equations, and graphs to help students analyze relations and to draw conclusions
- Real-life problems are provided to help students learn to apply the mathematics that they are learning to everyday life

Use appropriate tools strategically.
- *Graphic Organizers* support the thought process of what, when, and how to solve problems
- A variety of tool papers, such as graph paper, number lines, and manipulatives, are available as students consider how to approach a problem
- Opportunities to use the web, graphing calculators, and spreadsheets support student learning

Attend to precision.
- *On Your Own* questions encourage students to formulate consistent and appropriate reasoning
- Cooperative learning opportunities support precise communication

Look for and make use of structure.
- *Inductive Reasoning* activities provide students the opportunity to see patterns and structure in mathematics
- Real-world problems help students use the structure of mathematics to break down and solve more difficult problems

Look for and express regularity in repeated reasoning.
- Opportunities are provided to help students make generalizations
- Students are continually encouraged to check for reasonableness in their solutions

Go to *BigIdeasMath.com* for more information on the Common Core State Standards for Mathematical Practice.

Common Core State Standards for Mathematical Content for Grade 7 Accelerated

Chapter Coverage for Standards

1 · 2 · 3 · 4 · 5 · 6 · 7 · 8 · 9 · 10 · 11 · 12 · 13 · **14** · 15 · **16** · AT

Conceptual Category Number and Quantity

- The Real Number System
- Quantities

1 · 2 · **3** · 4 · 5 · **6** · 7 · 8 · 9 · 10 · 11 · 12 · **13** · 14 · 15 · 16 · AT

Conceptual Category Algebra

- Seeing Structure in Expressions
- Reasoning with Equations and Inequalities

1 · 2 · 3 · 4 · 5 · 6 · **7** · **8** · **9** · **10** · **11** · **12** · 13 · 14 · **15** · 16 · AT

Conceptual Category Geometry

- Congruence
- Similarity, Right Triangles, and Trigonometry
- Geometric Measurement and Dimension

1 · 2 · 3 · 4 · 5 · 6 · 7 · 8 · 9 · **10** · 11 · 12 · 13 · 14 · 15 · 16 · AT

Conceptual Category Statistics and Probability

- Making Inferences and Justifying Conclusions
- Conditional Probability and the Rules of Probability

Go to *BigIdeasMath.com* for more information on the Common Core State Standards for Mathematical Content.

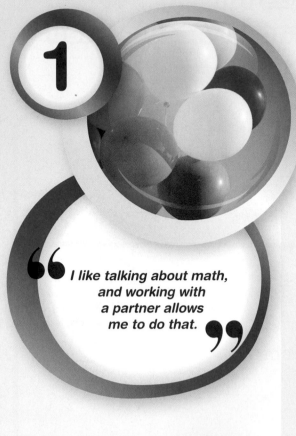

Integers

"I like talking about math, and working with a partner allows me to do that."

Rational Numbers

With my eBook, I get to decide when I use technology and when I use print.

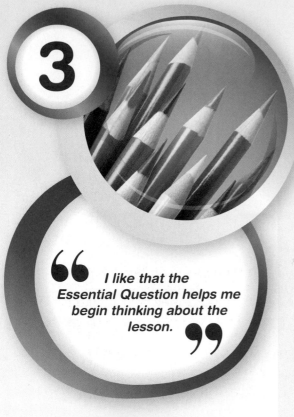

3 Expressions and Equations

" *I like that the Essential Question helps me begin thinking about the lesson.* "

Inequalities

> " *I really enjoy the projects at the end of the book because they help connect the math to other subjects, like science or art.* "

5 Ratios and Proportions

> *I like Newton and Descartes! The cartoons are funny and I like that they model the math that we are learning.*

Percents

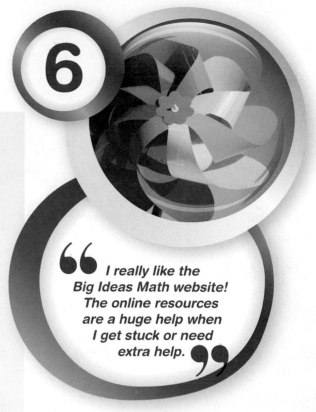

" *I really like the Big Ideas Math website! The online resources are a huge help when I get stuck or need extra help.* "

7 Constructions and Scale Drawings

I like the real-life application exercises because they show me how I can use the math in my own life.

Circles and Area

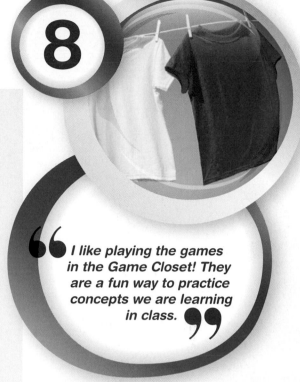

I like playing the games in the Game Closet! They are a fun way to practice concepts we are learning in class.

9 Surface Area and Volume

"With the BigIdeasMath.com website I don't have to worry if I forget my book or my workbook at school."

Probability and Statistics

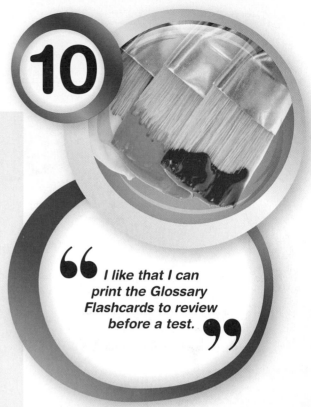

I like that I can print the Glossary Flashcards to review before a test.

11 Transformations

Angles and Triangles

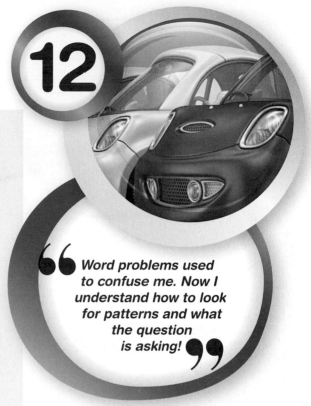

> *Word problems used to confuse me. Now I understand how to look for patterns and what the question is asking!*

13 Graphing and Writing Linear Equations

> I like the Big Ideas Math Tutorials because they help explain the math when I am at home.

Real Numbers and the Pythagorean Theorem

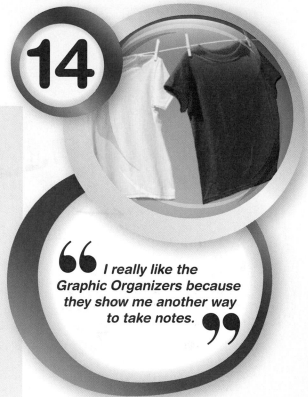

14

> " I really like the Graphic Organizers because they show me another way to take notes. "

15 Volume and Similar Solids

"Using the Interactive Manipulatives from the Dynamic Student Edition helps me to see the mathematics that I am learning."

Exponents and Scientific Notation

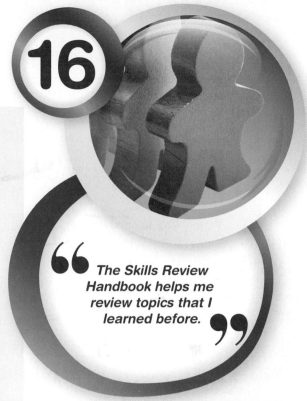

"The Skills Review Handbook helps me review topics that I learned before."

Additional Topics

How to Use Your Math Book

- Read the **Essential Question** in the activity.

  Discuss the question with your partner.

  Work with a partner to decide **What Is Your Answer?**

  Now you are ready to do the **Practice** problems.

- Find the **Key Vocabulary** words, **highlighted in yellow**.

  Read their definitions. Study the concepts in each **Key Idea**.
  If you forget a definition, you can look it up online in the

  Multi-Language Glossary at BigIdeasMath✓com.

- After you study each **EXAMPLE**, do the exercises in the ⬤ **On Your Own**.

  Now You're Ready to do the exercises that correspond to the example.

  As you study, look for a **Study Tip** or a **Common Error !**.

- The exercises are divided into 3 parts.

   Vocabulary and Concept Check

   Practice and Problem Solving

   Fair Game Review

  If an exercise has a ① next to it, look back at
  Example 1 for help with that exercise.

  More help is available at  **Check It Out**
  Lesson Tutorials
  BigIdeasMath✓com.

- To help study for your test, use the following.

  Quiz   **Study Help**

  Chapter Review   **Chapter Test**

SCAVENGER HUNT

Use this *Scavenger Hunt* to find where things are in **Chapter 1**.

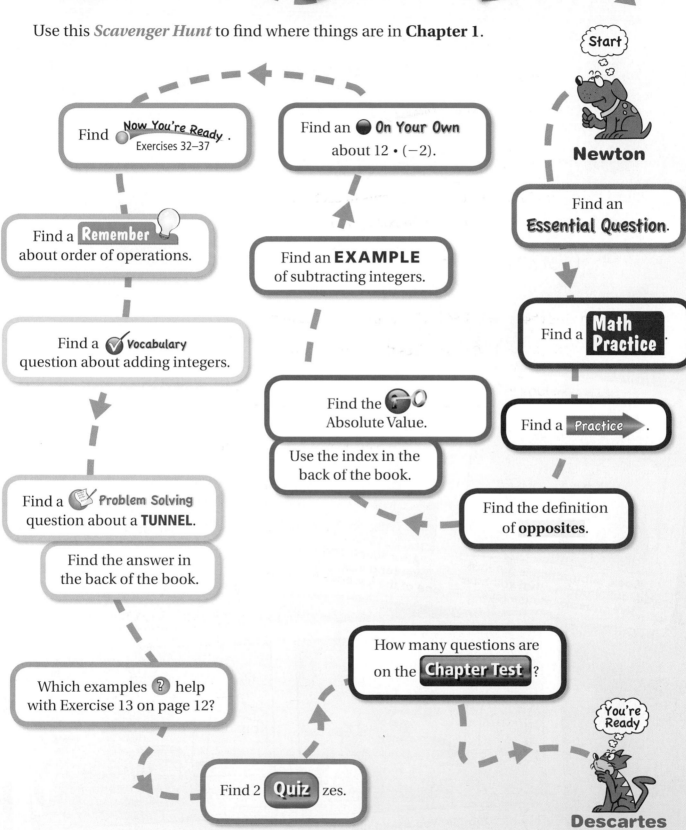

Start

Newton

Find **Now You're Ready**.
Exercises 32–37

Find an ● **On Your Own**
about 12 · (−2).

Find an **Essential Question**.

Find a **Remember**
about order of operations.

Find an **EXAMPLE**
of subtracting integers.

Find a **Math Practice**.

Find a **Vocabulary**
question about adding integers.

Find the **Absolute Value**.

Find a **Practice**.

Use the index in the
back of the book.

Find the definition
of **opposites**.

Find a **Problem Solving**
question about a **TUNNEL**.

Find the answer in
the back of the book.

How many questions are
on the **Chapter Test**?

Which examples ? help
with Exercise 13 on page 12?

Find 2 **Quiz**zes.

You're
Ready

Descartes

1 Integers

"Look, subtraction is not that difficult. Imagine that you have five squeaky mouse toys."

"After your friend Fluffy comes over for a visit, you notice that one of the squeaky toys is missing."

"Now, you go over to Fluffy's and retrieve the missing squeaky mouse toy. It's easy."

"Dear Sir: You asked me to 'find' the opposite of −1."

"I didn't know it was missing."

What You Learned Before

"I liked it because it is the opposite of the freezing point on the Fahrenheit temperature scale."

Commutative and Associative Properties (6.EE.3)

Example 1 **a. Simplify the expression 6 + (14 + x).**

$6 + (14 + x) = (6 + 14) + x$	Associative Property of Addition
$= 20 + x$	Add 6 and 14.

b. Simplify the expression (3.1 + x) + 7.4.

$(3.1 + x) + 7.4 = (x + 3.1) + 7.4$	Commutative Property of Addition
$= x + (3.1 + 7.4)$	Associative Property of Addition
$= x + 10.5$	Add 3.1 and 7.4.

c. Simplify the expression 5(12y).

$5(12y) = (5 \cdot 12)y$	Associative Property of Multiplication
$= 60y$	Multiply 5 and 12.

Try It Yourself
Simplify the expression. Explain each step.

1. $3 + (b + 8)$ 2. $(d + 4) + 6$ 3. $6(5p)$

Properties of Zero and One (6.EE.3)

Example 2 **a. Simplify the expression 6 • 0 • q.**

$6 \cdot 0 \cdot q = (6 \cdot 0) \cdot q$	Associative Property of Multiplication
$= 0 \cdot q = 0$	Multiplication Property of Zero

b. Simplify the expression 3.6 • s • 1.

$3.6 \cdot s \cdot 1 = 3.6 \cdot (s \cdot 1)$	Associative Property of Multiplication
$= 3.6 \cdot s$	Multiplication Property of One
$= 3.6s$	

Try It Yourself
Simplify the expression. Explain each step.

4. $13 \cdot m \cdot 0$ 5. $1 \cdot x \cdot 29$ 6. $(n + 14) + 0$

1.1 Integers and Absolute Value

Essential Question How can you use integers to represent the velocity and the speed of an object?

On these two pages, you will investigate vertical motion (up or down).

- Speed tells how fast an object is moving, but it does not tell the direction.
- Velocity tells how fast an object is moving, and it also tells the direction.

  When velocity is positive, the object is moving up.

  When velocity is negative, the object is moving down.

1 ACTIVITY: Falling Parachute

Work with a partner. You are gliding to the ground wearing a parachute. The table shows your height above the ground at different times.

Time (seconds)	0	1	2	3
Height (feet)	90	75	60	45

a. Describe the pattern in the table. How many feet do you move each second? After how many seconds will you land on the ground?

b. What integer represents your speed? Give the units.

c. Do you think your velocity should be represented by a positive or negative integer? Explain your reasoning.

d. What integer represents your velocity? Give the units.

2 ACTIVITY: Rising Balloons

Work with a partner. You release a group of balloons. The table shows the height of the balloons above the ground at different times.

Time (seconds)	0	1	2	3
Height (feet)	8	12	16	20

a. Describe the pattern in the table. How many feet do the balloons move each second? After how many seconds will the balloons be at a height of 40 feet?

b. What integer represents the speed of the balloons? Give the units.

c. Do you think the velocity of the balloons should be represented by a positive or negative integer? Explain your reasoning.

d. What integer represents the velocity of the balloons? Give the units.

COMMON CORE

Integers

In this lesson, you will
- define the absolute value of a number.
- find absolute values of numbers.
- solve real-life problems.

Preparing for Standards
7.NS.1
7.NS.2
7.NS.3

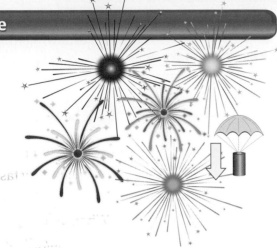

Work with a partner. The table shows the height of a firework's parachute above the ground at different times.

Math Practice 6

Use Clear Definitions

What information can you use to support your answer?

Time (seconds)	Height (feet)
0	480
1	360
2	240
3	120
4	0

a. Describe the pattern in the table. How many feet does the parachute move each second?

b. What integer represents the speed of the parachute? What integer represents the velocity? How are these integers similar in their relation to 0 on a number line?

Inductive Reasoning

4. Copy and complete the table.

Velocity (feet per second)	−14	20	−2	0	25	−15
Speed (feet per second)						

5. Find two different velocities for which the speed is 16 feet per second.

6. Which number is greater: −4 or 3? Use a number line to explain your reasoning.

7. One object has a velocity of −4 feet per second. Another object has a velocity of 3 feet per second. Which object has the greater speed? Explain your answer.

What Is Your Answer?

8. **IN YOUR OWN WORDS** How can you use integers to represent the velocity and the speed of an object?

9. **LOGIC** In this lesson, you will study *absolute value*. Here are some examples:

$$|-16| = 16 \qquad |16| = 16 \qquad |0| = 0 \qquad |-2| = 2$$

Which of the following is a true statement? Explain your reasoning.

$$|\text{velocity}| = \text{speed} \qquad\qquad |\text{speed}| = \text{velocity}$$

Practice

Use what you learned about absolute value to complete Exercises 4–11 on page 6.

Key Vocabulary 🔊
integer, *p. 4*
absolute value, *p. 4*

The following numbers are **integers:**

$$\ldots, -3, -2, -1, 0, 1, 2, 3, \ldots$$

🔑 Key Idea

Absolute Value

Words The **absolute value** of an integer is the distance between the number and 0 on a number line. The absolute value of a number a is written as $|a|$.

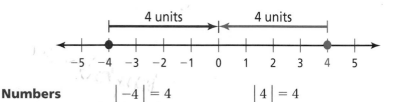

Numbers $|-4| = 4$ $|4| = 4$

EXAMPLE **1** **Finding Absolute Value**

Find the absolute value of 2.

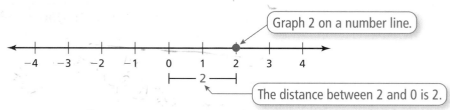

Graph 2 on a number line.

The distance between 2 and 0 is 2.

∴ So, $|2| = 2$.

EXAMPLE **2** **Finding Absolute Value**

Find the absolute value of −3.

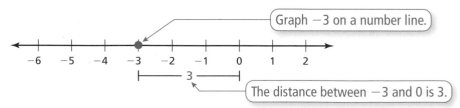

Graph −3 on a number line.

The distance between −3 and 0 is 3.

∴ So, $|-3| = 3$.

⚫ On Your Own

Now You're Ready
Exercises 4–19

Find the absolute value.

1. $|7|$ **2.** $|-1|$ **3.** $|-5|$ **4.** $|14|$

---

4   **Chapter 1**   Integers

🔊 Multi-Language Glossary at BigIdeasMath🗸com

EXAMPLE 3 Comparing Values

Compare 1 and $|-4|$.

Remember

A number line can be used to compare and order integers. Numbers to the left are less than numbers to the right. Numbers to the right are greater than numbers to the left.

Graph 1 on a number line.

Graph $|-4| = 4$ on a number line.

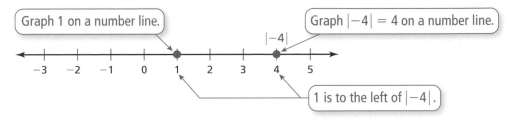

1 is to the left of $|-4|$.

So, $1 < |-4|$.

On Your Own

Now You're Ready
Exercises 20–25

Copy and complete the statement using <, >, or =.

5. $|-2|$ ☐ -1

6. -7 ☐ $|6|$

7. $|10|$ ☐ 11

8. 9 ☐ $|-9|$

EXAMPLE 4 Real-Life Application

Substance	Freezing Point (°C)
Butter	35
Airplane fuel	−53
Honey	−3
Mercury	−39
Candle wax	55

The *freezing point* is the temperature at which a liquid becomes a solid.

a. Which substance in the table has the lowest freezing point?

b. Is the freezing point of mercury or butter closer to the freezing point of water, 0°C?

a. Graph each freezing point.

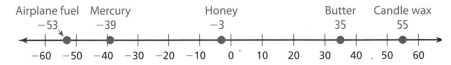

Airplane fuel has the lowest freezing point, −53°C.

b. The freezing point of water is 0°C, so you can use absolute values.

Mercury: $|-39| = 39$ **Butter:** $|35| = 35$

Because 35 is less than 39, the freezing point of butter is closer to the freezing point of water.

On Your Own

9. Is the freezing point of airplane fuel or candle wax closer to the freezing point of water? Explain your reasoning.

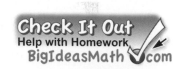

Vocabulary and Concept Check

1. **VOCABULARY** Which of the following numbers are integers?

$$9, 3.2, -1, \frac{1}{2}, -0.25, 15$$

2. **VOCABULARY** What is the absolute value of an integer?

3. **WHICH ONE DOESN'T BELONG?** Which expression does *not* belong with the other three? Explain your reasoning.

$$|6| \qquad 6 \qquad -6 \qquad |-6|$$

Practice and Problem Solving

Find the absolute value.

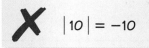 4. $|9|$ 5. $|-6|$ 6. $|-10|$ 7. $|10|$

8. $|-15|$ 9. $|13|$ 10. $|-7|$ 11. $|-12|$

12. $|5|$ 13. $|-8|$ 14. $|0|$ 15. $|18|$

16. $|-24|$ 17. $|-45|$ 18. $|60|$ 19. $|-125|$

Copy and complete the statement using <, >, or =.

 20. $2 \;\rule{1cm}{0.4pt}\; |-5|$ 21. $|-4| \;\rule{1cm}{0.4pt}\; 7$ 22. $-5 \;\rule{1cm}{0.4pt}\; |-9|$

23. $|-4| \;\rule{1cm}{0.4pt}\; -6$ 24. $|-1| \;\rule{1cm}{0.4pt}\; |-8|$ 25. $|5| \;\rule{1cm}{0.4pt}\; |-5|$

ERROR ANALYSIS Describe and correct the error.

26.
$$✗ \quad |10| = -10$$

27.
$$✗ \quad |-5| < 4$$

28. **SAVINGS** You deposit $50 in your savings account. One week later, you withdraw $20. Write each amount as an integer.

29. **ELEVATOR** You go down 8 floors in an elevator. Your friend goes up 5 floors in an elevator. Write each amount as an integer.

Order the values from least to greatest.

30. $8, |3|, -5, |-2|, -2$

31. $|-6|, -7, 8, |5|, -6$

32. $-12, |-26|, -15, |-12|, |10|$

33. $|-34|, 21, -17, |20|, |-11|$

Simplify the expression.

34. $|-30|$ 35. $-|4|$ 36. $-|-15|$

37. **PUZZLE** Use a number line.

    a. Graph and label the following points on a number line: $A = -3$, $E = 2$, $M = -6$, $T = 0$. What word do the letters spell?

    b. Graph and label the absolute value of each point in part (a). What word do the letters spell now?

38. **OPEN-ENDED** Write a negative integer whose absolute value is greater than 3.

REASONING Determine whether $n \geq 0$ or $n \leq 0$.

39. $n + |-n| = 2n$

40. $n + |-n| = 0$

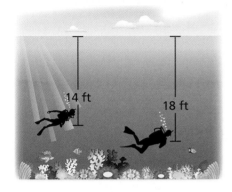

14 ft

18 ft

41. **CORAL REEF** The depths of two scuba divers exploring a living coral reef are shown.

    a. Write an integer for the position of each diver relative to sea level.

    b. Which integer in part (a) is greater?

    c. Which integer in part (a) has the greater absolute value? Compare this absolute value with the depth of that diver.

42. **VOLCANOES** The *summit elevation* of a volcano is the elevation of the top of the volcano relative to sea level. The summit elevation of the volcano Kilauea in Hawaii is 1277 meters. The summit elevation of the underwater volcano Loihi in the Pacific Ocean is -969 meters. Which summit is closer to sea level?

43. **MINIATURE GOLF** The table shows golf scores, relative to *par*.

    a. The player with the lowest score wins. Which player wins?

    b. Which player is at par?

    c. Which player is farthest from par?

Player	Score
1	+5
2	0
3	−4
4	−1
5	+2

True or False? Determine whether the statement is *true* or *false*. Explain your reasoning.

44. If $x < 0$, then $|x| = -x$.

45. The absolute value of every integer is positive.

Fair Game Review What you learned in previous grades & lessons

Add. *(Skills Review Handbook)*

46. $19 + 32$

47. $50 + 94$

48. $181 + 217$

49. $1149 + 2021$

50. **MULTIPLE CHOICE** Which value is *not* a whole number? *(Skills Review Handbook)*

    Ⓐ -5 **Ⓑ** 0 **Ⓒ** 4 **Ⓓ** 113

1.2 Adding Integers

Essential Question Is the sum of two integers *positive*, *negative*, or *zero*? How can you tell?

1 ACTIVITY: Adding Integers with the Same Sign

Work with a partner. Use integer counters to find −4 + (−3).

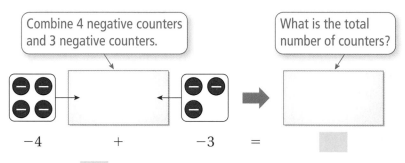

Combine 4 negative counters and 3 negative counters.

What is the total number of counters?

$$-4 \qquad + \qquad -3 \qquad =$$

∴ So, −4 + (−3) = ☐.

2 ACTIVITY: Adding Integers with Different Signs

Work with a partner. Use integer counters to find −3 + 2.

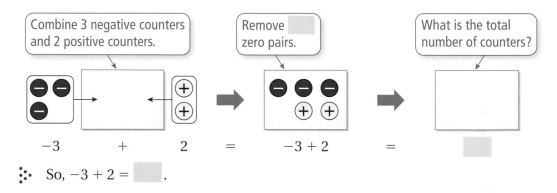

Combine 3 negative counters and 2 positive counters.

Remove ☐ zero pairs.

What is the total number of counters?

$$-3 \qquad + \qquad 2 \qquad = \qquad -3 + 2 \qquad =$$

∴ So, −3 + 2 = ☐.

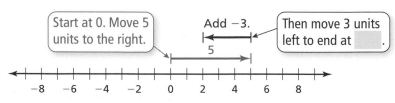

COMMON CORE

Integers
In this lesson, you will
- add integers.
- show that the sum of a number and its opposite is 0.
- solve real-life problems.

Learning Standards
7.NS.1a
7.NS.1b
7.NS.1d
7.NS.3

3 ACTIVITY: Adding Integers with Different Signs

Work with a partner. Use a number line to find 5 + (−3).

Start at 0. Move 5 units to the right.

Add −3.

Then move 3 units left to end at ☐.

5

```
←─┼──┼──┼──┼──┼──┼──┼──┼──┼──┼──┼──┼──┼──┼──┼──┼──→
  −8    −6    −4    −2    0    2    4    6    8
```

∴ So, 5 + (−3) = ☐.

Math Practice 3

Make Conjectures

How can the relationship between the integers help you write a rule?

Work with a partner. Write the addition expression shown. Then find the sum. How are the integers in the expression related to 0 on a number line?

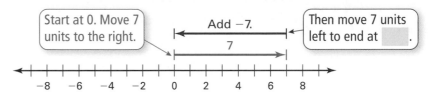

Start at 0. Move 7 units to the right.

Add −7.

7

Then move 7 units left to end at [ ].

Inductive Reasoning

Work with a partner. Use integer counters or a number line to complete the table.

	Exercise	Type of Sum	Sum	Sum: Positive, Negative, or Zero
1	**5.** $-4 + (-3)$	Integers with the same sign		
2	**6.** $-3 + 2$			
3	**7.** $5 + (-3)$			
4	**8.** $7 + (-7)$			
	9. $2 + 4$			
	10. $-6 + (-2)$			
	11. $-5 + 9$			
	12. $15 + (-9)$			
	13. $-10 + 10$			
	14. $-6 + (-6)$			
	15. $13 + (-13)$			

What Is Your Answer?

16. **IN YOUR OWN WORDS** Is the sum of two integers *positive*, *negative*, or *zero*? How can you tell?

17. **STRUCTURE** Write general rules for adding (a) two integers with the same sign, (b) two integers with different signs, and (c) two integers that vary only in sign.

Practice ➤ Use what you learned about adding integers to complete Exercises 8–15 on page 12.

Check It Out
Lesson Tutorials
BigIdeasMath com

Key Vocabulary
opposites, *p. 10*
additive inverse, *p. 10*

🔑 Key Idea

Adding Integers with the Same Sign

Words   Add the absolute values of the integers. Then use the common sign.

Numbers $2 + 5 = 7$ $-2 + (-5) = -7$

EXAMPLE ① **Adding Integers with the Same Sign**

Find $-2 + (-4)$. Use a number line to check your answer.

$$-2 + (-4) = -6 \qquad \text{Add } |-2| \text{ and } |-4|.$$

> Use the common sign.

∴ The sum is -6.

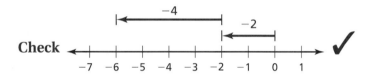

Check

The Meaning of a Word

Opposite

When you walk across a street, you are moving to the **opposite** side of the street.

● On Your Own

Add.

1. $7 + 13$ **2.** $-8 + (-5)$ **3.** $-20 + (-15)$

Two numbers that are the same distance from 0, but on opposite sides of 0, are called **opposites.** For example, -3 and 3 are opposites.

🔑 Key Ideas

Adding Integers with Different Signs

Words   Subtract the lesser absolute value from the greater absolute value. Then use the sign of the integer with the greater absolute value.

Numbers $8 + (-10) = -2$ $-13 + 17 = 4$

Additive Inverse Property

Words   The sum of an integer and its **additive inverse,** or opposite, is 0.

Numbers $6 + (-6) = 0$ $-25 + 25 = 0$ **Algebra** $a + (-a) = 0$

EXAMPLE 2 **Adding Integers with Different Signs**

a. **Find $5 + (-10)$.**

$$5 + (-10) = -5$$

$|-10| > |5|$. So, subtract $|5|$ from $|-10|$.

> Use the sign of -10.

⋮ The sum is -5.

b. **Find $-3 + 7$.**

$$-3 + 7 = 4$$

$|7| > |-3|$. So, subtract $|-3|$ from $|7|$.

> Use the sign of 7.

⋮ The sum is 4.

c. **Find $-12 + 12$.**

$$-12 + 12 = 0$$

The sum is 0 by the Additive Inverse Property.

> -12 and 12 are opposites.

⋮ The sum is 0.

EXAMPLE 3 **Adding More Than Two Integers**

The list shows four bank account transactions in July. Find the change C in the account balance.

JULY TRANSACTIONS	
Withdrawal	-$40
Deposit	$50
Deposit	$75
Withdrawal	-$50

Study Tip

A deposit of $50 and a withdrawal of $50 represent opposite quantities, +50 and −50, which have a sum of 0.

Find the sum of the four transactions.

$C = -40 + 50 + 75 + (-50)$	Write the sum.
$= -40 + 75 + 50 + (-50)$	Commutative Property of Addition
$= -40 + 75 + [50 + (-50)]$	Associative Property of Addition
$= -40 + 75 + 0$	Additive Inverse Property
$= 35 + 0$	Add -40 and 75.
$= 35$	Addition Property of Zero

⋮ Because $C = 35$, the account balance increased $35 in July.

On Your Own

Now You're Ready
Exercises 8–23
and 28–39

Add.

4. $-2 + 11$ **5.** $9 + (-10)$ **6.** $-31 + 31$

7. WHAT IF? In Example 3, the deposit amounts are $30 and $40. Find the change C in the account balance.

Check It Out
Help with Homework
BigIdeasMath.com

✓ Vocabulary and Concept Check

1. **WRITING** How do you find the additive inverse of an integer?

2. **NUMBER SENSE** Is $3 + (-4)$ the same as $-4 + 3$? Explain.

Tell whether the sum is *positive*, *negative*, or *zero* without adding. Explain your reasoning.

3. $-8 + 20$ 4. $30 + (-30)$ 5. $-10 + (-18)$

Tell whether the statement is *true* or *false*. Explain your reasoning.

6. The sum of two negative integers is always negative.

7. An integer and its absolute value are always opposites.

Practice and Problem Solving

Add.

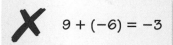 8. $6 + 4$ 9. $-4 + (-6)$ 10. $-2 + (-3)$ 11. $-5 + 12$

12. $5 + (-7)$ 13. $8 + (-8)$ 14. $9 + (-11)$ 15. $-3 + 13$

16. $-4 + (-16)$ 17. $-3 + (-1)$ 18. $14 + (-5)$ 19. $0 + (-11)$

20. $-10 + (-15)$ 21. $-13 + 9$ 22. $18 + (-18)$ 23. $-25 + (-9)$

ERROR ANALYSIS Describe and correct the error in finding the sum.

24.
$$✗ \quad 9 + (-6) = -3$$

25.
$$✗ \quad -10 + (-10) = 0$$

26. **TEMPERATURE** The temperature is $-3°F$ at 7:00 A.M. During the next 4 hours, the temperature increases $21°F$. What is the temperature at 11:00 A.M.?

27. **BANKING** Your bank account has a balance of $-\$12$. You deposit $\$60$. What is your new balance?

Tell how the Commutative and Associative Properties of Addition can help you find the sum mentally. Then find the sum.

 28. $9 + 6 + (-6)$ 29. $-8 + 13 + (-13)$ 30. $9 + (-17) + (-9)$

31. $7 + (-12) + (-7)$ 32. $-12 + 25 + (-15)$ 33. $6 + (-9) + 14$

Add.

34. $13 + (-21) + 16$ 35. $22 + (-14) + (-35)$ 36. $-13 + 27 + (-18)$

37. $-19 + 26 + 14$ 38. $-32 + (-17) + 42$ 39. $-41 + (-15) + (-29)$

40. SCIENCE A lithium atom has positively charged protons and negatively charged electrons. The sum of the charges represents the charge of the lithium atom. Find the charge of the atom.

Lithium Atom

41. OPEN-ENDED Write two integers with different signs that have a sum of -25. Write two integers with the same sign that have a sum of -25.

ALGEBRA Evaluate the expression when $a = 4$, $b = -5$, and $c = -8$.

42. $a + b$

43. $-b + c$

44. $|a + b + c|$

MENTAL MATH Use mental math to solve the equation.

45. $d + 12 = 2$

46. $b + (-2) = 0$

47. $-8 + m = -15$

48. PROBLEM SOLVING Starting at point A, the path of a dolphin jumping out of the water is shown.

    a. Is the dolphin deeper at point C or point E? Explain your reasoning.

    b. Is the dolphin higher at point B or point D? Explain your reasoning.

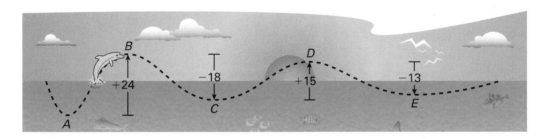

49. ✦**Puzzle**✦ According to a legend, the Chinese Emperor Yu-Huang saw a magic square on the back of a turtle. In a *magic square*, the numbers in each row and in each column have the same sum. This sum is called the *magic sum*.

Copy and complete the magic square so that each row and each column has a magic sum of 0. Use each integer from -4 to 4 exactly once.

Fair Game Review *What you learned in previous grades & lessons*

Subtract. *(Skills Review Handbook)*

50. $69 - 38$

51. $82 - 74$

52. $177 - 63$

53. $451 - 268$

54. MULTIPLE CHOICE What is the range of the numbers below? *(Skills Review Handbook)*

    12, 8, 17, 12, 15, 18, 30

    (A) 12   **(B)** 15   **(C)** 18   **(D)** 22

1.3 Subtracting Integers

Essential Question How are adding integers and subtracting integers related?

1 ACTIVITY: Subtracting Integers

Work with a partner. Use integer counters to find 4 − 2.

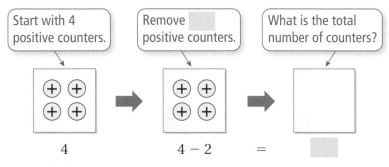

Start with 4 positive counters.

Remove [ ] positive counters.

What is the total number of counters?

4   4 − 2   =   [ ]

So, 4 − 2 = [ ].

2 ACTIVITY: Adding Integers

Work with a partner. Use integer counters to find 4 + (−2).

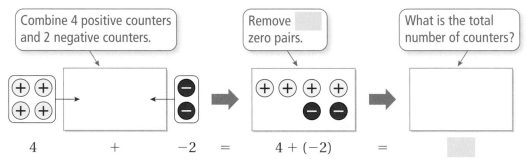

Combine 4 positive counters and 2 negative counters.

Remove [ ] zero pairs.

What is the total number of counters?

4   +   −2   =   4 + (−2)   =   [ ]

So, 4 + (−2) = [ ].

COMMON CORE

Integers
In this lesson, you will
• subtract integers.
• solve real-life problems.
Learning Standards
7.NS.1c
7.NS.1d
7.NS.3

3 ACTIVITY: Subtracting Integers

Work with a partner. Use a number line to find −3 − 1.

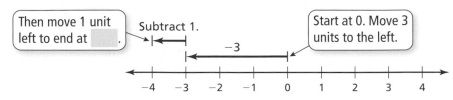

Then move 1 unit left to end at [ ].

Subtract 1.

Start at 0. Move 3 units to the left.

−3

So, −3 − 1 = [ ].

Math Practice 2

Make Sense of Quantities

What integers will you use in your addition expression?

Work with a partner. Write the addition expression shown. Then find the sum.

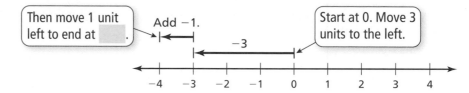

Then move 1 unit left to end at ___.

Add −1.

−3

Start at 0. Move 3 units to the left.

Inductive Reasoning

Work with a partner. Use integer counters or a number line to complete the table.

	Exercise	Operation: Add or Subtract	Answer
1	**5.** $4 - 2$	Subtract 2	
2	**6.** $4 + (-2)$		
3	**7.** $-3 - 1$		
4	**8.** $-3 + (-1)$		
	9. $3 - 8$		
	10. $3 + (-8)$		
	11. $9 - 13$		
	12. $9 + (-13)$		
	13. $-6 - (-3)$		
	14. $-6 + 3$		
	15. $-5 - (-12)$		
	16. $-5 + 12$		

What Is Your Answer?

17. IN YOUR OWN WORDS How are adding integers and subtracting integers related?

18. STRUCTURE Write a general rule for subtracting integers.

19. Use a number line to find the value of the expression $-4 + 4 - 9$. What property can you use to make your calculation easier? Explain.

Practice

Use what you learned about subtracting integers to complete Exercises 8–15 on page 18.

Key Idea

Subtracting Integers

Words   To subtract an integer, add its opposite.

Numbers $3 - 4 = 3 + (-4) = -1$

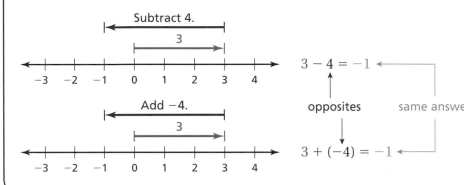

EXAMPLE **1** **Subtracting Integers**

a. **Find $3 - 12$.**

$$3 - 12 = 3 + (-12) \qquad \text{Add the opposite of 12.}$$
$$= -9 \qquad\qquad\quad \text{Add.}$$

∴ The difference is -9.

b. **Find $-8 - (-13)$.**

$$-8 - (-13) = -8 + 13 \qquad \text{Add the opposite of } -13.$$
$$= 5 \qquad\qquad\quad \text{Add.}$$

∴ The difference is 5.

c. **Find $5 - (-4)$.**

$$5 - (-4) = 5 + 4 \qquad \text{Add the opposite of } -4.$$
$$= 9 \qquad\qquad\quad \text{Add.}$$

∴ The difference is 9.

On Your Own

Now You're Ready
Exercises 8–23

Subtract.

1. $8 - 3$

2. $9 - 17$

3. $-3 - 3$

4. $-14 - 9$

5. $9 - (-8)$

6. $-12 - (-12)$

EXAMPLE **2** **Subtracting Integers**

Evaluate $-7 - (-12) - 14$.

$$-7 - (-12) - 14 = -7 + 12 - 14 \qquad \text{Add the opposite of } -12.$$
$$= 5 - 14 \qquad\qquad\quad \text{Add } -7 \text{ and } 12.$$
$$= 5 + (-14) \qquad\qquad \text{Add the opposite of } 14.$$
$$= -9 \qquad\qquad\qquad \text{Add.}$$

∴ So, $-7 - (-12) - 14 = -9$.

● **On Your Own**

Exercises 27–32

Evaluate the expression.

7. $-9 - 16 - 8$ **8.** $-4 - 20 - 9$

9. $0 - 9 - (-5)$ **10.** $-8 - (-6) - 0$

11. $15 - (-20) - 20$ **12.** $-14 - 9 - 36$

EXAMPLE **3** **Real-Life Application**

Which continent has the greater range of elevations?

	North America	Africa
Highest Elevation	6198 m	5895 m
Lowest Elevation	−86 m	−155 m

To find the range of elevations for each continent, subtract the lowest elevation from the highest elevation.

North America

$$\text{range} = 6198 - (-86)$$
$$= 6198 + 86$$
$$= 6284 \text{ m}$$

Africa

$$\text{range} = 5895 - (-155)$$
$$= 5895 + 155$$
$$= 6050 \text{ m}$$

∴ Because 6284 is greater than 6050, North America has the greater range of elevations.

● **On Your Own**

13. The highest elevation in Mexico is 5700 meters, on Pico de Orizaba. The lowest elevation in Mexico is −10 meters, in Laguna Salada. Find the range of elevations in Mexico.

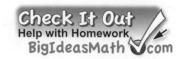
Check It Out
Help with Homework
BigIdeasMath ✓com

✓ Vocabulary and Concept Check

1. **WRITING** How do you subtract one integer from another?

2. **OPEN-ENDED** Write two integers that are opposites.

3. **DIFFERENT WORDS, SAME QUESTION** Which is different? Find "both" answers.

| Find the difference of 3 and -2. | What is 3 less than -2? |

| How much less is -2 than 3? | Subtract -2 from 3. |

MATCHING Match the subtraction expression with the corresponding addition expression.

4. $9 - (-5)$ 5. $-9 - 5$ 6. $-9 - (-5)$ 7. $9 - 5$

A. $-9 + 5$ B. $9 + (-5)$ C. $-9 + (-5)$ D. $9 + 5$

Practice and Problem Solving

Subtract.

 8. $4 - 7$ 9. $8 - (-5)$ 10. $-6 - (-7)$ 11. $-2 - 3$

12. $5 - 8$ 13. $-4 - 6$ 14. $-8 - (-3)$ 15. $10 - 7$

16. $-8 - 13$ 17. $15 - (-2)$ 18. $-9 - (-13)$ 19. $-7 - (-8)$

20. $-6 - (-6)$ 21. $-10 - 12$ 22. $32 - (-6)$ 23. $0 - 20$

24. **ERROR ANALYSIS** Describe and correct the error in finding the difference $7 - (-12)$.

$$\cancel{}\quad 7 - (-12) = 7 + (-12) = -5$$

25. **SWIMMING POOL** The floor of the shallow end of a swimming pool is at -3 feet. The floor of the deep end is 9 feet deeper. Which expression can be used to find the depth of the deep end?

| $-3 + 9$ | $-3 - 9$ | $9 - 3$ |

26. **SHARKS** A shark is at -80 feet. It swims up and jumps out of the water to a height of 15 feet. Write a subtraction expression for the vertical distance the shark travels.

Evaluate the expression.

② 27. $-2 - 7 + 15$ 28. $-9 + 6 - (-2)$ 29. $12 - (-5) - 8$

30. $-87 - 5 - 13$ 31. $-6 - (-8) + 6$ 32. $-15 - 7 - (-11)$

MENTAL MATH Use mental math to solve the equation.

33. $m - 5 = 9$ **34.** $w - (-3) = 7$ **35.** $6 - c = -9$

ALGEBRA Evaluate the expression when $k = -3$, $m = -6$, and $n = 9$.

36. $4 - n$ **37.** $m - (-8)$

38. $-5 + k - n$ **39.** $|m - k|$

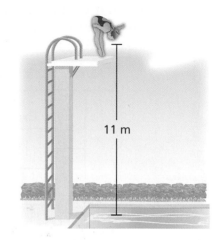

40. PLATFORM DIVING The figure shows a diver diving from a platform. The diver reaches a depth of 4 meters. What is the change in elevation of the diver?

11 m

41. OPEN-ENDED Write two different pairs of negative integers, x and y, that make the statement $x - y = -1$ true.

42. TEMPERATURE The table shows the record monthly high and low temperatures for a city in Alaska.

	Jan	Feb	Mar	Apr	May	Jun	Jul	Aug	Sep	Oct	Nov	Dec
High (°F)	56	57	56	72	82	92	84	85	73	64	62	53
Low (°F)	−35	−38	−24	−15	1	29	34	31	19	−6	−21	−36

   a. Find the range of temperatures for each month.

   b. What are the all-time high and all-time low temperatures?

   c. What is the range of the temperatures in part (b)?

REASONING Tell whether the difference between the two integers is *always*, *sometimes*, or *never* positive. Explain your reasoning.

43. two positive integers   **44.** two negative integers

45. a positive integer and a negative integer   **46.** a negative integer and a positive integer

Number Sense For what values of a and b is the statement true?

47. $|a - b| = |b - a|$ **48.** $|a + b| = |a| + |b|$ **49.** $|a - b| = |a| - |b|$

Fair Game Review What you learned in previous grades & lessons

Add. *(Section 1.2)*

50. $-5 + (-5) + (-5) + (-5)$ **51.** $-9 + (-9) + (-9) + (-9) + (-9)$

Multiply. *(Skills Review Handbook)*

52. 8×5 **53.** 6×78 **54.** 36×41 **55.** 82×29

56. MULTIPLE CHOICE Which value of n makes the value of the expression $4n + 3$ a composite number? *(Skills Review Handbook)*

   Ⓐ 1   Ⓑ 2   Ⓒ 3   Ⓓ 4

You can use an **idea and examples chart** to organize information about a concept. Here is an example of an idea and examples chart for absolute value.

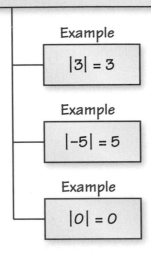

Absolute Value: the distance between a number and 0 on the number line

Example

$|3| = 3$

Example

$|-5| = 5$

Example

$|0| = 0$

On Your Own

Make idea and examples charts to help you study these topics.

1. integers

2. adding integers

   a. with the same sign

   b. with different signs

3. Additive Inverse Property

4. subtracting integers

After you complete this chapter, make idea and examples charts for the following topics.

5. multiplying integers

   a. with the same sign

   b. with different signs

6. dividing integers

   a. with the same sign

   b. with different signs

"I made an idea and examples chart to give my owner ideas for my birthday next week."

Copy and complete the statement using <, >, or =.   *(Section 1.1)*

1. $|-8|$ ☐ 3

2. 7 ☐ $|-7|$

Order the values from least to greatest.   *(Section 1.1)*

3. $-4, |-5|, |-4|, 3, -6$

4. $12, -8, |-15|, -10, |-9|$

Evaluate the expression.   *(Section 1.2 and Section 1.3)*

5. $-3 + (-8)$

6. $-4 + 16$

7. $3 - 9$

8. $-5 - (-5)$

Evaluate the expression when $a = -2$, $b = -8$, and $c = 5$.   *(Section 1.2 and Section 1.3)*

9. $4 - a - c$

10. $|b - c|$

11. EXPLORING Two climbers explore a cave.   *(Section 1.1)*

   a. Write an integer for the position of each climber relative to the surface.

   b. Which integer in part (a) is greater?

   c. Which integer in part (a) has the greater absolute value?

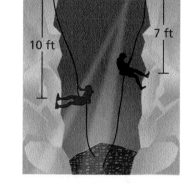

12. SCHOOL CARNIVAL The table shows the income and expenses for a school carnival. The school's goal was to raise $1100. Did the school reach its goal? Explain.   *(Section 1.2)*

Games	Concessions	Donations	Flyers	Decorations
$650	$530	$52	−$28	−$75

13. TEMPERATURE Temperatures in the Gobi Desert reach −40°F in the winter and 90°F in the summer. Find the range of the temperatures.   *(Section 1.3)*

1.4 Multiplying Integers

Essential Question
Is the product of two integers *positive*, *negative*, or *zero*? How can you tell?

1 ACTIVITY: Multiplying Integers with the Same Sign

Work with a partner. Use repeated addition to find 3 • 2.

Recall that multiplication is repeated addition. 3 • 2 means to add 3 groups of 2.

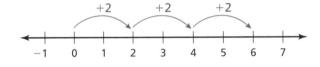

Now you can write

$3 • 2 = \boxed{} + \boxed{} + \boxed{}$

$= \boxed{}.$

So, $3 • 2 = \boxed{}$.

2 ACTIVITY: Multiplying Integers with Different Signs

Work with a partner. Use repeated addition to find 3 • (−2).

3 • (−2) means to add 3 groups of −2.

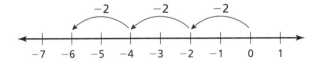

Now you can write

$3 • (−2) = \boxed{} + \boxed{} + \boxed{}$

$= \boxed{}.$

So, $3 • (−2) = \boxed{}$.

3 ACTIVITY: Multiplying Integers with Different Signs

Work with a partner. Use a table to find −3 • 2.

Describe the pattern of the products in the table. Then complete the table.

2	•	2	=	4
1	•	2	=	2
0	•	2	=	0
−1	•	2	=	
−2	•	2	=	
−3	•	2	=	

So, $−3 • 2 = \boxed{}$.

COMMON CORE

Integers
In this lesson, you will
- multiply integers.
- solve real-life problems.
Learning Standards
7.NS.2a
7.NS.2c
7.NS.3

Work with a partner. Use a table to find $-3 \cdot (-2)$**.**

Describe the pattern of the products in the table. Then complete the table.

Math Practice 7

Look for Patterns
How can you use the pattern to complete the table?

-3	$\cdot$	3	$=$	-9
-3	$\cdot$	2	$=$	-6
-3	$\cdot$	1	$=$	-3
-3	$\cdot$	0	$=$	
-3	$\cdot$	-1	$=$	
-3	$\cdot$	-2	$=$	

So, $-3 \cdot (-2) = $ ▢ .

Inductive Reasoning

Work with a partner. Complete the table.

	Exercise	Type of Product	Product	Product: Positive or Negative
1	**5.** $3 \cdot 2$	Integers with the same sign		
2	**6.** $3 \cdot (-2)$			
3	**7.** $-3 \cdot 2$			
4	**8.** $-3 \cdot (-2)$			
	9. $6 \cdot 3$			
	10. $2 \cdot (-5)$			
	11. $-6 \cdot 5$			
	12. $-5 \cdot (-3)$			

What Is Your Answer?

13. Write two integers whose product is 0.

14. **IN YOUR OWN WORDS** Is the product of two integers *positive*, *negative*, or *zero*? How can you tell?

15. **STRUCTURE** Write general rules for multiplying (a) two integers with the same sign and (b) two integers with different signs.

Practice ▶ Use what you learned about multiplying integers to complete Exercises 8–15 on page 26.

Check It Out
Lesson Tutorials
BigIdeasMath.com

Key Ideas

Multiplying Integers with the Same Sign

Words The product of two integers with the same sign is positive.

Numbers $2 \cdot 3 = 6$ $\qquad$ $-2 \cdot (-3) = 6$

Multiplying Integers with Different Signs

Words The product of two integers with different signs is negative.

Numbers $2 \cdot (-3) = -6$ $\qquad$ $-2 \cdot 3 = -6$

EXAMPLE **1** **Multiplying Integers with the Same Sign**

Find $-5 \cdot (-6)$.

The integers have the same sign.

$$-5 \cdot (-6) = 30$$

The product is positive.

$\vdots$ The product is 30.

EXAMPLE **2** **Multiplying Integers with Different Signs**

Multiply.

$\quad$ **a.** $3(-4)$ $\qquad\qquad\qquad$ **b.** $-7 \cdot 4$

The integers have different signs.

$$3(-4) = -12 \qquad\qquad -7 \cdot 4 = -28$$

The product is negative.

$\vdots$ The product is -12. $\qquad$ $\vdots$ The product is -28.

On Your Own

Now You're Ready
Exercises 8–23

Multiply.

1. $5 \cdot 5$ $\qquad\qquad\qquad$ **2.** $4(11)$

3. $-1(-9)$ $\qquad\qquad\qquad$ **4.** $-7 \cdot (-8)$

5. $12 \cdot (-2)$ $\qquad\qquad\qquad$ **6.** $4(-6)$

7. $-10(-6)(0)$ $\qquad\qquad\qquad$ **8.** $-7 \cdot (-5) \cdot (-4)$

EXAMPLE 3  Using Exponents

a. **Evaluate $(-2)^2$.**

$$(-2)^2 = (-2) \cdot (-2) \qquad \text{Write } (-2)^2 \text{ as repeated multiplication.}$$

$$= 4 \qquad \text{Multiply.}$$

b. **Evaluate -5^2.**

$$-5^2 = -(5 \cdot 5) \qquad \text{Write } 5^2 \text{ as repeated multiplication.}$$

$$= -25 \qquad \text{Multiply.}$$

c. **Evaluate $(-4)^3$.**

$$(-4)^3 = (-4) \cdot (-4) \cdot (-4) \qquad \text{Write } (-4)^3 \text{ as repeated multiplication.}$$

$$= 16 \cdot (-4) \qquad \text{Multiply.}$$

$$= -64 \qquad \text{Multiply.}$$

Study Tip

Place parentheses around a negative number to raise it to a power.

On Your Own

Evaluate the expression.

Now You're Ready
Exercises 32–37

9. $(-3)^2$ **10.** $(-2)^3$ **11.** -7^2 **12.** -6^3

EXAMPLE 4 **Real-Life Application**

Taxis in Service

The bar graph shows the number of taxis a company has in service. The number of taxis decreases by the same amount each year for 4 years. Find the total change in the number of taxis.

The bar graph shows that the number of taxis in service decreases by 50 each year. Use a model to solve the problem.

$$\boxed{\text{total change}} = \boxed{\text{change per year}} \cdot \boxed{\text{number of years}}$$

$$= -50 \cdot 4$$

$$= -200$$

Use -50 for the change per year because the number *decreases* each year.

∴ The total change in the number of taxis is -200.

On Your Own

13. A manatee population decreases by 15 manatees each year for 3 years. Find the total change in the manatee population.

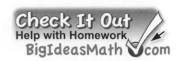

✓ Vocabulary and Concept Check

1. **WRITING** What can you conclude about the signs of two integers whose product is (a) positive and (b) negative?

2. **OPEN-ENDED** Write two integers whose product is negative.

Tell whether the product is *positive* or *negative* without multiplying. Explain your reasoning.

3. $4(-8)$

4. $-5(-7)$

5. $-3 \cdot 12$

Tell whether the statement is *true* or *false*. Explain your reasoning.

6. The product of three positive integers is positive.

7. The product of three negative integers is positive.

Practice and Problem Solving

Multiply.

8. $6 \cdot 4$

9. $7(-3)$

10. $-2(8)$

11. $-3(-4)$

12. $-6 \cdot 7$

13. $3 \cdot 9$

14. $8 \cdot (-5)$

15. $-1 \cdot (-12)$

16. $-5(10)$

17. $-13(0)$

18. $-9 \cdot 9$

19. $15(-2)$

20. $-10 \cdot 11$

21. $-6 \cdot (-13)$

22. $7(-14)$

23. $-11 \cdot (-11)$

24. **JOGGING** You burn 10 calories each minute you jog. What integer represents the change in your calories after you jog for 20 minutes?

25. **WETLANDS** About 60,000 acres of wetlands are lost each year in the United States. What integer represents the change in wetlands after 4 years?

Multiply.

26. $3 \cdot (-8) \cdot (-2)$

27. $6(-9)(-1)$

28. $-3(-5)(-4)$

29. $(-5)(-7)(-20)$

30. $-6 \cdot 3 \cdot (-2)$

31. $3 \cdot (-12) \cdot 0$

Evaluate the expression.

32. $(-4)^2$

33. $(-1)^3$

34. -8^2

35. -6^2

36. $-5^2 \cdot 4$

37. $-2 \cdot (-3)^3$

ERROR ANALYSIS Describe and correct the error in evaluating the expression.

38.

✗ $-2(-7) = -14$

39.

✗ $-10^2 = 100$

ALGEBRA Evaluate the expression when $a = -2$, $b = 3$, and $c = -8$.

40. ab

41. $\left| a^2 c \right|$

42. $-ab^3 - ac$

NUMBER SENSE Find the next two numbers in the pattern.

43. $-12, 60, -300, 1500, \ldots$

44. $7, -28, 112, -448, \ldots$

45. GYM CLASS You lose four points each time you attend gym class without sneakers. You forget your sneakers three times. What integer represents the change in your points?

46. MODELING The height of an airplane during a landing is given by $22{,}000 + (-480t)$, where t is the time in minutes.

   a. Copy and complete the table.

   b. Estimate how many minutes it takes the plane to land. Explain your reasoning.

Time (minutes)	5	10	15	20
Height (feet)				

47. INLINE SKATES In June, the price of a pair of inline skates is $165. The price changes each of the next 3 months.

   a. Copy and complete the table.

Month	Price of Skates
June	$165 \qquad = \$165$
July	$165 + (-12) = \$\underline{\quad}$
August	$165 + 2(-12) = \$\underline{\quad}$
September	$165 + 3(-12) = \$\underline{\quad}$

   b. Describe the change in the price of the inline skates for each month.

   c. The table at the right shows the amount of money you save each month to buy the inline skates. Do you have enough money saved to buy the inline skates in August? September? Explain your reasoning.

Amount Saved	
June	$35
July	$55
August	$45
September	$18

48. ✦**Reasoning** Two integers, a and b, have a product of 24. What is the least possible sum of a and b?

Fair Game Review   What you learned in previous grades & lessons

Divide. *(Skills Review Handbook)*

49. $27 \div 9$

50. $48 \div 6$

51. $56 \div 4$

52. $153 \div 9$

53. MULTIPLE CHOICE What is the prime factorization of 84? *(Skills Review Handbook)*

   (A) $2^2 \times 3^2$

   (B) $2^3 \times 7$

   (C) $3^3 \times 7$

   (D) $2^2 \times 3 \times 7$

1.5 Dividing Integers

Essential Question
Is the quotient of two integers *positive*, *negative*, or *zero*? How can you tell?

1 ACTIVITY: Dividing Integers with Different Signs

Work with a partner. Use integer counters to find −15 ÷ 3.

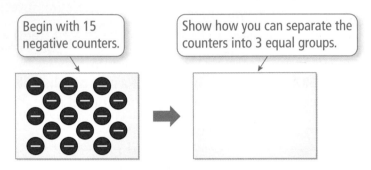

Begin with 15 negative counters.

Show how you can separate the counters into 3 equal groups.

:· Because there are ▢ negative counters in each group, −15 ÷ 3 = ▢.

2 ACTIVITY: Rewriting a Product as a Quotient

Work with a partner. Rewrite the product 3 · 4 = 12 as a quotient in two different ways.

First Way

12 is equal to 3 groups of ▢.

:· So, 12 ÷ 3 = ▢.

Second Way

12 is equal to 4 groups of ▢.

:· So, 12 ÷ 4 = ▢.

3 ACTIVITY: Dividing Integers with Different Signs

COMMON CORE

Integers
In this lesson, you will
- divide integers.
- solve real-life problems.
Learning Standards
7.NS.2b
7.NS.3

Work with a partner. Rewrite the product −3 · (−4) = 12 as a quotient in two different ways. What can you conclude?

First Way

$12 ÷ \left(▢ \right) = ▢$

Second Way

$12 ÷ \left(▢ \right) = ▢$

:· In each case, when you divide a ▢ integer by a ▢ integer, you get a ▢ integer.

Math Practice 8

Maintain Oversight

How do you know what the sign will be when you divide two integers?

Work with a partner. Rewrite the product $3 \cdot (-4) = -12$ as a quotient in two different ways. What can you conclude?

First Way

$-12 \div \left(\boxed{} \right) = \boxed{}$

Second Way

$-12 \div \left(\boxed{} \right) = \boxed{}$

∴ When you divide a [   ] integer by a [   ] integer, you get a [   ] integer. When you divide a [   ] integer by a [   ] integer, you get a [   ] integer.

Inductive Reasoning

Work with a partner. Complete the table.

Exercise	Type of Quotient	Quotient	Quotient: Positive, Negative, or Zero
① **5.** $-15 \div 3$	Integers with different signs		
② **6.** $12 \div 4$			
③ **7.** $12 \div (-3)$			
④ **8.** $-12 \div (-4)$			
9. $-6 \div 2$			
10. $-21 \div (-7)$			
11. $10 \div (-2)$			
12. $12 \div (-6)$			
13. $0 \div (-15)$			
14. $0 \div 4$			

What Is Your Answer?

15. IN YOUR OWN WORDS Is the quotient of two integers *positive*, *negative*, or *zero*? How can you tell?

16. STRUCTURE Write general rules for dividing (a) two integers with the same sign and (b) two integers with different signs.

Practice ▶ Use what you learned about dividing integers to complete Exercises 8–15 on page 32.

Check It Out
Lesson Tutorials
BigIdeasMath com

Key Ideas

Remember

Division by 0 is undefined.

Dividing Integers with the Same Sign

Words The quotient of two integers with the same sign is positive.

Numbers $8 \div 2 = 4$ $-8 \div (-2) = 4$

Dividing Integers with Different Signs

Words The quotient of two integers with different signs is negative.

Numbers $8 \div (-2) = -4$ $-8 \div 2 = -4$

EXAMPLE 1 **Dividing Integers with the Same Sign**

Find $-18 \div (-6)$.

The integers have the same sign.

$$-18 \div (-6) = 3$$

The quotient is positive.

∴ The quotient is 3.

EXAMPLE 2 **Dividing Integers with Different Signs**

Divide.

  a. $75 \div (-25)$ **b.** $\dfrac{-54}{6}$

The integers have different signs.

$$75 \div (-25) = -3 \qquad \dfrac{-54}{6} = -9$$

The quotient is negative.

∴ The quotient is -3.   ∴ The quotient is -9.

On Your Own

Now You're Ready
Exercises 8–23

Divide.

1. $14 \div 2$ **2.** $-32 \div (-4)$ **3.** $-40 \div (-8)$

4. $0 \div (-6)$ **5.** $\dfrac{-49}{7}$ **6.** $\dfrac{21}{-3}$

EXAMPLE **3** **Evaluating an Expression**

Evaluate $10 - x^2 \div y$ when $x = 8$ and $y = -4$.

$$
\begin{aligned}
10 - x^2 \div y &= 10 - 8^2 \div (-4) && \text{Substitute 8 for } x \text{ and } -4 \text{ for } y. \\
&= 10 - 8 \cdot 8 \div (-4) && \text{Write } 8^2 \text{ as repeated multiplication.} \\
&= 10 - 64 \div (-4) && \text{Multiply 8 and 8.} \\
&= 10 - (-16) && \text{Divide 64 by } -4. \\
&= 26 && \text{Subtract.}
\end{aligned}
$$

Remember

Use order of operations when evaluating an expression.

● **On Your Own**

Now You're Ready
Exercises 28–31

Evaluate the expression when $a = -18$ and $b = -6$.

7. $a \div b$

8. $\dfrac{a + 6}{3}$

9. $\dfrac{b^2}{a} + 4$

EXAMPLE **4** **Real-Life Application**

You measure the height of the tide using the support beams of a pier. Your measurements are shown in the picture. What is the mean hourly change in the height?

59 inches at 2 P.M.→
8 inches at 8 P.M.→

Use a model to solve the problem.

$$\text{mean hourly change} = \frac{\boxed{\text{final height}} \ - \ \boxed{\text{initial height}}}{\boxed{\text{elapsed time}}}$$

$$
\begin{aligned}
&= \frac{8 - 59}{6} && \text{Substitute. The elapsed time from} \\
& && \text{2 P.M. to 8 P.M. is 6 hours.} \\
&= \frac{-51}{6} && \text{Subtract.} \\
&= -8.5 && \text{Divide.}
\end{aligned}
$$

∴ The mean change in the height of the tide is -8.5 inches per hour.

● **On Your Own**

10. The height of the tide at the Bay of Fundy in New Brunswick decreases 36 feet in 6 hours. What is the mean hourly change in the height?

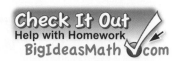
 Vocabulary and Concept Check

1. **WRITING** What can you tell about two integers when their quotient is positive? negative? zero?

2. **VOCABULARY** A quotient is undefined. What does this mean?

3. **OPEN-ENDED** Write two integers whose quotient is negative.

4. **WHICH ONE DOESN'T BELONG?** Which expression does *not* belong with the other three? Explain your reasoning.

$$\frac{10}{-5} \qquad \frac{-10}{5} \qquad \frac{-10}{-5} \qquad -\left(\frac{10}{5}\right)$$

Tell whether the quotient is *positive* or *negative* without dividing.

5. $-12 \div 4$

6. $\dfrac{-6}{-2}$

7. $15 \div (-3)$

 Practice and Problem Solving

Divide, if possible.

 8. $4 \div (-2)$

9. $21 \div (-7)$

10. $-20 \div 4$

11. $-18 \div (-3)$

12. $\dfrac{-14}{7}$

13. $\dfrac{0}{6}$

14. $\dfrac{-15}{-5}$

15. $\dfrac{54}{-9}$

16. $-33 \div 11$

17. $-49 \div (-7)$

18. $0 \div (-2)$

19. $60 \div (-6)$

20. $\dfrac{-56}{14}$

21. $\dfrac{18}{0}$

22. $\dfrac{65}{-5}$

23. $\dfrac{-84}{-7}$

ERROR ANALYSIS Describe and correct the error in finding the quotient.

24.

$$\times \quad \frac{-63}{-9} = -7$$

25.

$$\times \quad 0 \div (-5) = -5$$

26. **ALLIGATORS** An alligator population in a nature preserve in the Everglades decreases by 60 alligators over 5 years. What is the mean yearly change in the alligator population?

27. **READING** You read 105 pages of a novel over 7 days. What is the mean number of pages you read each day?

ALGEBRA Evaluate the expression when $x = 10$, $y = -2$, and $z = -5$.

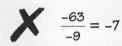 **28.** $x \div y$

29. $\dfrac{10y^2}{z}$

30. $\left| \dfrac{xz}{-y} \right|$

31. $\dfrac{-x^2 + 6z}{y}$

Find the mean of the integers.

32. $3, -10, -2, 13, 11$

33. $-26, 39, -10, -16, 12, 31$

Evaluate the expression.

34. $-8 - 14 \div 2 + 5$

35. $24 \div (-4) + (-2) \cdot (-5)$

36. **PATTERN** Find the next two numbers in the pattern $-128, 64, -32, 16, \ldots$. Explain your reasoning.

37. **SNOWBOARDING** A snowboarder descends a 1200-foot hill in 3 minutes. What is the mean change in elevation per minute?

38. **GOLF** The table shows a golfer's score for each round of a tournament.

    a. What was the golfer's total score?

    b. What was the golfer's mean score per round?

Scorecard	
Round 1	−2
Round 2	−6
Round 3	−7
Round 4	−3

39. **TUNNEL** The Detroit-Windsor Tunnel is an underwater highway that connects the cities of Detroit, Michigan, and Windsor, Ontario. How many times deeper is the roadway than the bottom of the ship?

0 ft
−15 ft
Detroit - Windsor Tunnel
−75 ft
Not drawn to scale

40. **AMUSEMENT PARK** The regular admission price for an amusement park is $72. For a group of 15 or more, the admission price is reduced by $25. How many people need to be in a group to save $500?

41. **Number Sense** Write five different integers that have a mean of -10. Explain how you found your answer.

✏️ **Fair Game Review** *What you learned in previous grades & lessons*

Graph the values on a number line. Then order the values from least to greatest. *(Section 1.1)*

42. $-6, 4, |2|, -1, |-10|$

43. $3, |0|, |-4|, -3, -8$

44. $|5|, -2, -5, |-2|, -7$

45. **MULTIPLE CHOICE** What is the value of $4 \cdot 3 + (12 \div 2)^2$? *(Skills Review Handbook)*

  Ⓐ 15   Ⓑ 48   Ⓒ 156   Ⓓ 324

Evaluate the expression. *(Section 1.4 and Section 1.5)*

1. $-7(6)$

2. $-1(-10)$

3. $\dfrac{-72}{-9}$

4. $-24 \div 3$

5. $-3 \cdot 4 \cdot (-6)$

6. $(-3)^3$

Evaluate the expression when $a = 4$, $b = -6$, **and** $c = -12$. *(Section 1.4 and Section 1.5)*

7. c^2

8. bc

9. $\dfrac{ab}{c}$

10. $\dfrac{|c - b|}{a}$

11. SPEECH In speech class, you lose 3 points for every 30 seconds you go over the time limit. Your speech is 90 seconds over the time limit. What integer represents the change in your points? *(Section 1.4)*

12. MOUNTAIN CLIMBING On a mountain, the temperature decreases by 18°F every 5000 feet. What integer represents the change in temperature at 20,000 feet? *(Section 1.4)*

13. GAMING You play a video game for 15 minutes. You lose 165 points. What is the mean change in points per minute? *(Section 1.5)*

14. DIVING You dive 21 feet from the surface of a lake in 7 seconds. *(Section 1.4 and Section 1.5)*

  a. What is the mean change in your position in feet per second?

  b. You continue diving. What is your position relative to the surface after 5 more seconds?

15. HIBERNATION A female grizzly bear weighs 500 pounds. After hibernating for 6 months, she weighs only 200 pounds. What is the mean change in weight per month? *(Section 1.5)*

1 Chapter Review

Check It Out
Vocabulary Help
BigIdeasMath ✓com

Review Key Vocabulary

integer, *p. 4*

opposites, *p. 10*

absolute value, *p. 4*

additive inverse, *p. 10*

Review Examples and Exercises

1.1 Integers and Absolute Value *(pp. 2–7)*

Find the absolute value of −2.

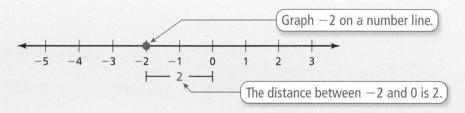

Graph −2 on a number line.

The distance between −2 and 0 is 2.

⋮ So, $|-2| = 2$.

Exercises

Find the absolute value.

1. $|3|$

2. $|-9|$

3. $|-17|$

4. $|8|$

5. **ELEVATION** The elevation of Death Valley, California, is −282 feet. The Mississippi River in Illinois has an elevation of 279 feet. Which is closer to sea level?

1.2 Adding Integers *(pp. 8–13)*

Find 6 + (−14).

$$6 + (-14) = -8$$ $|-14| > |6|$. So, subtract $|6|$ from $|-14|$.

Use the sign of −14.

⋮ The sum is −8.

Exercises

Add.

6. $-16 + (-11)$

7. $-15 + 5$

8. $100 + (-75)$

9. $-32 + (-2)$

1.3 **Subtracting Integers** *(pp. 14–19)*

Subtract.

a. $7 - 19 = 7 + (-19)$ Add the opposite of 19.

        $= -12$ Add.

  ∴ The difference is -12.

b. $-6 - (-10) = -6 + 10$ Add the opposite of -10.

         $= 4$ Add.

  ∴ The difference is 4.

Exercises

Subtract.

10. $8 - 18$ **11.** $-16 - (-5)$ **12.** $-18 - 7$ **13.** $-12 - (-27)$

14. GAME SHOW Your score on a game show is -300. You answer the final question incorrectly, so you lose 400 points. What is your final score?

1.4 **Multiplying Integers** *(pp. 22–27)*

a. Find $-7 \cdot (-9)$.

The integers have the same sign.

$$-7 \cdot (-9) = 63$$

The product is positive.

  ∴ The product is 63.

b. Find $-6(14)$.

The integers have different signs.

$$-6(14) = -84$$

The product is negative.

  ∴ The product is -84.

Exercises

Multiply.

15. $-8 \cdot 6$ **16.** $10(-7)$ **17.** $-3 \cdot (-6)$ **18.** $-12(5)$

1.5 **Dividing Integers** *(pp. 28–33)*

a. **Find $30 \div (-10)$.**

The integers have different signs.

$$30 \div (-10) = -3$$

The quotient is negative.

∴ The quotient is −3.

b. **Find $\dfrac{-72}{-9}$.**

The integers have the same sign.

$$\frac{-72}{-9} = 8$$

The quotient is positive.

∴ The quotient is 8.

Exercises

Divide.

19. $-18 \div 9$ **20.** $\dfrac{-42}{-6}$ **21.** $\dfrac{-30}{6}$ **22.** $84 \div (-7)$

Evaluate the expression when $x = 3$, $y = -4$, and $z = -6$.

23. $z \div x$ **24.** $\dfrac{xy}{z}$ **25.** $\dfrac{z - 2x}{y}$

Find the mean of the integers.

26. $-3, -8, 12, -15, 9$ **27.** $-54, -32, -70, -25, -65, -42$

28. **PROFITS** The table shows the weekly profits of a fruit vendor. What is the mean profit for these weeks?

Week	1	2	3	4
Profit	−$125	−$86	$54	−$35

29. **RETURNS** You return several shirts to a store. The receipt shows that the amount placed back on your credit card is −$30.60. Each shirt is −$6.12. How many shirts did you return?

Find the absolute value.

1. $|-9|$

2. $|64|$

3. $|-22|$

Copy and complete the statement using <, >, or =.

4. $4 \quad |-8|$

5. $|-7| \quad -12$

6. $-7 \quad |3|$

Evaluate the expression.

7. $-6 + (-11)$

8. $2 - (-9)$

9. $-9 \cdot 2$

10. $-72 \div (-3)$

Evaluate the expression when $x = 5$, $y = -3$, and $z = -2$.

11. $\dfrac{y + z}{x}$

12. $\dfrac{x - 5z}{y}$

Find the mean of the integers.

13. $11, -7, -14, 10, -5$

14. $-32, -41, -39, -27, -33, -44$

15. **NASCAR** A driver receives -25 points for each rule violation. What integer represents the change in points after 4 rule violations?

16. **GOLF** The table shows your scores, relative to *par*, for nine holes of golf. What is your total score for the nine holes?

Hole	1	2	3	4	5	6	7	8	9	Total
Score	+1	-2	-1	0	-1	+3	-1	-3	+1	?

17. **VISITORS** In a recent 10-year period, the change in the number of visitors to U.S. national parks was about $-11{,}150{,}000$ visitors.

a. What was the mean yearly change in the number of visitors?

b. During the seventh year, the change in the number of visitors was about $10{,}800{,}000$. Explain how the change for the 10-year period can be negative.

1. A football team gains 2 yards on the first play, loses 5 yards on the second play, loses 3 yards on the third play, and gains 4 yards on the fourth play. What is the team's overall gain or loss for all four plays? *(7.NS.1b)*

   A. a gain of 14 yards   C. a loss of 2 yards

   B. a gain of 2 yards   D. a loss of 14 yards

2. Which expression is *not* equal to the number 0? *(7.NS.1a)*

   F. $5 - 5$ H. $6 - (-6)$

   G. $-7 + 7$ I. $-8 - (-8)$

3. What is the value of the expression below when $a = -2$, $b = 3$, and $c = -5$? *(7.NS.3)*

   $$\left| a^2 - 2ac + 5b \right|$$

   A. -9 C. 1

   B. -1 D. 9

4. What is the value of the expression below? *(7.NS.1c)*

   $$17 - (-8)$$

5. Sam was evaluating an expression in the box below.

   $$(-2)^3 \cdot 3 - (-5) = 8 \cdot 3 - (-5)$$
   $$= 24 + 5$$
   $$= 29$$

   What should Sam do to correct the error that he made? *(7.NS.3)*

   F. Subtract 5 from 24 instead of adding.

   G. Rewrite $(-2)^3$ as -8.

   H. Subtract -5 from 3 before multiplying by $(-2)^3$.

   I. Multiply -2 by 3 before raising the quantity to the third power.

6. What is the value of the expression below when $x = 6$, $y = -4$, and $z = -2$?   *(7.NS.3)*

$$\frac{x - 2y}{-z}$$

A. -7

B. -1

C. 1

D. 7

7. What is the missing number in the sequence below?   *(7.NS.1c)*

$$39, 24, 9, \underline{}, -21$$

8. You are playing a game using the spinner shown. You start with a score of 0 and spin the spinner four times. When you spin blue or green, you add the number to your score. When you spin red or orange, you subtract the number from your score. Which sequence of colors represents the greatest score?   *(7.NS.3)*

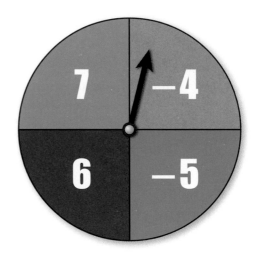

F. red, green, green, red

G. orange, orange, green, blue

H. red, blue, orange, green

I. blue, red, blue, red

9. Which expression represents a negative integer?   *(7.NS.3)*

A. $5 - (-6)$

B. $(-3)^3$

C. $-12 \div (-6)$

D. $(-2)(-4)$

10. Which expression has the greatest value when $x = -2$ and $y = -3$?   *(7.NS.3)*

F. $-xy$

G. xy

H. $x - y$

I. $-x - y$

11. What is the value of the expression below? *(7.NS.3)*

$$-5 \cdot (-4)^2 - (-3)$$

A. -83 **C.** 77

B. -77 **D.** 83

12. Which property does the equation below represent? *(7.NS.1d)*

$$-80 + 30 + (-30) = -80 + [30 + (-30)]$$

F. Commutative Property of Addition

G. Associative Property of Addition

H. Additive Inverse Property

I. Addition Property of Zero

13. What is the mean of the data set in the box below? *(7.NS.3)*

$$-8, -6, -2, 0, -6, -8, 4, -7, -8, 1$$

A. -8 **C.** -6

B. -7 **D.** -4

14. Consider the number line shown below. *(7.NS.1b, 7.NS.1c)*

Part A   Use the number line to explain how to add -2 and -3.

Part B   Use the number line to explain how to subtract 5 from 2.

15. What is the value of the expression below? *(7.NS.3)*

$$\frac{-3 - 2^2}{-1}$$

F. -25 **H.** 7

G. -1 **I.** 25

2 Rational Numbers

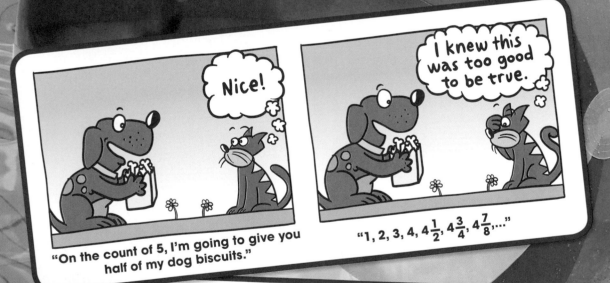

"On the count of 5, I'm going to give you half of my dog biscuits."

"1, 2, 3, 4, 4½, 4¾, 4⅞,..."

"I entered a contest for dog biscuits."

"I was notified that the number of biscuits I won was in the three-digit range."

What You Learned Before

"Let's play a game. The goal is to say a positive rational number that is less than the other pet's number... You go first."

This feels like a setup.

Writing Decimals and Fractions (4.NF.6)

Example 1   Write 0.37 as a fraction.

$$0.37 = \frac{37}{100}$$

Example 2   Write $\frac{2}{5}$ as a decimal.

$$\frac{2}{5} = \frac{2 \cdot 2}{5 \cdot 2} = \frac{4}{10} = 0.4$$

Try It Yourself
Write the decimal as a fraction or the fraction as a decimal.

1. 0.51
2. 0.731
3. $\frac{3}{5}$
4. $\frac{7}{8}$

Adding and Subtracting Fractions (5.NF.1)

Example 3   Find $\frac{1}{3} + \frac{1}{5}$.

$$\frac{1}{3} + \frac{1}{5} = \frac{1 \cdot 5}{3 \cdot 5} + \frac{1 \cdot 3}{5 \cdot 3}$$
$$= \frac{5}{15} + \frac{3}{15}$$
$$= \frac{8}{15}$$

Example 4   Find $\frac{1}{4} - \frac{2}{9}$.

$$\frac{1}{4} - \frac{2}{9} = \frac{1 \cdot 9}{4 \cdot 9} - \frac{2 \cdot 4}{9 \cdot 4}$$
$$= \frac{9}{36} - \frac{8}{36}$$
$$= \frac{1}{36}$$

Multiplying and Dividing Fractions (5.NF.4, 6.NS.1)

Example 5   Find $\frac{5}{6} \cdot \frac{3}{4}$.

$$\frac{5}{6} \cdot \frac{3}{4} = \frac{5 \cdot \overset{1}{\cancel{3}}}{\underset{2}{\cancel{6}} \cdot 4}$$
$$= \frac{5}{8}$$

Example 6   Find $\frac{2}{3} \div \frac{9}{10}$.

$$\frac{2}{3} \div \frac{9}{10} = \frac{2}{3} \cdot \frac{10}{9}$$

Multiply by the reciprocal of the divisor.

$$= \frac{2 \cdot 10}{3 \cdot 9}$$
$$= \frac{20}{27}$$

Try It Yourself
Evaluate the expression.

5. $\frac{1}{4} + \frac{13}{20}$
6. $\frac{14}{15} - \frac{1}{3}$
7. $\frac{3}{7} \cdot \frac{9}{10}$
8. $\frac{4}{5} \div \frac{16}{17}$

Essential Question

How can you use a number line to order rational numbers?

The Meaning of a Word • Rational

The word **rational** comes from the word *ratio*. Recall that you can write a ratio using fraction notation.

If you sleep for 8 hours in a day, then the ratio of your sleeping time to the total hours in a day can be written as $\dfrac{8 \text{ h}}{24 \text{ h}}$.

A **rational number** is a number that can be written as the ratio of two integers.

$$2 = \frac{2}{1} \qquad -3 = \frac{-3}{1} \qquad -\frac{1}{2} = \frac{-1}{2} \qquad 0.25 = \frac{1}{4}$$

---

1 ACTIVITY: Ordering Rational Numbers

Work in groups of five. Order the numbers from least to greatest.

- Use masking tape and a marker to make a number line on the floor similar to the one shown.

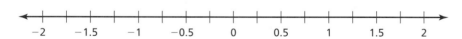

- Write the numbers on pieces of paper. Then each person should choose one.

- Stand on the location of your number on the number line.

- Use your positions to order the numbers from least to greatest.

a. $-0.5, 1.25, -\dfrac{1}{3}, 0.5, -\dfrac{5}{3}$

b. $-\dfrac{7}{4}, 1.1, \dfrac{1}{2}, -\dfrac{1}{10}, -1.3$

c. $-1.4, -\dfrac{3}{5}, \dfrac{9}{2}, \dfrac{1}{4}, 0.9$

d. $\dfrac{5}{4}, 0.75, -\dfrac{5}{4}, -0.8, -1.1$

COMMON CORE

Rational Numbers

In this lesson, you will

- understand that a rational number is an integer divided by an integer.
- convert rational numbers to decimals.

Learning Standards
7.NS.2b
7.NS.2d

2 **ACTIVITY: The Game of Math Card War**

Math Practice 1

Consider Similar Problems

What are some ways to determine which number is greater?

Preparation:

- Cut index cards to make 40 playing cards.
- Write each number in the table on a card.

To Play:

- Play with a partner.
- Deal 20 cards to each player facedown.
- Each player turns one card faceup. The player with the greater number wins. The winner collects both cards and places them at the bottom of his or her cards.
- Suppose there is a tie. Each player lays three cards facedown, then a new card faceup. The player with the greater of these new cards wins. The winner collects all ten cards and places them at the bottom of his or her cards.
- Continue playing until one player has all the cards. This player wins the game.

$-\dfrac{3}{2}$	$\dfrac{3}{10}$	$-\dfrac{3}{4}$	-0.6	1.25	-0.15	$\dfrac{5}{4}$	$\dfrac{3}{5}$	-1.6	-0.3
$\dfrac{3}{20}$	$\dfrac{8}{5}$	-1.2	$\dfrac{19}{10}$	0.75	-1.5	$-\dfrac{6}{5}$	$-\dfrac{3}{5}$	1.2	0.3
1.5	1.9	-0.75	-0.4	$\dfrac{3}{4}$	$-\dfrac{5}{4}$	-1.9	$\dfrac{2}{5}$	$-\dfrac{3}{20}$	$-\dfrac{19}{10}$
$\dfrac{6}{5}$	$-\dfrac{3}{10}$	1.6	$-\dfrac{2}{5}$	0.6	0.15	$\dfrac{3}{2}$	-1.25	0.4	$-\dfrac{8}{5}$

What Is Your Answer?

3. **IN YOUR OWN WORDS** How can you use a number line to order rational numbers? Give an example.

The numbers are in order from least to greatest. Fill in the blank spaces with rational numbers.

4. $-\dfrac{1}{2}$, ▢ , $\dfrac{1}{3}$, ▢ , $\dfrac{7}{5}$, ▢

5. $-\dfrac{5}{2}$, ▢ , -1.9, ▢ , $-\dfrac{2}{3}$, ▢

6. $-\dfrac{1}{3}$, ▢ , -0.1, ▢ , $\dfrac{4}{5}$, ▢

7. -3.4, ▢ , -1.5, ▢ , 2.2, ▢

Practice

Use what you learned about ordering rational numbers to complete Exercises 28–30 on page 48.

Check It Out
Lesson Tutorials
BigIdeasMath com

Key Vocabulary ◀))
rational number,
  p. 46
terminating decimal,
  p. 46
repeating decimal,
  p. 46

 **Key Idea**

Rational Numbers

A **rational number** is a number that can be written as $\frac{a}{b}$ where a and b are integers and $b \neq 0$.

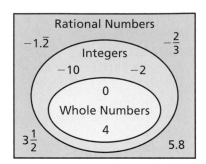

Because you can divide any integer by any nonzero integer, you can use long division to write fractions and mixed numbers as decimals. These decimals are also rational numbers and will either *terminate* or *repeat*.

A **terminating decimal** is a decimal that ends.

> $1.5, -0.25, 10.625$

A **repeating decimal** is a decimal that has a pattern that repeats.

> $-1.333\ldots = -1.\overline{3}$
> $0.151515\ldots = 0.\overline{15}$

Use *bar notation* to show which of the digits repeat.

EXAMPLE 1 **Writing Rational Numbers as Decimals**

a. Write $-2\frac{1}{4}$ as a decimal.

Notice that $-2\frac{1}{4} = -\frac{9}{4}$.

Divide 9 by 4.

$$\begin{array}{r} 2.25 \\ 4\overline{)9.00} \\ -8 \\ \hline 1\,0 \\ -8 \\ \hline 20 \\ -20 \\ \hline 0 \end{array}$$

The remainder is 0. So, it is a terminating decimal.

∴ So, $-2\frac{1}{4} = -2.25$.

b. Write $\frac{5}{11}$ as a decimal.

Divide 5 by 11.

$$\begin{array}{r} 0.4545 \\ 11\overline{)5.0000} \\ -4\,4 \\ \hline 60 \\ -55 \\ \hline 50 \\ -44 \\ \hline 60 \\ -55 \\ \hline 5 \end{array}$$

The remainder repeats. So, it is a repeating decimal.

∴ So, $\frac{5}{11} = 0.\overline{45}$.

● **On Your Own**

Now You're Ready
Exercises 11–18

Write the rational number as a decimal.

1. $-\frac{6}{5}$ **2.** $-7\frac{3}{8}$ **3.** $-\frac{3}{11}$ **4.** $1\frac{5}{27}$

◀)) Multi-Language Glossary at BigIdeasMath✓com

EXAMPLE **2** **Writing a Decimal as a Fraction**

Write -0.26 as a fraction in simplest form.

Study Tip

If p and q are integers, then $-\dfrac{p}{q} = \dfrac{-p}{q} = \dfrac{p}{-q}$.

$-0.26 = -\dfrac{26}{100}$ ← Write the digits after the decimal point in the numerator.

← The last digit is in the hundredths place. So, use 100 in the denominator.

$= -\dfrac{13}{50}$ Simplify.

On Your Own

Now You're Ready
Exercises 20–27

Write the decimal as a fraction or a mixed number in simplest form.

5. -0.7 **6.** 0.125 **7.** -3.1 **8.** -10.25

EXAMPLE **3** **Ordering Rational Numbers**

Creature	Elevation (kilometers)
Anglerfish	$-\dfrac{13}{10}$
Squid	$-2\dfrac{1}{5}$
Shark	$-\dfrac{2}{11}$
Whale	-0.8

The table shows the elevations of four sea creatures relative to sea level. Which of the sea creatures are deeper than the whale? Explain.

Write each rational number as a decimal.

$-\dfrac{13}{10} = -1.3$

$-2\dfrac{1}{5} = -2.2$

$-\dfrac{2}{11} = -0.\overline{18}$

Then graph each decimal on a number line.

⋮∙ Both -2.2 and -1.3 are less than -0.8. So, the squid and the anglerfish are deeper than the whale.

On Your Own

Now You're Ready
Exercises 28–33

9. WHAT IF? The elevation of a dolphin is $-\dfrac{1}{10}$ kilometer. Which of the sea creatures in Example 3 are deeper than the dolphin? Explain.

✓ Vocabulary and Concept Check

1. **VOCABULARY** Is the quotient of two integers always a rational number? Explain.

2. **WRITING** Are all terminating and repeating decimals rational numbers? Explain.

Tell whether the number belongs to each of the following number sets: *rational numbers, integers, whole numbers.*

3. -5

4. $-2.1\overline{6}$

5. 12

6. 0

Tell whether the decimal is *terminating* or *repeating*.

7. $-0.4848\ldots$

8. -0.151

9. 72.72

10. $-5.2\overline{36}$

Practice and Problem Solving

Write the rational number as a decimal.

 11. $\dfrac{7}{8}$

12. $\dfrac{1}{11}$

13. $-\dfrac{7}{9}$

14. $-\dfrac{17}{40}$

15. $1\dfrac{5}{6}$

16. $-2\dfrac{17}{18}$

17. $-5\dfrac{7}{12}$

18. $8\dfrac{15}{22}$

19. **ERROR ANALYSIS** Describe and correct the error in writing the rational number as a decimal.

$$\boldsymbol{X} \qquad -\dfrac{7}{11} = -0.6\overline{3}$$

Write the decimal as a fraction or a mixed number in simplest form.

20. -0.9

21. 0.45

22. -0.258

23. -0.312

24. -2.32

25. -1.64

26. 6.012

27. -12.405

Order the numbers from least to greatest.

28. $-\dfrac{3}{4}, 0.5, \dfrac{2}{3}, -\dfrac{7}{3}, 1.2$

29. $\dfrac{9}{5}, -2.5, -1.1, -\dfrac{4}{5}, 0.8$

30. $-1.4, -\dfrac{8}{5}, 0.6, -0.9, \dfrac{1}{4}$

31. $2.1, -\dfrac{6}{10}, -\dfrac{9}{4}, -0.75, \dfrac{5}{3}$

32. $-\dfrac{7}{2}, -2.8, -\dfrac{5}{4}, \dfrac{4}{3}, 1.3$

33. $-\dfrac{11}{5}, -2.4, 1.6, \dfrac{15}{10}, -2.25$

34. **COINS** You lose one quarter, two dimes, and two nickels.

   a. Write the amount as a decimal.

   b. Write the amount as a fraction in simplest form.

35. **HIBERNATION** A box turtle hibernates in sand at $-1\dfrac{5}{8}$ feet. A spotted turtle hibernates at $-1\dfrac{16}{25}$ feet. Which turtle is deeper?

Copy and complete the statement using <, >, or =.

36. -2.2 ☐ -2.42

37. -1.82 ☐ -1.81

38. $\dfrac{15}{8}$ ☐ $1\dfrac{7}{8}$

39. $-4\dfrac{6}{10}$ ☐ -4.65

40. $-5\dfrac{3}{11}$ ☐ $-5.\overline{2}$

41. $-2\dfrac{13}{16}$ ☐ $-2\dfrac{11}{14}$

42. OPEN-ENDED Find one terminating decimal and one repeating decimal between $-\dfrac{1}{2}$ and $-\dfrac{1}{3}$.

Player	Hits	At Bats
Eva	42	90
Michelle	38	80

43. SOFTBALL In softball, a batting average is the number of hits divided by the number of times at bat. Does Eva or Michelle have the higher batting average?

44. PROBLEM SOLVING You miss 3 out of 10 questions on a science quiz and 4 out of 15 questions on a math quiz. Which quiz has a higher percent of correct answers?

45. SKATING Is the half pipe deeper than the skating pool? Explain.

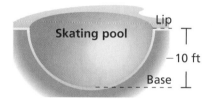

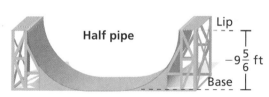

46. ENVIRONMENT The table shows the changes from the average water level of a pond over several weeks. Order the numbers from least to greatest.

Week	1	2	3	4
Change (inches)	$-\dfrac{7}{5}$	$-1\dfrac{5}{11}$	-1.45	$-1\dfrac{91}{200}$

47. **Critical Thinking** Given: a and b are integers.

a. When is $-\dfrac{1}{a}$ positive?

b. When is $\dfrac{1}{ab}$ positive?

Fair Game Review What you learned in previous grades & lessons

Add or subtract. *(Skills Review Handbook)*

48. $\dfrac{3}{5} + \dfrac{2}{7}$

49. $\dfrac{9}{10} - \dfrac{2}{3}$

50. $8.79 - 4.07$

51. $11.81 + 9.34$

52. MULTIPLE CHOICE In one year, a company has a profit of $-\$2$ million. In the next year, the company has a profit of $\$7$ million. How much more profit did the company make the second year? *(Section 1.3)*

(A) $2 million

(B) $5 million

(C) $7 million

(D) $9 million

2.2 Adding Rational Numbers

Essential Question How can you use what you know about adding integers to add rational numbers?

1 ACTIVITY: Adding Rational Numbers

Work with a partner. Use a number line to find the sum.

a. $2.7 + (-3.4)$

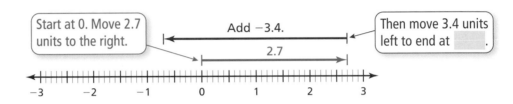

> Start at 0. Move 2.7 units to the right.

> Add −3.4.

> 2.7

> Then move 3.4 units left to end at [ ].

So, $2.7 + (-3.4) = $ [ ] .

b. $1.3 + (-1.5)$

c. $-2.1 + 0.8$

d. $-1\frac{1}{4} + \frac{3}{4}$

e. $\frac{3}{10} + \left(-\frac{3}{10}\right)$

2 ACTIVITY: Adding Rational Numbers

Work with a partner. Use a number line to find the sum.

a. $-1\frac{2}{5} + \left(-\frac{4}{5}\right)$

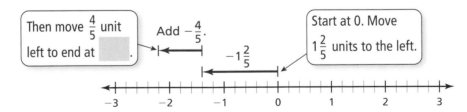

> Then move $\frac{4}{5}$ unit left to end at [ ].

> Add $-\frac{4}{5}$.

> $-1\frac{2}{5}$

> Start at 0. Move $1\frac{2}{5}$ units to the left.

So, $-1\frac{2}{5} + \left(-\frac{4}{5}\right) = $ [ ] .

b. $-\frac{7}{10} + \left(-1\frac{7}{10}\right)$

c. $-1\frac{2}{3} + \left(-1\frac{1}{3}\right)$

d. $-0.4 + (-1.9)$

e. $-2.3 + (-0.6)$

COMMON CORE

Rational Numbers

In this lesson, you will

- add rational numbers.
- solve real-life problems.

Learning Standards
7.NS.1a
7.NS.1b
7.NS.1d
7.NS.3

3 ACTIVITY: Writing Expressions

Work with a partner. Write the addition expression shown. Then find the sum.

Math Practice 2

Use Operations

What operation is represented in each number line? How does this help you write an expression?

a.

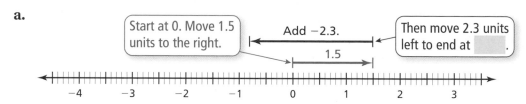

| Start at 0. Move 1.5 units to the right. | Add −2.3. | Then move 2.3 units left to end at ☐. |

1.5

b.

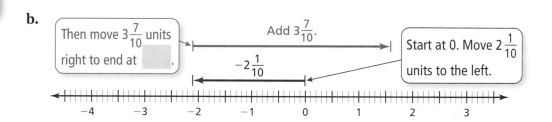

| Then move $3\frac{7}{10}$ units right to end at ☐. | Add $3\frac{7}{10}$. | Start at 0. Move $2\frac{1}{10}$ units to the left. |

$-2\frac{1}{10}$

c.

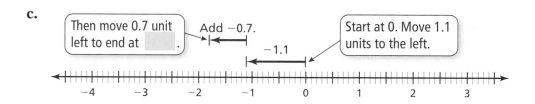

| Then move 0.7 unit left to end at ☐. | Add −0.7. | Start at 0. Move 1.1 units to the left. |

−1.1

What Is Your Answer?

4. IN YOUR OWN WORDS How can you use what you know about adding integers to add rational numbers?

PUZZLE Find a path through the table so that the numbers add up to the sum. You can move horizontally or vertically.

5. Sum: $\frac{3}{4}$

Start →

$\frac{1}{2}$	$\frac{2}{3}$	$-\frac{5}{7}$
$-\frac{1}{8}$	$-\frac{3}{4}$	$\frac{1}{3}$

← End

6. Sum: −0.07

Start →

2.43	1.75	−0.98
−1.09	3.47	−4.88

← End

Practice

Use what you learned about adding rational numbers to complete Exercises 4–6 on page 54.

Check It Out
Lesson Tutorials
BigIdeasMath com

🔑 Key Idea

Adding Rational Numbers

Words To add rational numbers, use the same rules for signs as you used for integers.

Numbers $-\dfrac{1}{3} + \dfrac{1}{6} = \dfrac{-2}{6} + \dfrac{1}{6} = \dfrac{-2+1}{6} = \dfrac{-1}{6} = -\dfrac{1}{6}$

EXAMPLE ① **Adding Rational Numbers**

Study Tip

In Example 1, notice how $-\dfrac{8}{3}$ is written as $-\dfrac{8}{3} = \dfrac{-8}{3} = \dfrac{-16}{6}$.

Find $-\dfrac{8}{3} + \dfrac{5}{6}$.

Estimate $-3 + 1 = -2$

$-\dfrac{8}{3} + \dfrac{5}{6} = \dfrac{-16}{6} + \dfrac{5}{6}$ Rewrite using the LCD (least common denominator).

$= \dfrac{-16 + 5}{6}$ Write the sum of the numerators over the common denominator.

$= \dfrac{-11}{6}$ Add.

$= -1\dfrac{5}{6}$ Write the improper fraction as a mixed number.

∴ The sum is $-1\dfrac{5}{6}$.   **Reasonable?** $-1\dfrac{5}{6} \approx -2$ ✓

EXAMPLE ② **Adding Rational Numbers**

Find $-4.05 + 7.62$.

$-4.05 + 7.62 = 3.57$ $|7.62| > |-4.05|$. So, subtract $|-4.05|$ from $|7.62|$.

Use the sign of 7.62.

∴ The sum is 3.57.

⬤ On Your Own

Now You're Ready
Exercises 4–12

Add.

1. $-\dfrac{7}{8} + \dfrac{1}{4}$

2. $-6\dfrac{1}{3} + \dfrac{20}{3}$

3. $2 + \left(-\dfrac{7}{2}\right)$

4. $-12.5 + 15.3$

5. $-8.15 + (-4.3)$

6. $0.65 + (-2.75)$

EXAMPLE 3 **Evaluating Expressions**

Evaluate $2x + y$ when $x = \dfrac{1}{4}$ and $y = -\dfrac{3}{2}$.

$$2x + y = 2\left(\dfrac{1}{4}\right) + \left(-\dfrac{3}{2}\right)$$ Substitute $\dfrac{1}{4}$ for x and $-\dfrac{3}{2}$ for y.

$$= \dfrac{1}{2} + \left(\dfrac{-3}{2}\right)$$ Multiply.

$$= \dfrac{1 + (-3)}{2}$$ Write the sum of the numerators over the common denominator.

$$= -1$$ Simplify.

EXAMPLE 4 **Real-Life Application**

Year	Profit (billions of dollars)
2008	−1.7
2009	−4.75
2010	1.7
2011	0.85
2012	3.6

The table shows the annual profits (in billions of dollars) of a financial company from 2008 to 2012. Positive numbers represent *gains*, and negative numbers represent *losses*. Which statement describes the profit over the five-year period?

(A) gain of $0.3 billion   (B) gain of $30 million

(C) loss of $3 million   (D) loss of $300 million

To determine whether there was a gain or a loss, find the sum of the profits.

five-year profit $= -1.7 + (-4.75) + 1.7 + 0.85 + 3.6$ Write the sum.

$= -1.7 + 1.7 + (-4.75) + 0.85 + 3.6$ Comm. Prop. of Add.

$= 0 + (-4.75) + 0.85 + 3.6$ Additive Inv. Prop.

$= -4.75 + 0.85 + 3.6$ Add. Prop. of Zero

$= -3.9 + 3.6$ Add −4.75 and 0.85.

$= -0.3$ Add −3.9 and 3.6.

The five-year profit is −$0.3 billion. So, the company has a five-year loss of $0.3 billion, or $300 million.

∴ The correct answer is (D).

● **On Your Own**

Now You're Ready
Exercises 15–17

Evaluate the expression when $a = \dfrac{1}{2}$ and $b = -\dfrac{5}{2}$.

7. $b + 4a$ 8. $|a + b|$

9. **WHAT IF?** In Example 4, the 2013 profit is $1.07 billion. State the company's gain or loss over the six-year period in millions of dollars.

 Vocabulary and Concept Check

1. **WRITING** Explain how to find the sum $-8.46 + 5.31$.

2. **OPEN-ENDED** Write an addition expression using fractions that equals $-\frac{1}{2}$.

3. **DIFFERENT WORDS, SAME QUESTION** Which is different? Find "both" answers.

Add -4.5 and 3.5.	What is the distance between -4.5 and 3.5?
What is -4.5 increased by 3.5?	Find the sum of -4.5 and 3.5.

 Practice and Problem Solving

Add. Write fractions in simplest form.

① ② 4. $\frac{11}{12} + \left(-\frac{7}{12}\right)$ 5. $-1\frac{1}{5} + \left(-\frac{3}{5}\right)$ 6. $-4.2 + 3.3$

7. $-\frac{9}{14} + \frac{2}{7}$ 8. $4 + \left(-1\frac{2}{3}\right)$ 9. $\frac{15}{4} + \left(-4\frac{1}{3}\right)$

10. $-3.1 + (-0.35)$ 11. $12.48 + (-10.636)$ 12. $20.25 + (-15.711)$

ERROR ANALYSIS Describe and correct the error in finding the sum.

13.
$$
\begin{array}{r}
-3.7 \\
+ \ (-0.25) \\
\hline
-0.62
\end{array}
$$

14.
$$-\frac{5}{8} + \frac{1}{8} = \frac{-5 + 1}{8} = \frac{-6}{8} = -\frac{3}{4}$$

Evaluate the expression when $x = \frac{1}{3}$ and $y = -\frac{7}{4}$.

③ 15. $x + y$ 16. $3x + y$ 17. $-x + |y|$

18. **BANKING** Your bank account balance is –$20.85. You deposit $15.50. What is your new balance?

19. **HOT DOGS** You eat $\frac{3}{10}$ of a pack of hot dogs.

Your friend eats $\frac{1}{5}$ of the pack of hot dogs.

What fraction of the pack of hot dogs do

you and your friend eat?

Add. Write fractions in simplest form.

20. $6 + \left(-4\frac{3}{4}\right) + \left(-2\frac{1}{8}\right)$

21. $-5\frac{2}{3} + 3\frac{1}{4} + \left(-7\frac{1}{3}\right)$

22. $10.9 + (-15.6) + 2.1$

23. NUMBER SENSE When is the sum of two negative mixed numbers an integer?

24. WRITING You are adding two rational numbers with different signs. How can you tell if the sum will be *positive*, *negative*, or *zero*?

25. RESERVOIR The table at the left shows the water level (in inches) of a reservoir for three months compared to the yearly average. Is the water level for the three-month period greater than or less than the yearly average? Explain.

June	July	August
$-2\frac{1}{8}$	$1\frac{1}{4}$	$-\frac{9}{16}$

26. BREAK EVEN The table at the right shows the annual profits (in thousands of dollars) of a county fair from 2008 to 2012. What must the 2012 profit be (in hundreds of dollars) to break even over the five-year period?

Year	Profit (thousands of dollars)
2008	2.5
2009	1.75
2010	-3.3
2011	-1.4
2012	?

27. REASONING Is $|a + b| = |a| + |b|$ for all rational numbers a and b? Explain.

28. **Repeated Reasoning** Evaluate the expression.

$$\frac{19}{20} + \left(\frac{-18}{20}\right) + \frac{17}{20} + \left(\frac{-16}{20}\right) + \cdots + \left(\frac{-4}{20}\right) + \frac{3}{20} + \left(\frac{-2}{20}\right) + \frac{1}{20}$$

 Fair Game Review *What you learned in previous grades & lessons*

Identify the property. Then simplify. *(Skills Review Handbook)*

29. $8 + (-3) + 2 = 8 + 2 + (-3)$

30. $2 \cdot (4.5 \cdot 9) = (2 \cdot 4.5) \cdot 9$

31. $\frac{1}{4} + \left(\frac{3}{4} + \frac{1}{8}\right) = \left(\frac{1}{4} + \frac{3}{4}\right) + \frac{1}{8}$

32. $\frac{3}{7} \cdot \frac{4}{5} \cdot \frac{14}{27} = \frac{3}{7} \cdot \frac{14}{27} \cdot \frac{4}{5}$

33. MULTIPLE CHOICE The regular price of a photo album is $18. You have a coupon for 15% off. How much is the discount? *(Skills Review Handbook)*

    Ⓐ $2.70   **Ⓑ** $3   **Ⓒ** $15   **Ⓓ** $15.30

You can use a **process diagram** to show the steps involved in a procedure. Here is an example of a process diagram for adding rational numbers.

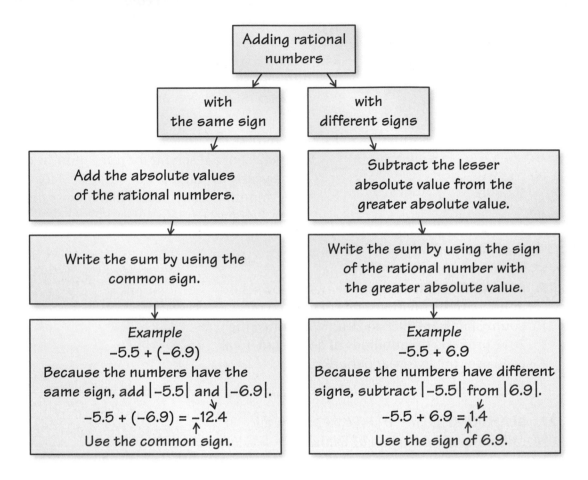

On Your Own

Make a process diagram with examples to help you study the topic.

1. writing rational numbers as decimals

After you complete this chapter, make process diagrams with examples for the following topics.

2. subtracting rational numbers

3. multiplying rational numbers

4. dividing rational numbers

"Does this process diagram accurately show how a cat claws furniture?"

Write the rational number as a decimal.   *(Section 2.1)*

1. $-\dfrac{3}{20}$

2. $-\dfrac{11}{6}$

Write the decimal as a fraction or a mixed number in simplest form.   *(Section 2.1)*

3. -0.325

4. -1.28

Order the numbers from least to greatest.   *(Section 2.1)*

5. $-\dfrac{1}{3}, -0.2, \dfrac{5}{3}, 0.4, 1.3$

6. $-\dfrac{4}{3}, -1.2, 0.3, \dfrac{4}{9}, -0.8$

Add. Write fractions in simplest form.   *(Section 2.2)*

7. $-\dfrac{4}{5} + \left(-\dfrac{3}{8}\right)$

8. $-\dfrac{13}{6} + \dfrac{7}{12}$

9. $-5.8 + 2.6$

10. $-4.28 + (-2.56)$

Evaluate the expression when $x = \dfrac{3}{4}$ **and** $y = -\dfrac{1}{2}$**.**   *(Section 2.2)*

11. $x + y$

12. $2x + y$

13. $x + |y|$

14. $|-x + y|$

15. STOCK The value of Stock A changes $-\$3.68$, and the value of Stock B changes $-\$3.72$. Which stock has the greater loss? Explain.   *(Section 2.1)*

16. LEMONADE You drink $\dfrac{2}{7}$ of a pitcher of lemonade. Your friend drinks $\dfrac{3}{14}$ of the pitcher. What fraction of the pitcher do you and your friend drink?   *(Section 2.2)*

17. FOOTBALL The table shows the statistics of a running back in a football game. Did he gain more than 50 yards total? Explain.   *(Section 2.2)*

Quarter	1	2	3	4	Total
Yards	$-8\dfrac{1}{2}$	23	$42\dfrac{1}{2}$	$-2\dfrac{1}{4}$	?

Essential Question How can you use what you know about subtracting integers to subtract rational numbers?

1 ACTIVITY: Subtracting Rational Numbers

Work with a partner. Use a number line to find the difference.

a. $-1\frac{1}{2} - \frac{1}{2}$

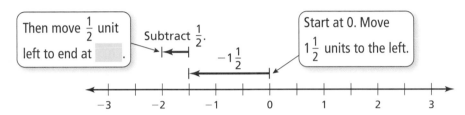

Then move $\frac{1}{2}$ unit left to end at ▢.

Subtract $\frac{1}{2}$.

Start at 0. Move $1\frac{1}{2}$ units to the left.

$-1\frac{1}{2}$

So, $-1\frac{1}{2} - \frac{1}{2} = $ ▢.

b. $\frac{6}{10} - 1\frac{3}{10}$

c. $-1\frac{1}{4} - 1\frac{3}{4}$

d. $-1.9 - 0.8$

e. $0.2 - 0.7$

2 ACTIVITY: Finding Distances on a Number Line

Work with a partner.

a. Plot -3 and 2 on the number line. Then find $-3 - 2$ and $2 - (-3)$. What do you notice about your results?

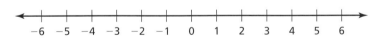

COMMON CORE

Rational Numbers

In this lesson, you will
- subtract rational numbers.
- solve real-life problems.

Learning Standards
7.NS.1c
7.NS.1d
7.NS.3

b. Plot $\frac{3}{4}$ and 1 on the number line. Then find $\frac{3}{4} - 1$ and $1 - \frac{3}{4}$. What do you notice about your results?

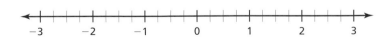

c. Choose any two points a and b on a number line. Find the values of $a - b$ and $b - a$. What do the absolute values of these differences represent? Is this true for any pair of rational numbers? Explain.

3 ACTIVITY: Financial Literacy

Work with a partner. The table shows the balance in a checkbook.

- Black numbers are amounts added to the account.
- Red numbers are amounts taken from the account.

Date	Check #	Transaction	Amount	Balance
––	––	Previous balance	––	100.00
1/02/2013	124	Groceries	34.57	
1/07/2013		Check deposit	875.50	
1/11/2013		ATM withdrawal	40.00	
1/14/2013	125	Electric company	78.43	
1/17/2013		Music store	10.55	
1/18/2013	126	Shoes	47.21	
1/22/2013		Check deposit	125.00	
1/24/2013		Interest	2.12	
1/25/2013	127	Cell phone	59.99	
1/26/2013	128	Clothes	65.54	
1/30/2013	129	Cable company	75.00	

Math Practice 4

Interpret Results

What does your answer represent? Does your answer make sense?

You can find the balance in the **second row** two different ways.

$$100.00 - 34.57 = 65.43 \qquad \text{Subtract 34.57 from 100.00.}$$
$$100.00 + (-34.57) = 65.43 \qquad \text{Add } -34.57 \text{ to 100.00.}$$

a. Copy the table. Then complete the balance column.

b. How did you find the balance in the **twelfth row**?

c. Use a different way to find the balance in part (b).

What Is Your Answer?

4. **IN YOUR OWN WORDS** How can you use what you know about subtracting integers to subtract rational numbers?

5. Give two real-life examples of subtracting rational numbers that are not integers.

Practice Use what you learned about subtracting rational numbers to complete Exercises 3–5 on page 62.

 **Key Idea**

Subtracting Rational Numbers

Words To subtract rational numbers, use the same rules for signs as you used for integers.

Numbers $\dfrac{2}{5} - \left(-\dfrac{1}{5}\right) = \dfrac{2}{5} + \dfrac{1}{5} = \dfrac{2+1}{5} = \dfrac{3}{5}$

EXAMPLE **1** **Subtracting Rational Numbers**

Find $-4\dfrac{1}{7} - \left(-\dfrac{6}{7}\right)$. **Estimate** $-4 - (-1) = -3$

$$-4\dfrac{1}{7} - \left(-\dfrac{6}{7}\right) = -4\dfrac{1}{7} + \dfrac{6}{7}$$ Add the opposite of $-\dfrac{6}{7}$.

$$= -\dfrac{29}{7} + \dfrac{6}{7}$$ Write the mixed number as an improper fraction.

$$= \dfrac{-29 + 6}{7}$$ Write the sum of the numerators over the common denominator.

$$= \dfrac{-23}{7}$$ Add.

$$= -3\dfrac{2}{7}$$ Write the improper fraction as a mixed number.

∴ The difference is $-3\dfrac{2}{7}$. **Reasonable?** $-3\dfrac{2}{7} \approx -3$ ✓

EXAMPLE **2** **Subtracting Rational Numbers**

Find $12.8 - 21.6$.

$12.8 - 21.6 = 12.8 + (-21.6)$ Add the opposite of 21.6.

$= -8.8$ $|-21.6| > |12.8|$. So, subtract $|12.8|$ from $|-21.6|$.

∴ The difference is -8.8.   Use the sign of -21.6.

On Your Own

Now You're Ready
Exercises 3–11

1. $\dfrac{1}{3} - \left(-\dfrac{1}{3}\right)$

2. $-3\dfrac{1}{3} - \dfrac{5}{6}$

3. $4\dfrac{1}{2} - 5\dfrac{1}{4}$

4. $-8.4 - 6.7$

5. $-20.5 - (-20.5)$

6. $0.41 - (-0.07)$

The distance between any two numbers on a number line is the absolute value of the difference of the numbers.

EXAMPLE ③ **Finding Distances Between Numbers on a Number Line**

Find the distance between the two numbers on the number line.

To find the distance between the numbers, first find the difference of the numbers.

$$-2\frac{2}{3} - 2\frac{1}{3} = -2\frac{2}{3} + \left(-2\frac{1}{3}\right) \quad \text{Add the opposite of } 2\frac{1}{3}.$$

$$= -\frac{8}{3} + \left(-\frac{7}{3}\right) \quad \text{Write the mixed numbers as improper fractions.}$$

$$= \frac{-15}{3} \quad \text{Add.}$$

$$= -5 \quad \text{Simplify.}$$

∴ Because $|-5| = 5$, the distance between $-2\frac{2}{3}$ and $2\frac{1}{3}$ is 5.

EXAMPLE ④ **Real-Life Application**

Clearance: 11 ft 8 in.

In the water, the bottom of a boat is 2.1 feet below the surface, and the top of the boat is 8.7 feet above it. Towed on a trailer, the bottom of the boat is 1.3 feet above the ground. Can the boat and trailer pass under the bridge?

Step 1: Find the height h of the boat.

$$h = 8.7 - (-2.1) \quad \text{Subtract the lowest point from the highest point.}$$

$$= 8.7 + 2.1 \quad \text{Add the opposite of } -2.1.$$

$$= 10.8 \quad \text{Add.}$$

Step 2: Find the height t of the boat and trailer.

$$t = 10.8 + 1.3 \quad \text{Add the trailer height to the boat height.}$$

$$= 12.1 \quad \text{Add.}$$

∴ Because 12.1 feet is greater than 11 feet 8 inches, the boat and trailer cannot pass under the bridge.

On Your Own

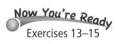
Now You're Ready
Exercises 13–15

7. Find the distance between -7.5 and -15.3 on a number line.

8. **WHAT IF?** In Example 4, the clearance is 12 feet 1 inch. Can the boat and trailer pass under the bridge?

✓ Vocabulary and Concept Check

1. **WRITING** Explain how to find the difference $-\dfrac{4}{5} - \dfrac{3}{5}$.

2. **WHICH ONE DOESN'T BELONG?** Which expression does *not* belong with the other three? Explain your reasoning.

$$-\dfrac{5}{8} - \dfrac{3}{4} \qquad -\dfrac{3}{4} + \dfrac{5}{8} \qquad -\dfrac{5}{8} + \left(-\dfrac{3}{4}\right) \qquad -\dfrac{3}{4} - \dfrac{5}{8}$$

Practice and Problem Solving

Subtract. Write fractions in simplest form.

① ② **3.** $\dfrac{5}{8} - \left(-\dfrac{7}{8}\right)$

4. $-1\dfrac{1}{3} - 1\dfrac{2}{3}$

5. $-1 - 2.5$

6. $-5 - \dfrac{5}{3}$

7. $-8\dfrac{3}{8} - 10\dfrac{1}{6}$

8. $-\dfrac{1}{2} - \left(-\dfrac{5}{9}\right)$

9. $5.5 - 8.1$

10. $-7.34 - (-5.51)$

11. $6.673 - (-8.29)$

12. ERROR ANALYSIS Describe and correct the error in finding the difference.

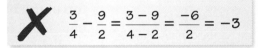

$$\times \qquad \dfrac{3}{4} - \dfrac{9}{2} = \dfrac{3-9}{4-2} = \dfrac{-6}{2} = -3$$

Find the distance between the two numbers on a number line.

③ **13.** $-2\dfrac{1}{2}, \ -5\dfrac{3}{4}$

14. $-2.2, \ 8.4$

15. $-7, \ -3\dfrac{2}{3}$

16. SPORTS DRINK Your sports drink bottle is $\dfrac{5}{6}$ full. After practice, the bottle is $\dfrac{3}{8}$ full. Write the difference of the amounts after practice and before practice.

17. SUBMARINE The figure shows the depths of a submarine.

a. Find the vertical distance traveled by the submarine.

b. Find the mean hourly vertical distance traveled by the submarine.

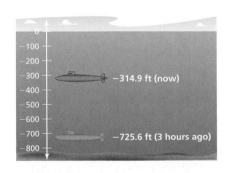

Evaluate.

18. $2\dfrac{1}{6} - \left(-\dfrac{8}{3}\right) + \left(-4\dfrac{7}{9}\right)$

19. $6.59 + (-7.8) - (-2.41)$

20. $-\dfrac{12}{5} + \left|-\dfrac{13}{6}\right| + \left(-3\dfrac{2}{3}\right)$

21. REASONING When is the difference of two decimals an integer? Explain.

22. RECIPE A cook has $2\frac{2}{3}$ cups of flour. A recipe calls for $2\frac{3}{4}$ cups of flour. Does the cook have enough flour? If not, how much more flour is needed?

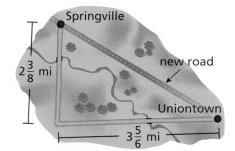

23. ROADWAY A new road that connects Uniontown to Springville is $4\frac{1}{3}$ miles long. What is the change in distance when using the new road instead of the dirt roads?

RAINFALL In Exercises 24–26, the bar graph shows the differences in a city's rainfall from the historical average.

24. What is the difference in rainfall between the wettest and the driest months?

25. Find the sum of the differences for the year.

26. What does the sum in Exercise 25 tell you about the rainfall for the year?

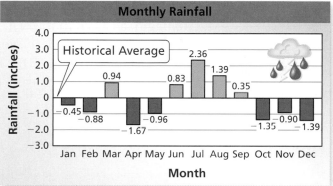

27. OPEN-ENDED Write two different pairs of negative decimals, x and y, that make the statement $x - y = 0.6$ true.

REASONING Tell whether the difference between the two numbers is *always*, *sometimes*, or *never* positive. Explain your reasoning.

28. two negative fractions

29. a positive decimal and a negative decimal

30. Structure Fill in the blanks to make the solution correct.

$$5.\ \boxed{}4 - \left(\boxed{}.8\boxed{}\right) = -3.61$$

 Fair Game Review *What you learned in previous grades & lessons*

Evaluate. *(Skills Review Handbook)*

31. 5.2×6.9

32. $7.2 \div 2.4$

33. $2\frac{2}{3} \times 3\frac{1}{4}$

34. $9\frac{4}{5} \div 3\frac{1}{2}$

35. MULTIPLE CHOICE A sports store has 116 soccer balls. Over 6 months, it sells 8 soccer balls per month. How many soccer balls are in inventory at the end of the 6 months? *(Section 1.3 and Section 1.4)*

(A) -48 (B) 48 (C) 68 (D) 108

2.4 Multiplying and Dividing Rational Numbers

Essential Question Why is the product of two negative rational numbers positive?

In Section 1.4, you used a table to see that the product of two negative integers is a positive integer. In this activity, you will find that same result another way.

1 ACTIVITY: Showing $(-1)(-1) = 1$

Work with a partner. How can you show that $(-1)(-1) = 1$?

To begin, assume that $(-1)(-1) = 1$ is a true statement. From the Additive Inverse Property, you know that $1 + (-1) = 0$. So, substitute $(-1)(-1)$ for 1 to get $(-1)(-1) + (-1) = 0$. If you can show that $(-1)(-1) + (-1) = 0$ is true, then you have shown that $(-1)(-1) = 1$.

Justify each step.

$$(-1)(-1) + (-1) = (-1)(-1) + 1(-1)$$

$$= (-1)[(-1) + 1]$$

$$= (-1)0$$

$$= 0$$

So, $(-1)(-1) = 1$.

2 ACTIVITY: Multiplying by -1

Work with a partner.

a. Graph each number below on three different number lines. Then multiply each number by -1 and graph the product on the appropriate number line.

$$2 \qquad 8 \qquad -1$$

b. How does multiplying by -1 change the location of the points in part (a)? What is the relationship between the number and the product?

c. Graph each number below on three different number lines. Where do you think the points will be after multiplying by -1? Plot the points. Explain your reasoning.

$$\frac{1}{2} \qquad 2.5 \qquad -\frac{5}{2}$$

d. What is the relationship between a rational number $-a$ and the product $-1(a)$? Explain your reasoning.

COMMON CORE

Rational Numbers
In this lesson, you will
- multiply and divide rational numbers.
- solve real-life problems.

Learning Standards
7.NS.2a
7.NS.2b
7.NS.2c
7.NS.3

3 ACTIVITY: Understanding the Product of Rational Numbers

Work with a partner. Let *a* and *b* be positive rational numbers.

a. Because *a* and *b* are positive, what do you know about $-a$ and $-b$?

b. Justify each step.

$$(-a)(-b) = (-1)(a)(-1)(b)$$

$$= (-1)(-1)(a)(b)$$

$$= (1)(a)(b)$$

$$= ab$$

c. Because *a* and *b* are positive, what do you know about the product *ab*?

d. What does this tell you about products of rational numbers? Explain.

4 ACTIVITY: Writing a Story

Work with a partner. Write a story that uses addition, subtraction, multiplication, or division of rational numbers.

- At least one of the numbers in the story has to be negative and *not* an integer.
- Draw pictures to help illustrate what is happening in the story.
- Include the solution of the problem in the story.

Math Practice 6

Specify Units

What units are in your story?

If you are having trouble thinking of a story, here are some common uses of negative numbers:

- A profit of $-\$15$ is a loss of $15.
- An elevation of -100 feet is a depth of 100 feet below sea level.
- A gain of -5 yards in football is a loss of 5 yards.
- A score of -4 in golf is 4 strokes under par.

What Is Your Answer?

5. IN YOUR OWN WORDS Why is the product of two negative rational numbers positive?

6. PRECISION Show that $(-2)(-3) = 6$.

7. How can you show that the product of a negative rational number and a positive rational number is negative?

 Use what you learned about multiplying rational numbers to complete Exercises 7–9 on page 68.

Check It Out
Lesson Tutorials
BigIdeasMath ✓com

🔑 Key Idea

Multiplying and Dividing Rational Numbers

Words To multiply or divide rational numbers, use the same rules for signs as you used for integers.

Remember

The *reciprocal* of $\frac{a}{b}$ is $\frac{b}{a}$.

Numbers $-\frac{2}{7} \cdot \frac{1}{3} = \frac{-2 \cdot 1}{7 \cdot 3} = \frac{-2}{21} = -\frac{2}{21}$

$-\frac{1}{2} \div \frac{4}{9} = \frac{-1}{2} \cdot \frac{9}{4} = \frac{-1 \cdot 9}{2 \cdot 4} = \frac{-9}{8} = -\frac{9}{8}$

EXAMPLE **1** **Dividing Rational Numbers**

Find $-5\frac{1}{5} \div 2\frac{1}{3}$.   **Estimate** $-5 \div 2 = -2\frac{1}{2}$

$-5\frac{1}{5} \div 2\frac{1}{3} = -\frac{26}{5} \div \frac{7}{3}$ Write mixed numbers as improper fractions.

$= \frac{-26}{5} \cdot \frac{3}{7}$ Multiply by the reciprocal of $\frac{7}{3}$.

$= \frac{-26 \cdot 3}{5 \cdot 7}$ Multiply the numerators and the denominators.

$= \frac{-78}{35}$, or $-2\frac{8}{35}$ Simplify.

∴ The quotient is $-2\frac{8}{35}$.   **Reasonable?** $-2\frac{8}{35} \approx -2\frac{1}{2}$ ✓

EXAMPLE **2** **Multiplying Rational Numbers**

Find $-2.5 \cdot 3.6$.

$$
\begin{array}{r}
-2.5 \\
\times\ 3.6 \\
\hline
1\,5\,0 \\
7\,5\,0 \\
\hline
-9.0\,0
\end{array}
$$

← The decimals have different signs.

← The product is negative.

∴ The product is -9.

EXAMPLE 3

Multiplying More Than Two Rational Numbers

Find $-\dfrac{1}{7} \cdot \left[\dfrac{4}{5} \cdot (-7) \right]$.

You can use properties of multiplication to make the product easier to find.

$$-\dfrac{1}{7} \cdot \left[\dfrac{4}{5} \cdot (-7) \right] = -\dfrac{1}{7} \cdot \left(-7 \cdot \dfrac{4}{5} \right)$$ Commutative Property of Multiplication

$$= -\dfrac{1}{7} \cdot (-7) \cdot \dfrac{4}{5}$$ Associative Property of Multiplication

$$= 1 \cdot \dfrac{4}{5}$$ Multiplicative Inverse Property

$$= \dfrac{4}{5}$$ Multiplication Property of One

∴ The product is $\dfrac{4}{5}$.

On Your Own

Now You're Ready
Exercises 10–30

Multiply or divide. Write fractions in simplest form.

1. $-\dfrac{6}{5} \div \left(-\dfrac{1}{2} \right)$

2. $\dfrac{1}{3} \div \left(-2\dfrac{2}{3} \right)$

3. $1.8(-5.1)$

4. $-6.3(-0.6)$

5. $-\dfrac{2}{3} \cdot 7\dfrac{7}{8} \cdot \dfrac{3}{2}$

6. $-7.2 \cdot 0.1 \cdot (-100)$

EXAMPLE 4 **Real-Life Application**

An investor owns Stocks A, B, and C. What is the mean change in the value of the stocks?

Account Positions	⟳		
Stock	**Original Value**	**Current Value**	**Change**
A	600.54	420.15	−180.39
B	391.10	518.38	127.28
C	380.22	99.70	−280.52

$$\text{mean} = \dfrac{-180.39 + 127.28 + (-280.52)}{3} = \dfrac{-333.63}{3} = -111.21$$

∴ The mean change in the value of the stocks is −$111.21.

On Your Own

7. **WHAT IF?** The change in the value of Stock D is $568.23. What is the mean change in the value of the four stocks?

Vocabulary and Concept Check

1. **WRITING** How is multiplying and dividing rational numbers similar to multiplying and dividing integers?

2. **NUMBER SENSE** Find the reciprocal of $-\dfrac{2}{5}$.

Tell whether the expression is *positive* or *negative* without evaluating.

3. $-\dfrac{3}{10} \times \left(-\dfrac{8}{15}\right)$

4. $1\dfrac{1}{2} \div \left(-\dfrac{1}{4}\right)$

5. -6.2×8.18

6. $\dfrac{-8.16}{-2.72}$

Practice and Problem Solving

Multiply.

7. $-1\left(\dfrac{4}{5}\right)$

8. $-1\left(-3\dfrac{1}{2}\right)$

9. $-0.25(-1)$

Divide. Write fractions in simplest form.

10. $-\dfrac{7}{10} \div \dfrac{2}{5}$

11. $\dfrac{1}{4} \div \left(-\dfrac{3}{8}\right)$

12. $-\dfrac{8}{9} \div \left(-\dfrac{8}{9}\right)$

13. $-\dfrac{1}{5} \div 20$

14. $-2\dfrac{4}{5} \div (-7)$

15. $-10\dfrac{2}{7} \div \left(-4\dfrac{4}{11}\right)$

16. $-9 \div 7.2$

17. $8 \div 2.2$

18. $-3.45 \div (-15)$

19. $-0.18 \div 0.03$

20. $8.722 \div (-3.56)$

21. $12.42 \div (-4.8)$

Multiply. Write fractions in simplest form.

22. $-\dfrac{1}{4} \times \left(-\dfrac{4}{3}\right)$

23. $\dfrac{5}{6}\left(-\dfrac{8}{15}\right)$

24. $-2\left(-1\dfrac{1}{4}\right)$

25. $-3\dfrac{1}{3} \cdot \left(-2\dfrac{7}{10}\right)$

26. $0.4 \times (-0.03)$

27. $-0.05 \times (-0.5)$

28. $-8(0.09)(-0.5)$

29. $\dfrac{5}{6} \cdot \left(-4\dfrac{1}{2}\right) \cdot \left(-2\dfrac{1}{5}\right)$

30. $\left(-1\dfrac{2}{3}\right)^3$

ERROR ANALYSIS Describe and correct the error.

31.
✗ $-2.2 \times 3.7 = 8.14$

32.
✗ $-\dfrac{1}{4} \div \dfrac{3}{2} = -\dfrac{4}{1} \times \dfrac{3}{2} = -\dfrac{12}{2} = -6$

33. **HOUR HAND** The hour hand of a clock moves $-30°$ every hour. How many degrees does it move in $2\dfrac{1}{5}$ hours?

34. **SUNFLOWER SEEDS** How many 0.75-pound packages can you make with 6 pounds of sunflower seeds?

Evaluate.

35. $-4.2 + 8.1 \times (-1.9)$

36. $2.85 - 6.2 \div 2^2$

37. $-3.64 \cdot |-5.3| - 1.5^3$

38. $1\frac{5}{9} \div \left(-\frac{2}{3}\right) + \left(-2\frac{3}{5}\right)$

39. $-3\frac{3}{4} \times \frac{5}{6} - 2\frac{1}{3}$

40. $\left(-\frac{2}{3}\right)^2 - \frac{3}{4}\left(2\frac{1}{3}\right)$

41. OPEN-ENDED Write two fractions whose product is $-\frac{3}{5}$.

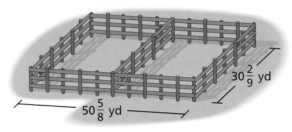

42. FENCING A farmer needs to enclose two adjacent rectangular pastures. How much fencing does the farmer need?

$30\frac{2}{9}$ yd

$50\frac{5}{8}$ yd

43. GASOLINE A 14.5-gallon gasoline tank is $\frac{3}{4}$ full. How many gallons will it take to fill the tank?

44. PRECISION A section of a boardwalk is made using 15 boards. Each board is $9\frac{1}{4}$ inches wide. The total width of the section is 144 inches. The spacing between each board is equal. What is the width of the spacing between each board?

45. RUNNING The table shows the changes in the times (in seconds) of four teammates. What is the mean change?

Teammate	Change
1	-2.43
2	-1.85
3	0.61
4	-1.45

46. **Critical Thinking** The daily changes in the barometric pressure for four days are -0.05, 0.09, -0.04, and -0.08 inches.

   a. What is the mean change?

   b. The mean change after five days is -0.01 inch. What is the change on the fifth day? Explain.

Fair Game Review What you learned in previous grades & lessons

Add or subtract. *(Section 2.2 and Section 2.3)*

47. $-6.2 + 4.7$

48. $-8.1 - (-2.7)$

49. $\frac{9}{5} - \left(-2\frac{7}{10}\right)$

50. $-4\frac{5}{6} + \left(-3\frac{4}{9}\right)$

51. MULTIPLE CHOICE What are the coordinates of the point in Quadrant IV? *(Skills Review Handbook)*

   (A) $(-4, 1)$

   (B) $(-3, -3)$

   (C) $(0, -2)$

   (D) $(3, -3)$

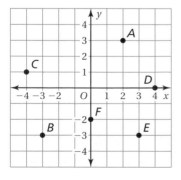

Subtract. Write fractions in simplest form. *(Section 2.3)*

1. $\dfrac{2}{7} - \left(\dfrac{6}{7}\right)$

2. $\dfrac{12}{7} - \left(-\dfrac{2}{9}\right)$

3. $9.1 - 12.9$

4. $5.647 - (-9.24)$

Find the distance between the two numbers on the number line. *(Section 2.3)*

5.

6.

Divide. Write fractions in simplest form. *(Section 2.4)*

7. $\dfrac{2}{3} \div \left(-\dfrac{5}{6}\right)$

8. $-8\dfrac{5}{9} \div \left(-1\dfrac{4}{7}\right)$

9. $-8.4 \div 2.1$

10. $32.436 \div (-4.24)$

Multiply. Write fractions in simplest form. *(Section 2.4)*

11. $\dfrac{5}{8} \times \left(-\dfrac{4}{15}\right)$

12. $-2\dfrac{3}{8} \times \dfrac{8}{5}$

13. $-9.4 \times (-4.7)$

14. $-100(-0.6)(0.01)$

15. PARASAILING A parasail is at 200.6 feet above the water. After 5 minutes, the parasail is at 120.8 feet above the water. What is the change in height of the parasail? *(Section 2.3)*

16. TEMPERATURE Use the thermometer shown. How much did the temperature drop from 5:00 P.M. to 10:00 P.M.? *(Section 2.3)*

17. LATE FEES You were overcharged $4.52 on your cell phone bill 3 months in a row. The cell phone company says that it will add −$4.52 to your next bill for each month you were overcharged. On the next bill, you see an adjustment of −13.28. Is this amount correct? Explain. *(Section 2.4)*

18. CASHEWS How many $1\dfrac{1}{4}$-pound packages can you make with $7\dfrac{1}{2}$ pounds of cashews? *(Section 2.4)*

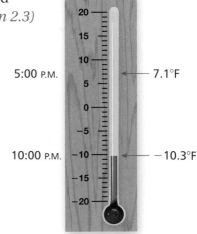

Check It Out
Vocabulary Help
BigIdeasMath ✓.com

Review Key Vocabulary

rational number, *p. 46*   repeating decimal, *p. 46*
terminating decimal, *p. 46*

Review Examples and Exercises

2.1 **Rational Numbers** *(pp. 44–49)*

a. Write $4\dfrac{3}{5}$ as a decimal.

Notice that $4\dfrac{3}{5} = \dfrac{23}{5}$.

Divide 23 by 5. → $\begin{array}{r} 4.6 \\ 5\overline{)23.0} \\ -20 \\ \hline 3\,0 \\ -3\,0 \\ \hline 0 \end{array}$

The remainder is 0. So, it is a terminating decimal. → 0

∴ So, $4\dfrac{3}{5} = 4.6$.

b. Write -0.14 as a fraction in simplest form.

$-0.14 = -\dfrac{14}{100}$ ← Write the digits after the decimal point in the numerator.

The last digit is in the hundredths place. So, use 100 in the denominator.

$= -\dfrac{7}{50}$ Simplify.

Exercises

Write the rational number as a decimal.

1. $-\dfrac{8}{15}$ **2.** $\dfrac{5}{8}$ **3.** $-\dfrac{13}{6}$ **4.** $1\dfrac{7}{16}$

Write the decimal as a fraction or a mixed number in simplest form.

5. -0.6 **6.** -0.35 **7.** -5.8 **8.** 24.23

2.2 Adding Rational Numbers (pp. 50–55)

Find $-\dfrac{7}{2} + \dfrac{5}{4}$.

$$-\dfrac{7}{2} + \dfrac{5}{4} = \dfrac{-14}{4} + \dfrac{5}{4}$$ Rewrite using the LCD (least common denominator).

$$= \dfrac{-14 + 5}{4}$$ Write the sum of the numerators over the common denominator.

$$= \dfrac{-9}{4}$$ Add.

$$= -2\dfrac{1}{4}$$ Write the improper fraction as a mixed number.

The sum is $-2\dfrac{1}{4}$.

Exercises

Add. Write fractions in simplest form.

9. $\dfrac{9}{10} + \left(-\dfrac{4}{5}\right)$ **10.** $-4\dfrac{5}{9} + \dfrac{8}{9}$ **11.** $-1.6 + (-2.4)$

2.3 Subtracting Rational Numbers (pp. 58–63)

Find $-4\dfrac{2}{5} - \left(-\dfrac{3}{5}\right)$.

$$-4\dfrac{2}{5} - \left(-\dfrac{3}{5}\right) = -4\dfrac{2}{5} + \dfrac{3}{5}$$ Add the opposite of $-\dfrac{3}{5}$.

$$= -\dfrac{22}{5} + \dfrac{3}{5}$$ Write the mixed number as an improper fraction.

$$= \dfrac{-22 + 3}{5}$$ Write the sum of the numerators over the common denominator.

$$= \dfrac{-19}{5}, \text{ or } -3\dfrac{4}{5}$$ Simplify.

The difference is $-3\dfrac{4}{5}$.

Exercises

Subtract. Write fractions in simplest form.

12. $-\dfrac{5}{12} - \dfrac{3}{10}$ **13.** $3\dfrac{3}{4} - \dfrac{7}{8}$ **14.** $3.8 - (-7.45)$

15. TURTLE A turtle is $20\dfrac{5}{6}$ inches below the surface of a pond. It dives to a depth of $32\dfrac{1}{4}$ inches. What is the change in the turtle's position?

2.4 Multiplying and Dividing Rational Numbers (pp. 64–69)

a. Find $-4\frac{1}{6} \div 1\frac{1}{3}$.

$$-4\frac{1}{6} \div 1\frac{1}{3} = -\frac{25}{6} \div \frac{4}{3} \qquad \text{Write mixed numbers as improper fractions.}$$

$$= \frac{-25}{6} \cdot \frac{3}{4} \qquad \text{Multiply by the reciprocal of } \frac{4}{3}.$$

$$= \frac{-25 \cdot 3}{6 \cdot 4} \qquad \text{Multiply the numerators and the denominators.}$$

$$= \frac{-25}{8}, \text{ or } -3\frac{1}{8} \qquad \text{Simplify.}$$

∴ The quotient is $-3\frac{1}{8}$.

b. Find $-1.6 \cdot 2.4$.

$$\begin{array}{r} -1.6 \\ \times\ 2.4 \\ \hline 64 \\ 320 \\ \hline -3.84 \end{array}$$

The decimals have different signs.

The product is negative.

∴ The product is -3.84.

Exercises

Divide. Write fractions in simplest form.

16. $\dfrac{9}{10} \div \left(-\dfrac{6}{5}\right)$ **17.** $-\dfrac{4}{11} \div \dfrac{2}{7}$ **18.** $6.4 \div (-3.2)$ **19.** $-15.4 \div (-2.5)$

Multiply. Write fractions in simplest form.

20. $-\dfrac{4}{9}\left(-\dfrac{7}{9}\right)$ **21.** $\dfrac{8}{15}\left(-\dfrac{2}{3}\right)$ **22.** $-5.9(-9.7)$

23. $4.5(-5.26)$ **24.** $-\dfrac{2}{3} \cdot \left(2\dfrac{1}{2}\right) \cdot (-3)$ **25.** $-1.6 \cdot (0.5) \cdot (-20)$

26. SUNKEN SHIP The elevation of a sunken ship is -120 feet. Your elevation is $\dfrac{5}{8}$ of the ship's elevation. What is your elevation?

Write the rational number as a decimal.

1. $\dfrac{7}{40}$

2. $-\dfrac{1}{9}$

3. $-\dfrac{21}{16}$

4. $\dfrac{36}{5}$

Write the decimal as a fraction or a mixed number in simplest form.

5. -0.122

6. 0.33

7. -4.45

8. -7.09

Add or subtract. Write fractions in simplest form.

9. $-\dfrac{4}{9} + \left(-\dfrac{23}{18}\right)$

10. $\dfrac{17}{12} - \left(-\dfrac{1}{8}\right)$

11. $9.2 + (-2.8)$

12. $2.86 - 12.1$

Multiply or divide. Write fractions in simplest form.

13. $3\dfrac{9}{10} \times \left(-\dfrac{8}{3}\right)$

14. $-1\dfrac{5}{6} \div 4\dfrac{1}{6}$

15. $-4.4 \times (-6.02)$

16. $-5 \div 1.5$

17. $-\dfrac{3}{5} \cdot \left(2\dfrac{2}{7}\right) \cdot \left(-3\dfrac{3}{4}\right)$

18. $-6 \cdot (-0.05) \cdot (-0.4)$

19. **ALMONDS** How many 2.25-pound containers can you make with 24.75 pounds of almonds?

20. **FISH** The elevation of a fish is -27 feet.

   a. The fish decreases its elevation by 32 feet, and then increases its elevation by 14 feet. What is its new elevation?

   b. Your elevation is $\dfrac{2}{5}$ of the fish's new elevation. What is your elevation?

21. **RAINFALL** The table shows the rainfall (in inches) for three months compared to the yearly average. Is the total rainfall for the three-month period greater than or less than the yearly average? Explain.

November	December	January
-0.86	2.56	-1.24

22. **BANK ACCOUNTS** Bank Account A has \$750.92, and Bank Account B has \$675.44. Account A changes by $-\$216.38$, and Account B changes by $-\$168.49$. Which account has the greater balance? Explain.

1. When José and Sean were each 5 years old, José was $1\frac{1}{2}$ inches taller than Sean. José grew at an average rate of $2\frac{3}{4}$ inches per year from the time that he was 5 years old until the time he was 13 years old. José was 63 inches tall when he was 13 years old. How tall was Sean when he was 5 years old? *(7.NS.3)*

A. $39\frac{1}{2}$ in.

C. $44\frac{3}{4}$ in.

B. $42\frac{1}{2}$ in.

D. $47\frac{3}{4}$ in.

Test-Taking Strategy
Estimate the Answer

One-fourth of the 36 cats in our town are tabbies. How many are not tabbies?

Ⓐ 9 Ⓑ 18 Ⓒ 27 Ⓓ 36

IC.

"Using estimation you can see that there are about 10 tabbies. So about 30 are not tabbies."

2. Which expression represents a positive integer? *(7.NS.2a)*

F. -6^2

H. $(-5)^2$

G. $(-3)^3$

I. -2^3

3. What is the missing number in the sequence below? *(7.NS.2a)*

$$\frac{9}{16}, \ -\frac{9}{8}, \ \frac{9}{4}, \ -\frac{9}{2}, \ 9, \ \underline{\quad}$$

4. What is the value of the expression below? *(7.NS.1c)*

$$\left| -2 - (-2.5) \right|$$

A. -4.5

C. 0.5

B. -0.5

D. 4.5

5. What is the distance between the two numbers on the number line? *(7.NS.1c)*

$-\frac{7}{4}$ $\frac{3}{8}$

$-2 \quad -1 \quad 0 \quad 1 \quad 2$

F. $-2\frac{1}{8}$

H. $1\frac{3}{8}$

G. $-1\frac{3}{8}$

I. $2\frac{1}{8}$

6. Sandra was evaluating an expression in the box below.

$$-4\frac{3}{4} \div 2\frac{1}{5} = -\frac{19}{4} \div \frac{11}{5}$$

$$= \frac{-4}{19} \cdot \frac{5}{11}$$

$$= \frac{-4 \cdot 5}{19 \cdot 11}$$

$$= \frac{-20}{209}$$

What should Sandra do to correct the error that she made? *(7.NS.3)*

A. Rewrite $-\frac{19}{4}$ as $-\frac{4}{19}$ and multiply by $\frac{11}{5}$.

B. Rewrite $\frac{11}{5}$ as $\frac{5}{11}$ and multiply by $-\frac{19}{4}$.

C. Rewrite $\frac{11}{5}$ as $-\frac{5}{11}$ and multiply by $-\frac{19}{4}$.

D. Rewrite $-4\frac{3}{4}$ as $-\frac{13}{4}$ and multiply by $\frac{5}{11}$.

7. What is the value of the expression below when $q = -2$, $r = -12$, and $s = 8$? *(7.NS.3)*

$$\frac{-q^2 - r}{s}$$

F. -2 **H.** 1

G. -1 **I.** 2

8. You are stacking wooden blocks with the dimensions shown below. How many blocks do you need to stack to build a block tower that is $7\frac{1}{2}$ inches tall? *(7.NS.3)*

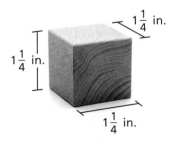

$1\frac{1}{4}$ in.

$1\frac{1}{4}$ in.

$1\frac{1}{4}$ in.

9. What is the area of a triangle with a base length of $2\frac{1}{2}$ inches and a height of 2 inches? *(7.NS.2c)*

   A. $2\frac{1}{4}$ in.^{2}

   B. $2\frac{1}{2}$ in.^{2}

   C. $4\frac{1}{2}$ in.^{2}

   D. 5 in.^{2}

10. What is the value of the expression below? *(7.NS.3)*
$$\frac{-4^2 - (-2)^3}{4}$$

   F. -6

   G. -2

   H. 2

   I. 6

11. Four points are graphed on the number line below. *(7.NS.3)*

   Part A   Choose the two points whose values have the greatest sum. Approximate this sum. Explain your reasoning.

   Part B   Choose the two points whose values have the greatest difference. Approximate this difference. Explain your reasoning.

   Part C   Choose the two points whose values have the greatest product. Approximate this product. Explain your reasoning.

   Part D   Choose the two points whose values have the greatest quotient. Approximate this quotient. Explain your reasoning.

12. What number belongs in the box to make the equation true? *(7.NS.3)*
$$\frac{-0.4}{\boxed{}} + 0.8 = -1.2$$

   A. -1

   B. -0.2

   C. 0.2

   D. 1

3 Expressions and Equations

"I can't find my algebra tiles, so I am painting some of my dog biscuits."

"Now I will be able to solve the equation $2x + (-2) = 2$."

"Descartes, if you solve for 🐭

in the equation, what do you get?"

What You Learned Before

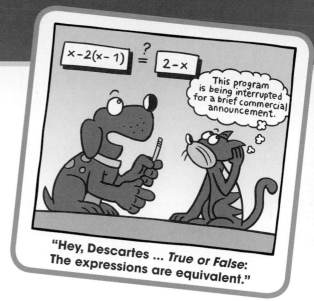

"Hey, Descartes ... True or False: The expressions are equivalent."

● Evaluating Expressions (7.NS.3)

Example 1 Evaluate $6x + 2y$ when $x = -3$ and $y = 5$.

$6x + 2y = 6(-3) + 2(5)$ Substitute -3 for x and 5 for y.

$\qquad = -18 + 10$ Using order of operations, multiply 6 and -3, and 2 and 5.

$\qquad = -8$ Add -18 and 10.

Example 2 Evaluate $6x^2 - 3(y + 2) + 8$ when $x = -2$ and $y = 4$.

$6x^2 - 3(y + 2) + 8 = 6(-2)^2 - 3(4 + 2) + 8$ Substitute -2 for x and 4 for y.

$\qquad = 6(-2)^2 - 3(6) + 8$ Using order of operations, evaluate within the parentheses.

$\qquad = 6(4) - 3(6) + 8$ Using order of operations, evaluate the exponent.

$\qquad = 24 - 18 + 8$ Using order of operations, multiply 6 and 4, and 3 and 6.

$\qquad = 14$ Subtract 18 from 24. Add the result to 8.

Try It Yourself

Evaluate the expression when $x = -\dfrac{1}{4}$ and $y = 3$.

1. $2xy$ **2.** $12x - 3y$ **3.** $-4x - y + 4$ **4.** $8x - y^2 - 3$

● Writing Algebraic Expressions (6.EE.2a)

Example 3 Write the phrase as an algebraic expression.

  a. the sum of twice a number m and four   **b.** eight less than three times a number x

        $2m + 4$ $3x - 8$

Try It Yourself

Write the phrase as an algebraic expression.

  5. five more than three times a number q **6.** nine less than a number n

  7. the product of a number p and six   **8.** the quotient of eight and a number h

  9. four more than three times a number t **10.** two less than seven times a number c

3.1 Algebraic Expressions

Essential Question
How can you simplify an algebraic expression?

1 ACTIVITY: Simplifying Algebraic Expressions

Work with a partner.

a. Evaluate each algebraic expression when $x = 0$ and when $x = 1$. Use the results to match each expression in the left table with its equivalent expression in the right table.

	Expression	Value When $x = 0$	Value When $x = 1$
A.	$3x + 2 - x + 4$		
B.	$5(x - 3) + 2$		
C.	$x + 3 - (2x + 1)$		
D.	$-4x + 2 - x + 3x$		
E.	$-(1 - x) + 3$		
F.	$2x + x - 3x + 4$		
G.	$4 - 3 + 2(x - 1)$		
H.	$2(1 - x + 4)$		
I.	$5 - (4 - x + 2x)$		
J.	$5x - (2x + 4 - x)$		

	Expression	Value When $x = 0$	Value When $x = 1$
a.	4		
b.	$-x + 1$		
c.	$4x - 4$		
d.	$2x + 6$		
e.	$5x - 13$		
f.	$-2x + 10$		
g.	$x + 2$		
h.	$2x - 1$		
i.	$-2x + 2$		
j.	$-x + 2$		

b. Compare each expression in the left table with its equivalent expression in the right table. In general, how do you think you obtain the equivalent expression in the right column?

COMMON CORE

Algebraic Expressions
In this lesson, you will
- apply properties of operations to simplify algebraic expressions.
- solve real-life problems.
Learning Standards
7.EE.1
7.EE.2

Math Practice 6

Communicate Precisely

What can you do to make sure that you are communicating exactly what is needed in the Key Idea?

Work with a partner. Use your results from Activity 1 to write a lesson on simplifying an algebraic expression.

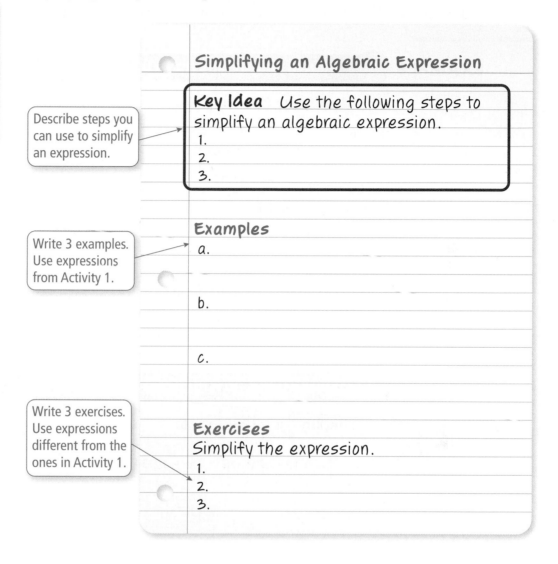

Describe steps you can use to simplify an expression.

Simplifying an Algebraic Expression

Key Idea   Use the following steps to simplify an algebraic expression.
1.
2.
3.

Examples
a.

b.

c.

Write 3 examples. Use expressions from Activity 1.

Write 3 exercises. Use expressions different from the ones in Activity 1.

Exercises
Simplify the expression.
1.
2.
3.

What Is Your Answer?

3. **IN YOUR OWN WORDS** How can you simplify an algebraic expression? Give an example that demonstrates your procedure.

4. **REASONING** Why would you want to simplify an algebraic expression? Discuss several reasons.

Practice

Use what you learned about simplifying algebraic expressions to complete Exercises 12–14 on page 84.

Key Vocabulary
like terms, *p. 82*
simplest form, *p. 82*

Parts of an algebraic expression are called *terms*. **Like terms** are terms that have the same variables raised to the same exponents. Constant terms are also like terms. To identify terms and like terms in an expression, first write the expression as a sum of its terms.

EXAMPLE 1 Identifying Terms and Like Terms

Identify the terms and like terms in each expression.

a. $9x - 2 + 7 - x$
Rewrite as a sum of terms.

$$9x + (-2) + 7 + (-x)$$

Terms: $9x,\ -2,\ 7,\ -x$

Like terms: $9x$ and $-x$, -2 and 7

b. $z^2 + 5z - 3z^2 + z$
Rewrite as a sum of terms.

$$z^2 + 5z + (-3z^2) + z$$

Terms: $z^2,\ 5z,\ -3z^2,\ z$

Like terms: z^2 and $-3z^2$, $5z$ and z

An algebraic expression is in **simplest form** when it has no like terms and no parentheses. To *combine* like terms that have variables, use the Distributive Property to add or subtract the coefficients.

EXAMPLE 2 Simplifying an Algebraic Expression

Simplify $\frac{3}{4}y + 12 - \frac{1}{2}y - 6$.

Study Tip

To subtract a variable term, add the term with the opposite coefficient.

$$\frac{3}{4}y + 12 - \frac{1}{2}y - 6 = \frac{3}{4}y + 12 + \left(-\frac{1}{2}y\right) + (-6) \quad \text{Rewrite as a sum.}$$

$$= \frac{3}{4}y + \left(-\frac{1}{2}y\right) + 12 + (-6) \quad \begin{array}{l}\text{Commutative Property}\\ \text{of Addition}\end{array}$$

$$= \left[\frac{3}{4} + \left(-\frac{1}{2}\right)\right]y + 12 + (-6) \quad \text{Distributive Property}$$

$$= \frac{1}{4}y + 6 \quad \text{Combine like terms.}$$

On Your Own

Now You're Ready
Exercises 5–10
and 12–17

Identify the terms and like terms in the expression.

1. $y + 10 - \frac{3}{2}y$ **2.** $2r^2 + 7r - r^2 - 9$ **3.** $7 + 4p - 5 + p + 2q$

Simplify the expression.

4. $14 - 3z + 8 + z$ **5.** $2.5x + 4.3x - 5$ **6.** $\frac{3}{8}b - \frac{3}{4}b$

Multi-Language Glossary at BigIdeasMath.com

EXAMPLE **3** **Simplifying an Algebraic Expression**

Simplify $-\frac{1}{2}(6n + 4) + 2n.$

$$-\frac{1}{2}(6n + 4) + 2n = -\frac{1}{2}(6n) + \left(-\frac{1}{2}\right)(4) + 2n \qquad \text{Distributive Property}$$

$$= -3n + (-2) + 2n \qquad \text{Multiply.}$$

$$= -3n + 2n + (-2) \qquad \text{Commutative Property of Addition}$$

$$= (-3 + 2)n + (-2) \qquad \text{Distributive Property}$$

$$= -n - 2 \qquad \text{Simplify.}$$

● **On Your Own**

Now You're Ready
Exercises 18–20

Simplify the expression.

7. $3(q + 1) - 4$ **8.** $-2(g + 4) + 7g$ **9.** $7 - 4\left(\frac{3}{4}x - \frac{1}{4}\right)$

EXAMPLE **4** **Real-Life Application**

Each person in a group buys a ticket, a medium drink, and a large popcorn. Write an expression in simplest form that represents the amount of money the group spends at the movies. Interpret the expression.

Evening Tickets $7.50
REFRESHMENTS
Drinks
Small $1.75
Medium $2.75
Large $3.50
Popcorn
Small $3.00
Large $4.00

Words Each ticket is $7.50, each medium drink is $2.75, and each large popcorn is $4.

Variable The same number of each item is purchased. So, x can represent the number of tickets, the number of medium drinks, and the number of large popcorns.

Expression $7.50\,x$ $+$ $2.75\,x$ $+$ $4\,x$

$$7.50x + 2.75x + 4x = (7.50 + 2.75 + 4)x \qquad \text{Distributive Property}$$

$$= 14.25x \qquad \text{Add coefficients.}$$

⋰ The expression $14.25x$ indicates that the total cost per person is $14.25.

● **On Your Own**

10. WHAT IF? Each person buys a ticket, a large drink, and a small popcorn. How does the expression change? Explain.

✓ Vocabulary and Concept Check

1. **WRITING** Explain how to identify the terms of $3y - 4 - 5y$.

2. **WRITING** Describe how to combine like terms in the expression $3n + 4n - 2$.

3. **VOCABULARY** Is the expression $3x + 2x - 4$ in simplest form? Explain.

4. **REASONING** Which algebraic expression is in simplest form? Explain.

$$5x - 4 + 6y \qquad 4x + 8 - x$$

$$3(7 + y) \qquad 12n - n$$

 Practice and Problem Solving

Identify the terms and like terms in the expression.

① **5.** $t + 8 + 3t$

6. $3z + 4 + 2 + 4z$

7. $2n - n - 4 + 7n$

8. $-x - 9x^2 + 12x^2 + 7$

9. $1.4y + 5 - 4.2 - 5y^2 + z$

10. $\frac{1}{2}s - 4 + \frac{3}{4}s + \frac{1}{8} - s^3$

11. **ERROR ANALYSIS** Describe and correct the error in identifying the like terms in the expression.

$$3x - 5 + 2x^2 + 9x = 3x + 2x^2 + 9x - 5$$

Like Terms: $3x$, $2x^2$, and $9x$

Simplify the expression.

② **12.** $12g + 9g$

13. $11x + 9 - 7$

14. $8s - 11s + 6$

15. $4.2v - 5 - 6.5v$

16. $8 + 4a + 6.2 - 9a$

17. $\frac{2}{5}y - 4 + 7 - \frac{9}{10}y$

③ **18.** $4(b - 6) + 19$

19. $4p - 5(p + 6)$

20. $-\frac{2}{3}(12c - 9) + 14c$

21. **HIKING** On a hike, each hiker carries the items shown. Write an expression in simplest form that represents the weight carried by x hikers. Interpret the expression.

4.6 lb

3.4 lb

2.2 lb

22. **STRUCTURE** Evaluate the expression $-8x + 5 - 2x - 4 + 5x$ when $x = 2$ before and after simplifying. Which method do you prefer? Explain.

23. **REASONING** Are the expressions $8x^2 + 3(x^2 + y)$ and $7x^2 + 7y + 4x^2 - 4y$ equivalent? Explain your reasoning.

24. **CRITICAL THINKING** Which solution shows a correct way of simplifying $6 - 4(2 - 5x)$? Explain the errors made in the other solutions.

   (A) $6 - 4(2 - 5x) = 6 - 4(-3x) = 6 + 12x$

   (B) $6 - 4(2 - 5x) = 6 - 8 + 20x = -2 + 20x$

   (C) $6 - 4(2 - 5x) = 2(2 - 5x) = 4 - 10x$

   (D) $6 - 4(2 - 5x) = 6 - 8 - 20x = -2 - 20x$

25. **BANNER** Write an expression in simplest form that represents the area of the banner.

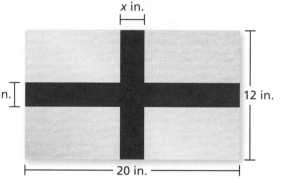

3 ft

We're #1 GO BIG BLUE!

$(3 + x)$ ft

26. **CAR WASH** Write an expression in simplest form that represents the earnings for washing and waxing x cars and y trucks.

	Car	Truck
Wash	$8	$10
Wax	$12	$15

MODELING Draw a diagram that shows how the expression can represent the area of a figure. Then simplify the expression.

27. $5(2 + x + 3)$

28. $(4 + 1)(x + 2x)$

29. **Critical Thinking** You apply gold foil to a piece of red poster board to make the design shown.

   a. Write an expression in simplest form that represents the area of the gold foil.
   b. Find the area of the gold foil when $x = 3$.
   c. The pattern at the right is called "St. George's Cross." Find a country that uses this pattern as its flag.

   x in.

   x in.

   12 in.

   20 in.

Fair Game Review *What you learned in previous grades & lessons*

Order the lengths from least to greatest. *(Skills Review Handbook)*

30. 15 in., 14.8 in., 15.8 in., 14.5 in., 15.3 in.

31. 0.65 m, 0.6 m, 0.52 m, 0.55 m, 0.545 m

32. **MULTIPLE CHOICE** A bird's nest is 12 feet above the ground. A mole's den is 12 inches below the ground. What is the difference in height of these two positions? *(Section 1.3)*

   (A) 24 in.   (B) 11 ft   (C) 13 ft   (D) 24 ft

Essential Question
How can you use algebra tiles to add or subtract algebraic expressions?

Key: ⬛ + = variable ⬛ − = −variable ⬛ + ⬛ − = zero pair

+ = 1 − = −1 + − = zero pair

---

1 ACTIVITY: Writing Algebraic Expressions

Work with a partner. Write an algebraic expression shown by the algebra tiles.

a. ⬛+ + + +

b. ⬛+ − −
   ⬛+

c. ⬛+ + + + + +
   ⬛+ − −

d. ⬛+ + + +
   ⬛+ − − − − −
   ⬛+ − −

---

2 ACTIVITY: Adding Algebraic Expressions

Work with a partner. Write the sum of two algebraic expressions modeled by the algebra tiles. Then use the algebra tiles to simplify the expression.

a. (⬛+ + +) + (⬛+ + + + +)

b. (⬛+ − − − − −) + (⬛+ − −)

c. (⬛+ + + + + +) + (⬛+ − − −
                          ⬛+)

d. (⬛+ − − − − −   + (⬛+ + + +
    ⬛+ − − −)   ⬛+ + +
                        ⬛+)

COMMON CORE

Linear Expressions
In this lesson, you will
● apply properties of operations to add and subtract linear expressions.
● solve real-life problems.
Learning Standards
7.EE.1
7.EE.2

ACTIVITY: Subtracting Algebraic Expressions

> **Math Practice** 2
>
> **Use Expressions**
> What do the tiles represent? How does this help you write an expression?

Work with a partner. Write the difference of two algebraic expressions modeled by the algebra tiles. Then use the algebra tiles to simplify the expression.

a. $\left(\boxed{+} \; \boxed{+}\boxed{+}\boxed{+} \right) - \left(\boxed{+} \; \boxed{+} \right)$

b. $\left(\boxed{+} \; \blacksquare\blacksquare\blacksquare\blacksquare \right) - \left(\boxed{+} \; \blacksquare\blacksquare\blacksquare \right)$

c. $\left(\begin{matrix} \boxed{+} \\ \boxed{+} \end{matrix} \; \boxed{+}\boxed{+}\boxed{+}\boxed{+}\boxed{+} \right) - \left(\boxed{+} \; \blacksquare \right)$

d. $\left(\begin{matrix} \boxed{+} \\ \boxed{+} \\ \boxed{+} \end{matrix} \; \begin{matrix}\blacksquare\blacksquare\blacksquare\blacksquare \\ \blacksquare\blacksquare \end{matrix} \right) - \left(\begin{matrix} \boxed{+} \\ \boxed{+} \end{matrix} \; \boxed{+}\boxed{+}\boxed{+} \right)$

ACTIVITY: Adding and Subtracting Algebraic Expressions

Work with a partner. Use algebra tiles to model the sum or difference. Then use the algebra tiles to simplify the expression.

a. $(2x + 1) + (x - 1)$

b. $(2x - 6) + (3x + 2)$

c. $(2x + 4) - (x + 2)$

d. $(4x + 3) - (2x - 1)$

What Is Your Answer?

5. **IN YOUR OWN WORDS** How can you use algebra tiles to add or subtract algebraic expressions?

6. Write the difference of two algebraic expressions modeled by the algebra tiles. Then use the algebra tiles to simplify the expression.

$\left(\blacksquare \; \boxed{+}\boxed{+}\boxed{+} \right) - \left(\begin{matrix} \blacksquare \\ \blacksquare \end{matrix} \; \blacksquare\blacksquare \right)$

> **Practice**
>
> Use what you learned about adding and subtracting algebraic expressions to complete Exercises 6 and 7 on page 90.

3.2 Lesson

Check It Out
Lesson Tutorials
BigIdeasMath com

Key Vocabulary
linear expression,
  p. 88

A **linear expression** is an algebraic expression in which the exponent of the variable is 1.

Linear Expressions	$-4x$	$3x + 5$	$5 - \dfrac{1}{6}x$
Nonlinear Expressions	x^2	$-7x^3 + x$	$x^5 + 1$

You can use a vertical or a horizontal method to add linear expressions.

EXAMPLE 1 **Adding Linear Expressions**

Find each sum.

a. $(x - 2) + (3x + 8)$

Vertical method: Align like terms vertically and add.

$$\begin{array}{r} x - 2 \\ + \ 3x + 8 \\ \hline 4x + 6 \end{array}$$

b. $(-4y + 3) + (11y - 5)$

Horizontal method: Use properties of operations to group like terms and simplify.

$(-4y + 3) + (11y - 5) = -4y + 3 + 11y - 5$ Rewrite the sum.

$\qquad\qquad\qquad\quad = -4y + 11y + 3 - 5$ Commutative Property of Addition

$\qquad\qquad\qquad\quad = (-4y + 11y) + (3 - 5)$ Group like terms.

$\qquad\qquad\qquad\quad = 7y - 2$ Combine like terms.

EXAMPLE 2 **Adding Linear Expressions**

Find $2(-7.5z + 3) + (5z - 2)$.

$2(-7.5z + 3) + (5z - 2) = -15z + 6 + 5z - 2$ Distributive Property

$\qquad\qquad\qquad\qquad = -15z + 5z + 6 - 2$ Commutative Property of Addition

$\qquad\qquad\qquad\qquad = -10z + 4$ Combine like terms.

On Your Own

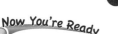
Now You're Ready
Exercises 8–16

Find the sum.

1. $(x + 3) + (2x - 1)$

2. $(-8z + 4) + (8z - 7)$

3. $(4 - n) + 2(-5n + 3)$

4. $\dfrac{1}{2}(w - 6) + \dfrac{1}{4}(w + 12)$

To subtract one linear expression from another, add the opposite of each term in the expression. You can use a vertical or a horizontal method.

EXAMPLE 3 **Subtracting Linear Expressions**

Find each difference.

a. $(5x + 6) - (-x + 6)$ **b.** $(7y + 5) - 2(4y - 3)$

a. **Vertical method:** Align like terms vertically and subtract.

Study Tip

To find the opposite of a linear expression, you can multiply the expression by −1.

$$
\begin{array}{r}
(5x + 6) \\
- (-x + 6)
\end{array}
$$

Add the opposite.

$$
\begin{array}{r}
5x + 6 \\
+ \quad x - 6 \\
\hline
6x
\end{array}
$$

b. **Horizontal method:** Use properties of operations to group like terms and simplify.

$(7y + 5) - 2(4y - 3) = 7y + 5 - 8y + 6$	Distributive Property
$= 7y - 8y + 5 + 6$	Commutative Property of Addition
$= (7y - 8y) + (5 + 6)$	Group like terms.
$= -y + 11$	Combine like terms.

EXAMPLE 4 **Real-Life Application**

The original price of a cowboy hat is d dollars. You use a coupon and buy the hat for $(d - 2)$ dollars. You decorate the hat and sell it for $(2d - 4)$ dollars. Write an expression that represents your earnings from buying and selling the hat. Interpret the expression.

earnings = selling price − purchase price	Use a model.
$= (2d - 4) - (d - 2)$	Write the difference.
$= (2d - 4) + (-d + 2)$	Add the opposite.
$= 2d - d - 4 + 2$	Group like terms.
$= d - 2$	Combine like terms.

⁘ You earn $(d - 2)$ dollars. You also paid $(d - 2)$ dollars, so you doubled your money by selling the hat for twice as much as you paid for it.

On Your Own

Find the difference.

5. $(m - 3) - (-m + 12)$ **6.** $-2(c + 2.5) - 5(1.2c + 4)$

7. **WHAT IF?** In Example 4, you sell the hat for $(d + 2)$ dollars. How much do you earn from buying and selling the hat?

3.2 Exercises

Check It Out
Help with Homework
BigIdeasMath.com

✓ Vocabulary and Concept Check

VOCABULARY Determine whether the algebraic expression is a linear expression. Explain.

1. $x^2 + x + 1$
2. $-2x - 8$
3. $x - x^4$

4. **WRITING** Describe two methods for adding or subtracting linear expressions.

5. **DIFFERENT WORDS, SAME QUESTION** Which is different? Find "both" answers.

Subtract x from $3x - 1$.

Find $3x - 1$ decreased by x.

What is x more than $3x - 1$?

What is the difference of $3x - 1$ and x?

Practice and Problem Solving

Write the sum or difference of two algebraic expressions modeled by the algebra tiles. Then use the algebra tiles to simplify the expression.

6.

7.

Find the sum.

8. $(n + 8) + (n - 12)$
9. $(7 - b) + (3b + 2)$
10. $(2w - 9) + (-4w - 5)$

11. $(2x - 6) + 4(x - 3)$
12. $5(-3.4k - 7) + (3k + 21)$
13. $(1 - 5q) + 2(2.5q + 8)$

14. $3(2 - 0.9h) + (-1.3h - 4)$
15. $\frac{1}{3}(9 - 6m) + \frac{1}{4}(12m - 8)$
16. $-\frac{1}{2}(7z + 4) + \frac{1}{5}(5z - 15)$

17. **BANKING** You start a new job. After w weeks, you have $(10w + 120)$ dollars in your savings account and $(45w + 25)$ dollars in your checking account. Write an expression that represents the total in both accounts.

18. **FIREFLIES** While catching fireflies, you and a friend decide to have a competition. After m minutes, you have $(3m + 13)$ fireflies and your friend has $(4m + 6)$ fireflies.

   a. Write an expression that represents the number of fireflies you and your friend caught together.

   b. The competition ends after 5 minutes. Who has more fireflies?

Find the difference.

③ 19. $(-2g + 7) - (g + 11)$

20. $(6d + 5) - (2 - 3d)$

21. $(4 - 5y) - 2(3.5y - 8)$

22. $(2n - 9) - 5(-2.4n + 4)$

23. $\dfrac{1}{8}(-8c + 16) - \dfrac{1}{3}(6 + 3c)$

24. $\dfrac{3}{4}(3x + 6) - \dfrac{1}{4}(5x - 24)$

25. ERROR ANALYSIS Describe and correct the error in finding the difference.

> ✗ $(4m + 9) - 3(2m - 5) = 4m + 9 - 6m - 15$
>
> $= 4m - 6m + 9 - 15$
>
> $= -2m - 6$

26. STRUCTURE Refer to the expressions in Exercise 18.

   a. How many fireflies are caught each minute during the competition?

   b. How many fireflies are caught before the competition starts?

27. LOGIC Your friend says the sum of two linear expressions is always a linear expression. Is your friend correct? Explain.

28. GEOMETRY The expression $17n + 11$ represents the perimeter (in feet) of the triangle. Write an expression that represents the measure of the third side.

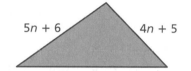

29. TAXI Taxi Express charges $2.60 plus $3.65 per mile, and Cab Cruiser charges $2.75 plus $3.90 per mile. Write an expression that represents how much more Cab Cruiser charges than Taxi Express.

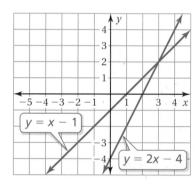

30. MODELING A rectangular room is 10 feet longer than it is wide. One-foot-by-one-foot tiles cover the entire floor. Write an expression that represents the number of tiles along the outside of the room.

31. Reasoning Write an expression in simplest form that represents the vertical distance between the two lines shown. What is the distance when $x = 3$? when $x = -3$?

Fair Game Review   What you learned in previous grades & lessons

Evaluate the expression when $x = -\dfrac{4}{5}$ and $y = \dfrac{1}{3}$. *(Section 2.2)*

32. $x + y$

33. $2x + 6y$

34. $-x + 4y$

35. MULTIPLE CHOICE What is the surface area of a cube that has a side length of 5 feet? *(Skills Review Handbook)*

   Ⓐ 25 ft^2 **Ⓑ** 75 ft^2 **Ⓒ** 125 ft^2 **Ⓓ** 150 ft^2

Check It Out
Lesson Tutorials
BigIdeasMath com

Key Vocabulary
factoring an expression, *p. 92*

When **factoring an expression**, you write the expression as a product of factors. You can use the Distributive Property to factor expressions.

EXAMPLE 1 Factoring Out the GCF

Factor 24x – 18 using the GCF.

Find the GCF of 24x and 18 by writing their prime factorizations.

$$24x = 2 \cdot 2 \cdot 2 \cdot 3 \cdot x$$
$$18 = 2 \cdot 3 \cdot 3$$

Circle the common prime factors.

So, the GCF of 24x and 18 is $2 \cdot 3 = 6$. Use the GCF to factor the expression.

$$24x - 18 = 6(4x) - 6(3)$$ Rewrite using GCF.
$$= 6(4x - 3)$$ Distributive Property

∴ So, $24x - 18 = 6(4x - 3)$.

You can also use the Distributive Property to factor out any rational number from an expression.

EXAMPLE 2 Factoring Out a Fraction

Factor $\frac{1}{2}$ out of $\frac{1}{2}x + \frac{3}{2}$.

Write each term as a product of $\frac{1}{2}$ and another factor.

$$\frac{1}{2}x = \frac{1}{2} \cdot x$$ Think: $\frac{1}{2}x$ is $\frac{1}{2}$ times what?

$$\frac{3}{2} = \frac{1}{2} \cdot 3$$ Think: $\frac{3}{2}$ is $\frac{1}{2}$ times what?

Use the Distributive Property to factor out $\frac{1}{2}$.

$$\frac{1}{2}x + \frac{3}{2} = \frac{1}{2} \cdot x + \frac{1}{2} \cdot 3$$ Rewrite the expression.

$$= \frac{1}{2}(x + 3)$$ Distributive Property

∴ So, $\frac{1}{2}x + \frac{3}{2} = \frac{1}{2}(x + 3)$.

COMMON CORE

Linear Expressions
In this extension, you will
● factor linear expressions.
Learning Standard
7.EE.1

◀ Multi-Language Glossary at BigIdeasMath.com

EXAMPLE **3** **Factoring Out a Negative Number**

Factor -2 out of $-4p + 10$.

Write each term as a product of -2 and another factor.

$$-4p = -2 \cdot 2p$$
Think: $-4p$ is -2 times what?

$$10 = -2 \cdot (-5)$$
Think: 10 is -2 times what?

Use the Distributive Property to factor out -2.

$$-4p + 10 = -2 \cdot 2p + (-2) \cdot (-5)$$
Rewrite the expression.

$$= -2[2p + (-5)]$$
Distributive Property

$$= -2(2p - 5)$$
Simplify.

⋰ So, $-4p + 10 = -2(2p - 5)$.

Math Practice **7**

View as Components

How does rewriting each term as a product help you see the common factor?

● Practice

Factor the expression using the GCF.

1. $9 + 21$ **2.** $32 - 48$ **3.** $8x + 2$ **4.** $3y - 24$

5. $20z - 8$ **6.** $15w + 65$ **7.** $36a + 16b$ **8.** $21m - 49n$

Factor out the coefficient of the variable.

9. $\frac{1}{3}b - \frac{1}{3}$ **10.** $\frac{3}{8}d + \frac{3}{4}$ **11.** $2.2x + 4.4$ **12.** $4h - 3$

13. Factor $-\frac{1}{2}$ out of $-\frac{1}{2}x + 6$.   **14.** Factor $-\frac{1}{4}$ out of $-\frac{1}{2}x - \frac{5}{4}y$.

15. WRESTLING A square wrestling mat has a perimeter of $(12x - 32)$ feet. Write an expression that represents the side length of the mat (in feet).

16. MAKING A DIAGRAM A table is 6 feet long and 3 feet wide. You extend the table by inserting two identical table *leaves*. The longest side length of each rectangular leaf is 3 feet. The extended table is rectangular with an area of $(18 + 6x)$ square feet.

    a. Make a diagram of the table and leaves.

    b. Write an expression that represents the length of the extended table. What does x represent?

17. STRUCTURE The area of the trapezoid is $\left(\frac{3}{4}x - \frac{1}{4}\right)$ square centimeters. Write two different pairs of expressions that represent possible lengths of the bases.

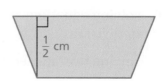

$\frac{1}{2}$ cm

You can use a **four square** to organize information about a topic. Each of the four squares can be a category, such as *definition, vocabulary, example, non-example, words, algebra, table, numbers, visual, graph,* or *equation.* Here is an example of a four square for like terms.

Definition	Examples
Terms that have the same variables raised to the same exponents	2 and −3, 3x and −7x, x^2 and $6x^2$

Like Terms

Words	Non-Examples
To *combine* like terms that have variables, use the Distributive Property to add or subtract the coefficients.	y and 4, 3x and −4y, $6x^2$ and 2x

On Your Own

Make four squares to help you study these topics.

1. simplest form

2. linear expression

3. factoring expressions

After you complete this chapter, make four squares for the following topics.

4. equivalent equations

5. solving equations using addition or subtraction

6. solving equations using multiplication or division

7. solving two-step equations

"My four square shows that my new red skateboard is faster than my old blue skateboard."

Check It Out
Progress Check
BigIdeasMath.com

Identify the terms and like terms in the expression.   *(Section 3.1)*

1. $11x + 2x$

2. $9x - 5x$

3. $21x + 6 - x - 5$

4. $8x + 14 - 3x + 1$

Simplify the expression.   *(Section 3.1)*

5. $2(3x + x)$

6. $-7 + 3x + 4x$

7. $2x + 4 - 3x + 2 + 3x$

8. $7x + 6 + 3x - 2 - 5x$

Find the sum or difference.   *(Section 3.2)*

9. $(s + 12) + (3s - 8)$

10. $(9t + 5) + (3t - 6)$

11. $(2 - k) + 3(-4k + 2)$

12. $\frac{1}{4}(q - 12) + \frac{1}{3}(q + 9)$

13. $(n - 8) - (-2n + 2)$

14. $-3(h - 4) - 2(-6h + 5)$

Factor out the coefficient of the variable.   *(Section 3.2)*

15. $5c - 15$ **16.** $\frac{2}{9}j + \frac{2}{3}$ **17.** $2.4n + 9.6$ **18.** $-6z + 12$

Paint
$21.79

Interior
Latex Paint
Orange
1 gallon

Brush
$3.99

Paint roller
$6.89

19. **PAINTING** You buy the same number of brushes, rollers, and paint cans. Write an expression in simplest form that represents the total amount of money you spend for painting supplies. *(Section 3.1)*

20. **APPLES** A basket holds n apples. You pick $2n - 3$ apples, and your friend picks $n + 4$ apples. Write an expression that represents the number of apples you and your friend picked. Interpret the expression.   *(Section 3.2)*

21. **EXERCISE** Write an expression in simplest form for the perimeter of the exercise mat. *(Section 3.1)*

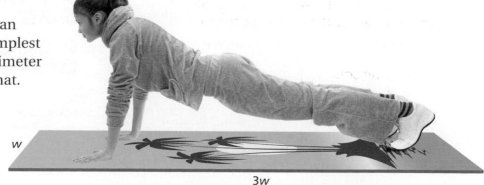

w

$3w$

3.3 Solving Equations Using Addition or Subtraction

Essential Question How can you use algebra tiles to solve addition or subtraction equations?

1 ACTIVITY: Solving Equations

Work with a partner. Use algebra tiles to model and solve the equation.

a. $x - 3 = -4$

Model the equation $x - 3 = -4$.

To get the variable tile by itself, remove the [ ] tiles on the left side by adding [ ] [ ] tiles to each side.

How many *zero pairs* can you remove from each side? [ ]
Circle them.

The remaining tile shows the value of x.

So, $x =$ [ ].

b. $z - 6 = 2$ **c.** $p - 7 = -3$ **d.** $-15 = t - 5$

2 ACTIVITY: Solving Equations

Work with a partner. Use algebra tiles to model and solve the equation.

a. $-5 = n + 2$

Model the equation $-5 = n + 2$.

Remove the [ ] tiles on the right side by adding [ ] [ ] tiles to each side.

How many *zero pairs* can you remove from the right side? [ ]
Circle them.

The remaining tiles show the value of n.

So, $n =$ [ ].

b. $y + 10 = -5$ **c.** $7 + b = -1$ **d.** $8 = 12 + z$

COMMON CORE

Solving Equations
In this lesson, you will
- write simple equations.
- solve equations using addition or subtraction.
- solve real-life problems.
Learning Standard
7.EE.4a

Math Practice 4

Interpret Results

How can you add tiles to make zero pairs? Explain how this helps you solve the equation.

③ ACTIVITY: Writing and Solving Equations

Work with a partner. Write an equation shown by the algebra tiles. Then solve.

a.

b.

c.

d.

④ ACTIVITY: Using a Different Method to Find a Solution

Work with a partner. The *melting point* of a solid is the temperature at which the solid melts to become a liquid. The melting point of the element bromine is about 19°F. This is about 57°F more than the melting point of mercury.

a. Which of the following equations can you use to find the melting point of mercury? What is the melting point of mercury?

$$x + 57 = 19 \qquad x - 57 = 19 \qquad x + 19 = 57 \qquad x + 19 = -57$$

b. **CHOOSE TOOLS** How can you solve this problem without using an equation? Explain. How are these two methods related?

What Is Your Answer?

5. **IN YOUR OWN WORDS** How can you use algebra tiles to solve addition or subtraction equations? Give an example of each.

6. **STRUCTURE** Explain how you could use inverse operations to solve addition or subtraction equations without using algebra tiles.

7. What makes the cartoon funny?

8. The word *variable* comes from the word *vary*. For example, the temperature in Maine varies a lot from winter to summer.

Write two other English sentences that use the word *vary*.

"To vary or not to vary." That is the question.

"Dear Sir: Yesterday you said x = 2. Today you are saying x = 3. Please make up your mind."

Practice ➤ Use what you learned about solving addition or subtraction equations to complete Exercises 5–8 on page 100.

Check It Out
Lesson Tutorials
BigIdeasMath ✓ com

Key Vocabulary 🔊
equivalent equations, p. 98

Two equations are **equivalent equations** if they have the same solutions. The Addition and Subtraction Properties of Equality can be used to write equivalent equations.

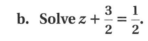 **Key Ideas**

Addition Property of Equality

Words Adding the same number to each side of an equation produces an equivalent equation.

Algebra If $a = b$, then $a + c = b + c$.

Remember

Addition and subtraction are inverse operations.

Subtraction Property of Equality

Words Subtracting the same number from each side of an equation produces an equivalent equation.

Algebra If $a = b$, then $a - c = b - c$.

EXAMPLE 1 **Solving Equations**

a. **Solve $x - 5 = -1$.**

$$x - 5 = -1$$ Write the equation.

Undo the subtraction. → $\underline{+5 \quad +5}$ Addition Property of Equality

$$x = 4$$ Simplify.

∴ The solution is $x = 4$.

Check

$$x - 5 = -1$$
$$4 - 5 \stackrel{?}{=} -1$$
$$-1 = -1 ✓$$

b. **Solve $z + \dfrac{3}{2} = \dfrac{1}{2}$.**

$$z + \frac{3}{2} = \frac{1}{2}$$ Write the equation.

Undo the addition. → $\underline{-\dfrac{3}{2} \quad -\dfrac{3}{2}}$ Subtraction Property of Equality

$$z = -1$$ Simplify.

∴ The solution is $z = -1$.

Check

$$z + \frac{3}{2} = \frac{1}{2}$$
$$-1 + \frac{3}{2} \stackrel{?}{=} \frac{1}{2}$$
$$\frac{1}{2} = \frac{1}{2} ✓$$

● **On Your Own**

Solve the equation. Check your solution.

Now You're Ready
Exercises 5–20

1. $p - 5 = -2$

2. $w + 13.2 = 10.4$

3. $x - \dfrac{5}{6} = -\dfrac{1}{6}$

EXAMPLE **2** **Writing an Equation**

A company has a profit of $750 this week. This profit is $900 more than the profit *P* last week. Which equation can be used to find *P*?

Ⓐ $750 = 900 - P$ Ⓑ $750 = P + 900$

Ⓒ $900 = P - 750$ Ⓓ $900 = P + 750$

Words   The profit this week is $900 more than the profit last week.

Equation   750   = *P*   +   900

∴ The equation is $750 = P + 900$. The correct answer is Ⓑ.

On Your Own

Now You're Ready
Exercises 22–25

4. A company has a profit of $120.50 today. This profit is $145.25 less than the profit *P* yesterday. Write an equation that can be used to find *P*.

EXAMPLE **3** **Real-Life Application**

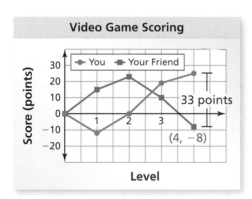

The line graph shows the scoring while you and your friend played a video game. Write and solve an equation to find your score after Level 4.

You can determine the following from the graph.

Words   Your friend's score is 33 points less than your score.

Variable Let *s* be your score after Level 4.

Equation   −8   =   *s*   −   33

$-8 = s - 33$ Write equation.

$\underline{+33 \quad +33}$ Addition Property of Equality

$25 = s$ Simplify.

∴ Your score after Level 4 is 25 points.

Reasonable? From the graph, your score after Level 4 is between 20 points and 30 points. So, 25 points is a reasonable answer.

On Your Own

5. **WHAT IF?** You have −12 points after Level 1. Your score is 27 points less than your friend's score. What is your friend's score?

3.3 Exercises

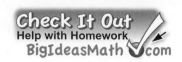

Check It Out
Help with Homework
BigIdeasMath.com

Vocabulary and Concept Check

1. **VOCABULARY** What property would you use to solve $m + 6 = -4$?
2. **VOCABULARY** Name two inverse operations.
3. **WRITING** Are the equations $m + 3 = -5$ and $m = -2$ equivalent? Explain.
4. **WHICH ONE DOESN'T BELONG?** Which equation does *not* belong with the other three? Explain your reasoning.

$x + 3 = -1$	$x + 1 = -5$	$x - 2 = -6$	$x - 9 = -13$

Practice and Problem Solving

Solve the equation. Check your solution.

① 5. $a - 6 = 13$ 6. $-3 = z - 8$ 7. $-14 = k + 6$ 8. $x + 4 = -14$

9. $c - 7.6 = -4$ 10. $-10.1 = w + 5.3$ 11. $\frac{1}{2} = q + \frac{2}{3}$ 12. $p - 3\frac{1}{6} = -2\frac{1}{2}$

13. $g - 9 = -19$ 14. $-9.3 = d - 3.4$ 15. $4.58 + y = 2.5$ 16. $x - 5.2 = -18.73$

17. $q + \frac{5}{9} = \frac{1}{6}$ 18. $-2\frac{1}{4} = r - \frac{4}{5}$ 19. $w + 3\frac{3}{8} = 1\frac{5}{6}$ 20. $4\frac{2}{5} + k = -3\frac{2}{11}$

21. **ERROR ANALYSIS** Describe and correct the error in finding the solution.

✗
$$\begin{aligned} x + 8 &= 10 \\ +8 & \quad +8 \\ x &= 18 \end{aligned}$$

Write the word sentence as an equation. Then solve.

② 22. 4 less than a number n is -15.
23. 10 more than a number c is 3.
24. The sum of a number y and -3 is -8.
25. The difference between a number p and 6 is -14.

In Exercises 26–28, write an equation. Then solve.

26. **DRY ICE** The temperature of dry ice is -109.3°F. This is 184.9°F less than the outside temperature. What is the outside temperature?
27. **PROFIT** A company makes a profit of \$1.38 million. This is \$2.54 million more than last year. What was the profit last year?
28. **HELICOPTER** The difference in elevation of a helicopter and a submarine is $18\frac{1}{2}$ meters. The elevation of the submarine is $-7\frac{3}{4}$ meters. What is the elevation of the helicopter?

GEOMETRY Write and solve an equation to find the unknown side length.

29. Perimeter = 12 cm

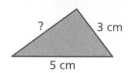

?   3 cm

5 cm

30. Perimeter = 24.2 in.

8.3 in.

?   3.8 in.

8.3 in.

31. Perimeter = 34.6 ft

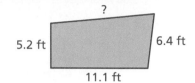

?

5.2 ft   6.4 ft

11.1 ft

In Exercises 32–36, write an equation. Then solve.

305 ft

32. STATUE OF LIBERTY The total height of the Statue of Liberty and its pedestal is 153 feet more than the height of the statue. What is the height of the statue?

33. BUNGEE JUMPING Your first jump is $50\frac{1}{6}$ feet higher than your second jump. Your first jump reaches $-200\frac{2}{5}$ feet. What is the height of your second jump?

34. TRAVEL Boatesville is $65\frac{3}{5}$ kilometers from Stanton. A bus traveling from Stanton is $24\frac{1}{3}$ kilometers from Boatesville. How far has the bus traveled?

35. GEOMETRY The sum of the measures of the angles of a triangle equals 180°. What is the measure of the missing angle?

$m°$

30.3°   40.8°

36. SKATEBOARDING The table shows your scores in a skateboarding competition. The leader has 311.62 points. What score do you need in the fourth round to win?

Round	1	2	3	4
Points	63.43	87.15	81.96	?

37. CRITICAL THINKING Find the value of $2x - 1$ when $x + 6 = 2$.

 Critical Thinking Find the values of x.

38. $|x| = 2$

39. $|x| - 2 = 4$

40. $|x| + 5 = 18$

 Fair Game Review What you learned in previous grades & lessons

Multiply or divide. *(Section 1.4 and Section 1.5)*

41. -7×8

42. $6 \times (-12)$

43. $18 \div (-2)$

44. $-26 \div 4$

45. MULTIPLE CHOICE A class of 144 students voted for a class president. Three-fourths of the students voted for you. Of the students who voted for you, $\frac{5}{9}$ are female. How many female students voted for you? *(Section 2.4)*

    Ⓐ 50   Ⓑ 60   Ⓒ 80   Ⓓ 108

Solving Equations Using Multiplication or Division

Essential Question How can you use multiplication or division to solve equations?

1 ACTIVITY: Using Division to Solve Equations

Work with a partner. Use algebra tiles to model and solve the equation.

a. $3x = -12$

Model the equation $3x = -12$.

Your goal is to get one variable tile by itself. Because there are ▢ variable tiles, divide the ▢ tiles into ▢ equal groups. Circle the groups.

Keep one of the groups. This shows the value of x.

∴ So, $x = $ ▢.

b. $2k = -8$

c. $-15 = 3t$

d. $-20 = 5m$

e. $4h = -16$

2 ACTIVITY: Writing and Solving Equations

Work with a partner. Write an equation shown by the algebra tiles. Then solve.

a.

b.

c.

d.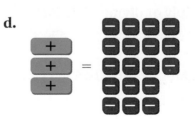

COMMON CORE

Solving Equations

In this lesson, you will
- solve equations using multiplication or division.
- solve real-life problems.

Learning Standard
7.EE.4a

Math Practice 1

Analyze Givens
How can you use the given information to decide which equation represents the situation?

Work with a partner. Choose the equation you can use to solve each problem. Solve the equation. Then explain how to solve the problem without using an equation. How are the two methods related?

a. For the final part of a race, a handcyclist travels 32 feet each second across a distance of 400 feet. How many seconds does it take for the handcyclist to travel the last 400 feet of the race?

$$32x = 400 \qquad 400x = 32$$

$$\frac{x}{32} = 400 \qquad \frac{x}{400} = 32$$

b. The melting point of the element radon is about $-96°F$. The melting point of nitrogen is about 3.6 times the melting point of radon. What is the melting point of nitrogen?

$$3.6x = -96 \qquad x + 96 = 3.6$$

$$\frac{x}{3.6} = -96 \qquad -96x = 3.6$$

c. This year, a hardware store has a profit of $-\$6.0$ million. This profit is $\frac{3}{4}$ of last year's profit. What is last year's profit?

$$\frac{x}{-6} = \frac{3}{4} \qquad -6x = \frac{3}{4}$$

$$\frac{3}{4} + x = -6 \qquad \frac{3}{4}x = -6$$

What Is Your Answer?

4. IN YOUR OWN WORDS How can you use multiplication or division to solve equations? Give an example of each.

Use what you learned about solving equations to complete Exercises 7–10 on page 106.

Key Ideas

Multiplication Property of Equality

Words   Multiplying each side of an equation by the same number produces an equivalent equation.

Algebra   If $a = b$, then $a \cdot c = b \cdot c$.

Remember

Multiplication and division are inverse operations.

Division Property of Equality

Words   Dividing each side of an equation by the same number produces an equivalent equation.

Algebra   If $a = b$, then $a \div c = b \div c$, $c \neq 0$.

EXAMPLE 1 | **Solving Equations**

a. Solve $\dfrac{x}{3} = -6$.

$$\dfrac{x}{3} = -6 \qquad \text{Write the equation.}$$

Undo the division. $\longrightarrow$ $3 \cdot \dfrac{x}{3} = 3 \cdot (-6) \qquad$ Multiplication Property of Equality

$$x = -18 \qquad \text{Simplify.}$$

∴ The solution is $x = -18$.

Check

$$\dfrac{x}{3} = -6$$

$$\dfrac{-18}{3} \stackrel{?}{=} -6$$

$$-6 = -6 \checkmark$$

b. Solve $18 = -4y$.

$$18 = -4y \qquad \text{Write the equation.}$$

Undo the multiplication. $\longrightarrow$ $\dfrac{18}{-4} = \dfrac{-4y}{-4} \qquad$ Division Property of Equality

$$-4.5 = y \qquad \text{Simplify.}$$

∴ The solution is $y = -4.5$.

Check

$$18 = -4y$$

$$18 \stackrel{?}{=} -4(-4.5)$$

$$18 = 18 \checkmark$$

On Your Own

Now You're Ready
Exercises 7–18

Solve the equation. Check your solution.

1. $\dfrac{x}{5} = -2$

2. $-a = -24$

3. $3 = -1.5n$

EXAMPLE 2 **Solving an Equation Using a Reciprocal**

Solve $-\dfrac{4}{5}x = -8$.

$$-\dfrac{4}{5}x = -8 \qquad\qquad \text{Write the equation.}$$

Multiply each side by $-\dfrac{5}{4}$, the reciprocal of $-\dfrac{4}{5}$.

$$-\dfrac{5}{4} \cdot \left(-\dfrac{4}{5}x\right) = -\dfrac{5}{4} \cdot (-8) \qquad \text{Multiplicative Inverse Property}$$

$$x = 10 \qquad\qquad \text{Simplify.}$$

:· The solution is $x = 10$.

On Your Own

Exercises 19–22

Solve the equation. Check your solution.

4. $-14 = \dfrac{2}{3}x$

5. $-\dfrac{8}{5}b = 5$

6. $\dfrac{3}{8}h = -9$

EXAMPLE 3 **Real-Life Application**

Record low temperature in Arizona

The record low temperature in Arizona is 1.6 times the record low temperature in Rhode Island. What is the record low temperature in Rhode Island?

Words   The record low in Arizona is 1.6 times the record low in Rhode Island.

Variable   Let t be the record low in Rhode Island.

Equation $-40 \qquad = 1.6 \quad \times \qquad t$

$$-40 = 1.6t \qquad \text{Write equation.}$$

$$-\dfrac{40}{1.6} = \dfrac{1.6t}{1.6} \qquad \text{Division Property of Equality}$$

$$-25 = t \qquad \text{Simplify.}$$

:· The record low temperature in Rhode Island is $-25°$F.

On Your Own

Exercises 24–27

7. The record low temperature in Hawaii is –0.15 times the record low temperature in Alaska. The record low temperature in Hawaii is 12°F. What is the record low temperature in Alaska?

Check It Out
Help with Homework
BigIdeasMath.com

 Vocabulary and Concept Check

1. **WRITING** Explain why you can use multiplication to solve equations involving division.

2. **OPEN-ENDED** Turning a light on and then turning the light off are considered to be inverse operations. Describe two other real-life situations that can be thought of as inverse operations.

Describe the inverse operation that will undo the given operation.

3. multiplying by 5
4. subtracting 12
5. dividing by -8
6. adding -6

 Practice and Problem Solving

Solve the equation. Check your solution.

① 7. $3h = 15$
   8. $-5t = -45$
   9. $\dfrac{n}{2} = -7$
   10. $\dfrac{k}{-3} = 9$

11. $5m = -10$
12. $8t = -32$
13. $-0.2x = 1.6$
14. $-10 = -\dfrac{b}{4}$

15. $-6p = 48$
16. $-72 = 8d$
17. $\dfrac{n}{1.6} = 5$
18. $-14.4 = -0.6p$

② 19. $\dfrac{3}{4}g = -12$
   20. $8 = -\dfrac{2}{5}c$
   21. $-\dfrac{4}{9}f = -3$
   22. $26 = -\dfrac{8}{5}y$

23. **ERROR ANALYSIS** Describe and correct the error in finding the solution.

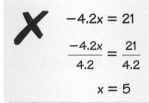
$$-4.2x = 21$$
$$\dfrac{-4.2x}{4.2} = \dfrac{21}{4.2}$$
$$x = 5$$

Write the word sentence as an equation. Then solve.

③ 24. A number divided by -9 is -16.
   25. A number multiplied by $\dfrac{2}{5}$ is $\dfrac{3}{20}$.

26. The product of 15 and a number is -75.

27. The quotient of a number and -1.5 is 21.

In Exercises 28 and 29, write an equation. Then solve.

28. **NEWSPAPERS** You make $0.75 for every newspaper you sell. How many newspapers do you have to sell to buy the soccer cleats?

29. **ROCK CLIMBING** A rock climber averages $12\dfrac{3}{5}$ feet per minute.

   How many feet does the rock climber climb in 30 minutes?

Soccer Cleats
$36⁰⁰

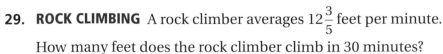

OPEN-ENDED (a) Write a multiplication equation that has the given solution.
(b) Write a division equation that has the same solution.

30. -3 **31.** -2.2 **32.** $-\dfrac{1}{2}$ **33.** $-1\dfrac{1}{4}$

34. REASONING Which of the methods can you use to solve $-\dfrac{2}{3}c = 16$?

> Multiply each side by $-\dfrac{2}{3}$.

> Multiply each side by $-\dfrac{3}{2}$.

> Divide each side by $-\dfrac{2}{3}$.

> Multiply each side by 3, then divide each side by -2.

35. STOCK A stock has a return of $-\$1.26$ per day. Write and solve an equation to find the number of days until the total return is $-\$10.08$.

36. ELECTION In a school election, $\dfrac{3}{4}$ of the students vote. There are 1464 ballots. Write and solve an equation to find the number of students.

37. OCEANOGRAPHY Aquarius is an underwater ocean laboratory located in the Florida Keys National Marine Sanctuary. Solve the equation $\dfrac{31}{25}x = -62$ to find the value of x.

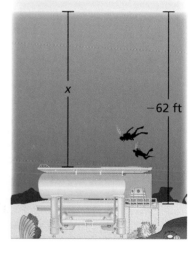

38. SHOPPING The price of a bike at Store A is $\dfrac{5}{6}$ the price at Store B. The price at Store A is $\$150.60$. Write and solve an equation to find how much you save by buying the bike at Store A.

39. CRITICAL THINKING Solve $-2|m| = -10$.

40. In four days, your family drives $\dfrac{5}{7}$ of a trip. Your rate of travel is the same throughout the trip. The total trip is 1250 miles. In how many more days will you reach your destination?

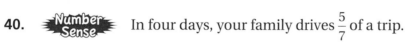

Fair Game Review What you learned in previous grades & lessons

Subtract. *(Section 1.3)*

41. $5 - 12$ **42.** $-7 - 2$ **43.** $4 - (-8)$ **44.** $-14 - (-5)$

45. MULTIPLE CHOICE Of the 120 apartments in a building, 75 have been scheduled to receive new carpet. What fraction of the apartments have not been scheduled to receive new carpet? *(Skills Review Handbook)*

Ⓐ $\dfrac{1}{4}$ Ⓑ $\dfrac{3}{8}$ Ⓒ $\dfrac{5}{8}$ Ⓓ $\dfrac{3}{4}$

3.5 Solving Two-Step Equations

Essential Question How can you use algebra tiles to solve a two-step equation?

1 ACTIVITY: Solving a Two-Step Equation

Work with a partner. Use algebra tiles to model and solve $2x - 3 = -5$.

Model the equation $2x - 3 = -5$.

Remove the ☐ red tiles on the left side by adding ☐ yellow tiles to each side.

How many *zero pairs* can you remove from each side? ☐
Circle them.

Because there are ☐ green tiles, divide the red tiles into ☐ equal groups. Circle the groups.

Keep one of the groups. This shows the value of x.

∴ So, $x = $ ☐ .

2 ACTIVITY: The Math behind the Tiles

Work with a partner. Solve $2x - 3 = -5$ without using algebra tiles. Complete each step. Then answer the questions.

Use the steps in Activity 1 as a guide.

$$2x - 3 = -5 \qquad \text{Write the equation.}$$

$$2x - 3 + \boxed{} = -5 + \boxed{} \qquad \text{Add } \boxed{} \text{ to each side.}$$

$$2x = \boxed{} \qquad \text{Simplify.}$$

$$\frac{2x}{\boxed{}} = \frac{\boxed{}}{\boxed{}} \qquad \text{Divide each side by } \boxed{}.$$

$$x = \boxed{} \qquad \text{Simplify.}$$

∴ So, $x = $ ☐ .

a. Which step is first, adding 3 to each side or dividing each side by 2?

b. How are the above steps related to the steps in Activity 1?

COMMON CORE

Solving Equations
In this lesson, you will
- solve two-step equations.
- solve real-life problems.
Learning Standard
7.EE.4a

3 **ACTIVITY: Solving Equations Using Algebra Tiles**

Work with a partner.

- Write an equation shown by the algebra tiles.
- Use algebra tiles to model and solve the equation.
- Check your answer by solving the equation without using algebra tiles.

a.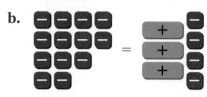

b.

4 **ACTIVITY: Working Backwards**

Work with a partner.

Math Practice 8

Maintain Oversight

How does working backwards help you decide which operation to do first? Explain.

a. **Sample:** Your friend pauses a video game to get a drink. You continue the game. You double the score by saving a princess. Then you lose 75 points because you do not collect the treasure. You finish the game with -25 points. How many points did you have when you started?

One way to solve the problem is to work backwards. To do this, start with the end result and retrace the events.

You have -25 points at the end of the game.	-25
You lost 75 points for not collecting the treasure, so add 75 to -25.	$-25 + 75 = 50$
You doubled your score for saving the princess, so find half of 50.	$50 \div 2 = 25$

∴ So, you started the game with 25 points.

b. You triple your account balance by making a deposit. Then you withdraw $127.32 to buy groceries. Your account is now overdrawn by $10.56. By working backwards, find your account balance before you made the deposit.

What Is Your Answer?

5. **IN YOUR OWN WORDS** How can you use algebra tiles to solve a two-step equation?

6. When solving the equation $4x + 1 = -11$, what is the first step?

7. **REPEATED REASONING** Solve the equation $2x - 75 = -25$. How do your steps compare with the strategy of working backwards in Activity 4?

Practice

Use what you learned about solving two-step equations to complete Exercises 6–11 on page 112.

3.5 Lesson

EXAMPLE 1 Solving a Two-Step Equation

Solve $-3x + 5 = 2$. Check your solution.

$$-3x + 5 = 2 \quad \text{Write the equation.}$$

Undo the addition. ⟶ $\underline{\quad -5 \qquad -5}$ Subtraction Property of Equality

$$-3x = -3 \quad \text{Simplify.}$$

Undo the multiplication. ⟶ $\dfrac{-3x}{-3} = \dfrac{-3}{-3}$ Division Property of Equality

$$x = 1 \quad \text{Simplify.}$$

Check

$$-3x + 5 = 2$$
$$-3(1) + 5 \overset{?}{=} 2$$
$$-3 + 5 \overset{?}{=} 2$$
$$2 = 2 \checkmark$$

∴ The solution is $x = 1$.

On Your Own

Now You're Ready
Exercises 6–17

Solve the equation. Check your solution.

1. $2x + 12 = 4$ **2.** $-5c + 9 = -16$ **3.** $3(x - 4) = 9$

EXAMPLE 2 Solving a Two-Step Equation

Solve $\dfrac{x}{8} - \dfrac{1}{2} = -\dfrac{7}{2}$. Check your solution.

$$\dfrac{x}{8} - \dfrac{1}{2} = -\dfrac{7}{2} \quad \text{Write the equation.}$$

$$\underline{+\dfrac{1}{2} \qquad +\dfrac{1}{2}} \quad \text{Addition Property of Equality}$$

$$\dfrac{x}{8} = -3 \quad \text{Simplify.}$$

$$8 \cdot \dfrac{x}{8} = 8 \cdot (-3) \quad \text{Multiplication Property of Equality}$$

$$x = -24 \quad \text{Simplify.}$$

Check

$$\dfrac{x}{8} - \dfrac{1}{2} = -\dfrac{7}{2}$$
$$\dfrac{-24}{8} - \dfrac{1}{2} \overset{?}{=} -\dfrac{7}{2}$$
$$-3 - \dfrac{1}{2} \overset{?}{=} -\dfrac{7}{2}$$
$$-\dfrac{7}{2} = -\dfrac{7}{2} \checkmark$$

Study Tip

You can simplify the equation in Example 2 before solving. Multiply each side by the LCD of the fractions, 8.

$$\dfrac{x}{8} - \dfrac{1}{2} = -\dfrac{7}{2}$$
$$x - 4 = -28$$
$$x = -24$$

∴ The solution is $x = -24$.

On Your Own

Now You're Ready
Exercises 20–25

Solve the equation. Check your solution.

4. $\dfrac{m}{2} + 6 = 10$ **5.** $-\dfrac{z}{3} + 5 = 9$ **6.** $\dfrac{2}{5} + 4a = -\dfrac{6}{5}$

Solve $3y - 8y = 25$.

$$3y - 8y = 25 \qquad \text{Write the equation.}$$
$$-5y = 25 \qquad \text{Combine like terms.}$$
$$y = -5 \qquad \text{Divide each side by } -5.$$

:· The solution is $y = -5$.

The height at the top of a roller coaster hill is 10 times the height h of the starting point. The height decreases 100 feet from the top to the bottom of the hill. The height at the bottom of the hill is −10 feet. Find h.

Location	Verbal Description	Expression
Start	The height at the start is h.	h
Top of hill	The height at the top of the hill is 10 times the starting height h.	$10h$
Bottom of hill	The height decreases by 100 feet. So, subtract 100.	$10h - 100$

The height at the bottom of the hill is −10 feet. Solve $10h - 100 = -10$ to find h.

$$10h - 100 = -10 \qquad \text{Write equation.}$$
$$10h = 90 \qquad \text{Add 100 to each side.}$$
$$h = 9 \qquad \text{Divide each side by 10.}$$

:· So, the height at the start is 9 feet.

● **On Your Own**

Now You're Ready
Exercises 29–34

Solve the equation. Check your solution.

7. $4 - 2y + 3 = -9$ **8.** $7x - 10x = 15$ **9.** $-8 = 1.3m - 2.1m$

10. WHAT IF? In Example 4, the height at the bottom of the hill is −5 feet. Find the height h.

 Vocabulary and Concept Check

1. **WRITING** How do you solve two-step equations?

Match the equation with the first step to solve it.

2. $4 + 4n = -12$ 3. $4n = -12$ 4. $\dfrac{n}{4} = -12$ 5. $\dfrac{n}{4} - 4 = -12$

A. Add 4.   **B.** Subtract 4.   **C.** Multiply by 4.   **D.** Divide by 4.

 Practice and Problem Solving

Solve the equation. Check your solution.

① 6. $2v + 7 = 3$ 7. $4b + 3 = -9$ 8. $17 = 5k - 2$

9. $-6t - 7 = 17$ 10. $8n + 16.2 = 1.6$ 11. $-5g + 2.3 = -18.8$

12. $2t - 5 = -10$ 13. $-4p + 9 = -5$ 14. $11 = -5x - 2$

15. $4 + 2.2h = -3.7$ 16. $-4.8f + 6.4 = -8.48$ 17. $7.3y - 5.18 = -51.9$

ERROR ANALYSIS Describe and correct the error in finding the solution.

18.
$$✗ \quad \begin{aligned} -6 + 2x &= -10 \\ -6 + \frac{2x}{2} &= -\frac{10}{2} \\ -6 + x &= -5 \\ x &= 1 \end{aligned}$$

19.
$$✗ \quad \begin{aligned} -3x + 2 &= -7 \\ -3x &= -9 \\ -\frac{3x}{3} &= \frac{-9}{3} \\ x &= -3 \end{aligned}$$

Solve the equation. Check your solution.

② 20. $\dfrac{3}{5}g - \dfrac{1}{3} = -\dfrac{10}{3}$ 21. $\dfrac{a}{4} - \dfrac{5}{6} = -\dfrac{1}{2}$ 22. $-\dfrac{1}{3} + 2z = -\dfrac{5}{6}$

23. $2 - \dfrac{b}{3} = -\dfrac{5}{2}$ 24. $-\dfrac{2}{3}x + \dfrac{3}{7} = \dfrac{1}{2}$ 25. $-\dfrac{9}{4}v + \dfrac{4}{5} = \dfrac{7}{8}$

In Exercises 26–28, write an equation. Then solve.

26. **WEATHER** Starting at 1:00 P.M., the temperature changes -4 degrees per hour. How long will it take to reach $-1°$?

27. **BOWLING** It costs $2.50 to rent bowling shoes. Each game costs $2.25. You have $9.25. How many games can you bowl?

28. **CELL PHONES** A cell phone company charges a monthly fee plus $0.25 for each text message. The monthly fee is $30.00 and you owe $59.50. How many text messages did you have?

Temperature
at 1:00 P.M.

35°F

Solve the equation. Check your solution.

③ 29. $3v - 9v = 30$

30. $12t - 8t = -52$

31. $-8d - 5d + 7d = 72$

32. $6(x - 2) = -18$

33. $-4(m + 3) = 24$

34. $-8(y + 9) = -40$

35. WRITING Write a real-world problem that can be modeled by $\frac{1}{2}x - 2 = 8$. Then solve the equation.

36. GEOMETRY The perimeter of the parallelogram is 102 feet. Find m.

REASONING Exercises 37 and 38 are missing information. Tell what information you need to solve the problem.

37. TAXI A taxi service charges an initial fee plus \$1.80 per mile. How far can you travel for \$12?

38. EARTH The coldest surface temperature on the Moon is 57 degrees colder than twice the coldest surface temperature on Earth. What is the coldest surface temperature on Earth?

39. PROBLEM SOLVING On Saturday, you catch insects for your science class. Five of the insects escape. The remaining insects are divided into three groups to share in class. Each group has nine insects. How many insects did you catch on Saturday?

   a. Solve the problem by working backwards.

   b. Solve the equation $\dfrac{x - 5}{3} = 9$. How does the answer compare with the answer to part (a)?

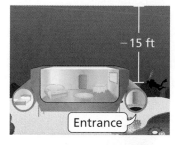

40. UNDERWATER HOTEL You must scuba dive to the entrance of your room at Jules' Undersea Lodge in Key Largo, Florida. The diver is 1 foot deeper than $\frac{2}{3}$ of the elevation of the entrance. What is the elevation of the entrance?

41. ✦Geometry✦ How much should you change the length of the rectangle so that the perimeter is 54 centimeters? Write an equation that shows how you found your answer.

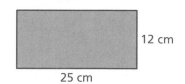

12 cm

25 cm

Fair Game Review *What you learned in previous grades & lessons*

Multiply or divide. *(Section 2.4)*

42. -6.2×5.6

43. $\dfrac{8}{3} \times \left(-2\dfrac{1}{2}\right)$

44. $\dfrac{5}{2} \div \left(-\dfrac{4}{5}\right)$

45. $-18.6 \div (-3)$

46. MULTIPLE CHOICE Which fraction is *not* equivalent to 0.75? *(Skills Review Handbook)*

  Ⓐ $\dfrac{15}{20}$ **Ⓑ** $\dfrac{9}{12}$ **Ⓒ** $\dfrac{6}{9}$ **Ⓓ** $\dfrac{3}{4}$

Solve the equation. Check your solution.   *(Section 3.3, Section 3.4, and Section 3.5)*

1. $-6.5 + x = -4.12$

2. $4\frac{1}{2} + p = -5\frac{3}{4}$

3. $-\dfrac{b}{7} = 4$

4. $-2w + 3.7 = -0.5$

Write the word sentence as an equation. Then solve.   *(Section 3.3 and Section 3.4)*

5. The difference between a number b and 7.4 is -6.8.

6. $5\frac{2}{5}$ more than a number a is $7\frac{1}{2}$.

7. A number x multiplied by $\dfrac{3}{8}$ is $-\dfrac{15}{32}$.

8. The quotient of two times a number k and -2.6 is 12.

Write and solve an equation to find the value of x.   *(Section 3.3 and Section 3.5)*

9. Perimeter = 26

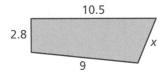
10.5
2.8
x
9

10. Perimeter = 23.59

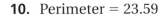
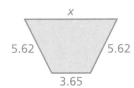
x
5.62   5.62
3.65

11. Perimeter = 33

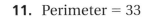
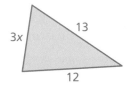
13
$3x$
12

12. BANKING You withdraw $29.79 from your bank account. Now your balance is $-$20.51. Write and solve an equation to find the amount of money in your bank account before you withdrew the money.   *(Section 3.3)*

13. WATER LEVEL During a drought, the water level of a lake changes $-3\frac{1}{5}$ feet per day. Write and solve an equation to find how long it takes for the water level to change -16 feet.   *(Section 3.4)*

14. BASKETBALL A basketball game has four quarters. The length of a game is 32 minutes. You play the entire game except for $4\frac{1}{2}$ minutes. Write and solve an equation to find the mean time you play per quarter.   *(Section 3.5)*

15. SCRAPBOOKING The mat needs to be cut to have a 0.5-inch border on all four sides.   *(Section 3.5)*

   a. How much should you cut from the left and right sides?

   b. How much should you cut from the top and bottom?

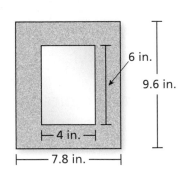
6 in.
9.6 in.
4 in.
7.8 in.

3 Chapter Review

Check It Out
Vocabulary Help
BigIdeasMath ✓.com

Review Key Vocabulary

like terms, *p. 82*
simplest form, *p. 82*
linear expression, *p. 88*

factoring an expression, *p. 92*
equivalent equations, *p. 98*

Review Examples and Exercises

3.1 Algebraic Expressions *(pp. 80–85)*

a. Identify the terms and like terms in the expression $6y + 9 + 3y - 7$.

Rewrite as a sum of terms.

$$6y + 9 + 3y + (-7)$$

Terms: $6y$, $\quad 9$, $\quad 3y$, $\quad -7$

Like terms: $6y$ and $3y$, 9 and -7

b. Simplify $\dfrac{2}{3}y + 14 - \dfrac{1}{6}y - 8$.

$\dfrac{2}{3}y + 14 - \dfrac{1}{6}y - 8 = \dfrac{2}{3}y + 14 + \left(-\dfrac{1}{6}y\right) + (-8)$ 　Rewrite as a sum.

$= \dfrac{2}{3}y + \left(-\dfrac{1}{6}y\right) + 14 + (-8)$ 　Commutative Property of Addition

$= \left[\dfrac{2}{3} + \left(-\dfrac{1}{6}\right)\right]y + 14 + (-8)$ 　Distributive Property

$= \dfrac{1}{2}y + 6$ 　Combine like terms.

Exercises

Identify the terms and like terms in the expression.

1. $z + 8 - 4z$　　　**2.** $3n + 7 - n - 3$　　　**3.** $10x^2 - y + 12 - 3x^2$

Simplify the expression.

4. $4h - 8h$

5. $6.4r - 7 - 2.9r$

6. $\dfrac{3}{5}x + 19 - \dfrac{3}{20}x - 7$

7. $3(2 + q) + 15$

8. $\dfrac{1}{8}(16m - 8) - 17$

9. $-1.5(4 - n) + 2.8$

Adding and Subtracting Linear Expressions *(pp. 86–93)*

a. **Find $(5z + 4) + (3z - 6)$.**

$$5z + 4$$
$$\underline{+\ 3z - 6} \qquad \text{Align like terms vertically and add.}$$
$$8z - 2$$

b. **Factor $\frac{1}{4}$ out of $\frac{1}{4}x - \frac{3}{4}$.**

Write each term as a product of $\frac{1}{4}$ and another factor.

$$\frac{1}{4}x = \frac{1}{4} \cdot x \qquad\qquad -\frac{3}{4} = \frac{1}{4} \cdot (-3)$$

Use the Distributive Property to factor out $\frac{1}{4}$.

$$\frac{1}{4}x - \frac{3}{4} = \frac{1}{4} \cdot x + \frac{1}{4} \cdot (-3) = \frac{1}{4}(x - 3)$$

∴ So, $\frac{1}{4}x - \frac{3}{4} = \frac{1}{4}(x - 3)$.

Exercises

Find the sum or difference.

10. $(c - 4) + (3c + 9)$

11. $\frac{2}{5}(d - 10) - \frac{2}{3}(d + 6)$

Factor out the coefficient of the variable.

12. $2b + 8$　　**13.** $\frac{1}{4}y + \frac{3}{8}$　　**14.** $1.7j - 3.4$　　**15.** $-5p + 20$

Solving Equations Using Addition or Subtraction *(pp. 96–101)*

Solve $x - 9 = -6$.

		Check
$x - 9 = -6$	Write the equation.	$x - 9 = -6$
Undo the subtraction. → $\underline{+9\quad +9}$	Addition Property of Equality	$3 - 9 \overset{?}{=} -6$
$x = 3$	Simplify.	$-6 = -6$ ✔

Exercises

Solve the equation. Check your solution.

16. $p - 3 = -4$　　**17.** $6 + q = 1$　　**18.** $-2 + j = -22$　　**19.** $b - 19 = -11$

20. $n + \frac{3}{4} = \frac{1}{4}$　　**21.** $v - \frac{5}{6} = -\frac{7}{8}$　　**22.** $t - 3.7 = 1.2$　　**23.** $\ell + 15.2 = -4.5$

3.4 **Solving Equations Using Multiplication or Division** *(pp. 102–107)*

Solve $\dfrac{x}{5} = -7$.

$\dfrac{x}{5} = -7$ Write the equation.

Undo the division. $\longrightarrow$ $5 \cdot \dfrac{x}{5} = 5 \cdot (-7)$ Multiplication Property of Equality

$x = -35$ Simplify.

Check

$\dfrac{x}{5} = -7$

$\dfrac{-35}{5} \overset{?}{=} -7$

$-7 = -7$ ✓

Exercises

Solve the equation. Check your solution.

24. $\dfrac{x}{3} = -8$ **25.** $-7 = \dfrac{y}{7}$ **26.** $-\dfrac{z}{4} = -\dfrac{3}{4}$ **27.** $-\dfrac{w}{20} = -2.5$

28. $4x = -8$ **29.** $-10 = 2y$ **30.** $-5.4z = -32.4$ **31.** $-6.8w = 3.4$

32. TEMPERATURE The mean temperature change is $-3.2°F$ per day for 5 days. What is the total change over the 5-day period?

3.5 **Solving Two-Step Equations** *(pp. 108–113)*

Solve $-6y + 7 = -5$. Check your solution.

$-6y + 7 = -5$ Write the equation.

$\underline{\; -7 \quad\; -7}$ Subtraction Property of Equality

$-6y = -12$ Simplify.

$\dfrac{-6y}{-6} = \dfrac{-12}{-6}$ Division Property of Equality

$y = 2$ Simplify.

Check

$-6y + 7 = -5$

$-6(2) + 7 \overset{?}{=} -5$

$-12 + 7 \overset{?}{=} -5$

$-5 = -5$ ✓

∴ The solution is $y = 2$.

Exercises

Solve the equation. Check your solution.

33. $-2c + 6 = -8$ **34.** $3(3w - 4) = -20$

35. $\dfrac{w}{6} + \dfrac{5}{8} = -1\dfrac{3}{8}$ **36.** $-3x - 4.6 = 5.9$

37. EROSION The floor of a canyon has an elevation of -14.5 feet. Erosion causes the elevation to change by -1.5 feet per year. How many years will it take for the canyon floor to have an elevation of -31 feet?

Check It Out
Test Practice
BigIdeasMath ✓com

Simplify the expression.

1. $8x - 5 + 2x$

2. $2.5w - 3y + 4w$

3. $3(5 - 2n) + 9n$

4. $\dfrac{5}{7}x + 15 - \dfrac{9}{14}x - 9$

Find the sum or difference.

5. $(3j + 11) + (8j - 7)$

6. $\dfrac{3}{4}(8p + 12) + \dfrac{3}{8}(16p - 8)$

7. $(2r - 13) - (-6r + 4)$

8. $-2.5(2s - 5) - 3(4.5s - 5.2)$

Factor out the coefficient of the variable.

9. $3n - 24$

10. $\dfrac{1}{2}q + \dfrac{5}{2}$

Solve the equation. Check your solution.

11. $7x = -3$

12. $2(x + 1) = -2$

13. $\dfrac{2}{9}g = -8$

14. $z + 14.5 = 5.4$

15. $-14 = 6c$

16. $\dfrac{2}{7}k - \dfrac{3}{8} = -\dfrac{19}{8}$

17. HAIR SALON Write an expression in simplest form that represents the income from w women and m men getting a haircut and a shampoo.

	Women	Men
Haircut	$45	$15
Shampoo	$12	$7

18. RECORD A runner is compared with the world record holder during a race. A negative number means the runner is ahead of the time of the world record holder. A positive number means that the runner is behind the time of the world record holder. The table shows the time difference between the runner and the world record holder for each lap. What time difference does the runner need for the fourth lap to match the world record?

Lap	Time Difference
1	−1.23
2	0.45
3	0.18
4	?

19. GYMNASTICS You lose 0.3 point for stepping out of bounds during a floor routine. Your final score is 9.124. Write and solve an equation to find your score before the penalty.

20. PERIMETER The perimeter of the triangle is 45. Find the value of x.

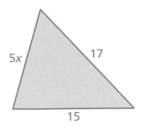

1. Which equation represents the word sentence shown below? *(7.EE.4a)*

   The quotient of a number b and 0.3 equals negative 10.

   A. $0.3b = 10$ **C.** $\dfrac{0.3}{b} = -10$

   B. $\dfrac{b}{0.3} = -10$ **D.** $\dfrac{b}{0.3} = 10$

2. What is the value of the expression below when $c = 0$ and $d = -6$? *(7.NS.2c)*

   $$\dfrac{cd - d^2}{4}$$

3. What is the value of the expression below? *(7.NS.1c)*

   $$-38 - (-14)$$

   F. -52 **H.** 24

   G. -24 **I.** 52

4. The daily low temperatures last week are shown below.

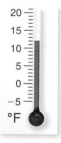

   What is the mean low temperature of last week? *(7.NS.3)*

   A. $-2°F$ **C.** $8°F$

   B. $6°F$ **D.** $10°F$

5. Which equation is equivalent to the equation shown below? *(7.EE.4a)*

$$-\frac{3}{4}x + \frac{1}{8} = -\frac{3}{8}$$

F. $-\frac{3}{4}x = -\frac{3}{8} - \frac{1}{8}$

G. $-\frac{3}{4}x = -\frac{3}{8} + \frac{1}{8}$

H. $x + \frac{1}{8} = -\frac{3}{8} \cdot \left(-\frac{4}{3}\right)$

I. $x + \frac{1}{8} = -\frac{3}{8} \cdot \left(-\frac{3}{4}\right)$

6. What is the value of the expression below? *(7.NS.2c)*

$$-0.28 \div -0.07$$

7. Karina was solving the equation in the box below.

$$-96 = -6(x - 15)$$
$$-96 = -6x - 90$$
$$-96 + 90 = -6x - 90 + 90$$
$$-6 = -6x$$
$$\frac{-6}{-6} = \frac{-6x}{-6}$$
$$1 = x$$

What should Karina do to correct the error that she made? *(7.EE.4a)*

A. First add 6 to both sides of the equation.

B. First subtract x from both sides of the equation.

C. Distribute the -6 to get $6x - 90$.

D. Distribute the -6 to get $-6x + 90$.

8. The perimeter of the rectangle is 400 inches. What is the value of j?
(All measurements are in inches.) *(7.EE.4a)*

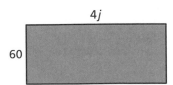

4*j*

60

F. 35

G. 85

H. 140

I. 200

9. Jacob was evaluating the expression below when $x = -2$ and $y = 4$.

$$3 + x^2 \div y$$

His work is in the box below.

$$3 + x^2 \div y = 3 + \left(-2^2\right) \div 4$$
$$= 3 - 4 \div 4$$
$$= 3 - 1$$
$$= 3$$

What should Jacob do to correct the error that he made?   *(7.NS.3)*

A. Divide 3 by 4 before subtracting.

B. Square -2, then divide.

C. Square then divide.

D. Subtract 4 from 3 before dividing.

10. Which number is equivalent to the expression shown below?   *(7.NS.3)*

$$-2\frac{1}{4} - \left(-8\frac{3}{8}\right)$$

F. $-10\frac{5}{8}$ **H.** $6\frac{1}{8}$

G. $-10\frac{1}{3}$ **I.** $6\frac{1}{2}$

11. You want to buy the bicycle. You already have $43.50 saved and plan to save an additional $7.25 every week.   *(7.EE.4a)*

Part A   Write and solve an equation to find the number of weeks you need to save before you can purchase the bicycle.

Part B   How much sooner could you purchase the bicycle if you had a coupon for $20 off and saved $8.75 every week? Explain your reasoning.

4 Inequalities

"If you reached into your water bowl and found more than $20..."

"And then reached into your cat food bowl and found more than $40..."

Someone else's bowls!

"What would you have?"

"Dear Precious Pet World: Your ad says 'Up to 75% off on selected items.'"

Hey, it didn't say who's doing the selecting.

"I select Yummy Tummy Bacon-Flavored Dog Biscuits."

What You Learned Before

"Move farther back. We still have an inequality."

Graphing Inequalities (6.EE.8)

Example 1 Graph $x \geq 2$.

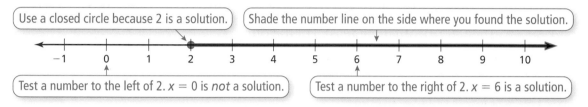

Use a closed circle because 2 is a solution.

Shade the number line on the side where you found the solution.

Test a number to the left of 2. $x = 0$ is *not* a solution.

Test a number to the right of 2. $x = 6$ is a solution.

Try It Yourself
Graph the inequality.

1. $x \geq 1$
2. $x < 5$
3. $x \leq 20$
4. $x > 13$

Comparing Numbers (6.NS.7a)

Example 2 Compare $-\dfrac{1}{3}$ and $-\dfrac{5}{6}$.

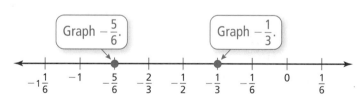

Graph $-\dfrac{5}{6}$.

Graph $-\dfrac{1}{3}$.

$-\dfrac{5}{6}$ is to the left of $-\dfrac{1}{3}$.

So, $-\dfrac{5}{6} < -\dfrac{1}{3}$.

Try It Yourself
Copy and complete the statement using < or >.

5. $-\dfrac{2}{3}$ ☐ $\dfrac{3}{8}$

6. $-\dfrac{1}{2}$ ☐ $-\dfrac{7}{8}$

7. $-\dfrac{1}{5}$ ☐ $\dfrac{1}{10}$

8. -1.4 ☐ 1.2

9. -2.2 ☐ -4.6

10. -1.9 ☐ -1.1

4.1 Writing and Graphing Inequalities

Essential Question How can you use a number line to represent solutions of an inequality?

1 ACTIVITY: Understanding Inequality Statements

Work with a partner. Read the statement. Circle each number that makes the statement true, and then answer the questions.

a. "You are in at least 5 of the photos."

$$-3 \quad -2 \quad -1 \quad 0 \quad 1 \quad 2 \quad 3 \quad 4 \quad 5 \quad 6$$

- What do you notice about the numbers that you circled?

- Is the number 5 included? Why or why not?

- Write four other numbers that make the statement true.

b. "The temperature is less than −4 degrees Fahrenheit."

$$-7 \quad -6 \quad -5 \quad -4 \quad -3 \quad -2 \quad -1 \quad 0 \quad 1 \quad 2$$

- What do you notice about the numbers that you circled?

- Can the temperature be exactly −4 degrees Fahrenheit? Explain.

- Write four other numbers that make the statement true.

c. "More than 3 students from our school are in the chess tournament."

$$-3 \quad -2 \quad -1 \quad 0 \quad 1 \quad 2 \quad 3 \quad 4 \quad 5 \quad 6$$

- What do you notice about the numbers that you circled?

- Is the number 3 included? Why or why not?

- Write four other numbers that make the statement true.

COMMON CORE

Inequalities
In this lesson, you will
- write and graph inequalities.
- use substitution to check whether a number is a solution of an inequality.
Preparing for Standard 7.EE.4b

d. "The balance in a yearbook fund is no more than −$5."

$$-7 \quad -6 \quad -5 \quad -4 \quad -3 \quad -2 \quad -1 \quad 0 \quad 1 \quad 2$$

- What do you notice about the numbers that you circled?

- Is the number −5 included? Why or why not?

- Write four other numbers that make the statement true.

2 ACTIVITY: Understanding Inequality Symbols

Work with a partner.

a. Consider the statement "x is a number such that $x > -1.5$."

- Can the number be exactly -1.5? Explain.

- Make a number line. Shade the part of the number line that shows the numbers that make the statement true.

- Write four other numbers that are not integers that make the statement true.

b. Consider the statement "x is a number such that $x \leq \dfrac{5}{2}$."

- Can the number be exactly $\dfrac{5}{2}$? Explain.

- Make a number line. Shade the part of the number line that shows the numbers that make the statement true.

- Write four other numbers that are not integers that make the statement true.

3 ACTIVITY: Writing and Graphing Inequalities

Math Practice **1**

Check Progress

All the graphs are similar. So, what can you do to make sure that you have correctly written each inequality?

Work with a partner. Write an inequality for each graph. Then, in words, describe all the values of x that make the inequality true.

a.
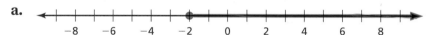

b.
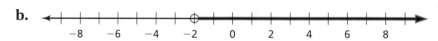

c.

d.

What Is Your Answer?

4. IN YOUR OWN WORDS How can you use a number line to represent solutions of an inequality?

5. STRUCTURE Is $x \geq -1.4$ the same as $-1.4 \leq x$? Explain.

Practice ▶ Use what you learned about writing and graphing inequalities to complete Exercises 4 and 5 on page 128.

Key Vocabulary
inequality, *p. 126*
solution of an
    inequality, *p. 126*
solution set, *p. 126*
graph of an
    inequality, *p. 127*

An **inequality** is a mathematical sentence that compares expressions. It contains the symbols <, >, ≤, or ≥. To write an inequality, look for the following phrases to determine where to place the inequality symbol.

Inequality Symbols				
Symbol	<	>	≤	≥
Key Phrases	• is less than • is fewer than	• is greater than • is more than	• is less than or equal to • is at most • is no more than	• is greater than or equal to • is at least • is no less than

EXAMPLE 1 Writing an Inequality

A number q plus 5 is greater than or equal to -7.9. Write this word sentence as an inequality.

A number q plus 5 is greater than or equal to -7.9.
$q + 5\geq-7.9$

An inequality is $q + 5 \geq -7.9$.

On Your Own

Now You're Ready
Exercises 6–9

Write the word sentence as an inequality.

1. A number x is at most -10.
2. Twice a number y is more than $-\dfrac{5}{2}$.

A **solution of an inequality** is a value that makes the inequality true. An inequality can have more than one solution. The set of all solutions of an inequality is called the **solution set**.

Reading

The symbol ≰ means *is not less than* or *equal to*.

Value of x	$x + 2 \leq -1$	Is the inequality true?
-2	$-2 + 2 \overset{?}{\leq} -1$ $0 \not\leq -1$ ✗	no
-3	$-3 + 2 \overset{?}{\leq} -1$ $-1 \leq -1$ ✓	yes
-4	$-4 + 2 \overset{?}{\leq} -1$ $-2 \leq -1$ ✓	yes

Multi-Language Glossary at BigIdeasMath.com

EXAMPLE ② **Checking Solutions**

Tell whether −2 is a solution of each inequality.

a. $y - 5 \geq -6$

$y - 5 \geq -6$	Write the inequality.
$-2 - 5 \overset{?}{\geq} -6$	Substitute −2 for y.
$-7 \ngeq -6$ ✗	Simplify.

−7 is *not* greater than or equal to −6.

∴ So, −2 is *not* a solution of the inequality.

b. $-5.5y < 14$

$-5.5y < 14$	
$-5.5(-2) \overset{?}{<} 14$	
$11 < 14$ ✓	

11 is less than 14.

∴ So, −2 is a solution of the inequality.

● **On Your Own**

Now You're Ready
Exercises 11–16

Tell whether −5 is a solution of the inequality.

3. $x + 12 > 7$ **4.** $1 - 2p \leq -9$ **5.** $n \div 2.5 \geq -3$

The **graph of an inequality** shows all the solutions of the inequality on a number line. An open circle ○ is used when a number is *not* a solution. A closed circle ● is used when a number is a solution. An arrow to the left or right shows that the graph continues in that direction.

EXAMPLE ③ **Graphing an Inequality**

Graph $y > -8$.

Use an open circle because −8 is *not* a solution.

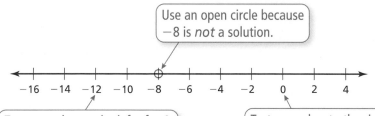

Test a number to the left of −8. $y = -12$ is *not* a solution.

Test a number to the right of −8. $y = 0$ is a solution.

Study Tip

The graph in Example 3 shows that the inequality has *infinitely many* solutions.

Shade the number line on the side where you found the solution.

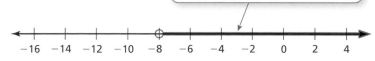

● **On Your Own**

Now You're Ready
Exercises 17–20

Graph the inequality on a number line.

6. $x < -1$ **7.** $z \geq 4$ **8.** $s \leq 1.4$ **9.** $-\dfrac{1}{2} < t$

Check It Out
Help with Homework
BigIdeasMath⊘com

✓ **Vocabulary and Concept Check**

1. **PRECISION** Should you use an open circle or a closed circle in the graph of the inequality $b \geq -42$? Explain.

2. **DIFFERENT WORDS, SAME QUESTION** Which is different? Write "both" inequalities.

k is less than or equal to -3.	k is no more than -3.
k is at most -3.	k is at least -3.

3. **REASONING** Do $x < 5$ and $5 < x$ represent the same inequality? Explain.

 Practice and Problem Solving

Write an inequality for the graph. Then, in words, describe all the values of x that make the inequality true.

4.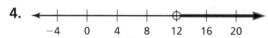

5.

Write the word sentence as an inequality.

① 6. A number y is no more than -8.

7. A number w added to 2.3 is more than 18.

8. A number t multiplied by -4 is at least $-\dfrac{2}{5}$.

9. A number b minus 4.2 is less than -7.5.

10. **ERROR ANALYSIS** Describe and correct the error in writing the word sentence as an inequality.

 Twice a number x is at most −24.

$2x \geq -24$

Tell whether the given value is a solution of the inequality.

② 11. $n + 8 \leq 13; n = 4$ 12. $5h > -15; h = -5$ 13. $p + 1.4 \leq 0.5; p = 0.1$

14. $\dfrac{a}{6} > -4; a = -18$ 15. $-\dfrac{2}{3}s \geq 6; s = -9$ 16. $\dfrac{7}{8} - 3k < -\dfrac{1}{2}; k = \dfrac{1}{4}$

Graph the inequality on a number line.

③ 17. $r \leq -9$ 18. $g > 2.75$ 19. $x \geq -3\dfrac{1}{2}$ 20. $z < 1\dfrac{1}{4}$

21. **FOOD TRUCK** Each day at lunchtime, at least 53 people buy food from a food truck. Write an inequality that represents this situation.

Tell whether the given value is a solution of the inequality.

22. $4k < k + 8$; $k = 3$

23. $\dfrac{w}{3} \geq w - 12$; $w = 15$

24. $7 - 2y > 3y + 13$; $y = -1$

25. $\dfrac{3}{4}b - 2 \leq 2b + 8$; $b = -4$

26. **MODELING** A subway ride for a student costs $1.25. A monthly pass costs $35.

    a. Write an inequality that represents the number of times you must ride the subway for the monthly pass to be a better deal.

    b. You ride the subway about 45 times per month. Should you buy the monthly pass? Explain.

27. **LOGIC** Consider the inequality $b > -2$.

    a. Describe the values of b that are solutions of the inequality.

    b. Describe the values of b that are *not* solutions of the inequality. Write an inequality for these values.

    c. What do all the values in parts (a) and (b) represent? Is this true for any inequality?

28. **Critical Thinking** A postal service says that a rectangular package can have a maximum combined length and *girth* of 108 inches. The girth of a package is the distance around the perimeter of a face that does not include the length.

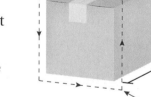

    a. Write an inequality that represents the allowable dimensions for the package.

    b. Find three different sets of allowable dimensions that are reasonable for the package. Find the volume of each package.

 Fair Game Review *What you learned in previous grades & lessons*

Solve the equation. Check your solution. *(Section 3.3)*

29. $p - 8 = 3$ **30.** $8.7 + w = 5.1$ **31.** $x - 2 = -9$

32. **MULTIPLE CHOICE** Which expression has a value less than -5? *(Section 1.2)*

    (A) $5 + 8$ (B) $-9 + 5$ (C) $1 + (-8)$ (D) $7 + (-2)$

Solving Inequalities Using Addition or Subtraction

Essential Question How can you use addition or subtraction to solve an inequality?

1 ACTIVITY: Writing an Inequality

Work with a partner. Members of the Boy Scouts must be less than 18 years old. In 4 years, your friend will still be eligible to be a scout.

a. Which of the following represents your friend's situation? What does x represent? Explain your reasoning.

$$x + 4 > 18$$ $$x + 4 < 18$$

$$x + 4 \geq 18$$ $$x + 4 \leq 18$$

b. Graph the possible ages of your friend on a number line. Explain how you decided what to graph.

2 ACTIVITY: Writing an Inequality

Work with a partner. Supercooling is the process of lowering the temperature of a liquid or a gas below its freezing point without it becoming a solid. Water can be supercooled to 86°F below its normal freezing point (32°F) and still not freeze.

a. Let x represent the temperature of water. Which inequality represents the temperature at which water can be a liquid or a gas? Explain your reasoning.

$$x - 32 > -86$$ $$x - 32 < -86$$

$$x - 32 \geq -86$$ $$x - 32 \leq -86$$

b. On a number line, graph the possible temperatures at which water can be a liquid or a gas. Explain how you decided what to graph.

COMMON CORE

Inequalities
In this lesson, you will
- solve inequalities using addition or subtraction.
- solve real-life problems.
Learning Standard
7.EE.4b

3 ACTIVITY: Solving Inequalities

Math Practice 4

Interpret Results

What does the solution of the inequality represent?

Work with a partner. Complete the following steps for Activity 1. Then repeat the steps for Activity 2.

- Use your inequality from part (a). Replace the inequality symbol with an equal sign.

- Solve the equation.

- Replace the equal sign with the original inequality symbol.

- Graph this new inequality.

- Compare the graph with your graph in part (b). What do you notice?

4 ACTIVITY: Temperatures of Continents

Work with a partner. The table shows the lowest recorded temperature on each continent. Write an inequality that represents each statement. Then solve and graph the inequality.

a. The temperature at a weather station in Asia is more than 150°F greater than the record low in Asia.

b. The temperature at a research station in Antarctica is at least 80°F greater than the record low in Antarctica.

Continent	Lowest Temperature
Africa	−11°F
Antarctica	−129°F
Asia	−90°F
Australia	−9.4°F
Europe	−67°F
North America	−81.4°F
South America	−27°F

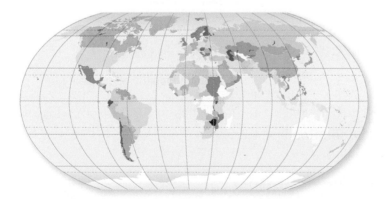

What Is Your Answer?

5. **IN YOUR OWN WORDS** How can you use addition or subtraction to solve an inequality?

6. Describe a real-life situation that you can represent with an inequality. Write the inequality. Graph the solution on a number line.

Practice

Use what you learned about solving inequalities to complete Exercises 3–5 on page 134.

Check It Out
Lesson Tutorials
BigIdeasMath ✓com

 **Key Ideas**

Study Tip

You can solve inequalities in the same way you solve equations. Use inverse operations to get the variable by itself.

Addition Property of Inequality

Words When you add the same number to each side of an inequality, the inequality remains true.

Numbers
$$-4 < 3$$
$$\underline{+2 \quad +2}$$
$$-2 < 5$$

Algebra If $a < b$, then $a + c < b + c$.

If $a > b$, then $a + c > b + c$.

Subtraction Property of Inequality

Words When you subtract the same number from each side of an inequality, the inequality remains true.

Numbers
$$-2 < 2$$
$$\underline{-3 \quad -3}$$
$$-5 < -1$$

Algebra If $a < b$, then $a - c < b - c$.

If $a > b$, then $a - c > b - c$.

These properties are also true for $\leq$ and $\geq$.

EXAMPLE 1 **Solving an Inequality Using Addition**

Solve $x - 5 < -3$. Graph the solution.

$$x - 5 < -3 \qquad \text{Write the inequality.}$$

Undo the subtraction. ⟶ $\underline{+5 \quad +5} \qquad$ Addition Property of Inequality

$$x < 2 \qquad \text{Simplify.}$$

∴ The solution is $x < 2$.

Check:

$x = 0$: $0 - 5 \overset{?}{<} -3$

$\qquad -5 < -3$ ✓

$x = 5$: $5 - 5 \overset{?}{<} -3$

$\qquad 0 \not< -3$ ✗

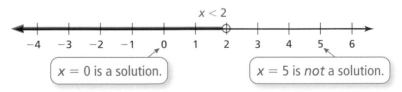

$x = 0$ is a solution.

$x = 5$ is *not* a solution.

On Your Own

Solve the inequality. Graph the solution.

1. $y - 6 > -7$

2. $b - 3.8 \leq 1.7$

3. $-\dfrac{1}{2} > z - \dfrac{1}{4}$

EXAMPLE **2** **Solving an Inequality Using Subtraction**

Solve $13 \leq x + 14$. Graph the solution.

	13	$\leq x + 14$	Write the inequality.
Undo the addition. →	-14	-14	Subtraction Property of Inequality
	-1	$\leq x$	Simplify.

The solution is $x \geq -1$.

Reading

The inequality $-1 \leq x$ is the same as $x \geq -1$.

$$x \geq -1$$

 Number line from -4 to 6 with a closed circle at -1 and shading to the right.

$-4 \quad -3 \quad -2 \quad -1 \quad 0 \quad 1 \quad 2 \quad 3 \quad 4 \quad 5 \quad 6$

● **On Your Own**

Now You're Ready
Exercises 3–17

Solve the inequality. Graph the solution.

4. $w - 7 \leq -10$ **5.** $-7.5 \geq d - 10$ **6.** $x + \dfrac{3}{4} > 1\dfrac{1}{2}$

EXAMPLE **3** **Real-Life Application**

A person can be no taller than 6.25 feet to become an astronaut pilot for NASA. Your friend is 5 feet 9 inches tall. Write and solve an inequality that represents how much your friend can grow and still meet the requirement.

Words   Current height   plus   amount your friend can grow   is no more than   the height limit.

Variable   Let h be the possible amounts your friend can grow.

Inequality 5.75 $+$ h $\leq$ 6.25

5 ft 9 in. = 60 + 9 = 69 in.

$69 \text{ in.} \times \dfrac{1 \text{ ft}}{12 \text{ in.}} = 5.75 \text{ ft}$

	$5.75 + h \leq$	6.25	Write the inequality.
	-5.75	-5.75	Subtraction Property of Inequality
	$h \leq$	0.5	Simplify.

So, your friend can grow no more than 0.5 foot, or 6 inches.

● **On Your Own**

7. Your cousin is 5 feet 3 inches tall. Write and solve an inequality that represents how much your cousin can grow and still meet the requirement.

Check It Out
Help with Homework
BigIdeasMath.com

 Vocabulary and Concept Check

1. **REASONING** Is the inequality $c + 3 > 5$ the same as $c > 5 - 3$? Explain.

2. **WHICH ONE DOESN'T BELONG?** Which inequality does *not* belong with the other three? Explain your reasoning.

$$w + \frac{7}{4} > \frac{3}{4}$$ $$w - \frac{3}{4} > -\frac{7}{4}$$ $$w + \frac{7}{4} < \frac{3}{4}$$ $$\frac{3}{4} < w + \frac{7}{4}$$

 Practice and Problem Solving

Solve the inequality. Graph the solution.

  3. $x + 7 \geq 18$ **4.** $a - 2 > 4$ **5.** $3 \leq 7 + g$

6. $8 + k \leq -3$ **7.** $-12 < y - 6$ **8.** $n - 4 < 5$

9. $t - 5 \leq -7$ **10.** $p + \frac{1}{4} \geq 2$ **11.** $\frac{2}{7} > b + \frac{5}{7}$

12. $z - 4.7 \geq -1.6$ **13.** $-9.1 < d - 6.3$ **14.** $\frac{8}{5} > s + \frac{12}{5}$

15. $-\frac{7}{8} \geq m - \frac{13}{8}$ **16.** $r + 0.2 < -0.7$ **17.** $h - 6 \leq -8.4$

ERROR ANALYSIS Describe and correct the error in solving the inequality or graphing the solution of the inequality.

18.
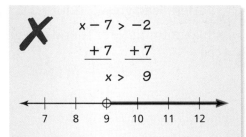

19.
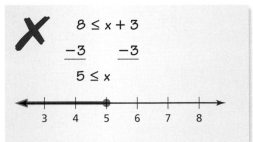

20. **AIRPLANE** A small airplane can hold 44 passengers. Fifteen passengers board the plane.

   a. Write and solve an inequality that represents the additional number of passengers that can board the plane.

   b. Can 30 more passengers board the plane? Explain.

Write and solve an inequality that represents *x*.

21. The perimeter is less than 28 feet.

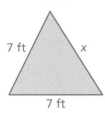

7 ft   *x*

7 ft

22. The base is greater than the height.

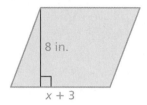

8 in.

x + 3

23. The perimeter is less than or equal to 51 meters.

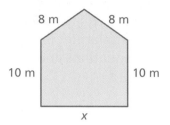

8 m   8 m

10 m   10 m

x

24. REASONING The solution of $d + s > -3$ is $d > -7$. What is the value of s?

25. BIRDFEEDER The hole for a birdfeeder post is 3 feet deep. The top of the post needs to be at least 5 feet above the ground. Write and solve an inequality that represents the required length of the post.

26. SHOPPING You can spend up to $35 on a shopping trip.

   a. You want to buy a shirt that costs $14. Write and solve an inequality that represents the amount of money you will have left if you buy the shirt.

   b. You notice that the shirt is on sale for 30% off. How does this change the inequality?

   c. Do you have enough money to buy the shirt that is on sale and a pair of pants that costs $23? Explain.

27. POWER A circuit overloads at 2400 watts of electricity. A portable heater that uses 1050 watts of electricity is plugged into the circuit.

   a. Write and solve an inequality that represents the additional number of watts you can plug in without overloading the circuit.

   b. In addition to the portable heater, what two other items in the table can you plug in at the same time without overloading the circuit? Is there more than one possibility? Explain.

Item	Watts
Aquarium	200
Hair dryer	1200
Television	150
Vacuum cleaner	1100

28. **Number Sense** The possible values of x are given by $x + 8 \le 6$. What is the greatest possible value of $7x$?

Fair Game Review What you learned in previous grades & lessons

Solve the equation. Check your solution. *(Section 3.4)*

29. $4x = 36$

30. $\dfrac{w}{3} = -9$

31. $-2b = 44$

32. $60 = \dfrac{3}{4}h$

33. MULTIPLE CHOICE Which fraction is equivalent to –2.4? *(Section 2.1)*

   Ⓐ $-\dfrac{12}{5}$ Ⓑ $-\dfrac{51}{25}$ Ⓒ $-\dfrac{8}{5}$ Ⓓ $-\dfrac{6}{25}$

You can use a **Y chart** to compare two topics. List differences in the branches and similarities in the base of the Y. Here is an example of a Y chart that compares solving equations and solving inequalities.

Solving Equations

- The sign between two expressions is an equal sign, =.
- One number is the solution.

Solving Inequalities

- The sign between two expressions is an inequality symbol: <, >, ≤, or ≥.
- More than one number can be a solution.

- Use inverse operations to group numbers on one side.
- Use inverse operations to group variables on one side.
- Solve for the variable.

On Your Own

Make Y charts to help you study and compare these topics.

1. writing equations and writing inequalities

2. graphing the solution of an equation and graphing the solution of an inequality

3. graphing inequalities that use > and graphing inequalities that use <

4. graphing inequalities that use > or < and graphing inequalities that use ≥ or ≤

5. solving inequalities using addition and solving inequalities using subtraction

"Hey Descartes, do you have any suggestions for the Y chart I am making?"

After you complete this chapter, make Y charts for the following topics.

6. solving inequalities using multiplication and solving inequalities using division

7. solving two-step equations and solving two-step inequalities

8. Pick two other topics that you studied earlier in this course and make a Y chart to compare them.

Write the word sentence as an inequality. *(Section 4.1)*

1. A number y plus 2 is greater than -5.

2. A number s minus 2.4 is at least 8.

Tell whether the given value is a solution of the inequality. *(Section 4.1)*

3. $8p < -3$; $p = -2$

4. $z + 2 > -4$; $z = -8$

Graph the inequality on a number line. *(Section 4.1)*

5. $x < -12$

6. $v > \dfrac{5}{4}$

7. $b \geq -\dfrac{1}{3}$

8. $q \leq 4.2$

Solve the inequality. Graph the solution. *(Section 4.2)*

9. $n + 2 \leq -6$

10. $t - \dfrac{3}{7} > \dfrac{6}{7}$

11. $-\dfrac{3}{4} \geq w + 1$

12. $y - 2.6 < -3.4$

13. STUDYING You plan to study at least 1.5 hours for a geography test. Write an inequality that represents this situation. *(Section 4.1)*

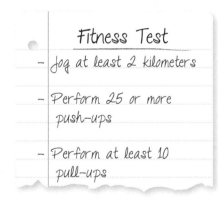

Fitness Test
- Jog at least 2 kilometers
- Perform 25 or more push-ups
- Perform at least 10 pull-ups

14. FITNESS TEST The three requirements to pass a fitness test are shown. *(Section 4.1)*

a. Write and graph three inequalities that represent the requirements.

b. You can jog 2500 meters, perform 30 push-ups, and perform 20 pull-ups. Do you satisfy the requirements of the test? Explain.

15. NUMBER LINE Use tape on the floor to make the number line shown. All units are in feet. You are standing at $-\dfrac{7}{2}$. You want to move to a number greater than $-\dfrac{3}{2}$. Write and solve an inequality that represents the distance you must move. *(Section 4.2)*

Solving Inequalities Using Multiplication or Division

Essential Question How can you use multiplication or division to solve an inequality?

1 ACTIVITY: Using a Table to Solve an Inequality

Work with a partner.

- Copy and complete the table.
- Decide which graph represents the solution of the inequality.
- Write the solution of the inequality.

a. $4x > 12$

x	−1	0	1	2	3	4	5
4x							
$4x \overset{?}{>} 12$							

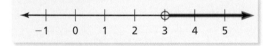 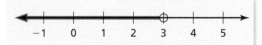

b. $-3x \le 9$

x	−5	−4	−3	−2	−1	0	1
−3x							
$-3x \overset{?}{\le} 9$							

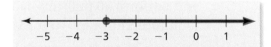 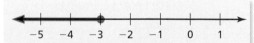

COMMON CORE

Inequalities

In this lesson, you will
- solve inequalities using multiplication or division.
- solve real-life problems.

Learning Standard
7.EE.4b

2 ACTIVITY: Solving an Inequality

Work with a partner.

a. Solve $-3x \le 9$ by adding $3x$ to each side of the inequality first. Then solve the resulting inequality.

b. Compare the solution in part (a) with the solution in Activity 1(b).

3 **ACTIVITY: Using a Table to Solve an Inequality**

Work with a partner.

- Copy and complete the table.
- Decide which graph represents the solution of the inequality.
- Write the solution of the inequality.

a. $\dfrac{x}{3} < 1$

x	−1	0	1	2	3	4	5
$\dfrac{x}{3}$							
$\dfrac{x}{3} \overset{?}{<} 1$							

b. $\dfrac{x}{-4} \geq \dfrac{3}{4}$

x	−5	−4	−3	−2	−1	0	1
$\dfrac{x}{-4}$							
$\dfrac{x}{-4} \overset{?}{\geq} \dfrac{3}{4}$							

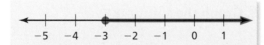

4 **ACTIVITY: Writing Rules**

Work with a partner. Use a table to solve each inequality.

a. $-2x \leq 10$ **b.** $-6x > 0$ **c.** $\dfrac{x}{-4} < 1$ **d.** $\dfrac{x}{-8} \geq \dfrac{1}{8}$

Write a set of rules that describes how to solve inequalities like those in Activities 1 and 3. Then use your rules to solve each of the four inequalities above.

Math Practice 3

Analyze Conjectures

When you apply your rules to parts (a)–(d), do you get the same solutions? Explain.

What Is Your Answer?

5. **IN YOUR OWN WORDS** How can you use multiplication or division to solve an inequality?

Practice

Use what you learned about solving inequalities using multiplication or division to complete Exercises 4–9 on page 143.

 **Key Idea**

Remember

Multiplication and division are inverse operations.

Multiplication and Division Properties of Inequality (Case 1)

Words When you multiply or divide each side of an inequality by the same *positive* number, the inequality remains true.

Numbers

$$-4 < 6 \qquad\qquad\qquad 4 > -6$$

$$2 \cdot (-4) < 2 \cdot 6 \qquad\qquad \frac{4}{2} > \frac{-6}{2}$$

$$-8 < 12 \qquad\qquad\qquad 2 > -3$$

Algebra If $a < b$ and c is positive, then

$$a \cdot c < b \cdot c \qquad \text{and} \qquad \frac{a}{c} < \frac{b}{c}.$$

If $a > b$ and c is positive, then

$$a \cdot c > b \cdot c \qquad \text{and} \qquad \frac{a}{c} > \frac{b}{c}.$$

These properties are also true for $\leq$ and $\geq$.

EXAMPLE 1   Solving an Inequality Using Multiplication

Solve $\dfrac{x}{5} \leq -3$. Graph the solution.

$$\frac{x}{5} \leq -3 \qquad\qquad \text{Write the inequality.}$$

Undo the division. → $\quad 5 \cdot \dfrac{x}{5} \leq 5 \cdot (-3) \qquad$ Multiplication Property of Inequality

$$x \leq -15 \qquad\qquad \text{Simplify.}$$

∴ The solution is $x \leq -15$.

$x \leq -15$

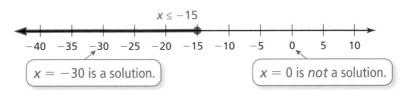

$x = -30$ is a solution. $x = 0$ is *not* a solution.

● **On Your Own**

Solve the inequality. Graph the solution.

1. $n \div 3 < 1$ **2.** $-0.5 \leq \dfrac{m}{10}$ **3.** $-3 > \dfrac{2}{3}p$

EXAMPLE 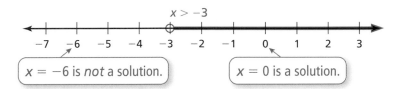 2 Solving an Inequality Using Division

Solve $6x > -18$. Graph the solution.

$$6x > -18$$ Write the inequality.

Undo the multiplication. $\longrightarrow$ $\dfrac{6x}{6} > \dfrac{-18}{6}$ Division Property of Inequality

$$x > -3$$ Simplify.

∴ The solution is $x > -3$.

$x > -3$

```
←——+————+————+————+————+————+————+————+————+————+————+——→
   -7   -6   -5   -4   -3   -2   -1    0    1    2    3
```

$x = -6$ is *not* a solution. $x = 0$ is a solution.

On Your Own

Now You're Ready
Exercises 10–18

Solve the inequality. Graph the solution.

4. $4b \geq 2$ **5.** $12k \leq -24$ **6.** $-15 < 2.5q$

Key Idea

Multiplication and Division Properties of Inequality (Case 2)

Words When you multiply or divide each side of an inequality by the same *negative* number, the direction of the inequality symbol must be reversed for the inequality to remain true.

Common Error ⚠

A negative sign in an inequality does not necessarily mean you must reverse the inequality symbol.

Only reverse the inequality symbol when you multiply or divide both sides by a negative number.

Numbers $-4 < 6$ $4 > -6$

$$-2 \cdot (-4) > -2 \cdot 6 \qquad \dfrac{4}{-2} < \dfrac{-6}{-2}$$

$$8 > -12 \qquad\qquad -2 < 3$$

Algebra If $a < b$ and c is negative, then

$$a \cdot c > b \cdot c \qquad \text{and} \qquad \dfrac{a}{c} > \dfrac{b}{c}.$$

If $a > b$ and c is negative, then

$$a \cdot c < b \cdot c \qquad \text{and} \qquad \dfrac{a}{c} < \dfrac{b}{c}.$$

These properties are also true for $\leq$ and $\geq$.

EXAMPLE 3 · **Solving an Inequality Using Multiplication**

Solve $-\frac{3}{2}n \le 6$. Graph the solution.

$$-\frac{3}{2}n \le 6 \qquad \text{Write the inequality.}$$

$$-\frac{2}{3} \cdot \left(-\frac{3}{2}n\right) \ge -\frac{2}{3} \cdot 6 \qquad \begin{array}{l}\text{Use the Multiplication Property of Inequality.} \\ \text{Reverse the inequality symbol.}\end{array}$$

$$n \ge -4 \qquad \text{Simplify.}$$

∴ The solution is $n \ge -4$.

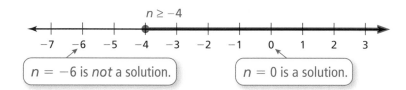

$n \ge -4$

$n = -6$ is *not* a solution.

$n = 0$ is a solution.

On Your Own

Solve the inequality. Graph the solution.

7. $\dfrac{x}{-3} > -4$

8. $0.5 \le -\dfrac{y}{2}$

9. $-12 \ge \dfrac{6}{5}m$

10. $-\dfrac{2}{5}h \le -8$

EXAMPLE 4 · **Solving an Inequality Using Division**

Solve $-3z > -4.5$. Graph the solution.

$$-3z > -4.5 \qquad \text{Write the inequality.}$$

Undo the multiplication. $\longrightarrow \quad \dfrac{-3z}{-3} < \dfrac{-4.5}{-3} \qquad \begin{array}{l}\text{Use the Division Property of Inequality.} \\ \text{Reverse the inequality symbol.}\end{array}$

$$z < 1.5 \qquad \text{Simplify.}$$

∴ The solution is $z < 1.5$.

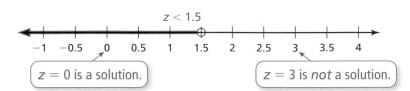

$z < 1.5$

$z = 0$ is a solution.

$z = 3$ is *not* a solution.

On Your Own

Now You're Ready
Exercises 27–35

Solve the inequality. Graph the solution.

11. $-5z < 35$

12. $-2a > -9$

13. $-1.5 < 3n$

14. $-4.2 \ge -0.7w$

Vocabulary and Concept Check

1. **WRITING** Explain how to solve $\frac{x}{3} < -2$.

2. **PRECISION** Explain how solving $4x < -16$ is different from solving $-4x < 16$.

3. **OPEN-ENDED** Write an inequality that you can solve using the Division Property of Inequality where the direction of the inequality symbol must be reversed.

Practice and Problem Solving

Use a table to solve the inequality.

4. $2x < 2$

5. $-3x \le 3$

6. $-6x > 18$

7. $\frac{x}{-5} \ge 7$

8. $\frac{x}{-1} > \frac{2}{5}$

9. $\frac{x}{3} \le \frac{1}{2}$

Solve the inequality. Graph the solution.

① ② 10. $2n > 20$

11. $\frac{c}{9} \le -4$

12. $2.2m < 11$

13. $-16 > x \div 2$

14. $\frac{1}{6}w \ge 2.5$

15. $7 < 3.5k$

16. $3x \le -\frac{5}{4}$

17. $4.2y \le -12.6$

18. $11.3 > \frac{b}{4.3}$

19. **ERROR ANALYSIS** Describe and correct the error in solving the inequality.

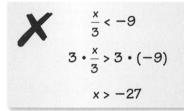

$$\frac{x}{3} < -9$$
$$3 \cdot \frac{x}{3} > 3 \cdot (-9)$$
$$x > -27$$

Write the word sentence as an inequality. Then solve the inequality.

20. The quotient of a number and 4 is at most 5.

21. A number divided by 7 is less than -3.

22. Six times a number is at least -24.

23. The product of -2 and a number is greater than 30.

24. **SMART PHONE** You earn $9.20 per hour at your summer job. Write and solve an inequality that represents the number of hours you need to work in order to buy a smart phone that costs $299.

25. **AVOCADOS** You have $9.60 to buy avocados for a guacamole recipe. Avocados cost $2.40 each.

   a. Write and solve an inequality that represents the number of avocados you can buy.

   b. Are there infinitely many solutions in this context? Explain.

26. **SCIENCE PROJECT** Students in a science class are divided into 6 equal groups with at least 4 students in each group for a project. Write and solve an inequality that represents the number of students in the class.

Solve the inequality. Graph the solution.

③ ④ 27. $-5n \le 15$

28. $-7w > 49$

29. $-\frac{1}{3}h \ge 8$

30. $-9 < -\frac{1}{5}x$

31. $-3y < -14$

32. $-2d \ge 26$

33. $4.5 > -\frac{m}{6}$

34. $\frac{k}{-0.25} \le 36$

35. $-2.4 > \frac{b}{-2.5}$

36. **ERROR ANALYSIS** Describe and correct the error in solving the inequality.

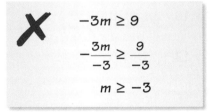

37. **TEMPERATURE** It is currently 0°C outside. The temperature is dropping 2.5°C every hour. Write and solve an inequality that represents the number of hours that must pass for the temperature to drop below −20°C.

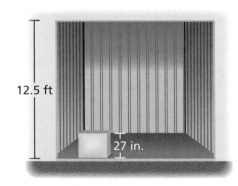

38. **STORAGE** You are moving some of your belongings into a storage facility.

   a. Write and solve an inequality that represents the number of boxes that you can stack vertically in the storage unit.

   b. Can you stack 6 boxes vertically in the storage unit? Explain.

Write and solve an inequality that represents x.

39. Area ≥ 120 cm²

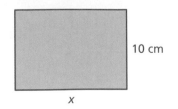

40. Area < 20 ft²

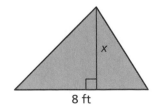

41. **AMUSEMENT PARK** You and four friends are planning a visit to an amusement park. You want to keep the cost below $100 per person. Write and solve an inequality that represents the total cost of visiting the amusement park.

42. **LOGIC** When you multiply or divide each side of an inequality by the same negative number, you must reverse the direction of the inequality symbol. Explain why.

43. **PROJECT** Choose two novels to research.

   a. Use the Internet or a magazine to complete the table.

   b. Use the table to find and compare the average number of copies sold per month for each novel. Which novel do you consider to be the most successful? Explain.

   c. Assume each novel continues to sell at the average rate. Write and solve an inequality that represents the number of months it will take for the total number of copies sold to exceed twice the current number sold.

	Author	Name of Novel	Release Date	Current Number of Copies Sold
1.				
2.				

  Describe all numbers that satisfy *both* inequalities. Include a graph with your description.

44. $4m > -4$ and $3m < 15$

45. $\dfrac{n}{3} \geq -4$ and $\dfrac{n}{-5} \geq 1$

46. $2x \geq -6$ and $2x \geq 6$

47. $-\dfrac{1}{2}s > -7$ and $\dfrac{1}{3}s < 12$

Fair Game Review What you learned in previous grades & lessons

Solve the equation. Check your solution. *(Section 3.5)*

48. $-2w + 4 = -12$

49. $\dfrac{v}{5} - 6 = 3$

50. $3(x - 1) = 18$

51. $\dfrac{m + 200}{4} = 51$

52. **MULTIPLE CHOICE** What is the value of $\dfrac{2}{3} + \left(-\dfrac{5}{7}\right)$? *(Section 2.2)*

   (A) $-\dfrac{3}{4}$ (B) $-\dfrac{1}{21}$ (C) $\dfrac{7}{10}$ (D) $1\dfrac{8}{21}$

4.4 Solving Two-Step Inequalities

Essential Question How can you use an inequality to describe the dimensions of a figure?

1 ACTIVITY: Areas and Perimeters of Figures

Work with a partner.

- **Use the given condition to choose the inequality that you can use to find the possible values of the variable. Justify your answer.**

- **Write four values of the variable that satisfy the inequality you chose.**

a. You want to find the values of x so that the area of the rectangle is more than 22 square units.

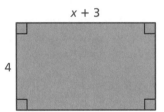

$4x + 12 > 22$ $4x + 3 > 22$

$4x + 12 \geq 22$ $2x + 14 > 22$

b. You want to find the values of x so that the perimeter of the rectangle is greater than or equal to 28 units.

$x + 7 \geq 28$ $4x + 12 \geq 28$ $2x + 14 \geq 28$ $2x + 14 \leq 28$

c. You want to find the values of y so that the area of the parallelogram is fewer than 41 square units.

$5y + 7 < 41$ $5y + 35 < 41$

$5y + 7 \leq 41$ $5y + 35 \leq 41$

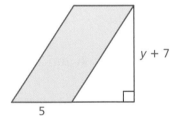

d. You want to find the values of z so that the area of the trapezoid is at most 100 square units.

$5z + 30 \leq 100$ $10z + 30 \leq 100$

$5z + 30 < 100$ $10z + 30 < 100$

COMMON CORE

Inequalities
In this lesson, you will
- solve multi-step inequalities.
- solve real-life problems.
Learning Standard
7.EE.4b

Work with a partner.

- Use the given condition to choose the inequality that you can use to find the possible values of the variable. Justify your answer.

- Write four values of the variable that satisfy the inequality you chose.

Math Practice 6

State the Meaning of Symbols

What inequality symbols do the phrases *at least* and *no more than* represent? Explain.

a. You want to find the values of x so that the volume of the rectangular prism is at least 50 cubic units.

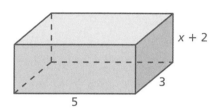

$x + 2$

3

5

| $15x + 30 > 50$ | $x + 10 \geq 50$ | $15x + 30 \geq 50$ | $15x + 2 \geq 50$ |

b. You want to find the values of x so that the volume of the rectangular prism is no more than 36 cubic units.

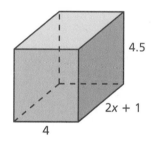

4.5

$2x + 1$

4

| $8x + 4 < 36$ | $36x + 18 < 36$ | $2x + 9.5 \leq 36$ | $36x + 18 \leq 36$ |

What Is Your Answer?

3. **IN YOUR OWN WORDS** How can you use an inequality to describe the dimensions of a figure?

4. Use what you know about solving equations and inequalities to describe how you can solve a two-step inequality. Give an example to support your explanation.

Practice

Use what you learned about solving two-step inequalities to complete Exercises 3 and 4 on page 150.

Check It Out
Lesson Tutorials
BigIdeasMath ✓com

You can solve two-step inequalities in the same way you solve two-step equations.

EXAMPLE 1   Solving Two-Step Inequalities

a. Solve $5x - 4 \geq 11$. Graph the solution.

$5x - 4 \geq 11$	Write the inequality.

Step 1: Undo the subtraction.

$\underline{+\ 4 \qquad +\ 4}$	Addition Property of Inequality
$5x \geq 15$	Simplify.

Step 2: Undo the multiplication.

$\dfrac{5x}{5} \geq \dfrac{15}{5}$	Division Property of Inequality
$x \geq 3$	Simplify.

∴ The solution is $x \geq 3$.

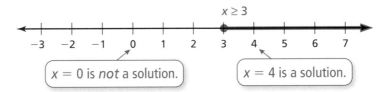

$x \geq 3$

$x = 0$ is *not* a solution.

$x = 4$ is a solution.

b. Solve $\dfrac{b}{-3} + 4 < 13$. Graph the solution.

$\dfrac{b}{-3} + 4 < 13$	Write the inequality.

Step 1: Undo the addition.

$\underline{-\ 4 \qquad -\ 4}$	Subtraction Property of Inequality
$\dfrac{b}{-3} < 9$	Simplify.

Step 2: Undo the division.

$-3 \cdot \dfrac{b}{-3} > -3 \cdot 9$	Use the Multiplication Property of Inequality. Reverse the inequality symbol.
$b > -27$	Simplify.

∴ The solution is $b > -27$.

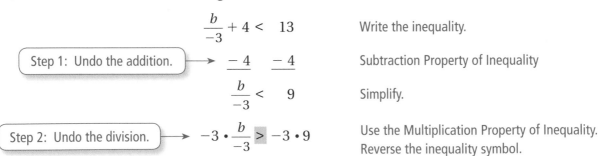

$b > -27$

On Your Own

Now You're Ready
Exercises 5–10

Solve the inequality. Graph the solution.

1. $6y - 7 > 5$

2. $4 - 3d \geq 19$

3. $\dfrac{w}{-4} + 8 > 9$

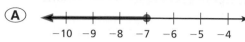

EXAMPLE 2 Graphing an Inequality

Which graph represents the solution of $-7(x + 3) \leq 28$?

Ⓐ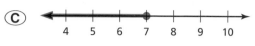
-10 -9 -8 -7 -6 -5 -4

Ⓑ
-10 -9 -8 -7 -6 -5 -4

Ⓒ
4 5 6 7 8 9 10

Ⓓ
4 5 6 7 8 9 10

$-7(x + 3) \leq 28$		Write the inequality.
$-7x - 21 \leq 28$		Distributive Property

Step 1: Undo the subtraction. ⟶ $\underline{+21 \quad +21}$ — Addition Property of Inequality

$-7x \leq 49$ — Simplify.

Step 2: Undo the multiplication. ⟶ $\dfrac{-7x}{-7} \geq \dfrac{49}{-7}$ — Use the Division Property of Inequality. Reverse the inequality symbol.

$x \geq -7$ — Simplify.

⁘ The correct answer is Ⓑ.

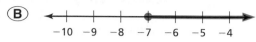

EXAMPLE 3 Real-Life Application

Progress Report	
Month	**Pounds Lost**
1	12
2	9
3	5
4	8

A contestant in a weight-loss competition wants to lose an average of at least 8 pounds per month during a 5-month period. How many pounds must the contestant lose in the fifth month to meet the goal?

Write and solve an inequality. Let x be the number of pounds lost in the fifth month.

$\dfrac{12 + 9 + 5 + 8 + x}{5} \geq 8$ — The phrase *at least* means *greater than or equal to.*

$\dfrac{34 + x}{5} \geq 8$ — Simplify.

$5 \cdot \dfrac{34 + x}{5} \geq 5 \cdot 8$ — Multiplication Property of Inequality

$34 + x \geq 40$ — Simplify.

$x \geq 6$ — Subtract 34 from each side.

Remember

In Example 3, the average is equal to the sum of the pounds lost divided by the number of months.

⁘ So, the contestant must lose at least 6 pounds to meet the goal.

On Your Own

Exercises 12–17

Solve the inequality. Graph the solution.

4. $2(k - 5) < 6$ **5.** $-4(n - 10) < 32$ **6.** $-3 \leq 0.5(8 + y)$

7. WHAT IF? In Example 3, the contestant wants to lose an average of at least 9 pounds per month. How many pounds must the contestant lose in the fifth month to meet the goal?

 Vocabulary and Concept Check

1. **WRITING** Compare and contrast solving two-step inequalities and solving two-step equations.

2. **OPEN-ENDED** Describe how to solve the inequality $3(a + 5) < 9$.

 Practice and Problem Solving

Match the inequality with its graph.

3. $\dfrac{t}{3} - 1 \geq -3$

A.
(number line −9 to −2, open circle at −6, shaded right)

B.
(number line −9 to −2, shaded left, closed circle at −6)

C.
(number line −9 to −2, closed circle at −7, shaded right)

4. $5x + 7 \leq 32$

A.
(number line 2 to 9, closed circle at 5, shaded left)

B.
(number line 2 to 9, open circle at 5, shaded left)

C.
(number line 2 to 9, closed circle at 5, shaded right)

Solve the inequality. Graph the solution.

① 5. $8y - 5 < 3$

6. $3p + 2 \geq -10$

7. $2 > 8 - \dfrac{4}{3}h$

8. $-2 > \dfrac{m}{6} - 7$

9. $-1.2b - 5.3 \geq 1.9$

10. $-1.3 \geq 2.9 - 0.6r$

11. **ERROR ANALYSIS** Describe and correct the error in solving the inequality.

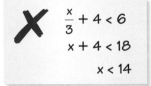
✗
$\dfrac{x}{3} + 4 < 6$
$x + 4 < 18$
$x < 14$

Solve the inequality. Graph the solution.

② 12. $5(g + 4) > 15$

13. $4(w - 6) \leq -12$

14. $-8 \leq \dfrac{2}{5}(k - 2)$

15. $-\dfrac{1}{4}(d + 1) < 2$

16. $7.2 > 0.9(n + 8.6)$

17. $20 \geq -3.2(c - 4.3)$

10 cm

18. **UNICYCLE** The first jump in a unicycle high-jump contest is shown. The bar is raised 2 centimeters after each jump. Solve the inequality $2n + 10 \geq 26$ to find the number of additional jumps needed to meet or exceed the goal of clearing a height of 26 centimeters.

Solve the inequality. Graph the solution.

19. $9x - 4x + 4 \geq 36 - 12$

20. $3d - 7d + 2.8 < 5.8 - 27$

21. SCUBA DIVER A scuba diver is at an elevation of -38 feet. The diver starts moving at a rate of -12 feet per minute. Write and solve an inequality that represents how long it will take the diver to reach an elevation deeper than -200 feet.

22. KILLER WHALES A killer whale has eaten 75 pounds of fish today. It needs to eat at least 140 pounds of fish each day.

   a. A bucket holds 15 pounds of fish. Write and solve an inequality that represents how many more buckets of fish the whale needs to eat.

   b. Should the whale eat *four* or *five* more buckets of fish? Explain.

23. REASONING A student theater charges $9.50 per ticket.

   a. The theater has already sold 70 tickets. Write and solve an inequality that represents how many more tickets the theater needs to sell to earn at least $1000.

   b. The theater increases the ticket price by $1. Without solving an inequality, describe how this affects the total number of tickets needed to earn at least $1000.

24. **Problem Solving** For what values of r will the area of the shaded region be greater than or equal to 12 square units?

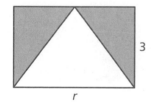

Fair Game Review What you learned in previous grades & lessons

Find the missing values in the ratio table. Then write the equivalent ratios.
(Skills Review Handbook)

25.

Flutes	7		28
Clarinets	4	12	

26.

Boys	6	3	
Girls	10		50

27. MULTIPLE CHOICE What is the volume of the cube?
(Skills Review Handbook)

   Ⓐ 8 ft^3 Ⓑ 16 ft^3

   Ⓒ 24 ft^3 Ⓓ 32 ft^3

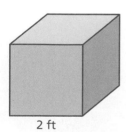

2 ft

Solve the inequality. Graph the solution. *(Section 4.3 and Section 4.4)*

1. $3p \le 18$

2. $2x > -\dfrac{3}{5}$

3. $\dfrac{r}{3} \ge -5$

4. $-\dfrac{z}{8} < 1.5$

5. $3n + 2 \le 11$

6. $-2 < 5 - \dfrac{k}{2}$

7. $1.3m - 3.8 < -1.2$

8. $4.8 \ge 0.3(12 - y)$

Write the word sentence as an inequality. Then solve the inequality. *(Section 4.3)*

9. The quotient of a number and 5 is less than 4.

10. Six times a number is at least -14.

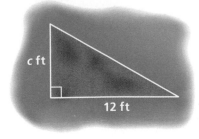

11. PEPPERS You have $18 to buy peppers. Peppers cost $1.50 each. Write and solve an inequality that represents the number of peppers you can buy. *(Section 4.3)*

12. MOVIES You have a gift card worth $90. You want to buy several movies that cost $12 each. Write and solve an inequality that represents the number of movies you can buy and still have at least $30 on the gift card. *(Section 4.4)*

13. CHOCOLATES Your class sells boxes of chocolates to raise $500 for a field trip. You earn $6.25 for each box of chocolates sold. Write and solve an inequality that represents the number of boxes your class must sell to meet or exceed the fundraising goal. *(Section 4.3)*

14. FENCE You want to put up a fence that encloses a triangular region with an area greater than or equal to 60 square feet. What is the least possible value of c? Explain. *(Section 4.3)*

Check It Out
Vocabulary Help
BigIdeasMath ✓com

Review Key Vocabulary

inequality, *p. 126*
solution of an inequality, *p. 126*

solution set, *p. 126*
graph of an inequality, *p. 127*

Review Examples and Exercises

4.1 Writing and Graphing Inequalities *(pp. 124–129)*

a. Six plus a number x is at most $-\dfrac{1}{4}$. Write this word sentence as an inequality.

Six plus a number x | is at most | $-\dfrac{1}{4}$.
$6 + x$ | $\leq$ | $-\dfrac{1}{4}$

∴ An inequality is $6 + x \leq -\dfrac{1}{4}$.

b. Graph $m > 3$.

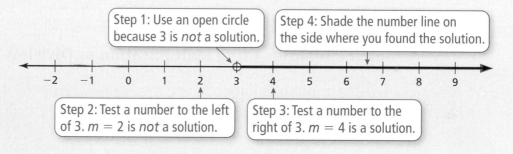

Step 1: Use an open circle because 3 is *not* a solution.

Step 4: Shade the number line on the side where you found the solution.

Step 2: Test a number to the left of 3. $m = 2$ is *not* a solution.

Step 3: Test a number to the right of 3. $m = 4$ is a solution.

Exercises

Write the word sentence as an inequality.

1. A number w is greater than -3.

2. A number y minus $\dfrac{1}{2}$ is no more than $-\dfrac{3}{2}$.

Tell whether the given value is a solution of the inequality.

3. $5 + j > 8; j = 7$

4. $6 \div n \leq -5; n = -3$

Graph the inequality on a number line.

5. $q > -1.3$

6. $s < 1\dfrac{3}{4}$

7. BUMPER CARS You must be at least 42 inches tall to ride the bumper cars at an amusement park. Write an inequality that represents this situation.

4.2 Solving Inequalities Using Addition or Subtraction   (pp. 130–135)

Solve $-5 < m - 3$. Graph the solution.

	$-5 < m - 3$	Write the inequality.
Undo the subtraction. →	$\underline{+\ 3 \qquad +\ 3}$	Addition Property of Inequality
	$-2 < m$	Simplify.

The solution is $m > -2$.

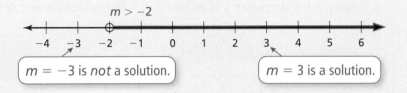

$m = -3$ is *not* a solution.

$m = 3$ is a solution.

Exercises

Solve the inequality. Graph the solution.

8. $d + 12 < 19$ **9.** $t - 4 \le -14$ **10.** $-8 \le z + 6.4$

4.3 Solving Inequalities Using Multiplication or Division   (pp. 138–145)

Solve $\dfrac{c}{-3} \ge -2$. Graph the solution.

	$\dfrac{c}{-3} \ge -2$	Write the inequality.
Undo the division. →	$-3 \cdot \dfrac{c}{-3} \le -3 \cdot (-2)$	Use the Multiplication Property of Inequality. Reverse the inequality symbol.
	$c \le 6$	Simplify.

The solution is $c \le 6$.

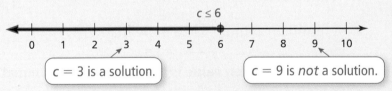

$c = 3$ is a solution.

$c = 9$ is *not* a solution.

Exercises

Solve the inequality. Graph the solution.

11. $6q < -18$ **12.** $-\dfrac{r}{3} \le 6$ **13.** $-4 > -\dfrac{4}{3}s$

4.4 **Solving Two-Step Inequalities** *(pp. 146–151)*

a. **Solve $6x - 8 \leq 10$. Graph the solution.**

$6x - 8 \leq 10$	Write the inequality.
$\underline{+\ 8 \quad +\ 8}$	Addition Property of Inequality
$6x \leq 18$	Simplify.
$\dfrac{6x}{6} \leq \dfrac{18}{6}$	Division Property of Inequality
$x \leq 3$	Simplify.

Step 1: Undo the subtraction.

Step 2: Undo the multiplication.

∴ The solution is $x \leq 3$.

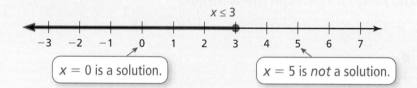

$x = 0$ is a solution. $x = 5$ is *not* a solution.

b. **Solve $\dfrac{q}{-4} + 7 < 11$. Graph the solution.**

$\dfrac{q}{-4} + 7 < 11$	Write the inequality.
$\underline{-\ 7 \quad -\ 7}$	Subtraction Property of Inequality
$\dfrac{q}{-4} < 4$	Simplify.
$-4 \cdot \dfrac{q}{-4} > -4 \cdot 4$	Use the Multiplication Property of Inequality. Reverse the inequality symbol.
$q > -16$	Simplify.

Step 1: Undo the addition.

Step 2: Undo the division.

∴ The solution is $q > -16$.

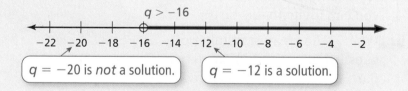

$q = -20$ is *not* a solution. $q = -12$ is a solution.

Exercises

Solve the inequality. Graph the solution.

14. $3x + 4 > 16$

15. $\dfrac{z}{-2} - 6 \leq -2$

16. $-2t - 5 < 9$

17. $7(q + 2) < -77$

18. $-\dfrac{1}{3}(p + 9) \leq 4$

19. $1.2(j + 3.5) \geq 4.8$

Write the word sentence as an inequality.

1. A number k plus 19.5 is less than or equal to 40.

2. A number q multiplied by $\frac{1}{4}$ is greater than -16.

Tell whether the given value is a solution of the inequality.

3. $n - 3 \leq 4; n = 7$

4. $-\frac{3}{7}m < 1; m = -7$

5. $-4c \geq 7; c = -2$

6. $-2.4m > -6.8; m = -3$

Solve the inequality. Graph the solution.

7. $w + 4 \leq 3$

8. $x - 4 > -6$

9. $-\frac{2}{9} + y \leq \frac{5}{9}$

10. $-6z \geq 36$

11. $-5.2 \geq \frac{p}{4}$

12. $4k - 8 \geq 20$

13. $\frac{4}{7} - b \geq -\frac{1}{7}$

14. $-0.6 > -0.3(d + 6)$

15. **GUMBALLS** You have $2.50. Each gumball in a gumball machine costs $0.25. Write and solve an inequality that represents the number of gumballs you can buy.

16. **PARTY** You can spend no more than $100 on a party you are hosting. The cost per guest is $8.

   a. Write and solve an inequality that represents the number of guests you can invite to the party.

   b. What is the greatest number of guests that you can invite to the party? Explain your reasoning.

17. **BASEBALL CARDS** You have $30 to buy baseball cards. Each pack of cards costs $5. Write and solve an inequality that represents the number of packs of baseball cards you can buy and still have at least $10 left.

4 Standards Assessment

You ate less than 3 cat treats for breakfast. Which number of cat treats could you have eaten?

Ⓐ 2　　Ⓑ 3　　Ⓒ 4　　Ⓓ 5

They were delicious!

"Scan the test and answer the easy questions first. Because $2 < 3$, A is correct."

1. What is the value of the expression below when $x = -5$, $y = 3$, and $z = -1$?　(7.NS.3)

$$\frac{x^2 - 3y}{z}$$

   A. -34　　　　**C.** 16

   B. -16　　　　**D.** 34

2. What is the value of the expression below? (7.NS.2a)

$$-\frac{3}{8} \cdot \left(\frac{2}{5}\right)$$

   F. $-\dfrac{20}{3}$　　　　**H.** $-\dfrac{15}{16}$

   G. $-\dfrac{16}{15}$　　　　**I.** $-\dfrac{3}{20}$

3. Which graph represents the inequality below?　(7.EE.4b)

$$\frac{x}{-4} - 8 \geq -9$$

   A.

   B.

   C.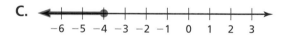

   D.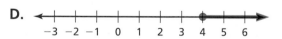

4. Which value of p makes the equation below true?　(7.EE.4a)

$$5(p + 6) = 25$$

   F. -1　　　　　　　　**H.** 11

   G. $3\dfrac{4}{5}$　　　　　　　　**I.** 14

5. You set up the lemonade stand. Your profit is equal to your revenue from lemonade sales minus your cost to operate the stand. Your cost is $8. How many cups of lemonade must you sell to earn a profit of $30? *(7.EE.4a)*

LEMONADE

Lemonade
50¢
per cup

   A. 4   **C.** 60

   B. 44   **D.** 76

6. Which value is a solution of the inequality below? *(7.EE.4b)*

$$3 - 2y < 7$$

   F. -6 **H.** -2

   G. -3 **I.** -1

7. What value of y makes the equation below true? *(7.EE.4a)*

$$12 - 3y = -6$$

8. What is the mean distance of the four points from -3? *(7.NS.3)*

   A. $-\dfrac{1}{2}$ **C.** 3

   B. $2\dfrac{1}{2}$ **D.** $7\dfrac{1}{8}$

9. Martin graphed the solution of the inequality $-4x + 18 > 6$ in the box below.

What should Martin do to correct the error that he made? *(7.EE.4b)*

 F. Use an open circle at 3 and shade to the left of 3.

 G. Use an open circle at 3 and shade to the right of 3.

 H. Use a closed circle and shade to the right of 3.

 I. Use an open circle and shade to the left of -3.

10. What is the value of the expression below? *(7.NS.1c)*

$$\frac{5}{12} - \frac{7}{8}$$

11. You are selling T-shirts to raise money for a charity. You sell the T-shirts for $10 each. *(7.EE.4b)*

Part A You have already sold 2 T-shirts. How many more T-shirts must you sell to raise at least $500? Explain.

Part B Your friend is raising money for the same charity. He sells the T-shirts for $8 each. What is the total number of T-shirts he must sell to raise at least $500? Explain.

Part C Who has to sell more T-shirts in total? How many more? Explain.

12. Which expression is equivalent to the expression below? *(7.NS.3)*

$$-\frac{2}{3} - \left(-\frac{4}{9}\right)$$

A. $-\dfrac{1}{3} + \dfrac{1}{9}$

C. $-\dfrac{1}{3} - \dfrac{7}{9}$

B. $-\dfrac{2}{3} \times \left(-\dfrac{1}{3}\right)$

D. $\dfrac{3}{2} \div \left(-\dfrac{1}{3}\right)$

5 Ratios and Proportions

"I am doing an experiment with slope. I want you to run up and down the board 10 times."

"This is what happens when you give a dog too many biscuits."

"Now with 2 more dog biscuits, do it again and we'll compare your rates."

"Dear Sir: I counted the number of bacon, cheese, and chicken dog biscuits in the box I bought."

"There were 16 bacon, 12 cheese, and only 8 chicken. That's a ratio of 4:3:2. Please go back to the original ratio of 1:1:1."

"And while you're at it, how about throwing in a couple of tuna and mouse flavors!"

What You Learned Before

"I wonder if our rate is proportional to the slope of the hill."

...or possibly proportional to our stupidity!

Simplifying Fractions (4.NF.1)

Example 1 Simplify $\frac{4}{8}$.

$$\frac{4 \div 4}{8 \div 4} = \frac{1}{2}$$

Example 2 Simplify $\frac{10}{15}$.

$$\frac{10 \div 5}{15 \div 5} = \frac{2}{3}$$

Identifying Equivalent Fractions (4.NF.1)

Example 3 Is $\frac{1}{4}$ equivalent to $\frac{13}{52}$?

$$\frac{13 \div 13}{52 \div 13} = \frac{1}{4}$$

∴ $\frac{1}{4}$ is equivalent to $\frac{13}{52}$.

Example 4 Is $\frac{30}{54}$ equivalent to $\frac{5}{8}$?

$$\frac{30 \div 6}{54 \div 6} = \frac{5}{9}$$

∴ $\frac{30}{54}$ is *not* equivalent to $\frac{5}{8}$.

Solving Equations (6.EE.7)

Example 5 Solve $12x = 168$.

$12x = 168$	Write the equation.
$\dfrac{12x}{12} = \dfrac{168}{12}$	Division Property of Equality
$x = 14$	Simplify.

Check

$$12x = 168$$
$$12(14) \stackrel{?}{=} 168$$
$$168 = 168 \checkmark$$

Try It Yourself

Simplify.

1. $\dfrac{12}{144}$

2. $\dfrac{15}{45}$

3. $\dfrac{75}{100}$

4. $\dfrac{16}{24}$

Are the fractions equivalent? Explain.

5. $\dfrac{15}{60} \stackrel{?}{=} \dfrac{3}{4}$

6. $\dfrac{2}{5} \stackrel{?}{=} \dfrac{24}{144}$

7. $\dfrac{15}{20} \stackrel{?}{=} \dfrac{3}{5}$

8. $\dfrac{2}{8} \stackrel{?}{=} \dfrac{16}{64}$

Solve the equation. Check your solution.

9. $\dfrac{y}{-5} = 3$

10. $0.6 = 0.2a$

11. $-2w = -9$

12. $\dfrac{1}{7}n = -4$

Essential Question How do rates help you describe real-life problems?

The Meaning of a Word ● Rate

When you rent snorkel gear at the beach, you should pay attention to the rental **rate**. The rental rate is in dollars per hour.

Snorkel Rentals
$8.75 per hour

Snorkel Rentals
$7.25 per hour

1 ACTIVITY: Finding Reasonable Rates

Work with a partner.

a. Match each description with a verbal rate.

b. Match each verbal rate with a numerical rate.

c. Give a reasonable numerical rate for each description. Then give an unreasonable rate.

Description	*Verbal Rate*	*Numerical Rate*
Your running rate in a 100-meter dash	Dollars per year	$\dfrac{\text{in.}}{\text{yr}}$
The fertilization rate for an apple orchard	Inches per year	$\dfrac{\text{lb}}{\text{acre}}$
The average pay rate for a professional athlete	Meters per second	$\dfrac{\$}{\text{yr}}$
The average rainfall rate in a rain forest	Pounds per acre	$\dfrac{\text{m}}{\text{sec}}$

COMMON CORE

Ratios and Rates

In this lesson, you will
● find ratios, rates, and unit rates.
● find ratios and rates involving ratios of fractions.

Learning Standards
7.RP.1
7.RP.3

2 ACTIVITY: Simplifying Expressions That Contain Fractions

Work with a partner. Describe a situation where the given expression may apply. Show how you can rewrite each expression as a division problem. Then simplify and interpret your result.

a. $\dfrac{\frac{1}{2}\text{ c}}{4\text{ fl oz}}$

b. $\dfrac{2\text{ in.}}{\frac{3}{4}\text{ sec}}$

c. $\dfrac{\frac{3}{8}\text{ c sugar}}{\frac{3}{5}\text{ c flour}}$

d. $\dfrac{\frac{5}{6}\text{ gal}}{\frac{2}{3}\text{ sec}}$

ACTIVITY: Using Ratio Tables to Find Equivalent Rates

Work with a partner. A communications satellite in orbit travels about 18 miles every 4 seconds.

a. Identify the rate in this problem.

b. Recall that you can use *ratio tables* to find and organize equivalent ratios and rates. Complete the ratio table below.

Time (seconds)	4	8	12	16	20
Distance (miles)					

c. How can you use a ratio table to find the speed of the satellite in miles per minute? miles per hour?

d. How far does the satellite travel in 1 second? Solve this problem (1) by using a ratio table and (2) by evaluating a quotient.

e. How far does the satellite travel in $\frac{1}{2}$ second? Explain your steps.

4 **ACTIVITY: Unit Analysis**

Math Practice 7

View as Components

What is the product of the numbers? What is the product of the units? Explain.

Work with a partner. Describe a situation where the product may apply. Then find each product and list the units.

a. $10 \text{ gal} \times \dfrac{22 \text{ mi}}{\text{gal}}$

b. $\dfrac{7}{2} \text{ lb} \times \dfrac{\$3}{\frac{1}{2} \text{ lb}}$

c. $\dfrac{1}{2} \text{ sec} \times \dfrac{30 \text{ ft}^2}{\text{sec}}$

What Is Your Answer?

5. IN YOUR OWN WORDS How do rates help you describe real-life problems? Give two examples.

6. To estimate the annual salary for a given hourly pay rate, multiply by 2 and insert "000" at the end.

Sample: $10 per hour is about $20,000 per year.

a. Explain why this works. Assume the person is working 40 hours a week.

b. Estimate the annual salary for an hourly pay rate of $8 per hour.

c. You earn $1 million per month. What is your annual salary?

d. Why is the cartoon funny?

"We had someone apply for the job. He says he would like $1 million a month, but will settle for $8 an hour."

Practice

Use what you discovered about ratios and rates to complete Exercises 7–10 on page 167.

Key Vocabulary
ratio, *p. 164*
rate, *p. 164*
unit rate, *p. 164*
complex fraction, *p. 165*

A **ratio** is a comparison of two quantities using division.

$$\frac{3}{4}, \ 3 \text{ to } 4, \ 3:4$$

A **rate** is a ratio of two quantities with different units.

$$\frac{60 \text{ miles}}{2 \text{ hours}}$$

A rate with a denominator of 1 is called a **unit rate**.

$$\frac{30 \text{ miles}}{1 \text{ hour}}$$

EXAMPLE ❶ **Finding Ratios and Rates**

There are 45 males and 60 females in a subway car. The subway car travels 2.5 miles in 5 minutes.

a. Find the ratio of males to females.

$$\frac{\text{males}}{\text{females}} = \frac{45}{60} = \frac{3}{4}$$

∴ The ratio of males to females is $\frac{3}{4}$.

b. Find the speed of the subway car.

$$2.5 \text{ miles in } 5 \text{ minutes} = \frac{2.5 \text{ mi}}{5 \text{ min}} = \frac{2.5 \text{ mi} \div 5}{5 \text{ min} \div 5} = \frac{0.5 \text{ mi}}{1 \text{ min}}$$

∴ The speed is 0.5 mile per minute.

EXAMPLE ❷ **Finding a Rate from a Ratio Table**

The ratio table shows the costs for different amounts of artificial turf. Find the unit rate in dollars per square foot.

	× 4	× 4	× 4	
Amount (square feet)	25	100	400	1600
Cost (dollars)	100	400	1600	6400
	× 4	× 4	× 4	

Use a ratio from the table to find the unit rate.

$$\frac{\text{cost}}{\text{amount}} = \frac{\$100}{25 \text{ ft}^2} \qquad \text{Use the first ratio in the table.}$$

$$= \frac{\$4}{1 \text{ ft}^2} \qquad \text{Simplify.}$$

Remember

The abbreviation ft² means *square feet*.

∴ So, the unit rate is $4 per square foot.

◀)) Multi-Language Glossary at BigIdeasMath✓.com

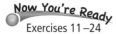

On Your Own

1. In Example 1, find the ratio of females to males.

2. In Example 1, find the ratio of females to total passengers.

3. The ratio table shows the distance that the *International Space Station* travels while orbiting Earth. Find the speed in miles per second.

Time (seconds)	3	6	9	12
Distance (miles)	14.4	28.8	43.2	57.6

A **complex fraction** has at least one fraction in the numerator, denominator, or both. You may need to simplify complex fractions when finding ratios and rates.

EXAMPLE ③ **Finding a Rate from a Graph**

The graph shows the speed of a subway car. Find the speed in miles per minute. Compare the speed to the speed of the subway car in Example 1.

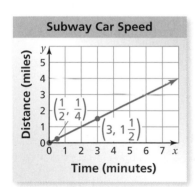

Subway Car Speed

Step 1: Choose and interpret a point on the line.

The point $\left(\frac{1}{2}, \frac{1}{4}\right)$ indicates that the subway car travels

$\frac{1}{4}$ mile in $\frac{1}{2}$ minute.

Step 2: Find the speed.

$$\frac{\text{distance traveled}}{\text{elapsed time}} = \frac{\frac{1}{4} \leftarrow \text{miles}}{\frac{1}{2} \leftarrow \text{minutes}}$$

$$= \frac{1}{4} \div \frac{1}{2} \qquad \text{Rewrite the quotient.}$$

$$= \frac{1}{4} \cdot 2 = \frac{1}{2} \qquad \text{Simplify.}$$

∴ The speed of the subway car is $\frac{1}{2}$ mile per minute.

Because $\frac{1}{2}$ mile per minute = 0.5 mile per minute, the speeds of the two subway cars are the same.

On Your Own

4. You use the point $\left(3, 1\frac{1}{2}\right)$ to find the speed of the subway car. Does your answer change? Explain your reasoning.

EXAMPLE 4 **Solving a Ratio Problem**

You mix $\frac{1}{2}$ cup of yellow paint for every $\frac{3}{4}$ cup of blue paint to make 15 cups of green paint. How much yellow paint and blue paint do you use?

Math Practice 1

Analyze Givens
What information is given in the problem? How does this help you know that the ratio table needs a "total" column? Explain.

Method 1: The ratio of yellow paint to blue paint is $\frac{1}{2}$ to $\frac{3}{4}$. Use a ratio table to find an equivalent ratio in which the total amount of yellow paint and blue paint is 15 cups.

Yellow (cups)	Blue (cups)	Total (cups)
$\frac{1}{2}$	$\frac{3}{4}$	$\frac{1}{2} + \frac{3}{4} = \frac{5}{4}$
2	3	5
6	9	15

$\times 4$ $\times 3$ $\times 4$ $\times 3$

∴ So, you use 6 cups of yellow paint and 9 cups of blue paint.

Method 2: Use the fraction of the green paint that is made from yellow paint and the fraction of the green paint that is made from blue paint. You use $\frac{1}{2}$ cup of yellow paint for every $\frac{3}{4}$ cup of blue paint, so the fraction of the green paint that is made from yellow paint is

yellow →
green →
$$\dfrac{\frac{1}{2}}{\frac{1}{2} + \frac{3}{4}} = \dfrac{\frac{1}{2}}{\frac{5}{4}} = \frac{1}{2} \cdot \frac{4}{5} = \frac{2}{5}.$$

Similarly, the fraction of the green paint that is made from blue paint is

blue →
green →
$$\dfrac{\frac{3}{4}}{\frac{1}{2} + \frac{3}{4}} = \dfrac{\frac{3}{4}}{\frac{5}{4}} = \frac{3}{4} \cdot \frac{4}{5} = \frac{3}{5}.$$

∴ So, you use $\frac{2}{5} \cdot 15 = 6$ cups of yellow paint and $\frac{3}{5} \cdot 15 = 9$ cups of blue paint.

On Your Own

Now You're Ready
Exercises 33 and 34

5. How much yellow paint and blue paint do you use to make 20 cups of green paint?

 ## Vocabulary and Concept Check

1. **VOCABULARY** How can you tell when a rate is a unit rate?

2. **WRITING** Why do you think rates are usually written as unit rates?

3. **OPEN-ENDED** Write a real-life rate that applies to you.

Estimate the unit rate.

4. $74.75

5. $1.19

6. $2.35

 ## Practice and Problem Solving

Find the product. List the units.

7. $8 \text{ h} \times \dfrac{\$9}{\text{h}}$

8. $8 \text{ lb} \times \dfrac{\$3.50}{\text{lb}}$

9. $\dfrac{29}{2} \text{ sec} \times \dfrac{60 \text{ MB}}{\text{sec}}$

10. $\dfrac{3}{4} \text{ h} \times \dfrac{19 \text{ mi}}{\frac{1}{4}\text{ h}}$

Write the ratio as a fraction in simplest form.

① 11. 25 to 45

12. $63 : 28$

13. $35 \text{ girls} : 15 \text{ boys}$

14. $51 \text{ correct} : 9 \text{ incorrect}$

15. 16 dogs to 12 cats

16. $2\dfrac{1}{3} \text{ feet} : 4\dfrac{1}{2} \text{ feet}$

Find the unit rate.

17. 180 miles in 3 hours

18. 256 miles per 8 gallons

19. $9.60 for 4 pounds

20. $4.80 for 6 cans

21. 297 words in 5.5 minutes

22. $21\dfrac{3}{4}$ meters in $2\dfrac{1}{2}$ hours

Use the ratio table to find the unit rate with the specified units.

② 23. servings per package

Packages	3	6	9	12
Servings	13.5	27	40.5	54

24. feet per year

Years	2	6	10	14
Feet	7.2	21.6	36	50.4

25. **DOWNLOAD** At 1:00 P.M., you have 24 megabytes of a movie. At 1:15 P.M., you have 96 megabytes. What is the download rate in megabytes per minute?

26. **POPULATION** In 2007, the U.S. population was 302 million people. In 2012, it was 314 million. What was the rate of population change per year?

27. **PAINTING** A painter can paint 350 square feet in 1.25 hours. What is the painting rate in square feet per hour?

3 28. **TICKETS** The graph shows the cost of buying tickets to a concert.

   a. What does the point (4, 122) represent?

   b. What is the unit rate?

   c. What is the cost of buying 10 tickets?

29. **CRITICAL THINKING** Are the two statements equivalent? Explain your reasoning.

   ● The ratio of boys to girls is 2 to 3.

   ● The ratio of girls to boys is 3 to 2.

30. **TENNIS** A sports store sells three different packs of tennis balls. Which pack is the best buy? Explain.

31. **FLOORING** It costs $68 for 16 square feet of flooring. How much does it cost for 12 square feet of flooring?

32. **OIL SPILL** An oil spill spreads 25 square meters every $\frac{1}{6}$ hour.

   How much area does the oil spill cover after 2 hours?

4 33. **JUICE** You mix $\frac{1}{4}$ cup of juice concentrate for every 2 cups of water to make 18 cups of juice. How much juice concentrate and water do you use?

34. **LANDSCAPING** A supplier sells $2\frac{1}{4}$ pounds of mulch for every $1\frac{1}{3}$ pounds of gravel. The supplier sells 172 pounds of mulch and gravel combined. How many pounds of each item does the supplier sell?

35. **HEART RATE** Your friend's heart beats 18 times in 15 seconds when at rest. While running, your friend's heart beats 25 times in 10 seconds.

   a. Find the heart rate in beats per minute at rest and while running.

   b. How many more times does your friend's heart beat in 3 minutes while running than while at rest?

36. PRECISION The table shows nutritional information for three beverages.

Beverage	Serving Size	Calories	Sodium
Whole milk	1 c	146	98 mg
Orange juice	1 pt	210	10 mg
Apple juice	24 fl oz	351	21 mg

a. Which has the most calories per fluid ounce?

b. Which has the least sodium per fluid ounce?

37. RESEARCH Fire hydrants are painted one of four different colors to indicate the rate at which water comes from the hydrant.

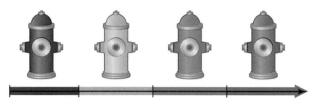

a. Use the Internet to find the ranges of the rates for each color.

b. Research why a firefighter needs to know the rate at which water comes out of a hydrant.

38. PAINT You mix $\frac{2}{5}$ cup of red paint for every $\frac{1}{4}$ cup of blue paint to make $1\frac{5}{8}$ gallons of purple paint.

a. How much red paint and blue paint do you use?

b. You decide that you want to make a lighter purple paint. You make the new mixture by adding $\frac{1}{10}$ cup of white paint for every $\frac{2}{5}$ cup of red paint and $\frac{1}{4}$ cup of blue paint. How much red paint, blue paint, and white paint do you use to make $\frac{3}{8}$ gallon of lighter purple paint?

39. **Critical Thinking** You and a friend start hiking toward each other from opposite ends of a 17.5-mile hiking trail. You hike $\frac{2}{3}$ mile every $\frac{1}{4}$ hour. Your friend hikes $2\frac{1}{3}$ miles per hour.

Big South Fork Trail 17.5 mi

a. Who hikes faster? How much faster?

b. After how many hours do you meet?

c. When you meet, who hiked farther? How much farther?

Fair Game Review *What you learned in previous grades & lessons*

Copy and complete the statement using <, >, or =. *(Section 2.1)*

40. $\frac{9}{2}$ ▢ $\frac{8}{3}$

41. $-\frac{8}{15}$ ▢ $\frac{10}{18}$

42. $\frac{-6}{24}$ ▢ $\frac{-2}{8}$

43. MULTIPLE CHOICE Which fraction is greater than $-\frac{2}{3}$ and less than $-\frac{1}{2}$? *(Section 2.1)*

Ⓐ $-\frac{3}{4}$　　Ⓑ $-\frac{7}{12}$　　Ⓒ $-\frac{5}{12}$　　Ⓓ $-\frac{3}{8}$

Essential Question How can proportions help you decide when things are "fair"?

The Meaning of a Word ● Proportional

When you work toward a goal, your success is usually **proportional** to the amount of work you put in.

An equation stating that two ratios are equal is a **proportion**.

---

1   ACTIVITY: Determining Proportions

Work with a partner. Tell whether the two ratios are equivalent. If they are not equivalent, change the next day to make the ratios equivalent. Explain your reasoning.

a. On the first day, you pay $5 for 2 boxes of popcorn. The next day, you pay $7.50 for 3 boxes.

First Day   Next Day
$$\frac{\$5.00}{2 \text{ boxes}} \overset{?}{=} \frac{\$7.50}{3 \text{ boxes}}$$

b. On the first day, it takes you $3\frac{1}{2}$ hours to drive 175 miles. The next day, it takes you 5 hours to drive 200 miles.

First Day   Next Day
$$\frac{3\frac{1}{2} \text{ h}}{175 \text{ mi}} \overset{?}{=} \frac{5 \text{ h}}{200 \text{ mi}}$$

COMMON CORE

Proportions
In this lesson, you will
● use equivalent ratios to determine whether two ratios form a proportion.
● use the Cross Products Property to determine whether two ratios form a proportion.
Learning Standard
7.RP.2a

c. On the first day, you walk 4 miles and burn 300 calories. The next day, you walk $3\frac{1}{3}$ miles and burn 250 calories.

First Day   Next Day
$$\frac{4 \text{ mi}}{300 \text{ cal}} \overset{?}{=} \frac{3\frac{1}{3} \text{ mi}}{250 \text{ cal}}$$

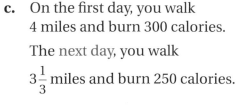

d. On the first day, you paint 150 square feet in $2\frac{1}{2}$ hours. The next day, you paint 200 square feet in 4 hours.

First Day   Next Day
$$\frac{150 \text{ ft}^2}{2\frac{1}{2} \text{ h}} \overset{?}{=} \frac{200 \text{ ft}^2}{4 \text{ h}}$$

2 ACTIVITY: Checking a Proportion

Work with a partner.

a. It is said that "one year in a dog's life is equivalent to seven years in a human's life." Explain why Newton thinks he has a score of 105 points. Did he solve the proportion correctly?

$$\frac{1 \text{ year}}{7 \text{ years}} \overset{?}{=} \frac{15 \text{ points}}{105 \text{ points}}$$

b. If Newton thinks his score is 98 points, how many points does he actually have? Explain your reasoning.

"I got 15 on my online test. That's 105 in dog points! Isn't that an A+?"

3 ACTIVITY: Determining Fairness

Math Practice 3

Justify Conclusions

What information can you use to justify your conclusion?

Work with a partner. Write a ratio for each sentence. Compare the ratios. If they are equal, then the answer is "It is fair." If they are not equal, then the answer is "It is not fair." Explain your reasoning.

a.

| You pay $184 for 2 tickets to a concert. | & | I pay $266 for 3 tickets to the same concert. |  Is this fair? |

b.

| You get 75 points for answering 15 questions correctly. | & | I get 70 points for answering 14 questions correctly. | Is this fair? |

c.

| You trade 24 football cards for 15 baseball cards. | & | I trade 20 football cards for 32 baseball cards. | Is this fair? |

What Is Your Answer?

4. Find a recipe for something you like to eat. Then show how two of the ingredient amounts are proportional when you double or triple the recipe.

5. **IN YOUR OWN WORDS** How can proportions help you decide when things are "fair"? Give an example.

Practice

Use what you discovered about proportions to complete Exercises 15–20 on page 174.

 **Key Idea**

Proportions

Words A **proportion** is an equation stating that two ratios are equivalent. Two quantities that form a proportion are **proportional**.

Numbers $\dfrac{2}{3} = \dfrac{4}{6}$ The proportion is read "2 is to 3 as 4 is to 6."

EXAMPLE **1** **Determining Whether Ratios Form a Proportion**

Tell whether $\dfrac{6}{4}$ and $\dfrac{8}{12}$ form a proportion.

Compare the ratios in simplest form.

$$\dfrac{6}{4} = \dfrac{6 \div 2}{4 \div 2} = \dfrac{3}{2} \longleftarrow$$

The ratios are *not* equivalent.

$$\dfrac{8}{12} = \dfrac{8 \div 4}{12 \div 4} = \dfrac{2}{3} \longleftarrow$$

∴ So, $\dfrac{6}{4}$ and $\dfrac{8}{12}$ do *not* form a proportion.

EXAMPLE **2** **Determining Whether Two Quantities Are Proportional**

Tell whether *x* and *y* are proportional.

Compare each ratio *x* to *y* in simplest form.

$$\dfrac{\frac{1}{2}}{3} = \dfrac{1}{6} \qquad \dfrac{1}{6} \qquad \dfrac{\frac{3}{2}}{9} = \dfrac{1}{6} \qquad \dfrac{2}{12} = \dfrac{1}{6}$$

The ratios are equivalent.

x	*y*
$\frac{1}{2}$	3
1	6
$\frac{3}{2}$	9
2	12

Reading

Two quantities that are proportional are in a *proportional relationship*.

∴ So, *x* and *y* are proportional.

● **On Your Own**

Now You're Ready
Exercises 5–14

Tell whether the ratios form a proportion.

1. $\dfrac{1}{2}, \dfrac{5}{10}$ 2. $\dfrac{4}{6}, \dfrac{18}{24}$ 3. $\dfrac{10}{3}, \dfrac{5}{6}$ 4. $\dfrac{25}{20}, \dfrac{15}{12}$

5. Tell whether *x* and *y* are proportional.

Birdhouses Built, *x*	1	2	4	6
Nails Used, *y*	12	24	48	72

 **Key Ideas**

Cross Products

In the proportion $\dfrac{a}{b} = \dfrac{c}{d}$, the products $a \cdot d$ and $b \cdot c$ are called **cross products**.

Cross Products Property

Words The cross products of a proportion are equal.

<table>
<tr><td align="center">**Numbers**</td><td align="center">**Algebra**</td></tr>
<tr><td align="center"></td><td align="center">$\dfrac{a}{b} \Large\times \dfrac{c}{d}$</td></tr>
<tr><td align="center">$2 \cdot 6 = 3 \cdot 4$</td><td align="center">$ad = bc,$
where $b \neq 0$ and $d \neq 0$</td></tr>
</table>

Study Tip

You can use the Multiplication Property of Equality to show that the cross products are equal.

$$\frac{a}{b} = \frac{c}{d}$$

$$\cancel{b}d \cdot \frac{a}{\cancel{b}} = b\cancel{d} \cdot \frac{c}{\cancel{d}}$$

$$ad = bc$$

EXAMPLE 3 | **Identifying Proportional Relationships**

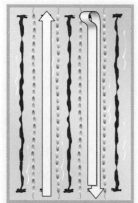

1 length   1 lap

You swim your first 4 laps in 2.4 minutes. You complete 16 laps in 12 minutes. Is the number of laps proportional to your time?

Method 1: Compare unit rates.

$$\overset{\div 4}{\frac{2.4 \text{ min}}{4 \text{ laps}} = \frac{0.6 \text{ min}}{1 \text{ lap}}} \qquad \overset{\div 16}{\frac{12 \text{ min}}{16 \text{ laps}} = \frac{0.75 \text{ min}}{1 \text{ lap}}}$$

$$\div 4 \qquad\qquad\qquad \div 16$$

> The unit rates are *not* equivalent.

So, the number of laps is *not* proportional to the time.

Method 2: Use the Cross Products Property.

$$\frac{2.4 \text{ min}}{4 \text{ laps}} \overset{?}{=} \frac{12 \text{ min}}{16 \text{ laps}} \qquad \text{Test to see if the rates are equivalent.}$$

$$2.4 \cdot 16 \overset{?}{=} 4 \cdot 12 \qquad \text{Find the cross products.}$$

$$38.4 \neq 48 \qquad \text{The cross products are } not \text{ equal.}$$

So, the number of laps is *not* proportional to the time.

On Your Own

Now You're Ready
Exercises 15–20

6. You read the first 20 pages of a book in 25 minutes. You read 36 pages in 45 minutes. Is the number of pages read proportional to your time?

Vocabulary and Concept Check

1. **VOCABULARY** What does it mean for two ratios to form a proportion?

2. **VOCABULARY** What are two ways you can tell that two ratios form a proportion?

3. **OPEN-ENDED** Write two ratios that are equivalent to $\frac{3}{5}$.

4. **WHICH ONE DOESN'T BELONG?** Which ratio does *not* belong with the other three? Explain your reasoning.

$$\frac{4}{10} \qquad \frac{2}{5} \qquad \frac{3}{5} \qquad \frac{6}{15}$$

Practice and Problem Solving

Tell whether the ratios form a proportion.

① 5. $\frac{1}{3}, \frac{7}{21}$ **6.** $\frac{1}{5}, \frac{6}{30}$ **7.** $\frac{3}{4}, \frac{24}{18}$ **8.** $\frac{2}{5}, \frac{40}{16}$

9. $\frac{48}{9}, \frac{16}{3}$ **10.** $\frac{18}{27}, \frac{33}{44}$ **11.** $\frac{7}{2}, \frac{16}{6}$ **12.** $\frac{12}{10}, \frac{14}{12}$

Tell whether x and y are proportional.

② 13.

x	1	2	3	4
y	7	8	9	10

14.

x	2	4	6	8
y	5	10	15	20

Tell whether the two rates form a proportion.

③ 15. 7 inches in 9 hours; 42 inches in 54 hours

16. 12 players from 21 teams; 15 players from 24 teams

17. 440 calories in 4 servings; 300 calories in 3 servings

18. 120 units made in 5 days; 88 units made in 4 days

19. 66 wins in 82 games; 99 wins in 123 games

20. 68 hits in 172 at bats; 43 hits in 123 at bats

21. FITNESS You can do 90 sit-ups in 2 minutes. Your friend can do 135 sit-ups in 3 minutes. Do these rates form a proportion? Explain.

22. HEART RATES Find the heart rates of you and your friend. Do these rates form a proportion? Explain.

	Heartbeats	Seconds
You	22	20
Friend	18	15

Tell whether the ratios form a proportion.

23. $\dfrac{2.5}{4}, \dfrac{7}{11.2}$

24. 2 to 4, 11 to $\dfrac{11}{2}$

25. $2 : \dfrac{4}{5}, \dfrac{3}{4} : \dfrac{3}{10}$

26. PAY RATE You earn $56 walking your neighbor's dog for 8 hours. Your friend earns $36 painting your neighbor's fence for 4 hours.

    a. What is your pay rate?

    b. What is your friend's pay rate?

    c. Are the pay rates equivalent? Explain.

27. GEOMETRY Are the heights and bases of the two triangles proportional? Explain.

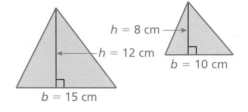

$h = 8$ cm

$h = 12$ cm

$b = 10$ cm

$b = 15$ cm

28. BASEBALL A pitcher coming back from an injury limits the number of pitches thrown in bull pen sessions as shown.

    a. Which quantities are proportional?

    b. How many pitches that are not curveballs do you think the pitcher will throw in Session 5?

Session Number, x	Pitches, y	Curveballs, z
1	10	4
2	20	8
3	30	12
4	40	16

29. NAIL POLISH A specific shade of red nail polish requires 7 parts red to 2 parts yellow. A mixture contains 35 quarts of red and 8 quarts of yellow. How can you fix the mixture to make the correct shade of red?

30. COIN COLLECTION The ratio of quarters to dimes in a coin collection is $5 : 3$. You add the same number of new quarters as dimes to the collection.

    a. Is the ratio of quarters to dimes still $5 : 3$?

    b. If so, illustrate your answer with an example. If not, show why with a "counterexample."

31. AGE You are 13 years old, and your cousin is 19 years old. As you grow older, is your age proportional to your cousin's age? Explain your reasoning.

32. **Critical Thinking** Ratio A is equivalent to Ratio B. Ratio B is equivalent to Ratio C. Is Ratio A equivalent to Ratio C? Explain.

Fair Game Review What you learned in previous grades & lessons

Add or subtract. *(Section 1.2 and Section 1.3)*

33. $-28 + 15$

34. $-6 + (-11)$

35. $-10 - 8$

36. $-17 - (-14)$

37. MULTIPLE CHOICE Which fraction is not equivalent to $\dfrac{2}{6}$? *(Skills Review Handbook)*

    Ⓐ $\dfrac{1}{3}$ **Ⓑ** $\dfrac{12}{36}$ **Ⓒ** $\dfrac{4}{12}$ **Ⓓ** $\dfrac{6}{9}$

Recall that you can graph the values from a ratio table.

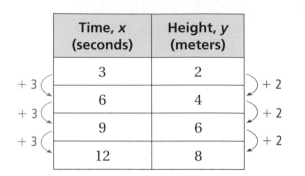

Time, x (seconds)	Height, y (meters)
3	2
6	4
9	6
12	8

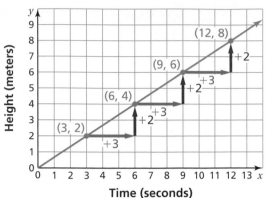

The structure in the ratio table shows why the graph has a constant *rate of change*. You can use the constant rate of change to show that the graph passes through the origin. The graph of every proportional relationship is a line through the origin.

EXAMPLE 1 Determining Whether Two Quantities Are Proportional

Use a graph to tell whether x and y are in a proportional relationship.

a.

x	2	4	6
y	6	8	10

Plot (2, 6), (4, 8), and (6, 10). Draw a line through the points.

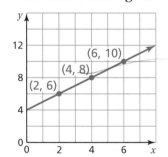

The graph is a line that does not pass through the origin.

∴ So, x and y are not in a proportional relationship.

b.

x	1	2	3
y	2	4	6

Plot (1, 2), (2, 4), and (3, 6). Draw a line through the points.

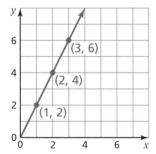

The graph is a line that passes through the origin.

∴ So, x and y are in a proportional relationship.

COMMON CORE

Proportions

In this extension, you will
- use graphs to determine whether two ratios form a proportion.
- interpret graphs of proportional relationships.

Learning Standards
7.RP.2a
7.RP.2b
7.RP.2d

 **Practice**

Use a graph to tell whether x and y are in a proportional relationship.

1.

x	1	2	3	4
y	3	4	5	6

2.

x	1	3	5	7
y	0.5	1.5	2.5	3.5

The graph shows that the distance traveled by the Mars rover *Curiosity* is proportional to the time traveled. Interpret each plotted point in the graph.

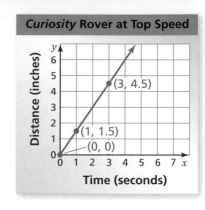

Curiosity **Rover at Top Speed**

(0, 0):   The rover travels 0 inches in 0 seconds.

(1, 1.5):   The rover travels 1.5 inches in 1 second. So, the unit rate is 1.5 inches per second.

Study Tip

In the graph of a proportional relationship, you can find the unit rate from the point (1, y).

(3, 4.5):   The rover travels 4.5 inches in 3 seconds. Because the relationship is proportional, you can also use this point to find the unit rate.

$$\frac{4.5 \text{ in.}}{3 \text{ sec}} = \frac{1.5 \text{ in.}}{1 \text{ sec}}, \text{ or } 1.5 \text{ inches per second}$$

Practice

Interpret each plotted point in the graph of the proportional relationship.

3.

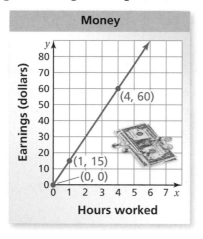

4.

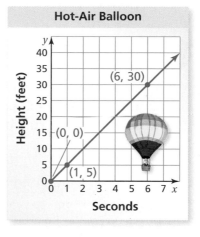

Tell whether x and y are in a proportional relationship. If so, find the unit rate.

5.

x (hours)	1	4	7	10
y (feet)	5	20	35	50

6.   Let y be the temperature x hours after midnight. The temperature is 60°F at midnight and decreases 2°F every $\frac{1}{2}$ hour.

7.   **REASONING** The graph of a proportional relationship passes through (12, 16) and (1, y). Find y.

8.   **MOVIE RENTAL** You pay $1 to rent a movie plus an additional $0.50 per day until you return the movie. Your friend pays $1.25 per day to rent a movie.

   a.   Make tables showing the costs to rent a movie up to 5 days.

   b.   Which person pays an amount proportional to the number of days rented?

5.3 Writing Proportions

Essential Question How can you write a proportion that solves a problem in real life?

1 ACTIVITY: Writing Proportions

Work with a partner. A rough rule for finding the correct bat length is "the bat length should be half of the batter's height." So, a 62-inch-tall batter uses a bat that is 31 inches long. Write a proportion to find the bat length for each given batter height.

a. 58 inches

b. 60 inches

c. 64 inches

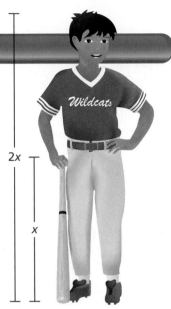

2 ACTIVITY: Bat Lengths

Work with a partner. Here is a more accurate table for determining the bat length for a batter. Find all the batter heights and corresponding weights for which the rough rule in Activity 1 is exact.

Weight of Batter (pounds)	Height of Batter (inches)							
	45–48	49–52	53–56	57–60	61–64	65–68	69–72	Over 72
Under 61	28	29	29					
61–70	28	29	30	30				
71–80	28	29	30	30	31			
81–90	29	29	30	30	31	32		
91–100	29	30	30	31	31	32		
101–110	29	30	30	31	31	32		
111–120	29	30	30	31	31	32		
121–130	29	30	30	31	32	33	33	
131–140	30	30	31	31	32	33	33	
141–150	30	30	31	31	32	33	33	
151–160	30	31	31	32	32	33	33	33
161–170		31	31	32	32	33	33	34
171–180				32	33	33	34	34
Over 180					33	33	34	34

COMMON CORE

Proportions
In this lesson, you will
• write proportions.
• solve proportions using mental math.
Learning Standards
7.RP.2c
7.RP.3

3 **ACTIVITY: Writing Proportions**

Work with a partner. The batting average of a baseball player is the number of "hits" divided by the number of "at bats."

$$\text{batting average} = \frac{\text{hits }(H)}{\text{at bats }(A)}$$

A player whose batting average is 0.250 is said to be "batting 250."

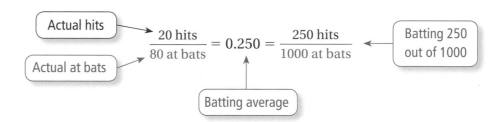

Write a proportion to find how many hits *H* a player needs to achieve the given batting average. Then solve the proportion.

a. 50 times at bat; batting average is 0.200.

b. 84 times at bat; batting average is 0.250.

c. 80 times at bat; batting average is 0.350.

d. 1 time at bat; batting average is 1.000.

What Is Your Answer?

4. **IN YOUR OWN WORDS** How can you write a proportion that solves a problem in real life?

5. Two players have the same batting average.

	At Bats	Hits	Batting Average
Player 1	132	45	
Player 2	132	45	

Player 1 gets four hits in the next five at bats. Player 2 gets three hits in the next three at bats.

a. Who has the higher batting average?

b. Does this seem fair? Explain your reasoning.

Practice Use what you discovered about proportions to complete Exercises 4–7 on page 182.

One way to write a proportion is to use a table.

	Last Month	This Month
Purchase	2 ringtones	3 ringtones
Total Cost	6 dollars	x dollars

Use the columns or the rows to write a proportion.

Use columns:

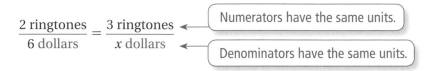

$$\frac{2 \text{ ringtones}}{6 \text{ dollars}} = \frac{3 \text{ ringtones}}{x \text{ dollars}}$$

← Numerators have the same units.

← Denominators have the same units.

Use rows:

$$\frac{2 \text{ ringtones}}{3 \text{ ringtones}} = \frac{6 \text{ dollars}}{x \text{ dollars}}$$

The units are the same on each side of the proportion.

EXAMPLE 1 Writing a Proportion

Black Bean Soup

1.5 cups black beans
0.5 cup salsa
2 cups water
1 tomato
2 teaspoons seasoning

A chef increases the amounts of ingredients in a recipe to make a proportional recipe. The new recipe has 6 cups of black beans. Write a proportion that gives the number x of tomatoes in the new recipe.

Organize the information in a table.

	Original Recipe	New Recipe
Black Beans	1.5 cups	6 cups
Tomatoes	1 tomato	x tomatoes

One proportion is $\dfrac{1.5 \text{ cups beans}}{1 \text{ tomato}} = \dfrac{6 \text{ cups beans}}{x \text{ tomatoes}}$.

On Your Own

Exercises 8–11

1. Write a different proportion that gives the number x of tomatoes in the new recipe.

2. Write a proportion that gives the amount y of water in the new recipe.

EXAMPLE **2** **Solving Proportions Using Mental Math**

Solve $\dfrac{3}{2} = \dfrac{x}{8}$.

Step 1: Think: The product of 2 and what number is 8?

$$\dfrac{3}{2} = \dfrac{x}{8}$$

$$2 \times \; ? = 8$$

Step 2: Because the product of 2 and 4 is 8, multiply the numerator by 4 to find x.

$$3 \times 4 = 12$$

$$\dfrac{3}{2} = \dfrac{x}{8}$$

$$2 \times 4 = 8$$

∴ The solution is $x = 12$.

EXAMPLE **3** **Solving Proportions Using Mental Math**

In Example 1, how many tomatoes are in the new recipe?

Solve the proportion $\dfrac{1.5}{1} = \dfrac{6}{x}$. ← cups black beans
← tomatoes

Step 1: Think: The product of 1.5 and what number is 6?

$$1.5 \times \; ? = 6$$

$$\dfrac{1.5}{1} = \dfrac{6}{x}$$

Step 2: Because the product of 1.5 and 4 is 6, multiply the denominator by 4 to find x.

$$1.5 \times 4 = 6$$

$$\dfrac{1.5}{1} = \dfrac{6}{x}$$

$$1 \times 4 = 4$$

∴ So, there are 4 tomatoes in the new recipe.

● **On Your Own**

Now You're Ready
Exercises 16–21

Solve the proportion.

3. $\dfrac{5}{8} = \dfrac{20}{d}$

4. $\dfrac{7}{z} = \dfrac{14}{10}$

5. $\dfrac{21}{24} = \dfrac{x}{8}$

6. A school has 950 students. The ratio of female students to all students is $\dfrac{48}{95}$. Write and solve a proportion to find the number f of students who are female.

Vocabulary and Concept Check

1. **WRITING** Describe two ways you can use a table to write a proportion.

2. **WRITING** What is your first step when solving $\dfrac{x}{15} = \dfrac{3}{5}$? Explain.

3. **OPEN-ENDED** Write a proportion using an unknown value x and the ratio 5 : 6. Then solve it.

Practice and Problem Solving

Write a proportion to find how many points a student needs to score on the test to get the given score.

4. test worth 50 points; test score of 40%

5. test worth 50 points; test score of 78%

6. test worth 80 points; test score of 80%

7. test worth 150 points; test score of 96%

Use the table to write a proportion.

8.

	Game 1	Game 2
Points	12	18
Shots	14	w

9.

	May	June
Winners	n	34
Entries	85	170

10.

	Today	Yesterday
Miles	15	m
Hours	2.5	4

11.

	Race 1	Race 2
Meters	100	200
Seconds	x	22.4

12. **ERROR ANALYSIS** Describe and correct the error in writing the proportion.

	Monday	Tuesday
Dollars	2.08	d
Ounces	8	16

$$\dfrac{2.08}{16} = \dfrac{d}{8}$$

13. **T-SHIRTS** You can buy 3 T-shirts for $24. Write a proportion that gives the cost c of buying 7 T-shirts.

14. **COMPUTERS** A school requires 2 computers for every 5 students. Write a proportion that gives the number c of computers needed for 145 students.

15. **SWIM TEAM** The school team has 80 swimmers. The ratio of seventh-grade swimmers to all swimmers is 5 : 16. Write a proportion that gives the number s of seventh-grade swimmers.

Solve the proportion.

② ③ 16. $\dfrac{1}{4} = \dfrac{z}{20}$

17. $\dfrac{3}{4} = \dfrac{12}{y}$

18. $\dfrac{35}{k} = \dfrac{7}{3}$

19. $\dfrac{15}{8} = \dfrac{45}{c}$

20. $\dfrac{b}{36} = \dfrac{5}{9}$

21. $\dfrac{1.4}{2.5} = \dfrac{g}{25}$

22. **ORCHESTRA** In an orchestra, the ratio of trombones to violas is 1 to 3.

    a. There are 9 violas. Write a proportion that gives the number t of trombones in the orchestra.

    b. How many trombones are in the orchestra?

23. **ATLANTIS** Your science teacher has a 1 : 200 scale model of the space shuttle *Atlantis*. Which of the proportions can you use to find the actual length x of *Atlantis*? Explain.

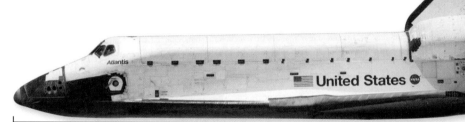

$\dfrac{1}{200} = \dfrac{19.5}{x}$ $\dfrac{1}{200} = \dfrac{x}{19.5}$ $\dfrac{200}{19.5} = \dfrac{x}{1}$ $\dfrac{x}{200} = \dfrac{1}{19.5}$

19.5 cm

24. **YOU BE THE TEACHER** Your friend says "$48x = 6 \cdot 12$." Is your friend right? Explain.

> Solve $\dfrac{6}{x} = \dfrac{12}{48}$.

25. **Reasoning** There are 180 white lockers in the school. There are 3 white lockers for every 5 blue lockers. How many lockers are in the school?

Fair Game Review What you learned in previous grades & lessons

Solve the equation. *(Section 3.4)*

26. $\dfrac{x}{6} = 25$

27. $8x = 72$

28. $150 = 2x$

29. $35 = \dfrac{x}{4}$

30. **MULTIPLE CHOICE** What is the value of $-\dfrac{9}{4} + \left| -\dfrac{8}{5} \right| - 2\dfrac{1}{2}$? *(Section 2.3)*

   Ⓐ $-6\dfrac{7}{20}$ Ⓑ $-5\dfrac{7}{20}$ Ⓒ $-3\dfrac{3}{20}$ Ⓓ $-2\dfrac{3}{20}$

You can use an **information wheel** to organize information about a concept. Here is an example of an information wheel for ratio.

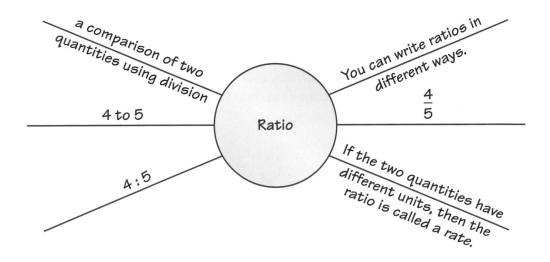

a comparison of two quantities using division

4 to 5

4 : 5

Ratio

You can write ratios in different ways.

$\dfrac{4}{5}$

If the two quantities have different units, then the ratio is called a rate.

On Your Own

Make information wheels to help you study these topics.

1. rate

2. unit rate

3. proportion

4. cross products

5. graphing proportional relationships

After you complete this chapter, make information wheels for the following topics.

6. solving proportions

7. slope

8. direct variation

"My information wheel summarizes how cats act when they get baths."

5.1–5.3　Quiz

Check It Out
Progress Check
BigIdeasMath ✓com

Write the ratio as a fraction in simplest form.　*(Section 5.1)*

1. 18 red buttons : 12 blue buttons

2. $\frac{5}{4}$ inches to $\frac{2}{3}$ inch

Use the ratio table to find the unit rate with the specified units.　*(Section 5.1)*

3. cost per song

Songs	0	2	4	6
Cost	$0	$1.98	$3.96	$5.94

4. gallons per hour

Hours	3	6	9	12
Gallons	10.5	21	31.5	42

Tell whether the ratios form a proportion.　*(Section 5.2)*

5. $\frac{1}{8}, \frac{4}{32}$

6. $\frac{2}{3}, \frac{10}{30}$

7. $\frac{7}{4}, \frac{28}{16}$

Tell whether the two rates form a proportion.　*(Section 5.2)*

8. 75 miles in 3 hours; 140 miles in 4 hours

9. 12 gallons in 4 minutes; 21 gallons in 7 minutes

10. 150 steps in 50 feet; 72 steps in 24 feet

11. 3 rotations in 675 days; 2 rotations in 730 days

Use the table to write a proportion.　*(Section 5.3)*

12.

	Monday	Tuesday
Dollars	42	56
Hours	6	h

13.

	Series 1	Series 2
Games	g	6
Wins	4	3

14. MUSIC DOWNLOAD The amount of time needed to download music is shown in the table. Find the unit rate in megabytes per second. *(Section 5.1)*

Seconds	6	12	18	24
Megabytes	2	4	6	8

15. SOUND The graph shows the distance that sound travels through steel. Interpret each plotted point in the graph of the proportional relationship. *(Section 5.2)*

16. GAMING You advance 3 levels in 15 minutes. Your friend advances 5 levels in 20 minutes. Do these rates form a proportion? Explain. *(Section 5.2)*

17. CLASS TIME You spend 150 minutes in 3 classes. Write and solve a proportion to find how many minutes you spend in 5 classes. *(Section 5.3)*

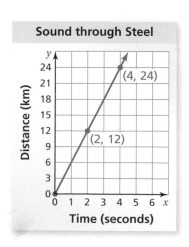

Sound through Steel

(4, 24)

(2, 12)

Distance (km)

Time (seconds)

5.4 Solving Proportions

Essential Question How can you use ratio tables and cross products to solve proportions?

1 **ACTIVITY: Solving a Proportion in Science**

Work with a partner. You can use ratio tables to determine the amount of a compound (like salt) that is dissolved in a solution. Determine the unknown quantity. Explain your procedure.

a. **Salt Water**

Salt Water	1 L	3 L
Salt	250 g	x g

1 liter   3 liters

$$\frac{1\,L}{} = \frac{}{}$$ Write proportion.

$1 \cdot \boxed{} = \boxed{} \cdot \boxed{}$ Set cross products equal.

$\boxed{} = \boxed{}$ Simplify.

∴ There are [ ] grams of salt in the 3-liter solution.

b. **White Glue Solution**

Water	½ cup	1 cup
White Glue	½ cup	x cups

Recipe for SLIME

1. Add ½ cup of water and ½ cup white glue. Mix thoroughly. This is your white glue solution.

2. Add a couple drops of food coloring to the white glue solution. Mix thoroughly.

c. **Borax Solution**

Borax	1 tsp	2 tsp
Water	1 cup	x cups

3. Add 1 teaspoon of borax to 1 cup of water. Mix thoroughly. This is your borax solution (about 1 cup).

d. **Slime (See recipe.)**

Borax Solution	½ cup	1 cup
White Glue Solution	y cups	x cups

4. Pour the borax solution and the glue solution into a separate bowl.

5. Place the slime that forms into a plastic bag. Squeeze the mixture repeatedly to mix it up.

COMMON CORE

Proportions
In this lesson, you will
- solve proportions using multiplication or the Cross Products Property.
- use a point on a graph to write and solve proportions.
Learning Standards
7.RP.2b
7.RP.2c

② ACTIVITY: The Game of Criss Cross

Math Practice 2

Use Operations
How can you use the name of the game to determine which operation to use?

CRISS CROSS

Preparation:

- Cut index cards to make 48 playing cards.
- Write each number on a card.

  1, 1, 1, 2, 2, 2, 3, 3, 3, 4, 4, 4, 5, 5, 5, 6, 6, 6, 7, 7,

  7, 8, 8, 8, 9, 9, 9, 10, 10, 10, 12, 12, 12, 13, 13,

  13, 14, 14, 14, 15, 15, 15, 16, 16, 16, 18, 20, 25

- Make a copy of the game board.

To Play:

- Play with a partner.
- Deal eight cards to each player.
- Begin by drawing a card from the remaining cards. Use four of your cards to try to form a proportion.
- Lay the four cards on the game board. If you form a proportion, then say "Criss Cross." You earn 4 points. Place the four cards in a discard pile. Now it is your partner's turn.
- If you cannot form a proportion, then it is your partner's turn.
- When the original pile of cards is empty, shuffle the cards in the discard pile. Start again.
- The first player to reach 20 points wins.

What Is Your Answer?

3. **IN YOUR OWN WORDS** How can you use ratio tables and cross products to solve proportions? Give an example.

4. **PUZZLE** Use each number once to form three proportions.

1	2	10	4	12	20
15	5	16	6	8	3

Practice ➤ Use what you discovered about solving proportions to complete Exercises 10–13 on page 190.

Check It Out
Lesson Tutorials
BigIdeasMath ✓com

🔑 Key Idea

Solving Proportions

Method 1   Use mental math.   *(Section 5.3)*

Method 2   Use the Multiplication Property of Equality.   *(Section 5.4)*

Method 3   Use the Cross Products Property.   *(Section 5.4)*

EXAMPLE 1   Solving Proportions Using Multiplication

Solve $\dfrac{5}{7} = \dfrac{x}{21}$.

$$\dfrac{5}{7} = \dfrac{x}{21}$$ Write the proportion.

$$21 \cdot \dfrac{5}{7} = 21 \cdot \dfrac{x}{21}$$ Multiplication Property of Equality

$$15 = x$$ Simplify.

∴· The solution is 15.

🔵 On Your Own

Now You're Ready
Exercises 4–9

Use multiplication to solve the proportion.

1. $\dfrac{w}{6} = \dfrac{6}{9}$

2. $\dfrac{12}{10} = \dfrac{a}{15}$

3. $\dfrac{y}{6} = \dfrac{2}{4}$

EXAMPLE 2   Solving Proportions Using the Cross Products Property

Solve each proportion.

a. $\dfrac{x}{8} = \dfrac{7}{10}$

$x \cdot 10 = 8 \cdot 7$ Cross Products Property

$10x = 56$ Multiply.

$x = 5.6$ Divide.

∴· The solution is 5.6.

b. $\dfrac{9}{y} = \dfrac{3}{17}$

$9 \cdot 17 = y \cdot 3$

$153 = 3y$

$51 = y$

∴· The solution is 51.

On Your Own

Now You're Ready
Exercises 10–21

Use the Cross Products Property to solve the proportion.

4. $\dfrac{2}{7} = \dfrac{x}{28}$ **5.** $\dfrac{12}{5} = \dfrac{6}{y}$ **6.** $\dfrac{40}{z+1} = \dfrac{15}{6}$

EXAMPLE **3** **Real-Life Application**

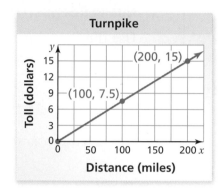

The graph shows the toll y due on a turnpike for driving x miles. Your toll is $7.50. How many *kilometers* did you drive?

The point (100, 7.5) on the graph shows that the toll is $7.50 for driving 100 miles. Convert 100 miles to kilometers.

Method 1: Convert using a ratio.

1 mi ≈ 1.61 km

$$100 \ \cancel{mi} \times \frac{1.61 \ km}{1 \ \cancel{mi}} = 161 \ km$$

So, you drove about 161 kilometers.

Method 2: Convert using a proportion.

Let x be the number of kilometers equivalent to 100 miles.

kilometers → $\dfrac{1.61}{1} = \dfrac{x}{100}$ ← kilometers

miles → ⠀⠀⠀⠀⠀ ← miles   Write a proportion. Use 1.61 km ≈ 1 mi.

$1.61 \cdot 100 = 1 \cdot x$ Cross Products Property

$161 = x$ Simplify.

So, you drove about 161 kilometers.

On Your Own

Now You're Ready
Exercises 28–30

Write and solve a proportion to complete the statement. Round to the nearest hundredth, if necessary.

7. 7.5 in. ≈ ▢ cm   **8.** 100 g ≈ ▢ oz

9. 2 L ≈ ▢ qt   **10.** 4 m ≈ ▢ ft

Vocabulary and Concept Check

1. **WRITING** What are three ways you can solve a proportion?

2. **OPEN-ENDED** Which way would you choose to solve $\dfrac{3}{x} = \dfrac{6}{14}$? Explain your reasoning.

3. **NUMBER SENSE** Does $\dfrac{x}{4} = \dfrac{15}{3}$ have the same solution as $\dfrac{x}{15} = \dfrac{4}{3}$? Use the Cross Products Property to explain your answer.

Practice and Problem Solving

Use multiplication to solve the proportion.

❶ 4. $\dfrac{9}{5} = \dfrac{z}{20}$

5. $\dfrac{h}{15} = \dfrac{16}{3}$

6. $\dfrac{w}{4} = \dfrac{42}{24}$

7. $\dfrac{35}{28} = \dfrac{n}{12}$

8. $\dfrac{7}{16} = \dfrac{x}{4}$

9. $\dfrac{y}{9} = \dfrac{44}{54}$

Use the Cross Products Property to solve the proportion.

❷ 10. $\dfrac{a}{6} = \dfrac{15}{2}$

11. $\dfrac{10}{7} = \dfrac{8}{k}$

12. $\dfrac{3}{4} = \dfrac{v}{14}$

13. $\dfrac{5}{n} = \dfrac{16}{32}$

14. $\dfrac{36}{42} = \dfrac{24}{r}$

15. $\dfrac{9}{10} = \dfrac{d}{6.4}$

16. $\dfrac{x}{8} = \dfrac{3}{12}$

17. $\dfrac{8}{m} = \dfrac{6}{15}$

18. $\dfrac{4}{24} = \dfrac{c}{36}$

19. $\dfrac{20}{16} = \dfrac{d}{12}$

20. $\dfrac{30}{20} = \dfrac{w}{14}$

21. $\dfrac{2.4}{1.8} = \dfrac{7.2}{k}$

22. **ERROR ANALYSIS** Describe and correct the error in solving the proportion $\dfrac{m}{8} = \dfrac{15}{24}$.

$$\boxed{\times} \quad \dfrac{m}{8} = \dfrac{15}{24}$$
$$8 \cdot m = 24 \cdot 15$$
$$m = 45$$

23. **PENS** Forty-eight pens are packaged in 4 boxes. How many pens are packaged in 9 boxes?

24. **PIZZA PARTY** How much does it cost to buy 10 medium pizzas?

3 Medium Pizzas for $10.50

Solve the proportion.

25. $\dfrac{2x}{5} = \dfrac{9}{15}$

26. $\dfrac{5}{2} = \dfrac{d-2}{4}$

27. $\dfrac{4}{k+3} = \dfrac{8}{14}$

Write and solve a proportion to complete the statement. Round to the nearest hundredth if necessary.

③ 28. 6 km ≈ ▓ mi

29. 2.5 L ≈ ▓ gal

30. 90 lb ≈ ▓ kg

31. TRUE OR FALSE? Tell whether the statement is *true* or *false*. Explain.

If $\dfrac{a}{b} = \dfrac{2}{3}$, then $\dfrac{3}{2} = \dfrac{b}{a}$.

32. CLASS TRIP It costs $95 for 20 students to visit an aquarium. How much does it cost for 162 students?

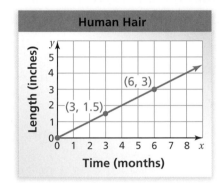

Human Hair

Length (inches) vs. Time (months)

(3, 1.5) (6, 3)

33. GRAVITY A person who weighs 120 pounds on Earth weighs 20 pounds on the Moon. How much does a 93-pound person weigh on the Moon?

34. HAIR The length of human hair is proportional to the number of months it has grown.

   a. What is the hair length in *centimeters* after 6 months?

   b. How long does it take hair to grow 8 inches?

   c. Use a different method than the one in part (b) to find how long it takes hair to grow 20 inches.

35. SWING SET It takes 6 hours for 2 people to build a swing set. Can you use the proportion $\dfrac{2}{6} = \dfrac{5}{h}$ to determine the number of hours h it will take 5 people to build the swing set? Explain.

36. REASONING There are 144 people in an audience. The ratio of adults to children is 5 to 3. How many are adults?

37. PROBLEM SOLVING Three pounds of lawn seed covers 1800 square feet. How many bags are needed to cover 8400 square feet?

38. **Critical Thinking** Consider the proportions $\dfrac{m}{n} = \dfrac{1}{2}$ and $\dfrac{n}{k} = \dfrac{2}{5}$. What is the ratio $\dfrac{m}{k}$? Explain your reasoning.

 Fair Game Review *What you learned in previous grades & lessons*

Plot the ordered pair in a coordinate plane. *(Skills Review Handbook)*

39. $A(-5, -2)$

40. $B(-3, 0)$

41. $C(-1, 2)$

42. $D(1, 4)$

43. MULTIPLE CHOICE What is the value of $(3w - 8) - 4(2w + 3)$? *(Section 3.2)*

  Ⓐ $11w + 4$ Ⓑ $-5w - 5$ Ⓒ $-5w + 4$ Ⓓ $-5w - 20$

Essential Question How can you compare two rates graphically?

1 ACTIVITY: Comparing Unit Rates

Work with a partner. The table shows the maximum speeds of several animals.

a. Find the missing speeds. Round your answers to the nearest tenth.

b. Which animal is fastest? Which animal is slowest?

c. Explain how you convert between the two units of speed.

Animal	Speed (miles per hour)	Speed (feet per second)
Antelope	61.0	
Black mamba snake		29.3
Cheetah		102.6
Chicken		13.2
Coyote	43.0	
Domestic pig		16.0
Elephant		36.6
Elk		66.0
Giant tortoise	0.2	
Giraffe	32.0	
Gray fox		61.6
Greyhound	39.4	
Grizzly bear		44.0
Human		41.0
Hyena	40.0	
Jackal	35.0	
Lion		73.3
Peregrine falcon	200.0	
Quarter horse	47.5	
Spider		1.76
Squirrel	12.0	
Thomson's gazelle	50.0	
Three-toed sloth		0.2
Tuna	47.0	

COMMON CORE

Slope

In this lesson, you will
- find the slopes of lines.
- interpret the slopes of lines as rates.

Learning Standard
7.RP.2b

Math Practice 4

Apply Mathematics

How can you use the graph to determine which animal has the greater speed?

Work with a partner. A cheetah and a Thomson's gazelle run at maximum speed.

a. Use the table in Activity 1 to calculate the missing distances.

| | Cheetah | Gazelle |
Time (seconds)	Distance (feet)	Distance (feet)
0		
1		
2		
3		
4		
5		
6		
7		

b. Use the table to write ordered pairs. Then plot the ordered pairs and connect the points for each animal. What do you notice about the graphs?

c. Which graph is steeper? The speed of which animal is greater?

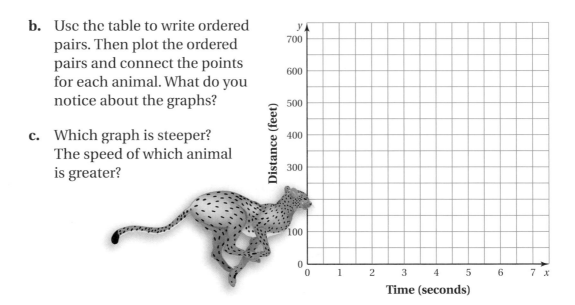

What Is Your Answer?

3. **IN YOUR OWN WORDS** How can you compare two rates graphically? Explain your reasoning. Give some examples with your answer.

4. **REPEATED REASONING** Choose 10 animals from Activity 1.

   a. Make a table for each animal similar to the table in Activity 2.

   b. Sketch a graph of the distances for each animal.

   c. Compare the steepness of the 10 graphs. What can you conclude?

Key Vocabulary
slope, *p. 194*

Study Tip

The slope of a line is the same between any two points on the line because lines have a *constant* rate of change.

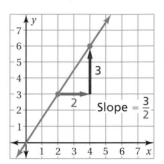

Key Idea

Slope

Slope is the rate of change between any two points on a line. It is a measure of the *steepness* of a line.

To find the slope of a line, find the ratio of the change in *y* (vertical change) to the change in *x* (horizontal change).

$$\text{slope} = \frac{\text{change in } y}{\text{change in } x}$$

$$\text{Slope} = \frac{3}{2}$$

EXAMPLE 1 Finding Slopes

Find the slope of each line.

a.

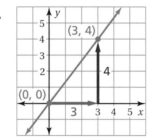

$$\text{slope} = \frac{\text{change in } y}{\text{change in } x}$$

$$= \frac{4}{3}$$

⋮⋮ The slope of the line is $\frac{4}{3}$.

b.

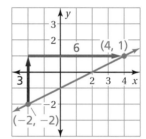

$$\text{slope} = \frac{\text{change in } y}{\text{change in } x}$$

$$= \frac{3}{6} = \frac{1}{2}$$

⋮⋮ The slope of the line is $\frac{1}{2}$.

On Your Own

Now You're Ready
Exercises 4–9

Find the slope of the line.

1.

2.

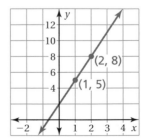

◀) Multi-Language Glossary at BigIdeasMath com

EXAMPLE 2 Interpreting a Slope

The table shows your earnings for babysitting.

a. Graph the data.

b. Find and interpret the slope of the line through the points.

Hours, x	0	2	4	6	8	10
Earnings, y (dollars)	0	10	20	30	40	50

a. Graph the data. Draw a line through the points.

b. Choose any two points to find the slope of the line.

$$\text{slope} = \frac{\text{change in } y}{\text{change in } x}$$

$$= \frac{20}{4} \quad \leftarrow \boxed{\text{dollars}}$$
$$\qquad \leftarrow \boxed{\text{hours}}$$

$$= 5$$

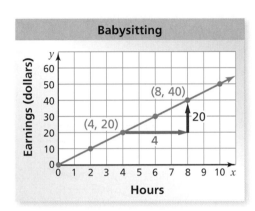

Babysitting

∴ The slope of the line represents the unit rate.
The slope is 5. So, you earn $5 per hour babysitting.

On Your Own

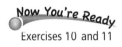

Now You're Ready
Exercises 10 and 11

3. In Example 2, use two other points to find the slope. Does the slope change?

4. The graph shows the amounts you and your friend earn babysitting.

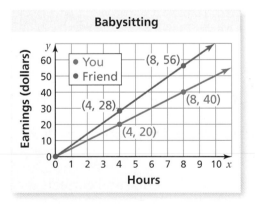

Babysitting

a. Compare the steepness of the lines. What does this mean in the context of the problem?

b. Find and interpret the slope of the blue line.

Check It Out
Help with Homework
BigIdeasMath ✓com

 Vocabulary and Concept Check

1. **VOCABULARY** Is there a connection between rate and slope? Explain.

2. **REASONING** Which line has the greatest slope?

3. **REASONING** Is it more difficult to run up a ramp with a slope of $\frac{1}{5}$ or a ramp with a slope of 5? Explain.

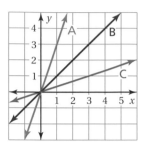

 Practice and Problem Solving

Find the slope of the line.

④ 4.
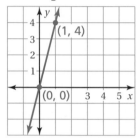

5.
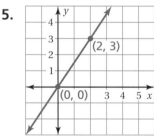

6.
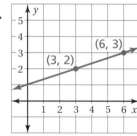

7.

8.

9.

Graph the data. Then find and interpret the slope of the line through the points.

② 10.
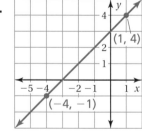

Minutes, x	3	5	7	9
Words, y	135	225	315	405

11.
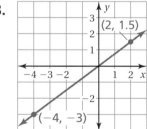

Gallons, x	5	10	15	20
Miles, y	162.5	325	487.5	650

12. **ERROR ANALYSIS** Describe and correct the error in finding the slope of the line passing through (0, 0) and (4, 5).

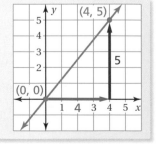

$$\text{slope} = \frac{4}{5}$$

Graph the line that passes through the two points. Then find the slope of the line.

13. $(0, 0)$, $\left(\dfrac{1}{3}, \dfrac{7}{3}\right)$

14. $\left(-\dfrac{3}{2}, -\dfrac{3}{2}\right)$, $\left(\dfrac{3}{2}, \dfrac{3}{2}\right)$

15. $\left(1, \dfrac{5}{2}\right)$, $\left(-\dfrac{1}{2}, -\dfrac{1}{4}\right)$

16. CAMPING The graph shows the amount of money you and a friend are saving for a camping trip.

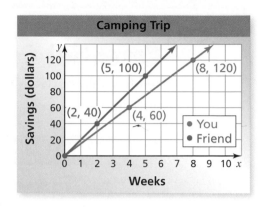

a. Compare the steepness of the lines. What does this mean in the context of the problem?

b. Find the slope of each line.

c. How much more money does your friend save each week than you?

d. The camping trip costs $165. How long will it take you to save enough money?

17. MAPS An atlas contains a map of Ohio. The table shows data from the key on the map.

Distance on Map (mm), x	10	20	30	40
Actual Distance (mi), y	25	50	75	100

a. Graph the data.

b. Find the slope of the line. What does this mean in the context of the problem?

c. The map distance between Toledo and Columbus is 48 millimeters. What is the actual distance?

d. Cincinnati is about 225 miles from Cleveland. What is the distance between these cities on the map?

18. CRITICAL THINKING What is the slope of a line that passes through the points $(2, 0)$ and $(5, 0)$? Explain.

19. A line has a slope of 2. It passes through the points $(1, 2)$ and $(3, y)$. What is the value of y?

Fair Game Review *What you learned in previous grades & lessons*

Multiply. *(Section 2.4)*

20. $-\dfrac{3}{5} \times \dfrac{8}{6}$

21. $1\dfrac{1}{2} \times \left(-\dfrac{6}{15}\right)$

22. $-2\dfrac{1}{4} \times \left(-1\dfrac{1}{3}\right)$

23. MULTIPLE CHOICE You have 18 stamps from Mexico in your stamp collection. These stamps represent $\dfrac{3}{8}$ of your collection. The rest of the stamps are from the United States. How many stamps are from the United States? *(Section 3.4)*

   A 12   **B** 24   **C** 30   **D** 48

Essential Question How can you use a graph to show the relationship between two quantities that vary directly? How can you use an equation?

1 ACTIVITY: Math in Literature

Gulliver's Travels was written by Jonathan Swift and published in 1726. Gulliver was shipwrecked on the island Lilliput, where the people were only 6 inches tall. When the Lilliputians decided to make a shirt for Gulliver, a Lilliputian tailor stated that he could determine Gulliver's measurements by simply measuring the distance around Gulliver's thumb. He said "Twice around the thumb equals once around the wrist. Twice around the wrist is once around the neck. Twice around the neck is once around the waist."

Work with a partner. Use the tailor's statement to complete the table.

COMMON CORE

Direct Variation

In this lesson, you will
- identify direct variation from graphs or equations.
- use direct variation models to solve problems.

Learning Standards
7.RP.2a
7.RP.2b
7.RP.2c
7.RP.2d

Thumb, t	Wrist, w	Neck, n	Waist, x
0 in.			
1 in.			
	4 in.		
		12 in.	
			32 in.
	10 in.		

2 ACTIVITY: Drawing a Graph

Work with a partner. Use the information from Activity 1.

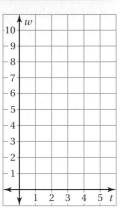

a. In your own words, describe the relationship between t and w.

b. Use the table to write the ordered pairs (t, w). Then plot the ordered pairs.

c. What do you notice about the graph of the ordered pairs?

d. Choose two points and find the slope of the line between them.

e. The quantities t and w are said to *vary directly*. An equation that describes the relationship is

$$w = \boxed{} \; t.$$

3 ACTIVITY: Drawing a Graph and Writing an Equation

Math Practice 6

Label Axes

How do you know which labels to use for the axes? Explain.

Work with a partner. Use the information from Activity 1 to draw a graph of the relationship. Write an equation that describes the relationship between the two quantities.

a. Thumb t and neck n $(n = \boxed{} \; t)$

b. Wrist w and waist x $(x = \boxed{} \; w)$

c. Wrist w and thumb t $(t = \boxed{} \; w)$

d. Waist x and wrist w $(w = \boxed{} \; x)$

What Is Your Answer?

4. IN YOUR OWN WORDS How can you use a graph to show the relationship between two quantities that vary directly? How can you use an equation?

5. STRUCTURE How are all the graphs in Activity 3 alike?

6. Give a real-life example of two variables that vary directly.

7. Work with a partner. Use string to find the distance around your thumb, wrist, and neck. Do your measurements agree with the tailor's statement in *Gulliver's Travels*? Explain your reasoning.

Practice

Use what you learned about quantities that vary directly to complete Exercises 4 and 5 on page 202.

Check It Out
Lesson Tutorials
BigIdeasMath $\checkmark$com

Key Vocabulary 🔊
direct variation,
  p. 200
constant of
  proportionality,
  p. 200

 **Key Idea**

Direct Variation

Words Two quantities x and y show **direct variation** when $y = kx$, where k is a number and $k \neq 0$. The number k is called the **constant of proportionality**.

Graph The graph of $y = kx$ is a line with a slope of k that passes through the origin. So, two quantities that show direct variation are in a proportional relationship.

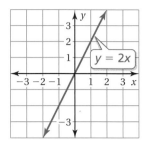

EXAMPLE **1** **Identifying Direct Variation**

Tell whether x and y show direct variation. Explain your reasoning.

a.

x	1	2	3	4
y	−2	0	2	4

b.

x	0	2	4	6
y	0	2	4	6

Study Tip

Other ways to say that x and y show direct variation are "y varies directly with x" and "x and y are directly proportional."

Plot the points. Draw a line through the points.

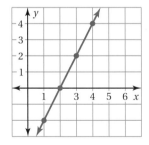

Plot the points. Draw a line through the points.

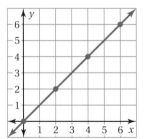

∴ The line does *not* pass through the origin. So, x and y do *not* show direct variation.

∴ The line passes through the origin. So, x and y show direct variation.

EXAMPLE **2** **Identifying Direct Variation**

Tell whether x and y show direct variation. Explain your reasoning.

a. $y + 1 = 2x$

    $y = 2x - 1$ Solve for y.

b. $\frac{1}{2}y = x$

    $y = 2x$ Solve for y.

∴ The equation *cannot* be written as $y = kx$. So, x and y do *not* show direct variation.

∴ The equation can be written as $y = kx$. So, x and y show direct variation.

🔊 Multi-Language Glossary at BigIdeasMath $\checkmark$com

On Your Own

Tell whether x and y show direct variation. Explain your reasoning.

1.

x	y
0	-2
1	1
2	4
3	7

2.

x	y
1	4
2	8
3	12
4	16

3.

x	y
-2	4
-1	2
0	0
1	2

4. $xy = 3$

5. $x = \frac{1}{3}y$

6. $y + 1 = x$

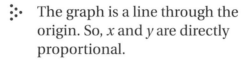

EXAMPLE 3 **Real-Life Application**

x	y
$\frac{1}{2}$	8
1	16
$\frac{3}{2}$	24
2	32

The table shows the area y (in square feet) that a robotic vacuum cleans in x minutes.

a. **Graph the data. Tell whether x and y are directly proportional.**

Graph the data. Draw a line through the points.

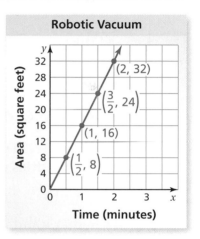

Robotic Vacuum

•• The graph is a line through the origin. So, x and y are directly proportional.

b. **Write an equation that represents the line.**

Choose any two points to find the slope of the line.

$$\text{slope} = \frac{\text{change in } y}{\text{change in } x} = \frac{16}{1} = 16$$

•• The slope of the line is the constant of proportionality, k. So, an equation of the line is $y = 16x$.

c. **Use the equation to find the area cleaned in 10 minutes.**

$y = 16x$ — Write the equation.

$= 16(10)$ — Substitute 10 for x.

$= 160$ — Multiply.

•• So, the vacuum cleans 160 square feet in 10 minutes.

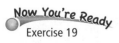

On Your Own

7. **WHAT IF?** The battery weakens and the robot begins cleaning less and less area each minute. Do x and y show direct variation? Explain.

 Vocabulary and Concept Check

1. **VOCABULARY** What does it mean for x and y to vary directly?

2. **WRITING** What point is on the graph of every direct variation equation?

3. **DIFFERENT WORDS, SAME QUESTION** Which is different? Find "both" answers.

Do x and y show direct variation?

Are x and y in a proportional relationship?

Is the graph of the relationship a line?

Does y vary directly with x?

 Practice and Problem Solving

Graph the ordered pairs in a coordinate plane. Do you think that graph shows that the quantities vary directly? Explain your reasoning.

4. $(-1, -1), (0, 0), (1, 1), (2, 2)$

5. $(-4, -2), (-2, 0), (0, 2), (2, 4)$

Tell whether x and y show direct variation. Explain your reasoning. If so, find k.

① 6.

x	1	2	3	4
y	2	4	6	8

7.

x	−2	−1	0	1
y	0	2	4	6

8.

x	−1	0	1	2
y	−2	−1	0	1

9.

x	3	6	9	12
y	2	4	6	8

② 10. $y - x = 4$

11. $x = \dfrac{2}{5}y$

12. $y + 3 = x + 6$

13. $y - 5 = 2x$

14. $x - y = 0$

15. $\dfrac{x}{y} = 2$

16. $8 = xy$

17. $x^2 = y$

18. **ERROR ANALYSIS** Describe and correct the error in telling whether x and y show direct variation.

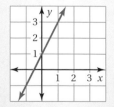
The graph is a line, so it shows direct variation.

③ 19. **RECYCLING** The table shows the profit y for recycling x pounds of aluminum. Graph the data. Tell whether x and y show direct variation. If so, write an equation that represents the line.

Aluminum (lb), x	10	20	30	40
Profit, y	$4.50	$9.00	$13.50	$18.00

The variables x and y vary directly. Use the values to find the constant of proportionality. Then write an equation that relates x and y.

20. $y = 72; x = 3$

21. $y = 20; x = 12$

22. $y = 45; x = 40$

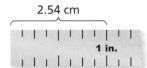
2.54 cm
1 in.

23. MEASUREMENT Write a direct variation equation that relates x inches to y centimeters.

24. MODELING Design a waterskiing ramp. Show how you can use direct variation to plan the heights of the vertical supports.

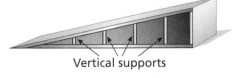
Vertical supports

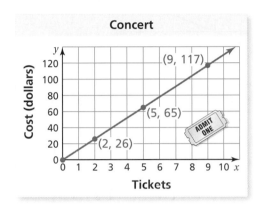
Concert

25. REASONING Use $y = kx$ to show why the graph of a proportional relationship always passes through the origin.

26. TICKETS The graph shows the cost of buying concert tickets. Tell whether x and y show direct variation. If so, find and interpret the constant of proportionality. Then write an equation and find the cost of 14 tickets.

27. CELL PHONE PLANS Tell whether x and y show direct variation. If so, write an equation of direct variation.

Minutes, x	500	700	900	1200
Cost, y	$40	$50	$60	$75

28. CHLORINE The amount of chlorine in a swimming pool varies directly with the volume of water. The pool has 2.5 milligrams of chlorine per liter of water. How much chlorine is in the pool?

29. **Critical Thinking** Is the graph of every direct variation equation a line? Does the graph of every line represent a direct variation equation? Explain your reasoning.

8000 gallons

 Fair Game Review *What you learned in previous grades & lessons*

Write the fraction as a decimal. *(Section 2.1)*

30. $\dfrac{13}{20}$

31. $\dfrac{9}{16}$

32. $\dfrac{21}{40}$

33. $\dfrac{24}{25}$

34. MULTIPLE CHOICE Which rate is *not* equivalent to 180 feet per 8 seconds? *(Section 5.1)*

(A) $\dfrac{225 \text{ ft}}{10 \text{ sec}}$

(B) $\dfrac{45 \text{ ft}}{2 \text{ sec}}$

(C) $\dfrac{135 \text{ ft}}{6 \text{ sec}}$

(D) $\dfrac{180 \text{ ft}}{1 \text{ sec}}$

Solve the proportion. *(Section 5.4)*

1. $\dfrac{7}{n} = \dfrac{42}{48}$

2. $\dfrac{x}{2} = \dfrac{40}{16}$

3. $\dfrac{3}{11} = \dfrac{27}{z}$

Find the slope of the line. *(Section 5.5)*

4.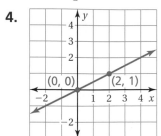

5.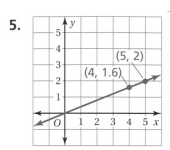

Graph the data. Then find and interpret the slope of the line through the points.
(Section 5.5)

6.

Hours, x	2	4	6	8
Miles, y	10	20	30	40

7.

Packages, x	6	10	14	18
Servings, y	9	15	21	27

Tell whether x and y show direct variation. Explain your reasoning. *(Section 5.6)*

8. $y - 9 = 6 + x$

9. $x = \dfrac{5}{8}y$

10. CONCERT A benefit concert with three performers lasts 8 hours. At this rate, how many hours is a concert with four performers? *(Section 5.4)*

11. LAWN MOWING The graph shows how much you and your friend each earn mowing lawns. *(Section 5.5)*

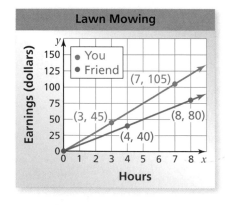

a. Compare the steepness of the lines. What does this mean in the context of the problem?

b. Find and interpret the slope of each line.

c. How much more money do you earn per hour than your friend?

12. PIE SALE The table shows the profits of a pie sale. Tell whether x and y show direct variation. If so, write the equation of direct variation. *(Section 5.6)*

Pies Sold, x	10	12	14	16
Profit, y	$79.50	$95.40	$111.30	$127.20

Check It Out
Vocabulary Help
BigIdeasMath ✓com

Review Key Vocabulary

ratio, *p. 164*	proportion, *p. 172*	direct variation, *p. 200*
rate, *p. 164*	proportional, *p. 172*	constant of proportionality,
unit rate, *p. 164*	cross products, *p. 173*	*p. 200*
complex fraction, *p. 165*	slope, *p. 194*	

Review Examples and Exercises

5.1 Ratios and Rates (pp. 162–169)

**There are 15 orangutans and 25 gorillas in a nature preserve.
One of the orangutans swings 75 feet in 15 seconds on a rope.**

a. Find the ratio of orangutans to gorillas.

b. How fast is the orangutan swinging?

a. $\dfrac{\text{orangutans}}{\text{gorillas}} = \dfrac{15}{25} = \dfrac{3}{5}$

∴ The ratio of orangutans to gorillas is $\dfrac{3}{5}$.

b. 75 feet in 15 seconds $= \dfrac{75 \text{ ft}}{15 \text{ sec}}$

$= \dfrac{75 \text{ ft} \div 15}{15 \text{ sec} \div 15}$

$= \dfrac{5 \text{ ft}}{1 \text{ sec}}$

∴ The orangutan is swinging 5 feet per second.

Exercises

Find the unit rate.

1. 289 miles on 10 gallons

2. $6\dfrac{2}{5}$ revolutions in $2\dfrac{2}{3}$ seconds

3. calories per serving

Servings	2	4	6	8
Calories	240	480	720	960

5.2 Proportions (pp. 170–177)

Tell whether the ratios $\dfrac{9}{12}$ and $\dfrac{6}{8}$ form a proportion.

$\dfrac{9}{12} = \dfrac{9 \div 3}{12 \div 3} = \dfrac{3}{4}$ ←

$\dfrac{6}{8} = \dfrac{6 \div 2}{8 \div 2} = \dfrac{3}{4}$ ←

The ratios are equivalent.

∴ So, $\dfrac{9}{12}$ and $\dfrac{6}{8}$ form a proportion.

Exercises

Tell whether the ratios form a proportion.

4. $\dfrac{4}{9}, \dfrac{2}{3}$ **5.** $\dfrac{12}{22}, \dfrac{18}{33}$ **6.** $\dfrac{8}{50}, \dfrac{4}{10}$ **7.** $\dfrac{32}{40}, \dfrac{12}{15}$

8. Use a graph to determine whether x and y are in a proportional relationship.

x	1	3	6	8
y	4	12	24	32

5.3 Writing Proportions *(pp. 178–183)*

Write a proportion that gives the number r of returns on Saturday.

	Friday	Saturday
Sales	40	85
Returns	32	r

One proportion is $\dfrac{40 \text{ sales}}{32 \text{ returns}} = \dfrac{85 \text{ sales}}{r \text{ returns}}$.

Exercises

Use the table to write a proportion.

9.

	Game 1	Game 2
Penalties	6	8
Minutes	16	m

10.

	Concert 1	Concert 2
Songs	15	18
Hours	2.5	h

5.4 Solving Proportions *(pp. 186–191)*

Solve $\dfrac{15}{2} = \dfrac{30}{y}$.

$15 \cdot y = 2 \cdot 30$ Cross Products Property

$15y = 60$ Multiply.

$y = 4$ Divide.

The solution is 4.

Exercises

Solve the proportion.

11. $\dfrac{x}{4} = \dfrac{2}{5}$ **12.** $\dfrac{5}{12} = \dfrac{y}{15}$ **13.** $\dfrac{8}{20} = \dfrac{6}{w}$ **14.** $\dfrac{s+1}{4} = \dfrac{4}{8}$

5.5 Slope (pp. 192–197)

The graph shows the number of visits your website received over the past 6 months. Find and interpret the slope.

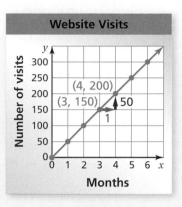

Website Visits

Choose any two points to find the slope of the line.

$$\text{slope} = \frac{\text{change in } y}{\text{change in } x}$$

$$= \frac{50}{1} \leftarrow \boxed{\text{visits}}$$
$$\phantom{= \frac{50}{1}} \leftarrow \boxed{\text{months}}$$

$$= 50$$

∴ The slope of the line represents the unit rate. The slope is 50. So, the number of visits increased by 50 each month.

Exercises

Find the slope of the line.

15.

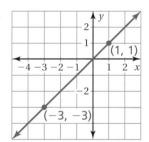

16.

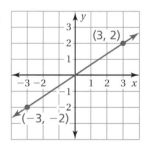

17.

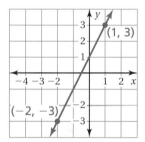

5.6 Direct Variation (pp. 198–203)

Tell whether x and y show direct variation. Explain your reasoning.

a. $x + y - 1 = 3$

$y = 4 - x$ Solve for y.

∴ The equation *cannot* be written as $y = kx$. So, x and y do *not* show direct variation.

b. $x = 8y$

$\dfrac{1}{8}x = y$ Solve for y.

∴ The equation can be written as $y = kx$. So, x and y show direct variation.

Exercises

Tell whether x and y show direct variation. Explain your reasoning.

18. $x + y = 6$ **19.** $y - x = 0$ **20.** $\dfrac{x}{y} = 20$ **21.** $x = y + 2$

Check It Out
Test Practice
BigIdeasMath ✓com

Find the unit rate.

1. 84 miles in 12 days

2. $2\frac{2}{5}$ kilometers in $3\frac{3}{4}$ minutes

Tell whether the ratios form a proportion.

3. $\frac{1}{9}, \frac{6}{54}$

4. $\frac{9}{12}, \frac{8}{72}$

Use a graph to tell whether x and y are in a proportional relationship.

5.

x	2	4	6	8
y	10	20	30	40

6.

x	1	3	5	7
y	3	7	11	15

Use the table to write a proportion.

7.

	Monday	Tuesday
Gallons	6	8
Miles	180	m

8.

	Thursday	Friday
Classes	6	c
Hours	8	4

Solve the proportion.

9. $\frac{x}{8} = \frac{9}{4}$

10. $\frac{17}{3} = \frac{y}{6}$

Graph the line that passes through the two points. Then find the slope of the line.

11. $(15, 9), (-5, -3)$

12. $(2, 9), (4, 18)$

Tell whether x and y show direct variation. Explain your reasoning.

13. $xy - 11 = 5$

14. $x = \frac{3}{y}$

15. $\frac{y}{x} = 8$

16. MOVIE TICKETS Five movie tickets cost $36.25. What is the cost of 8 movie tickets?

17. CROSSWALK The graph shows the number of cycles of a crosswalk signal during the day and during the night.

   a. Compare the steepness of the lines. What does this mean in the context of the problem?

   b. Find and interpret the slope of each line.

Don't Walk

Walk

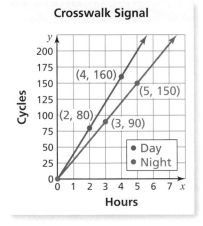

Crosswalk Signal

(4, 160)
(5, 150)
(2, 80)
(3, 90)
● Day
● Night

18. GLAZE A specific shade of green glaze requires 5 parts blue to 3 parts yellow. A glaze mixture contains 25 quarts of blue and 9 quarts of yellow. How can you fix the mixture to make the specific shade of green glaze?

1. The school store sells 4 pencils for $0.80. What is the unit cost of a pencil? *(7.RP.1)*

    A. $0.20

    B. $0.80

    C. $3.20

    D. $5.00

2. Which expressions do *not* have a value of 3? *(7.NS.3)*

    I. $2 + (-1)$

    II. $2 - (-1)$

    III. $-3 \times (-1)$

    IV. $-3 \div (-1)$

    F. I only

    G. III and IV

    H. II only

    I. I, III, and IV

3. What is the value of the expression below? *(7.NS.3)*

    $$-4 \times (-6) - (-5)$$

4. What is the slope of the line shown? *(7.RP.2b)*

    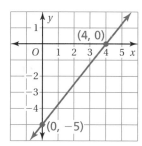

    A. $\dfrac{4}{5}$

    B. $\dfrac{5}{4}$

    C. 4

    D. 5

5. The graph below represents which inequality? *(7.EE.4b)*

    F. $-3 - 6x < -27$

    G. $2x + 6 \geq 14$

    H. $5 - 3x > -7$

    I. $2x + 3 \leq 11$

6. The quantities *x* and *y* are proportional. What is the missing value in the table? *(7.RP.2a)*

x	y
$\frac{2}{3}$	6
$\frac{4}{3}$	12
$\frac{8}{3}$	24
5	

A. 30

B. 36

C. 45

D. 48

7. You are selling tomatoes. You have already earned $16 today. How many additional pounds of tomatoes do you need to sell to earn a total of $60? *(7.EE.4a)*

F. 4

G. 11

H. 15

I. 19

$4 per pound

8. The distance traveled by the a high-speed train is proportional to the number of hours traveled. Which of the following is *not* a valid interpretation of the graph below? *(7.RP.2d)*

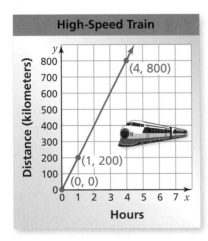

High-Speed Train

(4, 800)

(1, 200)

(0, 0)

Distance (kilometers)

Hours

A. The train travels 0 kilometers in 0 hours.

B. The unit rate is 200 kilometers per hour.

C. After 4 hours, the train is traveling 800 kilometers per hour.

D. The train travels 800 kilometers in 4 hours.

9. Regina was evaluating the expression below. What should Regina do to correct the error she made? *(7.NS.3)*

$$-\frac{3}{2} \div \left(-\frac{8}{7}\right) = -\frac{2}{3} \times \left(-\frac{7}{8}\right)$$

$$= \frac{2 \times 7}{3 \times 8}$$

$$= \frac{14}{24}$$

$$= \frac{7}{12}$$

F. Rewrite $-\frac{3}{2} \div \left(-\frac{8}{7}\right)$ as $-\frac{2}{3} \times \left(-\frac{8}{7}\right)$.

G. Rewrite $-\frac{3}{2} \div \left(-\frac{8}{7}\right)$ as $-\frac{3}{2} \times \left(-\frac{7}{8}\right)$.

H. Rewrite $-\frac{3}{2} \div \left(-\frac{8}{7}\right)$ as $-\frac{3}{7} \times \left(-\frac{8}{2}\right)$.

I. Rewrite $-\frac{2}{3} \times \left(-\frac{7}{8}\right)$ as $-\frac{2 \times 7}{3 \times 8}$.

10. What is the least value of t for which the inequality is true? *(7.EE.4b)*

$$3 - 6t \le -15$$

11. You can mow 800 square feet of lawn in 15 minutes. At this rate, how many minutes will you take to mow a lawn that measures 6000 square feet? *(7.RP.2c)*

Part A Write a proportion to represent the problem. Use m to represent the number of minutes. Explain your reasoning.

Part B Solve the proportion you wrote in Part A. Then use it to answer the problem. Show your work.

12. What value of p makes the equation below true? *(7.EE.4a)*

$$6 - 2p = -48$$

A. -27

B. -21

C. 21

D. 27

6 Percents

"Here's my sales strategy. I buy each dog bone for $0.05."

"Then I mark each one up to $1. Then, I have a 75% off sale. Cool, huh?"

"At 4 a day, I have chewed 17,536 dog biscuits. At only 99.9% pure, that means that..."

"I have swallowed seventeen and a half contaminated dog biscuits during the past twelve years."

What You Learned Before

"The fact that these two percents do not total 100 is a sad commentary on humans."

Writing Percents as Fractions
(6.RP.3C)

Example 1 Write 45% as a fraction in simplest form.

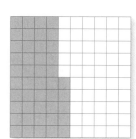

$$45\% = \frac{45}{100}$$ Write as a fraction with a denominator of 100.

$$= \frac{9}{20}$$ Simplify.

So, $45\% = \frac{9}{20}$.

Try It Yourself

Write the percent as a fraction or mixed number in simplest form.

1. 16%   **2.** 40%   **3.** 68%   **4.** 85%

5. 148%   **6.** 150%   **7.** 105%   **8.** 276%

Writing Fractions as Percents (6.RP.3C)

Example 2 Write $\frac{3}{25}$ as a percent.

$$\frac{3}{25} \overset{\times 4}{=} \frac{12}{100} = 12\%$$

Because $25 \times 4 = 100$, multiply the numerator and denominator by 4. Write the numerator with a percent symbol.

Try It Yourself

Write the fraction or mixed number as a percent.

9. $\frac{9}{25}$ **10.** $\frac{43}{50}$ **11.** $\frac{11}{20}$ **12.** $\frac{3}{5}$

13. $1\frac{1}{4}$ **14.** $1\frac{12}{25}$ **15.** $1\frac{4}{5}$ **16.** $2\frac{3}{10}$

6.1 Percents and Decimals

Essential Question

How does the decimal point move when you rewrite a percent as a decimal and when you rewrite a decimal as a percent?

1 ACTIVITY: Writing Percents as Decimals

Work with a partner. Write the percent shown by the model. Write the percent as a decimal.

a.

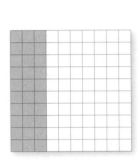

$$\boxed{}\% = \frac{\boxed{}}{\boxed{}} \leftarrow \text{per} \leftarrow \text{cent}$$

$$= \frac{\boxed{}}{\boxed{}} \qquad \text{Simplify.}$$

$$= \boxed{} \qquad \text{Write fraction as a decimal.}$$

b.

c.

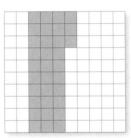

d.

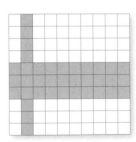

e.

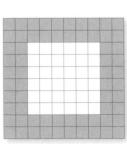

f.

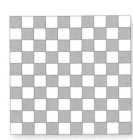

g.

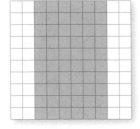

COMMON CORE

Percents and Decimals

In this lesson, you will
- write percents as decimals.
- write decimals as percents.
- solve real-life problems.

Learning Standard
7.EE.3

2 ACTIVITY: Writing Percents as Decimals

Math Practice 6

Communicate Precisely

How can reading the fraction aloud help you write it as a decimal?

Work with a partner. Write the percent as a decimal.

a. 13.5%

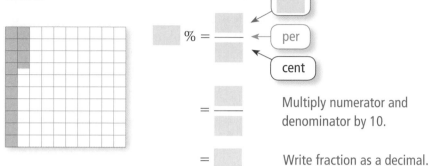

$$\boxed{}\% = \cfrac{\boxed{}}{\boxed{}} \leftarrow \text{per} \quad \leftarrow \boxed{}$$
$$\leftarrow \text{cent}$$

$$= \cfrac{\boxed{}}{\boxed{}} \quad \text{Multiply numerator and denominator by 10.}$$

$$= \boxed{} \quad \text{Write fraction as a decimal.}$$

b. 12.5%   **c.** 3.8%   **d.** 0.5%

3 ACTIVITY: Writing Decimals as Percents

Work with a partner. Draw a model to represent the decimal. Write the decimal as a percent.

a. 0.1

$$0.1 \quad = \quad 0.10 = \cfrac{\boxed{}}{\boxed{}} \quad = \quad \boxed{}\%$$

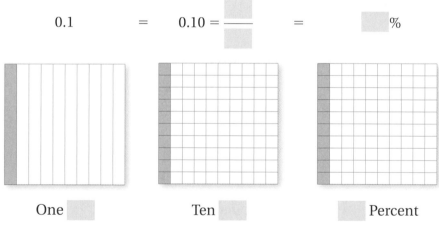

One ▢   Ten ▢   ▢ Percent

b. 0.24   **c.** 0.58   **d.** 0.05

What Is Your Answer?

4. IN YOUR OWN WORDS How does the decimal point move when you rewrite a percent as a decimal and when you rewrite a decimal as a percent?

5. Explain why the decimal point moves when you rewrite a percent as a decimal and when you rewrite a decimal as a percent.

Practice Use what you learned about percents and decimals to complete Exercises 7–12 and 19–24 on page 218.

Check It Out
Lesson Tutorials
BigIdeasMath.com

Key Idea

Writing Percents as Decimals

Words Remove the percent symbol. Then divide by 100, or just move the decimal point two places to the left.

Numbers $23\% = 23.\% = 0.23$

EXAMPLE **1** **Writing Percents as Decimals**

a. Write 52% as a decimal.

$$52\% = 52.\% = 0.52$$

b. Write 7% as a decimal.

$$7\% = 07.\% = 0.07$$

Study Tip

When moving the decimal point, you may need to place one or more zeros in the number.

Check

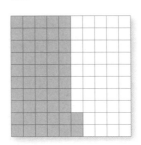

Check

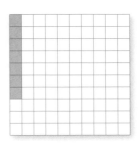

On Your Own

Now You're Ready
Exercises 7–18

Write the percent as a decimal. Use a model to check your answer.

1. 24%   **2.** 3%   **3.** 107%   **4.** 92.7%

Key Idea

Writing Decimals as Percents

Words Multiply by 100, or just move the decimal point two places to the right. Then add a percent symbol.

Numbers $0.36 = 0.36 = 36\%$

EXAMPLE **2** **Writing Decimals as Percents**

a. Write 0.47 as a percent.

$$0.47 = 0.47 = 47\%$$

b. Write 0.663 as a percent.

$$0.663 = 0.663 = 66.3\%$$

c. Write 1.8 as a percent.

$$1.8 = 1.80 = 180\%$$

d. Write 0.009 as a percent.

$$0.009 = 0.009 = 0.9\%$$

On Your Own

Write the decimal as a percent. Use a model to check your answer.

5. 0.94   **6.** 1.2   **7.** 0.316   **8.** 0.005

EXAMPLE 3   **Writing a Fraction as a Percent and a Decimal**

On a math test, you get 92 out of a possible 100 points. Which of the
following is *not* another way of expressing 92 out of 100?

Ⓐ $\dfrac{23}{25}$ **Ⓑ** 92%   **Ⓒ** $\dfrac{17}{20}$ **Ⓓ** 0.92

$$92 \text{ out of } 100 = \frac{92}{100} \begin{cases} = 92\% & \text{Eliminate Choice B.} \\ = \dfrac{23}{25} & \text{Eliminate Choice A.} \\ = 0.92 & \text{Eliminate Choice D.} \end{cases}$$

⋰• So, the correct answer is **Ⓒ**.

EXAMPLE 4   **Real-Life Application**

The figure shows the portions of ultraviolet (UV) rays reflected by
four different surfaces. How many times more UV rays are reflected
by water than by sea foam?

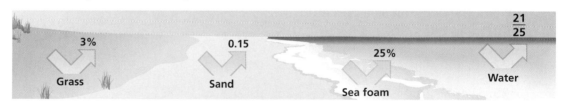

Write 25% and $\dfrac{21}{25}$ as decimals.

Sea foam: $25\% = 25.\% = 0.25$ **Water:** $\dfrac{21}{25} = \dfrac{84}{100} = 0.84$

Divide 0.84 by 0.25: $0.25\overline{)0.84} \longrightarrow 25\overline{)84.00}$ $\overset{3.36}{}$

⋰• So, water reflects about 3.4 times more UV rays than sea foam.

On Your Own

9. Write "18 out of 100" as a percent, a fraction, and a decimal.

10. In Example 4, how many times more UV rays are reflected
by water than by sand?

Check It Out
Help with Homework
BigIdeasMath .com

 Vocabulary and Concept Check

MATCHING Match the decimal with its equivalent percent.

1. 0.42   **2.** 4.02   **3.** 0.042   **4.** 0.0402

   A. 4.02%   **B.** 42%   **C.** 4.2%   **D.** 402%

5. OPEN-ENDED Write three different decimals that are between 10% and 20%.

6. WHICH ONE DOESN'T BELONG? Which one does *not* belong with the other three? Explain your reasoning.

| 70% | 0.7 | $\dfrac{7}{10}$ | 0.07 |

 Practice and Problem Solving

Write the percent as a decimal.

7. 78%   **8.** 55%   **9.** 18.5%

10. 57.4%   **11.** 33%   **12.** 9%

13. 47.63%   **14.** 91.25%   **15.** 166%

16. 217%   **17.** 0.06%   **18.** 0.034%

Write the decimal as a percent.

19. 0.74   **20.** 0.52   **21.** 0.89

22. 0.768   **23.** 0.99   **24.** 0.49

25. 0.487   **26.** 0.128   **27.** 3.68

28. 5.12   **29.** 0.0371   **30.** 0.0046

31. ERROR ANALYSIS Describe and correct the error in writing 0.86 as a percent.

> ✗ 0.86 = 00.86 = 0.0086%

32. MUSIC Thirty-six percent of the songs on your MP3 player are pop songs. Write this percent as a decimal.

33. CAT About 0.34 of the length of a cat is its tail. Write this decimal as a percent.

34. COMPUTER Write the percent of free space on the computer as a decimal.

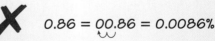

Volume	Capacity	Free Space	% Free Space
(C:)	149 GB	133 GB	89 %

Write the percent as a fraction in simplest form and as a decimal.

35. 36%   **36.** 23.5%   **37.** 16.24%

38. SCHOOL The percents of students who travel to school by car, bus, and bicycle are shown for a school of 825 students.

Car: 20%   School bus: 48%   Bicycle: 8%

   a. Write the percents as decimals.

   b. Write the percents as fractions.

   c. What percent of students use another method to travel to school?

   d. **RESEARCH** Make a bar graph that represents how the students in your class travel to school.

39. ELECTIONS In an election, the winning candidate receives 60% of the votes. What percent of the votes does the other candidate receive?

40. COLORS Students in a class were asked to tell their favorite color.

   a. What percent said red, blue, or yellow?

   b. How many times more students said red than yellow?

   c. Use two methods to find the percent of students who said green. Which method do you prefer?

Favorite Color

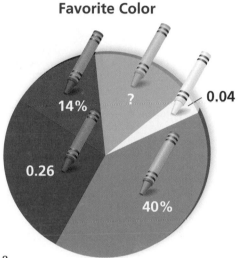

41. *Problem Solving* In the first 42 Super Bowls, $0.1\overline{6}$ of the MVPs (most valuable players) were running backs.

   a. What percent of the MVPs were running backs?

   b. What fraction of the MVPs were *not* running backs?

 Fair Game Review What you learned in previous grades & lessons

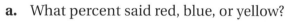

Write the decimal as a fraction or mixed number in simplest form.
(Skills Review Handbook)

42. 0.46   **43.** 0.31   **44.** 2.2   **45.** 4.32

Simplify the expression. *(Section 3.1)*

46. $4x + 3 - 9x$ **47.** $5 + 3.2n - 6 - 4.8n$

48. $2y - 5(y - 3)$ **49.** $-\dfrac{1}{2}(8b + 3) + 3b$

50. MULTIPLE CHOICE Ham costs $4.48 per pound. Cheese costs $6.36 per pound. You buy 1.5 pounds of ham and 0.75 pound of cheese. How much more do you pay for the ham? *(Skills Review Handbook)*

   Ⓐ $1.41   Ⓑ $1.95   Ⓒ $4.77   Ⓓ $6.18

Comparing and Ordering Fractions, Decimals, and Percents

Essential Question How can you order numbers that are written as fractions, decimals, and percents?

1 ACTIVITY: Using Fractions, Decimals, and Percents

Work with a partner. Decide which number form (fraction, decimal, or percent) is more common. Then find which is greater.

a. 7% sales tax   or $\dfrac{1}{20}$ sales tax

b. 0.37 cup of flour   or $\dfrac{1}{3}$ cup of flour

c. $\dfrac{5}{8}$-inch wrench   or   0.375-inch wrench

d. $12\dfrac{3}{5}$ dollars   or   12.56 dollars

e. 93% test score   or $\dfrac{7}{8}$ test score

f. $5\dfrac{5}{6}$ fluid ounces   or   5.6 fluid ounces

COMMON CORE

Fractions, Decimals, and Percents

In this lesson, you will
• compare and order fractions, decimals, and percents.
• solve real-life problems.

Learning Standard
7.EE.3

2 ACTIVITY: Ordering Numbers

Work with a partner to order the following numbers.

$$\dfrac{1}{8} \qquad 11\% \qquad \dfrac{3}{20} \qquad 0.172 \qquad 0.32 \qquad 43\% \qquad 7\% \qquad 0.7 \qquad \dfrac{5}{6}$$

a. Decide on a strategy for ordering the numbers. Will you write them all as fractions, decimals, or percents?

b. Use your strategy and a number line to order the numbers from least to greatest. (Note: Label the number line appropriately.)

Math
Practice **2**

Make Sense of Quantities
What strategies can you use to determine which number is greater?

Preparation:

- Cut index cards to make 40 playing cards.
- Write each number in the table onto a card.

To Play:

- Play with a partner.
- Deal 20 cards facedown to each player.
- Each player turns one card faceup. The player with the greater number wins. The winner collects both cards and places them at the bottom of his or her cards.
- Suppose there is a tie. Each player lays three cards facedown, then a new card faceup. The player with the greater of these new cards wins. The winner collects all 10 cards and places them at the bottom of his or her cards.
- Continue playing until one player has all the cards. This player wins the game.

75%	$\frac{3}{4}$	$\frac{1}{3}$	$\frac{3}{10}$	0.3	25%	0.4	0.25	100%	0.27
0.75	$66\frac{2}{3}\%$	12.5%	40%	$\frac{1}{4}$	4%	0.5%	0.04	$\frac{1}{100}$	$\frac{2}{3}$
0	30%	5%	$\frac{27}{100}$	0.05	$33\frac{1}{3}\%$	$\frac{2}{5}$	0.333...	27%	1%
1	0.01	$\frac{1}{20}$	$\frac{1}{8}$	0.125	$\frac{1}{25}$	$\frac{1}{200}$	0.005	0.666...	0%

What Is Your Answer?

4. **IN YOUR OWN WORDS** How can you order numbers that are written as fractions, decimals, and percents? Give an example with your answer.

5. All but one of the U.S. coins shown has a name that is related to its value. Which one is it? How are the names of the others related to their values?

Practice ▶   Use what you learned about ordering numbers to complete Exercises 4–7, 16, and 17 on page 224.

6.2 Lesson

Check It Out
Lesson Tutorials
BigIdeasMath ✓com

When comparing and ordering fractions, decimals, and percents, write the numbers as all fractions, all decimals, or all percents.

EXAMPLE 1 **Comparing Fractions, Decimals, and Percents**

a. Which is greater, $\frac{3}{20}$ or 16%?

Study Tip

It is usually easier to order decimals or percents than to order fractions.

Write $\frac{3}{20}$ as a percent:
$$\frac{3}{20} \xrightarrow{\times 5} \frac{15}{100} = 15\%$$
$\xrightarrow{\times 5}$

⋮ 15% is less than 16%. So, 16% is the greater number.

b. Which is greater, 79% or 0.08?

Write 79% as a decimal: $79\% = 79.\% = 0.79$

⋮ 0.79 is greater than 0.08. So, 79% is the greater number.

⬤ On Your Own

Now You're Ready
Exercises 4–15

1. Which is greater, 25% or $\frac{7}{25}$?   **2.** Which is greater, 0.49 or 94%?

EXAMPLE 2 **Real-Life Application**

You, your sister, and a friend each take the same number of shots at a soccer goal. You make 72% of your shots, your sister makes $\frac{19}{25}$ of her shots, and your friend makes 0.67 of his shots. Who made the fewest shots?

Remember

To order numbers from least to greatest, write them as they appear on a number line from left to right.

Write 72% and $\frac{19}{25}$ as decimals.

You: $72\% = 72.\% = 0.72$ **Sister:** $\frac{19}{25} \xrightarrow{\times 4} \frac{76}{100} = 0.76$
$\xrightarrow{\times 4}$

Graph the decimals on a number line.

Friend: 0.67   You: 72% = 0.72   Sister: $\frac{19}{25}$ = 0.76

0.66 0.68 0.70 0.72 0.74 0.76 0.78

⋮ 0.67 is the least number. So, your friend made the fewest shots.

On Your Own

3. You make 75% of your shots, your sister makes $\frac{13}{20}$ of her shots, and your friend makes 0.7 of his shots. Who made the most shots?

EXAMPLE **3** **Real-Life Application**

Washington: $\frac{1}{50}$ Michigan: 0.03

New York: 6%

California: 0.12

Ohio: $\frac{1}{25}$

The map shows the portions of the U.S. population that live in five states.

List the five states in order by population from least to greatest.

Begin by writing each portion as a fraction, a decimal, and a percent.

State	Fraction	Decimal	Percent
Michigan	$\frac{3}{100}$	0.03	3%
New York	$\frac{6}{100}$	0.06	6%
Washington	$\frac{1}{50}$	0.02	2%
California	$\frac{12}{100}$	0.12	12%
Ohio	$\frac{1}{25}$	0.04	4%

Graph the percent for each state on a number line.

Michigan: 3%   New York: 6%

Washington: 2%   Ohio: 4%   California: 12%

0%   2%   4%   6%   8%   10%   12%   14%

∴ The states in order by population from least to greatest are Washington, Michigan, Ohio, New York, and California.

On Your Own

4. The portion of the U.S. population that lives in Texas is $\frac{2}{25}$. The portion that lives in Illinois is 0.042. Reorder the states in Example 3 including Texas and Illinois.

Section 6.2 Comparing and Ordering Fractions, Decimals, and Percents **223**

 ## Vocabulary and Concept Check

1. **NUMBER SENSE** Copy and complete the table.

2. **NUMBER SENSE** How would you decide whether $\frac{3}{5}$ or 59% is greater? Explain.

3. **WHICH ONE DOESN'T BELONG?** Which one does *not* belong with the other three? Explain your reasoning.

40%	$\frac{2}{5}$
0.4	0.04

Fraction	Decimal	Percent
$\frac{18}{25}$	0.72	
$\frac{17}{20}$		85%
$\frac{13}{50}$		
	0.62	
		45%

 ## Practice and Problem Solving

Tell which number is greater.

① 4. 0.9, 95%
5. 20%, 0.02
6. $\frac{37}{50}$, 37%
7. 50%, $\frac{13}{25}$

8. 0.086, 86%
9. 76%, 0.67
10. 60%, $\frac{5}{8}$
11. 0.12, 1.2%

12. 17%, $\frac{4}{25}$
13. 140%, 0.14
14. $\frac{1}{3}$, 30%
15. 80%, $\frac{7}{9}$

Use a number line to order the numbers from least to greatest.

② 16. 38%, $\frac{8}{25}$, 0.41
17. 68%, 0.63, $\frac{13}{20}$

18. $\frac{43}{50}$, 0.91, $\frac{7}{8}$, 84%
19. 0.15%, $\frac{3}{20}$, 0.015

20. 2.62, $2\frac{2}{5}$, 26.8%, 2.26, 271%
21. $\frac{87}{200}$, 0.44, 43.7%, $\frac{21}{50}$

22. **TEST** You answered 21 out of 25 questions correctly on a test. Did you reach your goal of getting at least 80%?

23. **POPULATION** The table shows the portions of the world population that live in four countries. Order the countries by population from least to greatest.

Country	Brazil	India	Russia	United States
Portion of World Population	2.8%	$\frac{7}{40}$	$\frac{1}{50}$	0.044

PRECISION Order the numbers from least to greatest.

24. 66.1%, 0.66, $\dfrac{2}{3}$, 0.667

25. $\dfrac{2}{9}$, 21%, $0.2\overline{1}$, $\dfrac{11}{50}$

Tell which letter shows the graph of the number.

26. $\dfrac{2}{5}$ **27.** 45.2% **28.** 0.435 **29.** $\dfrac{4}{9}$

30. TOUR DE FRANCE The Tour de France is a bicycle road race. The whole race is made up of 21 small races called *stages*. The table shows how several stages compare to the whole Tour de France in a recent year. Order the stages from shortest to longest.

Stage	1	7	8	17	21
Portion of Total Distance	$\dfrac{11}{200}$	0.044	$\dfrac{6}{125}$	0.06	4%

31. SLEEP The table shows the portions of the day that several animals sleep.

  a. Order the animals by sleep time from least to greatest.

  b. Estimate the portion of the day that you sleep.

  c. Where do you fit on the ordered list?

Animal	Portion of Day Sleeping
Dolphin	0.433
Lion	56.3%
Rabbit	$\dfrac{19}{40}$
Squirrel	$\dfrac{31}{50}$
Tiger	65.8%

32. **Number Sense** Tell what whole number you can substitute for *a* in each list so the numbers are ordered from least to greatest. If there is none, explain why.

  a. $\dfrac{2}{a}$, $\dfrac{a}{22}$, 33%

  b. $\dfrac{1}{a}$, $\dfrac{a}{8}$, 33%

Fair Game Review What you learned in previous grades & lessons

Tell whether the ratios form a proportion. *(Section 5.2)*

33. $\dfrac{6}{10}$, $\dfrac{9}{15}$

34. $\dfrac{7}{16}$, $\dfrac{28}{80}$

35. $\dfrac{20}{12}$, $\dfrac{35}{21}$

36. MULTIPLE CHOICE What is the solution of $2n - 4 > -12$? *(Section 4.4)*

  Ⓐ $n < -10$ **Ⓑ** $n < -4$ **Ⓒ** $n > -2$ **Ⓓ** $n > -4$

Essential Question How can you use models to estimate percent questions?

The statement "25% of 12 is 3" has three numbers. In real-life problems, any one of these numbers can be unknown.

Question	Which number is missing?	Type of Question
What is 25% of 12?	3	Find a part of a number.
3 is what percent of 12?	25%	Find a percent.
3 is 25% of what?	12	Find the whole.

1 ACTIVITY: Estimating a Part

Work with a partner. Use a model to estimate the answer to each question.

a. What number is 50% of 30?

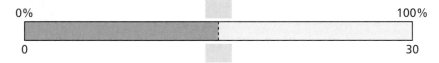

⋮⋮ So, from the model, ▢ is 50% of 30.

b. What number is 75% of 30?   **c.** What number is 40% of 30?

d. What number is 6% of 30?   **e.** What number is 65% of 30?

2 ACTIVITY: Estimating a Percent

COMMON CORE

Percent Proportion

In this lesson, you will
• use the percent proportion to find parts, wholes, and percents.

Learning Standard
7.RP.3

Work with a partner. Use a model to estimate the answer to each question.

a. 15 is what percent of 75?

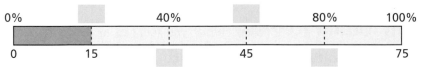

⋮⋮ So, from the model, 15 is ▢ of 75.

b. 5 is what percent of 20?   **c.** 18 is what percent of 40?

d. 50 is what percent of 80?   **e.** 75 is what percent of 50?

3   ACTIVITY: Estimating a Whole

Math Practice 4

Use a Model

What quantities are given? How can you use the model to find the unknown quantity?

Work with a partner. Use a model to estimate the answer to each question.

a. 24 is $33\frac{1}{3}$% of what number?

| 0% | | $33\frac{1}{3}$% | | $66\frac{2}{3}$% | | 100% |

| 0 | | 24 | | | | |

∴ So, from the model, 24 is $33\frac{1}{3}$% of ▢.

b. 13 is 25% of what number?   **c.** 110 is 20% of what number?

d. 75 is 75% of what number?   **e.** 81 is 45% of what number?

4   ACTIVITY: Using Ratio Tables

Work with a partner. Use a ratio table to answer each question. Then compare your answer to the estimate you found using the model.

1d a. What number is 6% of 30?

Part	6		
Whole	100		30

1e b. What number is 65% of 30?

Part	65		
Whole	100		30

2c c. 18 is what percent of 40?

Part	18		
Whole	40		100

3e d. 81 is 45% of what number?

Part	45		81
Whole	100		

What Is Your Answer?

5. IN YOUR OWN WORDS How can you use models to estimate percent questions? Give examples to support your answer.

6. Complete the proportion below using the given labels.

percent
whole
100
part

$$\frac{}{} = \frac{}{}$$

Practice

Use what you learned about estimating percent questions to complete Exercises 5–10 on page 230.

 **Key Idea**

The Percent Proportion

Words You can represent "a is p percent of w" with the proportion

$$\frac{a}{w} = \frac{p}{100}$$

where a is part of the whole w, and $p\%$, or $\dfrac{p}{100}$, is the percent.

Study Tip

In percent problems, the word *of* is usually followed by the whole.

Numbers 3 out of 4 is 75%.

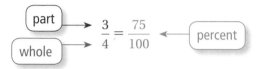

part $\longrightarrow$ $\dfrac{3}{4} = \dfrac{75}{100}$ $\longleftarrow$ percent

whole $\longrightarrow$

EXAMPLE **1** **Finding a Percent**

What percent of 15 is 12?

$\dfrac{a}{w} = \dfrac{p}{100}$	Write the percent proportion.
$\dfrac{12}{15} = \dfrac{p}{100}$	Substitute 12 for a and 15 for w.
$100 \cdot \dfrac{12}{15} = 100 \cdot \dfrac{p}{100}$	Multiplication Property of Equality
$80 = p$	Simplify.

So, 80% of 15 is 12.

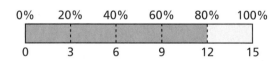

EXAMPLE **2** **Finding a Part**

What number is 36% of 50?

$\dfrac{a}{w} = \dfrac{p}{100}$	Write the percent proportion.
$\dfrac{a}{50} = \dfrac{36}{100}$	Substitute 50 for w and 36 for p.
$50 \cdot \dfrac{a}{50} = 50 \cdot \dfrac{36}{100}$	Multiplication Property of Equality
$a = 18$	Simplify.

So, 18 is 36% of 50.

EXAMPLE **3**   **Finding a Whole**

150% of what number is 24?

$$\frac{a}{w} = \frac{p}{100}$$ Write the percent proportion.

$$\frac{24}{w} = \frac{150}{100}$$ Substitute 24 for a and 150 for p.

$24 \cdot 100 = w \cdot 150$ Cross Products Property

$2400 = 150w$ Multiply.

$16 = w$ Divide each side by 150.

:• So, 150% of 16 is 24.

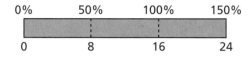

On Your Own

Now You're Ready
Exercises 11–18

Write and solve a proportion to answer the question.

1. What percent of 5 is 3?

2. 25 is what percent of 20?

3. What number is 80% of 60?

4. 10% of 40.5 is what number?

5. 0.1% of what number is 4?

6. $\frac{1}{2}$ is 25% of what number?

EXAMPLE **4**   **Real-Life Application**

2011 Alabama Tornadoes

Number of tornadoes

70
60 58
50
40 36
30   29
20   13
10   7
0   2
  EF0 EF1 EF2 EF3 EF4 EF5
      Strength

The bar graph shows the strengths of tornadoes that occurred in Alabama in 2011. What percent of the tornadoes were EF1s?

The total number of tornadoes, 145, is the *whole*, and the number of EF1 tornadoes, 58, is the *part*.

$$\frac{a}{w} = \frac{p}{100}$$ Write the percent proportion.

$$\frac{58}{145} = \frac{p}{100}$$ Substitute 58 for a and 145 for w.

$$100 \cdot \frac{58}{145} = 100 \cdot \frac{p}{100}$$ Multiplication Property of Equality

$$40 = p$$ Simplify.

:• So, 40% of the tornadoes were EF1s.

On Your Own

7. Twenty percent of the tornadoes occurred in central Alabama on April 27. How many tornadoes does this represent?

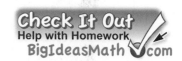

 Vocabulary and Concept Check

1. **VOCABULARY** Write the percent proportion in words.

2. **WRITING** Explain how to use a proportion to find 30% of a number.

3. **NUMBER SENSE** Write and solve the percent proportion represented by the model.

0%	20%	40%	60%	80%	100%

0   40

4. **WHICH ONE DOESN'T BELONG?** Which proportion does *not* belong with the other three? Explain your reasoning.

$$\frac{15}{w} = \frac{50}{100}$$ $$\frac{12}{15} = \frac{40}{n}$$ $$\frac{15}{25} = \frac{p}{100}$$ $$\frac{a}{20} = \frac{35}{100}$$

 Practice and Problem Solving

Use a model to estimate the answer to the question. Use a ratio table to check your answer.

5. What number is 24% of 80?

6. 15 is what percent of 40?

7. 15 is 30% of what number?

8. What number is 120% of 70?

9. 20 is what percent of 52?

10. 48 is 75% of what number?

Write and solve a proportion to answer the question.

11. What percent of 25 is 12?

12. 14 is what percent of 56?

13. 25% of what number is 9?

14. 36 is 0.9% of what number?

15. 75% of 124 is what number?

16. 110% of 90 is what number?

17. What number is 0.4% of 40?

18. 72 is what percent of 45?

$$\times \quad \frac{a}{w} = \frac{p}{100}$$
$$\frac{a}{34} = \frac{40}{100}$$
$$a = 13.6$$

19. **ERROR ANALYSIS** Describe and correct the error in using the percent proportion to answer the question below.

"40% of what number is 34?"

20. **FITNESS** Of 140 seventh-grade students, 15% earn the Presidential Physical Fitness Award. How many students earn the award?

21. **COMMISSION** A salesperson receives a 3% commission on sales. The salesperson receives $180 in commission. What is the amount of sales?

Write and solve a proportion to answer the question.

22. 0.5 is what percent of 20?

23. 14.2 is 35.5% of what number?

24. $\frac{3}{4}$ is 60% of what number?

25. What number is 25% of $\frac{7}{8}$?

26. HOMEWORK You are assigned 32 math exercises for homework. You complete 87.5% of these before dinner. How many do you have left to do after dinner?

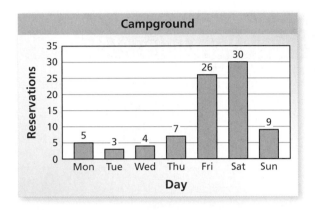

27. HOURLY WAGE Your friend earns $10.50 per hour. This is 125% of her hourly wage last year. How much did your friend earn per hour last year?

28. CAMPSITE The bar graph shows the numbers of reserved campsites at a campground for one week. What percent of the reservations were for Friday or Saturday?

29. PROBLEM SOLVING A classmate displays the results of a class president election in the bar graph shown.

a. What is missing from the bar graph?

b. What percent of the votes does the last-place candidate receive? Explain your reasoning.

c. There are 124 votes total. How many votes does Chloe receive?

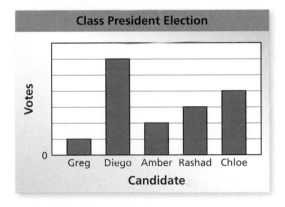

30. REASONING 20% of a number is x. What is 100% of the number? Assume $x > 0$.

31. Structure Answer each question. Assume $x > 0$.

a. What percent of $8x$ is $5x$?

b. What is 65% of $80x$?

 Fair Game Review *What you learned in previous grades & lessons*

Evaluate the expression when $a = -15$ and $b = -5$. *(Section 1.5)*

32. $a \div b$

33. $\dfrac{b + 14}{a}$

34. $\dfrac{b^2}{a + 5}$

35. MULTIPLE CHOICE What is the solution of $9x = -1.8$? *(Section 3.4)*

Ⓐ $x = -5$ Ⓑ $x = -0.2$ Ⓒ $x = 0.2$ Ⓓ $x = 5$

6.4 The Percent Equation

Essential Question How can you use an equivalent form of the percent proportion to solve a percent problem?

1 ACTIVITY: Solving Percent Problems Using Different Methods

Work with a partner. The circle graph shows the number of votes received by each candidate during a school election. So far, only half the students have voted.

a. Complete the table.

Candidate	Number of votes received / Total number of votes
Sue	
Miguel	
Leon	
Hong	

Votes Received by Each Candidate

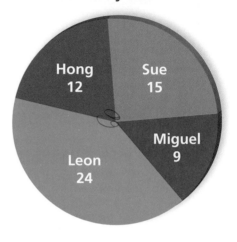

b. Find the percent of students who voted for each candidate. Explain the method you used to find your answers.

c. Compare the method you used in part (b) with the methods used by other students in your class. Which method do you prefer? Explain.

2 ACTIVITY: Finding Parts Using Different Methods

COMMON CORE

Percent Equation
In this lesson, you will
• use the percent equation to find parts, wholes, and percents.
• solve real-life problems.
Learning Standards
7.RP.3
7.EE.3

Work with a partner. The circle graph shows the final results of the election.

a. Find the number of students who voted for each candidate. Explain the method you used to find your answers.

b. Compare the method you used in part (a) with the methods used by other students in your class. Which method do you prefer? Explain.

Final Results

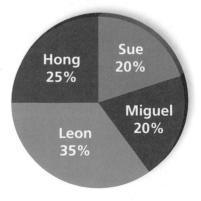

ACTIVITY: Deriving the Percent Equation

Work with a partner. In Section 6.3, you used the percent proportion to find the missing percent, part, or whole. You can also use the *percent equation* to find these missing values.

  a. Complete the steps below to find the percent equation.

$$\frac{\text{part}}{\text{whole}} = \text{percent}$$ Definition of percent

$$\frac{\text{part}}{\text{whole}} \cdot \boxed{} = \boxed{} \cdot \boxed{}$$ Multiply each side by the $\boxed{}$.

$$\text{part} = \boxed{} \cdot \boxed{}$$ Divide out common factors.
This is the percent equation.

  b. Use the percent equation to find the number of students who voted for each candidate in Activity 2. How does this method compare to the percent proportion?

4 **ACTIVITY: Identifying Different Equations**

Work with a partner. Without doing any calculations, choose the equation that you cannot use to answer each question.

Math Practice 3

Justify Conclusions

How can you justify the equations that you chose?

  a. What number is 55% of 80?

$$a = 0.55 \cdot 80 \qquad a = \frac{11}{20} \cdot 80 \qquad 80a = 0.55 \qquad \frac{a}{80} = \frac{55}{100}$$

  b. 24 is 60% of what number?

$$\frac{24}{w} = \frac{60}{100} \qquad 24 = 0.6 \cdot w \qquad \frac{24}{60} = w \qquad 24 = \frac{3}{5} \cdot w$$

What Is Your Answer?

  5. **IN YOUR OWN WORDS** How can you use an equivalent form of the percent proportion to solve a percent problem?

  6. Write a percent proportion and a percent equation that you can use to answer the question below.

16 is what percent of 250?

Practice → Use what you learned about solving percent problems to complete Exercises 4–9 on page 236.

Key Idea

The Percent Equation

Words   To represent "*a* is *p* percent of *w*," use an equation.

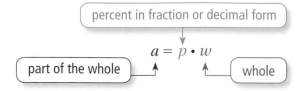

$$a = p \cdot w$$

Numbers $15 = 0.5 \cdot 30$

EXAMPLE 1   **Finding a Part of a Number**

What number is 24% of 50?   **Estimate**

Common Error

Remember to convert a percent to a fraction or a decimal before using the percent equation. For Example 1, write 24% as $\frac{24}{100}$.

$a = p \cdot w$ Write percent equation.

$= \frac{24}{100} \cdot 50$ Substitute $\frac{24}{100}$ for *p* and 50 for *w*.

$= 12$ Simplify.

∴ So, 12 is 24% of 50.   **Reasonable?** $12 \approx 12.5$ ✓

EXAMPLE 2   **Finding a Percent**

9.5 is what percent of 25?   **Estimate**

$a = p \cdot w$ Write percent equation.

$9.5 = p \cdot 25$ Substitute 9.5 for *a* and 25 for *w*.

$\frac{9.5}{25} = \frac{p \cdot 25}{25}$ Division Property of Equality

$0.38 = p$ Simplify.

∴ Because 0.38 equals 38%,   **Reasonable?** $38\% \approx 40\%$ ✓
9.5 is 38% of 25.

EXAMPLE **3** **Finding a Whole**

39 is 52% of what number?   **Estimate**

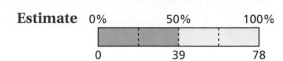

$a = p \cdot w$	Write percent equation.
$39 = 0.52 \cdot w$	Substitute 39 for a and 0.52 for p.
$75 = w$	Divide each side by 0.52.

So, 39 is 52% of 75.   **Reasonable?** $75 \approx 78$ ✓

On Your Own

Now You're Ready
Exercises 10–17

Write and solve an equation to answer the question.

1. What number is 10% of 20?

2. What number is 150% of 40?

3. 3 is what percent of 600?

4. 18 is what percent of 20?

5. 8 is 80% of what number?

6. 90 is 18% of what number?

EXAMPLE **4** **Real-Life Application**

8th Street Cafe

DATE: MAY04'13 05:45PM
TABLE: 29
SERVER: JANE

Food Total   27.50
Tax   1.65
Subtotal   29.15

TIP: _____

TOTAL: _____

Thank You

a. Find the percent of sales tax on the food total.

Answer the question: $1.65 is what percent of $27.50?

$a = p \cdot w$	Write percent equation.
$1.65 = p \cdot 27.50$	Substitute 1.65 for a and 27.50 for w.
$0.06 = p$	Divide each side by 27.50.

Because 0.06 equals 6%, the percent of sales tax is 6%.

b. Find the amount of a 16% tip on the food total.

Answer the question: What tip amount is 16% of $27.50?

$a = p \cdot w$	Write percent equation.
$= 0.16 \cdot 27.50$	Substitute 0.16 for p and 27.50 for w.
$= 4.40$	Multiply.

So, the amount of the tip is $4.40.

On Your Own

7. **WHAT IF?** Find the amount of a 20% tip on the food total.

 Vocabulary and Concept Check

1. **VOCABULARY** Write the percent equation in words.

2. **REASONING** A number n is 150% of number m. Is n *greater than*, *less than*, or *equal to m*? Explain your reasoning.

3. **DIFFERENT WORDS, SAME QUESTION** Which is different? Find "both" answers.

What number is 20% of 55?	55 is 20% of what number?
20% of 55 is what number?	0.2 • 55 is what number?

 Practice and Problem Solving

Answer the question. Explain the method you chose.

4. What number is 24% of 80?

5. 15 is what percent of 40?

6. 15 is 30% of what number?

7. What number is 120% of 70?

8. 20 is what percent of 52?

9. 48 is 75% of what number?

Write and solve an equation to answer the question.

①
10. 20% of 150 is what number?

11. 45 is what percent of 60?

②
12. 35% of what number is 35?

13. 0.8% of 150 is what number?

③
14. 29 is what percent of 20?

15. 0.5% of what number is 12?

16. What percent of 300 is 51?

17. 120% of what number is 102?

ERROR ANALYSIS Describe and correct the error in using the percent equation.

18. What number is 35% of 20?

$$\times \quad a = p \cdot w$$
$$= 35 \cdot 20$$
$$= 700$$

19. 30 is 60% of what number?

$$\times \quad a = p \cdot w$$
$$= 0.6 \cdot 30$$
$$= 18$$

20. **COMMISSION** A salesperson receives a 2.5% commission on sales. What commission does the salesperson receive for $8000 in sales?

21. **FUNDRAISING** Your school raised 125% of its fundraising goal. The school raised $6750. What was the goal?

SALE
$240

22. **SURFBOARD** The sales tax on a surfboard is $12. What is the percent of sales tax?

PUZZLE There were w signers of the Declaration of Independence. The youngest was Edward Rutledge, who was x years old. The oldest was Benjamin Franklin, who was y years old.

23. x is 25% of 104. What was Rutledge's age?

24. 7 is 10% of y. What was Franklin's age?

25. w is 80% of y. How many signers were there?

26. y is what percent of $(w + y - x)$?

Favorite Sport

Other

40.0%

37.5%

27. LOGIC How can you tell whether the percent of a number will be *greater than*, *less than*, or *equal to* the number? Give examples to support your answer.

28. SURVEY In a survey, a group of students were asked their favorite sport. Eighteen students chose "other" sports.

    a. How many students participated?

    b. How many chose football?

29. WATER TANK Water tank A has a capacity of 550 gallons and is 66% full. Water tank B is 53% full. The ratio of the capacity of Tank A to Tank B is $11:15$.

    a. How much water is in Tank A?

    b. What is the capacity of Tank B?

    c. How much water is in Tank B?

30. TRUE OR FALSE? Tell whether the statement is *true* or *false*. Explain your reasoning.

    If W is 25% of Z, then $Z:W$ is $75:25$.

31. **Reasoning** The table shows your test results for math class. What test score do you need on the last exam to earn 90% of the total points?

Test Score	Point Value
83%	100
91.6%	250
88%	150
?	300

 Fair Game Review What you learned in previous grades & lessons

Simplify. Write the answer as a decimal. *(Skills Review Handbook)*

32. $\dfrac{10 - 4}{10}$ **33.** $\dfrac{25 - 3}{25}$ **34.** $\dfrac{105 - 84}{84}$ **35.** $\dfrac{170 - 125}{125}$

36. MULTIPLE CHOICE There are 160 people in a grade. The ratio of boys to girls is 3 to 5. Which proportion can you use to find the number x of boys? *(Section 5.3)*

    Ⓐ $\dfrac{3}{8} = \dfrac{x}{160}$ **Ⓑ** $\dfrac{3}{5} = \dfrac{x}{160}$ **Ⓒ** $\dfrac{5}{8} = \dfrac{x}{160}$ **Ⓓ** $\dfrac{3}{5} = \dfrac{160}{x}$

You can use a **summary triangle** to explain a concept. Here is an example of a summary triangle for writing a percent as a decimal.

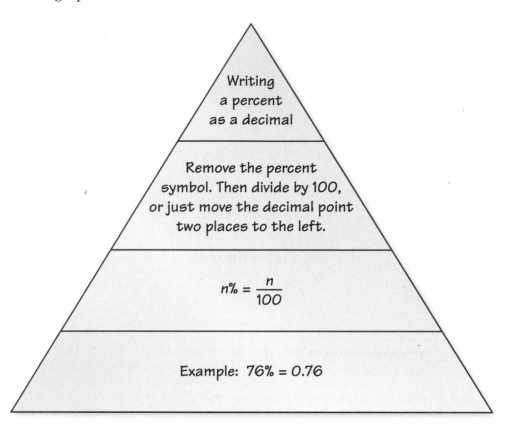

Writing a percent as a decimal

Remove the percent symbol. Then divide by 100, or just move the decimal point two places to the left.

$$n\% = \frac{n}{100}$$

Example: 76% = 0.76

On Your Own

Make summary triangles to help you study these topics.

1. writing a decimal as a percent

2. comparing and ordering fractions, decimals, and percents

3. the percent proportion

4. the percent equation

After you complete this chapter, make summary triangles for the following topics.

5. percent of change
6. discount

7. markup
8. simple interest

"I found this great summary triangle in my *Beautiful Beagle Magazine.*"

Write the percent as a decimal.　*(Section 6.1)*

1. 34%

2. 0.12%

3. 62.5%

Write the decimal as a percent.　*(Section 6.1)*

4. 0.67

5. 5.35

6. 0.685

Tell which number is greater.　*(Section 6.2)*

7. $\frac{11}{15}$, 74%

8. 3%, 0.3

Use a number line to order the numbers from least to greatest.　*(Section 6.2)*

9. 125%, $\frac{6}{5}$, 1.22

10. 42%, 0.43, $\frac{17}{40}$

Write and solve a proportion to answer the question.　*(Section 6.3)*

11. What percent of 15 is 6?

12. 35 is what percent of 25?

13. What number is 40% of 50?

14. 0.5% of what number is 5?

Write and solve an equation to answer the question.　*(Section 6.4)*

15. What number is 28% of 75?

16. 42 is 21% of what number?

17. FISHING On a fishing trip, 38% of the fish that you catch are perch. Write this percent as a decimal.　*(Section 6.1)*

18. SCAVENGER HUNT The table shows the results of 8 teams competing in a scavenger hunt. Which team collected the most items? Which team collected the fewest items?　*(Section 6.2)*

Team	1	2	3	4	5	6	7	8
Portion Collected	$\frac{3}{4}$	0.8	77.5%	0.825	$\frac{29}{40}$	76.25%	$\frac{63}{80}$	81.25%

19. COMPLETIONS A quarterback completed 68% of his passes in a game. He threw 25 passes. How many passes did the quarterback complete?　*(Section 6.3)*

20. TEXT MESSAGES You have 44 text messages in your inbox. How many messages can your cell phone hold?　*(Section 6.4)*

6.5 Percents of Increase and Decrease

Essential Question What is a percent of decrease? What is a percent of increase?

1 ACTIVITY: Percent of Decrease

Work with a partner.

Each year in the Columbia River Basin, adult salmon swim upriver to streams to lay eggs and hatch their young.

To go up the river, the adult salmon use fish ladders. But to go down the river, the young salmon must pass through several dams.

At one time, there were electric turbines at each of the eight dams on the main stem of the Columbia and Snake Rivers. About 88% of the young salmon passed through these turbines unharmed.

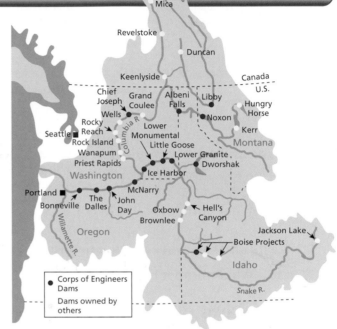

a. Copy and complete the table to show the number of young salmon that made it through the dams.

Dam	0	1	2	3	4	5	6	7	8
Salmon	1000	880	774						

88% of $1000 = 0.88 \cdot 1000$
$= 880$

88% of $880 = 0.88 \cdot 880$
$= 774.4$
≈ 774

b. Display the data in a bar graph.

c. By what percent did the number of young salmon decrease when passing through each dam?

COMMON CORE

Percents

In this lesson, you will
- find percents of increase.
- find percents of decrease.

Learning Standard
7.RP.3

Work with a partner. In 2013, the population of a city was 18,000 people.

a. An organization projects that the population will increase by 2% each year for the next 7 years. Copy and complete the table to find the populations of the city for 2014 through 2020. Then display the data in a bar graph.

For 2014:

$$2\% \text{ of } 18,000 = 0.02 \cdot 18,000$$
$$= 360$$

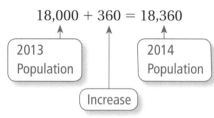

$$18,000 + 360 = 18,360$$

2013 Population

2014 Population

Increase

Year	Population
2013	18,000
2014	18,360
2015	
2016	
2017	
2018	
2019	
2020	

b. Another organization projects that the population will increase by 3% each year for the next 7 years. Repeat part (a) using this percent.

c. Which organization projects the larger populations? How many more people do they project for 2020?

What Is Your Answer?

3. **IN YOUR OWN WORDS** What is a percent of decrease? What is a percent of increase?

4. Describe real-life examples of a percent of decrease and a percent of increase.

Practice Use what you learned about percent of increase and percent of decrease to complete Exercises 4–7 on page 244.

A **percent of change** is the percent that a quantity changes from the original amount.

$$\text{percent of change} = \frac{\text{amount of change}}{\text{original amount}}$$

 **Key Idea**

Percents of Increase and Decrease

When the original amount increases, the percent of change is called a **percent of increase**.

$$\text{percent of increase} = \frac{\text{new amount} - \text{original amount}}{\text{original amount}}$$

When the original amount decreases, the percent of change is called a **percent of decrease**.

$$\text{percent of decrease} = \frac{\text{original amount} - \text{new amount}}{\text{original amount}}$$

EXAMPLE 1 Finding a Percent of Increase

The table shows the numbers of hours you spent online last weekend. What is the percent of change in your online time from Saturday to Sunday?

Day	Hours Online
Saturday	2
Sunday	4.5

The number of hours on Sunday is greater than the number of hours on Saturday. So, the percent of change is a percent of increase.

$$\text{percent of increase} = \frac{\text{new amount} - \text{original amount}}{\text{original amount}}$$

$$= \frac{4.5 - 2}{2} \qquad \text{Substitute.}$$

$$= \frac{2.5}{2} \qquad \text{Subtract.}$$

$$= 1.25, \text{ or } 125\% \qquad \text{Write as a percent.}$$

∴ So, your online time increased 125% from Saturday to Sunday.

On Your Own

Find the percent of change. Round to the nearest tenth of a percent if necessary.

1. 10 inches to 25 inches

2. 57 people to 65 people

Multi-Language Glossary at BigIdeasMath.com

EXAMPLE **2** **Finding a Percent of Decrease**

The bar graph shows a softball player's home run totals. What was the percent of change from 2012 to 2013?

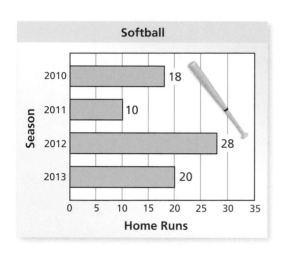

Softball

The number of home runs decreased from 2012 to 2013. So, the percent of change is a percent of decrease.

$$\text{percent of decrease} = \frac{\text{original amount} - \text{new amount}}{\text{original amount}}$$

$$= \frac{28 - 20}{28} \qquad \text{Substitute.}$$

$$= \frac{8}{28} \qquad \text{Subtract.}$$

$$\approx 0.286, \text{ or } 28.6\% \qquad \text{Write as a percent.}$$

∴ So, the number of home runs decreased about 28.6%.

 Key Idea

Study Tip

The amount of error is always positive.

Percent Error

A **percent error** is the percent that an estimated quantity differs from the actual amount.

$$\text{percent error} = \frac{\text{amount of error}}{\text{actual amount}}$$

EXAMPLE **3** **Finding a Percent Error**

You estimate that the length of your classroom is 16 feet. The actual length is 21 feet. Find the percent error.

The amount of error is $21 - 16 = 5$ feet.

$$\text{percent error} = \frac{\text{amount of error}}{\text{actual amount}} \qquad \text{Write percent error equation.}$$

$$= \frac{5}{21} \qquad \text{Substitute.}$$

$$\approx 0.238, \text{ or } 23.8\% \qquad \text{Write as a percent.}$$

∴ The percent error is about 23.8%.

On Your Own

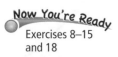

Exercises 8–15 and 18

3. In Example 2, what was the percent of change from 2010 to 2011?

4. **WHAT IF?** In Example 3, your friend estimates that the length of the classroom is 23 feet. Who has the greater percent error? Explain.

✓ Vocabulary and Concept Check

1. **VOCABULARY** How do you know whether a percent of change is a *percent of increase* or a *percent of decrease*?

2. **NUMBER SENSE** Without calculating, which has a greater percent of increase?
   - 5 bonus points on a 50-point exam
   - 5 bonus points on a 100-point exam

3. **WRITING** What does it mean to have a 100% decrease?

Practice and Problem Solving

Find the new amount.

4. 8 meters increased by 25%

5. 15 liters increased by 60%

6. 50 points decreased by 26%

7. 25 penalties decreased by 32%

Identify the percent of change as an *increase* or a *decrease*. Then find the percent of change. Round to the nearest tenth of a percent if necessary.

8. 12 inches to 36 inches

9. 75 people to 25 people

10. 50 pounds to 35 pounds

11. 24 songs to 78 songs

12. 10 gallons to 24 gallons

13. 72 paper clips to 63 paper clips

14. 16 centimeters to 44.2 centimeters

15. 68 miles to 42.5 miles

16. **ERROR ANALYSIS** Describe and correct the error in finding the percent increase from 18 to 26.

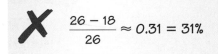
$$\frac{26 - 18}{26} \approx 0.31 = 31\%$$

17. **VIDEO GAME** Last week, you finished Level 2 of a video game in 32 minutes. Today, you finish Level 2 in 28 minutes. What is your percent of change?

18. **PIG** You estimate that a baby pig weighs 20 pounds. The actual weight of the baby pig is 16 pounds. Find the percent error.

19. **CONCERT** You estimate that 200 people attended a school concert. The actual attendance was 240 people.

   a. Find the percent error.

   b. What other estimate gives the same percent error? Explain your reasoning.

Identify the percent of change as an *increase* or a *decrease*. Then find the percent of change. Round to the nearest tenth of a percent if necessary.

20. $\frac{1}{4}$ to $\frac{1}{2}$

21. $\frac{4}{5}$ to $\frac{3}{5}$

22. $\frac{3}{8}$ to $\frac{7}{8}$

23. $\frac{5}{4}$ to $\frac{3}{8}$

24. CRITICAL THINKING Explain why a change from 20 to 40 is a 100% increase, but a change from 40 to 20 is a 50% decrease.

25. POPULATION The table shows population data for a community.

Year	Population
2007	118,000
2013	138,000

   a. What is the percent of change from 2007 to 2013?

   b. Use this percent of change to predict the population in 2019.

26. GEOMETRY Suppose the length and the width of the sandbox are doubled.

   a. Find the percent of change in the perimeter.

   b. Find the percent of change in the area.

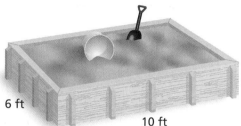

6 ft

10 ft

27. CEREAL A cereal company fills boxes with 16 ounces of cereal. The acceptable percent error in filling a box is 2.5%. Find the least and the greatest acceptable weights.

June   September

28. PRECISION Find the percent of change from June to September in the time to run a mile.

29. CRITICAL THINKING A number increases by 10%, and then decreases by 10%. Will the result be *greater than*, *less than*, or *equal to* the original number? Explain.

30. DONATIONS Donations to an annual fundraiser are 15% greater this year than last year. Last year, donations were 10% greater than the year before. The amount raised this year is $10,120. How much was raised 2 years ago?

31. Reasoning Forty students are in the science club. Of those, 45% are girls. This percent increases to 56% after new girls join the club. How many new girls join?

 Fair Game Review *What you learned in previous grades & lessons*

Write and solve an equation to answer the question. *(Section 6.4)*

32. What number is 25% of 64?

33. 39.2 is what percent of 112?

34. 5 is 5% of what number?

35. 18 is 32% of what number?

36. MULTIPLE CHOICE Which set of ratios does *not* form a proportion? *(Section 5.2)*

   Ⓐ $\frac{1}{4}, \frac{6}{24}$

   Ⓑ $\frac{4}{7}, \frac{7}{10}$

   Ⓒ $\frac{16}{24}, \frac{2}{3}$

   Ⓓ $\frac{36}{10}, \frac{18}{5}$

6.6 Discounts and Markups

Essential Question How can you find discounts and selling prices?

1 **ACTIVITY: Comparing Discounts**

Work with a partner. The same pair of sneakers is on sale at three stores. Which one is the best buy? Explain.

a. Regular Price: $45   **b.** Regular Price: $49   **c.** Regular Price: $39

40% off

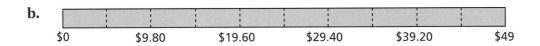
50% off

up to **70%** off

a.
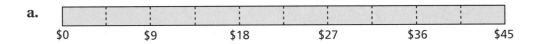
$0 $9 $18 $27 $36 $45

b.
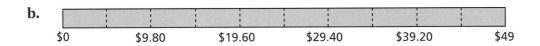
$0 $9.80 $19.60 $29.40 $39.20 $49

c.
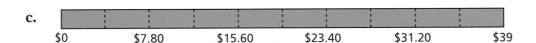
$0 $7.80 $15.60 $23.40 $31.20 $39

2 **ACTIVITY: Finding the Original Price**

Work with a partner.

clearance **30%** off ORIGINAL PRICE

a. You buy a shirt that is on sale for 30% off. You pay $22.40. Your friend wants to know the original price of the shirt. Show how you can use the model below to find the original price.

b. Explain how you can use the percent proportion to find the original price.

$0 $22.40   Original Price

COMMON CORE

Percents

In this lesson, you will
- use percent of discounts to find prices of items.
- use percent of markups to find selling prices of items.

Learning Standard
7.RP.3

Math Practice 2

Make Sense of Quantities

What do the quantities represent? What is the relationship between the quantities?

You own a small jewelry store. You increase the price of the jewelry by 125%.

Work with a partner. Use a model to estimate the selling price of the jewelry. Then use a calculator to find the selling price.

a. Your cost is $250.

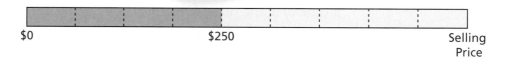

b. Your cost is $50.

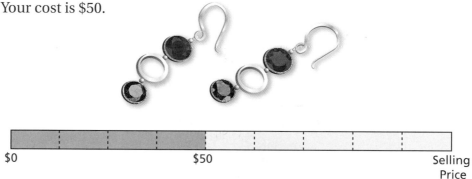

c. Your cost is $170.

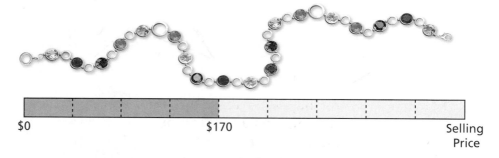

What Is Your Answer?

4. **IN YOUR OWN WORDS** How can you find discounts and selling prices? Give examples of each.

Practice

Use what you learned about discounts to complete Exercises 4, 9, and 14 on page 250.

Check It Out
Lesson Tutorials
BigIdeasMath ✓com

Key Vocabulary 🔊
discount, *p. 248*
markup, *p. 248*

 **Key Ideas**

Discounts

A **discount** is a decrease in the original price of an item.

Markups

To make a profit, stores charge more than what they pay. The increase from what the store pays to the selling price is called a **markup**.

EXAMPLE 1 **Finding a Sale Price**

The original price of the shorts is $35. What is the sale price?

Method 1: First, find the discount. The discount is 25% of $35.

$$a = p \cdot w \qquad \text{Write percent equation.}$$
$$= 0.25 \cdot 35 \qquad \text{Substitute 0.25 for } p \text{ and 35 for } w.$$
$$= 8.75 \qquad \text{Multiply.}$$

Next, find the sale price.

sale price	=	original price	−	discount
	=	35	−	8.75
	=	26.25		

∴ So, the sale price is $26.25.

Method 2: First, find the percent of the original price.

$$100\% - 25\% = 75\%$$

Next, find the sale price.

$$\text{sale price} = 75\% \text{ of } \$35$$
$$= 0.75 \cdot 35$$
$$= 26.25$$

Study Tip

A 25% discount is the same as paying 75% of the original price.

∴ So, the sale price is $26.25.

Check

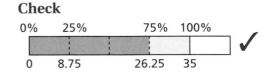

On Your Own

Now You're Ready
Exercises 4−8

1. The original price of a skateboard is $50. The sale price includes a 20% discount. What is the sale price?

🔊 **Multi-Language Glossary at** BigIdeasMath ✓com

EXAMPLE 2 **Finding an Original Price**

What is the original price of the shoes?

The sale price is
100% − 40% = 60%
of the original price.

Answer the question: 33 is 60% of what number?

$a = p \cdot w$ Write percent equation.

$33 = 0.6 \cdot w$ Substitute 33 for a and 0.6 for p.

$55 = w$ Divide each side by 0.6.

∴ So, the original price
of the shoes is $55.

Check

| 0% | | 60% | 100% |

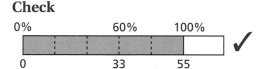

0   33   55   ✓

EXAMPLE 3 **Finding a Selling Price**

A store pays $70 for a bicycle. The percent of markup is 20%. What is the selling price?

Method 1: First, find the markup.
The markup is 20% of $70.

$a = p \cdot w$

$= 0.20 \cdot 70$

$= 14$

Next, find the selling price.

$$\text{selling price} = \text{cost to store} + \text{markup}$$

$$= 70 + 14$$

$$= 84$$

∴ So, the selling price is $84.

Method 2: Use a ratio table.
The selling price is 120% of the cost to the store.

Percent	Dollars
100%	$70
20%	$14
120%	$84

÷ 5   × 6   ÷ 5   × 6

∴ So, the selling price is $84.

Check

| 0% | 40% | 80% | 120% |

0   14   28   42   56   70   84   ✓

● **On Your Own**

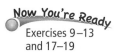
Exercises 9–13
and 17–19

2. The discount on a DVD is 50%. It is on sale for $10. What is the original price of the DVD?

3. A store pays $75 for an aquarium. The markup is 20%. What is the selling price?

✓ Vocabulary and Concept Check

1. **WRITING** Describe how to find the sale price of an item that has been discounted 25%.

2. **WRITING** Describe how to find the selling price of an item that has been marked up 110%.

3. **REASONING** Which would you rather pay? Explain your reasoning.

   a. | 6% tax on a discounted price | or | 6% tax on the original price |

   b. | 30% markup on a $30 shirt | or | $30 markup on a $30 shirt |

Practice and Problem Solving

Copy and complete the table.

		Original Price	Percent of Discount	Sale Price
①	4.	$80	20%	
	5.	$42	15%	
	6.	$120	80%	
	7.	$112	32%	
	8.	$69.80	60%	
②	9.		25%	$40
	10.		5%	$57
	11.		80%	$90
	12.		64%	$72
	13.		15%	$146.54
	14.	$60		$45
	15.	$82		$65.60
	16.	$95		$61.75

Find the selling price.

③ 17. Cost to store: $50
Markup: 10%

18. Cost to store: $80
Markup: 60%

19. Cost to store: $140
Markup: 25%

20. YOU BE THE TEACHER The cost to a store for an MP3 player is $60. The selling price is $105. A classmate says that the markup is 175% because $\frac{\$105}{\$60} = 1.75$. Is your classmate correct? If not, explain how to find the correct percent of markup.

21. SCOOTER The scooter is on sale for 90% off the original price. Which of the methods can you use to find the sale price? Which method do you prefer? Explain.

| Multiply $45.85 by 0.9. | Multiply $45.85 by 0.1. |

| Multiply $45.85 by 0.9, then add to $45.85. | Multiply $45.85 by 0.9, then subtract from $45.85. |

22. GAMING You are shopping for a video game system.

a. At which store should you buy the system?

b. Store A has a weekend sale. What discount must Store A offer for you to buy the system there?

Store	Cost to Store	Markup
A	$162	40%
B	$155	30%
C	$160	25%

23. STEREO A $129.50 stereo is discounted 40%. The next month, the sale price is discounted 60%. Is the stereo now "free"? If not, what is the sale price?

24. CLOTHING You buy a pair of jeans at a department store.

a. What is the percent of discount to the nearest percent?

b. What is the percent of sales tax to the nearest tenth of a percent?

c. The price of the jeans includes a 60% markup. After the discount, what is the percent of markup to the nearest percent?

Department Store

Jeans	39.99
Discount	-10.00
Subtotal	29.99
Sales Tax	1.95
Total	31.94

Thank You

25. You buy a bicycle helmet for $22.26, which includes 6% sales tax. The helmet is discounted 30% off the selling price. What is the original price?

Fair Game Review What you learned in previous grades & lessons

Evaluate. *(Skills Review Handbook)*

26. 2000(0.085)　　　**27.** 1500(0.04)(3)　　　**28.** 3200(0.045)(8)

29. MULTIPLE CHOICE Which measurement is greater than 1 meter? *(Skills Review Handbook)*

Ⓐ 38 inches　　　Ⓑ 1 yard　　　Ⓒ 3.4 feet　　　Ⓓ 98 centimeters

Essential Question How can you find the amount of simple interest earned on a savings account? How can you find the amount of interest owed on a loan?

Simple interest is money earned on a savings account or an investment. It can also be money you pay for borrowing money.

> Write the annual interest rate in decimal form.

Simple interest	=	Principal	×	Annual interest rate	×	Time
($)		($)		(% per yr)		(Years)

$$I = Prt$$

1 ACTIVITY: Finding Simple Interest

Work with a partner. You put $100 in a savings account. The account earns 6% simple interest per year. **(a)** Find the interest earned and the balance at the end of 6 months. **(b)** Copy and complete the table. Then make a bar graph that shows how the balance grows in 6 months.

a. $I = Prt$ Write simple interest formula.

　　$=$ 　　　　　 Substitute values.

　　$=$ 　　　　　 Multiply.

∴ At the end of 6 months, you earn $ 　　 in interest. So, your balance is $ 　　.

b.

Time	Interest	Balance
0 month	$0	$100
1 month		
2 months		
3 months		
4 months		
5 months		
6 months		

Account Balance

COMMON CORE

Percents

In this lesson, you will
- use the simple interest formula to find interest earned or paid, annual interest rates, and amounts paid on loans.

Learning Standard
7.RP.3

2 ACTIVITY: Financial Literacy

Work with a partner. Use the following information to write a report about credit cards. In the report, describe how a credit card works. Include examples that show the amount of interest paid each month on a credit card.

Math Practice 5

Use Other Resources

What resources can you use to find more information about credit cards?

> **U.S. Credit Card Data**
>
> • A typical household with credit card debt in the United States owes about $16,000 to credit card companies.
>
> • A typical credit card interest rate is 14% to 16% per year. This is called the annual percentage rate.

3 ACTIVITY: The National Debt

Work with a partner. In 2012, the United States owed about $16 trillion in debt. The interest rate on the national debt is about 1% per year.

a. Write $16 trillion in decimal form. How many zeros does this number have?

b. How much interest does the United States pay each year on its national debt?

c. How much interest does the United States pay each day on its national debt?

d. The United States has a population of about 314 million people. Estimate the amount of interest that each person pays per year toward interest on the national debt.

What Is Your Answer?

4. **IN YOUR OWN WORDS** How can you find the amount of simple interest earned on a savings account? How can you find the amount of interest owed on a loan? Give examples with your answer.

Practice Use what you learned about simple interest to complete Exercises 4–7 on page 256.

6.7 Lesson

Check It Out
Lesson Tutorials
BigIdeasMath.com

Key Vocabulary 🔊
interest, *p. 254*
principal, *p. 254*
simple interest,
   p. 254

Interest is money paid or earned for the use of money. The **principal** is the amount of money borrowed or deposited.

🔑 Key Idea

Simple Interest

Words **Simple interest** is money paid or earned only on the principal.

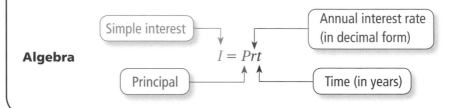

Algebra $I = Prt$

EXAMPLE ❶ **Finding Interest Earned**

You put $500 in a savings account. The account earns 3% simple interest per year. (a) What is the interest earned after 3 years? (b) What is the balance after 3 years?

a. $I = Prt$ Write simple interest formula.

     $= 500(0.03)(3)$ Substitute 500 for *P*, 0.03 for *r*, and 3 for *t*.

     $= 45$ Multiply.

   ∴ So, the interest earned is $45 after 3 years.

b. To find the balance, add the interest to the principal.

   ∴ So, the balance is $500 + $45 = $545 after 3 years.

EXAMPLE ❷ **Finding an Annual Interest Rate**

You put $1000 in an account. The account earns $100 simple interest in 4 years. What is the annual interest rate?

       $I = Prt$ Write simple interest formula.

    $100 = 1000(r)(4)$ Substitute 100 for *I*, 1000 for *P*, and 4 for *t*.

    $100 = 4000r$ Simplify.

  $0.025 = r$ Divide each side by 4000.

  ∴ So, the annual interest rate of the account is 0.025, or 2.5%.

🔊 Multi-Language Glossary at BigIdeasMath.com

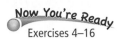
On Your Own

1. In Example 1, what is the balance of the account after 9 months?

2. You put $350 in an account. The account earns $17.50 simple interest in 2.5 years. What is the annual interest rate?

EXAMPLE ③ Finding an Amount of Time

A bank offers three savings accounts. The simple interest rate is determined by the principal. How long does it take an account with a principal of $800 to earn $100 in interest?

3.0%
More than $5000

2.0%
$500-$5000

1.5%
Less than $500

The pictogram shows that the interest rate for a principal of $800 is 2%.

$I = Prt$	Write simple interest formula.
$100 = 800(0.02)(t)$	Substitute 100 for I, 800 for P, and 0.02 for r.
$100 = 16t$	Simplify.
$6.25 = t$	Divide each side by 16.

∴ So, the account earns $100 in interest in 6.25 years.

EXAMPLE ④ Finding an Amount Paid on a Loan

You borrow $600 to buy a violin. The simple interest rate is 15%. You pay off the loan after 5 years. How much do you pay for the loan?

$I = Prt$	Write simple interest formula.
$= 600(0.15)(5)$	Substitute 600 for P, 0.15 for r, and 5 for t.
$= 450$	Multiply.

To find the amount you pay, add the interest to the loan amount.

∴ So, you pay $600 + $450 = $1050 for the loan.

On Your Own

3. In Example 3, how long does it take an account with a principal of $10,000 to earn $750 in interest?

4. **WHAT IF?** In Example 4, you pay off the loan after 2 years. How much money do you save?

Vocabulary and Concept Check

1. **VOCABULARY** Define each variable in $I = Prt$.

2. **WRITING** In each situation, tell whether you would want a *higher* or *lower* interest rate. Explain your reasoning.

   a. you borrow money   **b.** you open a savings account

3. **REASONING** An account earns 6% simple interest. You want to find the interest earned on $200 after 8 months. What conversions do you need to make before you can use the formula $I = Prt$?

Practice and Problem Solving

An account earns simple interest. (a) Find the interest earned. (b) Find the balance of the account.

4. $600 at 5% for 2 years   **5.** $1500 at 4% for 5 years

6. $350 at 3% for 10 years   **7.** $1800 at 6.5% for 30 months

8. $700 at 8% for 6 years   **9.** $1675 at 4.6% for 4 years

10. $925 at 2% for 2.4 years   **11.** $5200 at 7.36% for 54 months

12. **ERROR ANALYSIS** Describe and correct the error in finding the simple interest earned on $500 at 6% for 18 months.

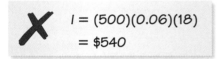
$I = (500)(0.06)(18)$
$= \$540$

Find the annual interest rate.

13. $I = \$24$, $P = \$400$, $t = 2$ years   **14.** $I = \$562.50$, $P = \$1500$, $t = 5$ years

15. $I = \$54$, $P = \$900$, $t = 18$ months   **16.** $I = \$160.67$, $P = \$2000$, $t = 8$ months

Find the amount of time.

17. $I = \$30$, $P = \$500$, $r = 3\%$ **18.** $I = \$720$, $P = \$1000$, $r = 9\%$

19. $I = \$54$, $P = \$800$, $r = 4.5\%$ **20.** $I = \$450$, $P = \$2400$, $r = 7.5\%$

21. **BANKING** A savings account earns 5% simple interest per year. The principal is $1200. What is the balance after 4 years?

22. **SAVINGS** You put $400 in an account. The account earns $18 simple interest in 9 months. What is the annual interest rate?

23. **CD** You put $3000 in a CD (certificate of deposit) at the promotional rate. How long will it take to earn $336 in interest?

Certificate of Deposit

This certificate is the original Spectrum and valid document from the treasury and Security department of this here trust financial group & associates. The agreement herein contained are thorough, correct and binding on the parties. Alterations made on the certificate after it has been legally issued.

Promotional Rate 5.6%
Simple Interest

DIRECTOR'S SIGNATURE

Find the amount paid for the loan.

(4) 24. $1500 at 9% for 2 years

25. $2000 at 12% for 3 years

26. $2400 at 10.5% for 5 years

27. $4800 at 9.9% for 4 years

Copy and complete the table.

	Principal	Interest Rate	Time	Simple Interest
28.	$12,000	4.25%	5 years	
29.		6.5%	18 months	$828.75
30.	$15,500	8.75%		$5425.00
31.	$18,000		54 months	$4252.50

32. ZOO A family charges a trip to the zoo on a credit card. The simple interest rate is 12%. The charges are paid after 3 months. What is the total amount paid for the trip?

Zoo Trip	
Tickets	67.70
Food	62.34
Gas	45.50
Total Cost	?

33. MONEY MARKET You deposit $5000 in an account earning 7.5% simple interest. How long will it take for the balance of the account to be $6500?

11.8% Simple Interest
Equal monthly
payments for 2 years

34. LOANS A music company offers a loan to buy a drum set for $1500. What is the monthly payment?

35. REASONING How many years will it take for $2000 to double at a simple interest rate of 8%? Explain how you found your answer.

36. PROBLEM SOLVING You have two loans, for 2 years each. The total interest for the two loans is $138. On the first loan, you pay 7.5% simple interest on a principal of $800. On the second loan, you pay 3% simple interest. What is the principal for the second loan?

37. Critical Thinking You put $500 in an account that earns 4% annual interest. The interest earned each year is added to the principal to create a new principal. Find the total amount in your account after each year for 3 years.

Fair Game Review *What you learned in previous grades & lessons*

Solve the inequality. Graph the solution. *(Section 4.2)*

38. $x + 5 < 2$

39. $b - 2 \geq -1$

40. $w + 6 \leq -3$

41. MULTIPLE CHOICE What is the solution of $4x + 5 = -11$? *(Section 3.5)*

Ⓐ $x = -4$ Ⓑ $x = -1.5$ Ⓒ $x = 1.5$ Ⓓ $x = 4$

Check It Out
Progress Check
BigIdeasMath.com

Identify the percent of change as an *increase* or a *decrease*. Then find the percent of change. Round to the nearest tenth of a percent if necessary.   (*Section 6.5*)

1. 8 inches to 24 inches

2. 300 miles to 210 miles

Find the original price, discount, sale price, or selling price.   (*Section 6.6*)

3. Original price: $30
Discount: 10%
Sale price: ?

4. Original price: $55
Discount: ?
Sale price: $46.75

5. Original price: ?
Discount: 75%
Sale price: $74.75

6. Cost to store: $152
Markup: 50%
Selling price: ?

An account earns simple interest. Find the interest earned, principal, interest rate, or time.   (*Section 6.7*)

7. Interest earned: ?
Principal: $1200
Interest rate: 2%
Time: 5 years

8. Interest earned: $25
Principal: $500
Interest rate: 5%
Time: ?

9. Interest earned: $76
Principal: $800
Interest rate: ?
Time: 2 years

10. Interest earned: $119.88
Principal: ?
Interest rate: 3.6%
Time: 3 years

11. HEIGHT You estimate that your friend is 50 inches tall. The actual height of your friend is 54 inches. Find the percent error.   (*Section 6.5*)

12. DIGITAL CAMERA A digital camera costs $230. The camera is on sale for 30% off, and you have a coupon for an additional 15% off the sale price. What is the final price?   (*Section 6.6*)

13. WATER SKIS The original price of the water skis was $200. What is the percent of discount?   (*Section 6.6*)

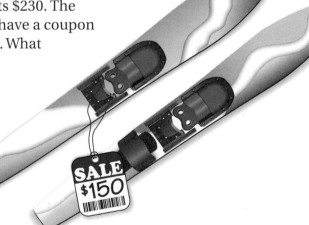

2 Ways to Own:
1. $75 cash back with 3.5% simple interest
2. No interest for 2 years

14. SAXOPHONE A saxophone costs $1200. A store offers two loan options. Which option saves more money if you pay the loan in 2 years?   (*Section 6.7*)

15. LOAN You borrow $200. The simple interest rate is 12%. You pay off the loan after 2 years. How much do you pay for the loan?   (*Section 6.7*)

Check It Out
Vocabulary Help
BigIdeasMath ✓com

Review Key Vocabulary

percent of change, *p. 242*   percent error, *p. 243*   interest, *p. 254*
percent of increase, *p. 242*   discount, *p. 248*   principal, *p. 254*
percent of decrease, *p. 242*   markup, *p. 248*   simple interest, *p. 254*

Review Examples and Exercises

 Percents and Decimals   *(pp. 214–219)*

   a. Write 64% as a decimal.   **b.** Write 0.023 as a percent.

     64% = 64.% = 0.64   0.023 = 0.023 = 2.3%

Exercises

Write the percent as a decimal. Use a model to check your answer.

  1. 76%   **2.** 6%   **3.** 334%

Write the decimal as a percent. Use a model to check your answer.

  4. 0.15   **5.** 1.24   **6.** 0.097

 Comparing and Ordering Fractions, Decimals, and Percents
(pp. 220–225)

Which is greater, $\dfrac{9}{10}$ or 88%?

Write $\dfrac{9}{10}$ as a percent: $\dfrac{9}{10} = \dfrac{90}{100} = 90\%$

  88% is less than 90%. So, $\dfrac{9}{10}$ is the greater number.

Exercises

Tell which number is greater.

  7. $\dfrac{1}{2}$, 52%   **8.** $\dfrac{12}{5}$, 245%

  9. 0.46, 43%   **10.** 0.023, 22%

Use a number line to order the numbers from least to greatest.

  11. $\dfrac{41}{50}$, 0.83, 80%   **12.** $\dfrac{9}{4}$, 220%, 2.15

  13. 0.67, 66%, $\dfrac{2}{3}$ **14.** 0.88, $\dfrac{7}{8}$, 90%

The Percent Proportion (pp. 226–231)

a. **What percent of 24 is 9?**

$$\frac{a}{w} = \frac{p}{100}$$ Write the percent proportion.

$$\frac{9}{24} = \frac{p}{100}$$ Substitute 9 for a and 24 for w.

$$100 \cdot \frac{9}{24} = 100 \cdot \frac{p}{100}$$ Multiplication Property of Equality

$$37.5 = p$$ Simplify.

⋮ So, 37.5% of 24 is 9.

b. **What number is 15% of 80?**

$$\frac{a}{w} = \frac{p}{100}$$ Write the percent proportion.

$$\frac{a}{80} = \frac{15}{100}$$ Substitute 80 for w and 15 for p.

$$80 \cdot \frac{a}{80} = 80 \cdot \frac{15}{100}$$ Multiplication Property of Equality

$$a = 12$$ Simplify.

⋮ So, 12 is 15% of 80.

c. **120% of what number is 54?**

$$\frac{a}{w} = \frac{p}{100}$$ Write the percent proportion.

$$\frac{54}{w} = \frac{120}{100}$$ Substitute 54 for a and 120 for p.

$$54 \cdot 100 = w \cdot 120$$ Cross Products Property

$$5400 = 120w$$ Multiply.

$$45 = w$$ Divide each side by 120.

⋮ So, 120% of 45 is 54.

Exercises

Write and solve a proportion to answer the question.

15. What percent of 60 is 18?

16. 40 is what percent of 32?

17. What number is 70% of 70?

18. $\frac{3}{4}$ is 75% of what number?

The Percent Equation *(pp. 232–237)*

a. What number is 72% of 25?

$a = p \cdot w$ Write percent equation.

$ = 0.72 \cdot 25$ Substitute 0.72 for p and 25 for w.

$ = 18$ Multiply.

So, 72% of 25 is 18.

b. 28 is what percent of 70?

$a = p \cdot w$ Write percent equation.

$28 = p \cdot 70$ Substitute 28 for a and 70 for w.

$\dfrac{28}{70} = \dfrac{p \cdot 70}{70}$ Division Property of Equality

$0.4 = p$ Simplify.

Because 0.4 equals 40%, 28 is 40% of 70.

c. 22.1 is 26% of what number?

$a = p \cdot w$ Write percent equation.

$22.1 = 0.26 \cdot w$ Substitute 22.1 for a and 0.26 for p.

$85 = w$ Divide each side by 0.26.

So, 22.1 is 26% of 85.

Exercises

Write and solve an equation to answer the question.

19. What number is 24% of 25?

20. 9 is what percent of 20?

21. 60.8 is what percent of 32?

22. 91 is 130% of what number?

23. 85% of what number is 10.2?

24. 83% of 20 is what number?

25. PARKING 15% of the school parking spaces are handicap spaces. The school has 18 handicap spaces. How many parking spaces are there?

26. FIELD TRIP Of the 25 students on a field trip, 16 students bring cameras. What percent of the students bring cameras?

6.5 Percents of Increase and Decrease (pp. 240–245)

The table shows the numbers of skim boarders at a beach on Saturday and Sunday. What was the percent of change in boarders from Saturday to Sunday?

The number of skim boarders on Sunday is less than the number of skim boarders on Saturday. So, the percent of change is a percent of decrease.

percent of decrease = $\dfrac{\text{original amount} - \text{new amount}}{\text{original amount}}$

$= \dfrac{12 - 9}{12}$ Substitute.

$= \dfrac{3}{12}$ Subtract.

$= 0.25 = 25\%$ Write as a percent.

Day	Number of Skim Boarders
Saturday	12
Sunday	9

So, the number of skim boarders decreased by 25% from Saturday to Sunday.

Exercises

Identify the percent of change as an *increase* or a *decrease*. Then find the percent of change. Round to the nearest tenth of a percent if necessary.

27. 6 yards to 36 yards

28. 120 meals to 52 meals

29. MARBLES You estimate that a jar contains 68 marbles. The actual number of marbles is 60. Find the percent error.

6.6 Discounts and Markups (pp. 246–251)

What is the original price of the tennis racquet?

The sale price is 100% − 30% = 70% of the original price.

Answer the question: 21 is 70% of what number?

$a = p \cdot w$ Write percent equation.

$21 = 0.7 \cdot w$ Substitute 21 for a and 0.7 for p.

$30 = w$ Divide each side by 0.7.

So, the original price of the tennis racquet is $30.

SALE 30% off Now $21

Exercises

Find the sale price or original price.

30. Original price: $50
Discount: 15%
Sale price: ?

31. Original price: ?
Discount: 20%
Sale price: $75

Simple Interest *(pp. 252–257)*

You put $200 in a savings account. The account earns 2% simple interest per year.

a. What is the interest earned after 4 years?

b. What is the balance after 4 years?

a. $I = Prt$ Write simple interest formula.

    $= 200(0.02)(4)$ Substitute 200 for P, 0.02 for r, and 4 for t.

    $= 16$ Multiply.

    ⋮ So, the interest earned is $16 after 4 years.

b. To find the balance, add the interest to the principal.

    ⋮ So, the balance is $200 + $16 = $216 after 4 years.

You put $500 in an account. The account earns $55 simple interest in 5 years. What is the annual interest rate?

        $I = Prt$ Write simple interest formula.

    $55 = 500(r)(5)$ Substitute 55 for I, 500 for P, and 5 for t.

    $55 = 2500r$ Simplify.

    $0.022 = r$ Divide each side by 2500.

⋮ So, the annual interest rate of the account is 0.022, or 2.2%.

Exercises

An account earns simple interest.

a. Find the interest earned.

b. Find the balance of the account.

32. $300 at 4% for 3 years   **33.** $2000 at 3.5% for 4 years

Find the annual simple interest rate.

34. $I = \$17$, $P = \$500$, $t = 2$ years   **35.** $I = \$426$, $P = \$1200$, $t = 5$ years

Find the amount of time.

36. $I = \$60$, $P = \$400$, $r = 5\%$ **37.** $I = \$237.90$, $P = \$1525$, $r = 2.6\%$

38. SAVINGS You put $100 in an account. The account earns $2 simple interest in 6 months. What is the annual interest rate?

Write the percent as a decimal.

1. 0.96% **2.** 65% **3.** 25.7%

Write the decimal as a percent.

4. 0.42 **5.** 7.88 **6.** 0.5854

Tell which number is greater.

7. $\frac{16}{25}$, 65% **8.** 56%, 5.6

Use a number line to order the numbers from least to greatest.

9. 85%, $\frac{15}{18}$, 0.84 **10.** 58.3%, 0.58, $\frac{7}{12}$

Answer the question.

11. What percent of 28 is 21? **12.** 64 is what percent of 40?

13. What number is 80% of 45? **14.** 0.8% of what number is 6?

Identify the percent of change as an *increase* or a *decrease*. Then find the percent of change. Round to the nearest tenth of a percent if necessary.

15. 4 strikeouts to 10 strikeouts **16.** $24 to $18

Find the sale price or selling price.

17. Original price: $15
Discount: 5%
Sale price: ?

18. Cost to store: $5.50
Markup: 75%
Selling price: ?

An account earns simple interest. Find the interest earned or the principal.

19. Interest earned: ?
Principal: $450
Interest rate: 6%
Time: 8 years

20. Interest earned: $27
Principal: ?
Interest rate: 1.5%
Time: 2 years

21. BASKETBALL You, your cousin, and a friend each take the same number of free throws at a basketball hoop. Who made the most free throws?

22. PARKING LOT You estimate that there are 66 cars in a parking lot. The actual number of cars is 75.

a. Find the percent error.

b. What other estimate gives the same percent error? Explain your reasoning.

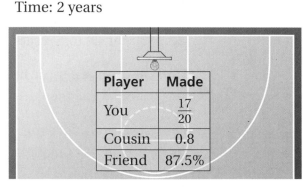

Player	Made
You	$\frac{17}{20}$
Cousin	0.8
Friend	87.5%

23. INVESTMENT You put $800 in an account that earns 4% simple interest. Find the total amount in your account after each year for 3 years.

Test-Taking Strategy
Read All Choices Before Answering

Which amount of increase in your catnip allowance do you want?
Ⓐ 50%   Ⓑ 75%   Ⓒ 98%   Ⓓ 10%

I get it. C for catnip.

"Reading all choices before answering can really pay off!"

1. A movie theater offers 30% off the price of a movie ticket to students from your school. The regular price of a movie ticket is $8.50. What is the discounted price that you would pay for a ticket? *(7.RP.3)*

   A. $2.55   **C.** $5.95

   B. $5.50   **D.** $8.20

2. You are comparing the prices of four boxes of cereal. Two of the boxes contain free extra cereal.

   - Box F costs $3.59 and contains 16 ounces.

   - Box G costs $3.79 and contains 16 ounces, plus an additional 10% for free.

   - Box H costs $4.00 and contains 500 grams.

   - Box I costs $4.69 and contains 500 grams, plus an additional 20% for free.

   Which box has the least unit cost? (1 ounce = 28.35 grams)   *(7.RP.3)*

   F. Box F   **H.** Box H

   G. Box G   **I.** Box I

3. What value makes the equation $11 - 3x = -7$ true?   *(7.EE.4a)*

4. Which proportion represents the problem below?   *(7.RP.3)*

   "17% of a number is 43. What is the number?"

   A. $\dfrac{17}{43} = \dfrac{n}{100}$ **C.** $\dfrac{n}{43} = \dfrac{17}{100}$

   B. $\dfrac{n}{17} = \dfrac{43}{100}$ **D.** $\dfrac{43}{n} = \dfrac{17}{100}$

5. Which list of numbers is in order from least to greatest? *(7.EE.3)*

F. 0.8, $\frac{5}{8}$, 70%, 0.09

H. $\frac{5}{8}$, 70%, 0.8, 0.09

G. 0.09, $\frac{5}{8}$, 0.8, 70%

I. 0.09, $\frac{5}{8}$, 70%, 0.8

6. What is the value of $\frac{9}{8} \div \left(-\frac{11}{4}\right)$? *(7.NS.2b)*

7. A pair of running shoes is on sale for 25% off the original price.

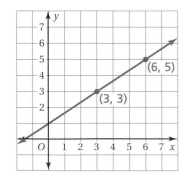

ORIGINAL PRICE
$123.75

Which price is closest to the sale price of the running shoes? *(7.RP.3)*

A. $93

C. $124

B. $99

D. $149

8. What is the slope of the line? *(7.RP.2b)*

F. $\frac{2}{3}$

H. 2

G. $\frac{3}{2}$

I. 3

9. Brad solved the equation in the box shown.

What should Brad do to correct the error that he made? *(7.EE.4a)*

$$-3(2 + w) = -45$$
$$2 + w = -15$$
$$w = -17$$

A. Multiply -45 by -3 to get $2 + w = 135$.

B. Add 3 to -45 to get $2 + w = -42$.

C. Add 2 to -15 to get $w = -13$.

D. Divide -45 by -3 to get 15.

10. You are comparing the costs of a certain model of ladder at a hardware store and at an online store. *(7.RP.3)*

Think
Solve
Explain

HARDWARE STORE

Ladders: $350/ea.

Sales tax: 6% of cost of ladder

BUY HARDWARE ONLINE

Ladder $320

Sales tax: 6% of cost of ladder

Shipping & handling: 5% of cost of ladder

Part A What is the cost of the ladder at each of the stores? Show your work and explain your reasoning.

Part B Suppose that the hardware store is offering 10% off the price of the ladder and that the online store is offering free shipping and handling. Which store offers the better final cost? by how much? Show your work and explain your reasoning.

11. Which graph represents the inequality below? *(7.EE.4b)*

$$-5 - 3x \geq -11$$

F.
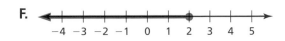

H.
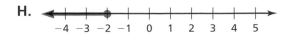

G.

I.

7 Constructions and Scale Drawings

"Move 4 of the lines to make 3 equilateral triangles."

"Well done, Descartes!"

"I'm at 3rd base. You are running to 1st base, and Fluffy is running to 2nd base."

"Should I throw the ball to 2nd to get Fluffy out or throw it to 1st to get you out?"

What You Learned Before

"Look at this baby crocodile! Isn't it cute?"

Yes, it's very acute.

Measuring Angles (4.MD.6)

Example 1 Use a protractor to find the measure of each angle. Then classify the angle as *acute*, *obtuse*, *right*, or *straight*.

a.

b.

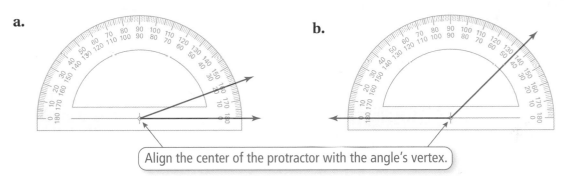

Align the center of the protractor with the angle's vertex.

∴ The angle measure is 20°. So, the angle is acute.

∴ The angle measure is 135°. So, the angle is obtuse.

Drawing Angles (4.G.1)

Example 2 Use a protractor to draw a 45° angle.

Draw a ray. Place the center of the protractor on the endpoint of the ray and align the protractor so the ray passes through the 0° mark. Make a mark at 45°. Then draw a ray from the endpoint at the center of the protractor through the mark at 45°.

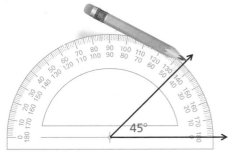

45°

Try It Yourself

Use a protractor to find the measure of the angle. Then classify the angle as *acute*, *obtuse*, *right*, or *straight*.

1.

2.

3.

Use a protractor to draw an angle with the given measure.

4. 55°   **5.** 160°   **6.** 85°   **7.** 180°

Essential Question What can you conclude about the angles formed by two intersecting lines?

Classification of Angles

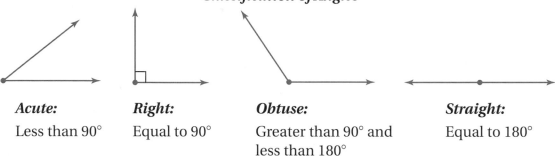

Acute:
Less than 90°

Right:
Equal to 90°

Obtuse:
Greater than 90° and less than 180°

Straight:
Equal to 180°

1 ACTIVITY: Drawing Angles

Work with a partner.

a. Draw the hands of the clock to represent the given type of angle.

| Acute | Straight | Right | Obtuse |

b. What is the measure of the angle formed by the hands of the clock at the given time?

9:00   6:00   12:00

The Meaning of a Word ● Adjacent

When two states are **adjacent,**   they are next to each other and they share a common border.

Maine

New Hampshire

Maine

New Hampshire

COMMON CORE

Geometry

In this lesson, you will
- identify adjacent and vertical angles.
- find angle measures using adjacent and vertical angles.

Learning Standard 7.G.5

Work with a partner. Some angles, such as ∠A, can be named by a single letter. When this does not clearly identify an angle, you should use three letters, as shown.

Math Practice 3

Justify Conclusions

When you name an angle, does the order in which you write the letters matter? Explain.

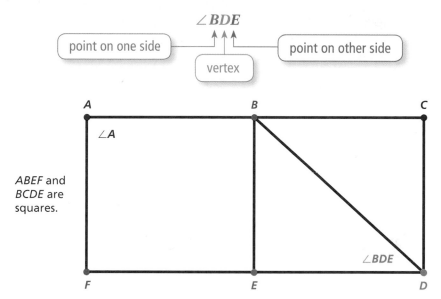

∠BDE

point on one side → ← point on other side

vertex

ABEF and *BCDE* are squares.

∠A

∠BDE

a. Name all the right angles, acute angles, and obtuse angles.

b. Which pairs of angles do you think are *adjacent*? Explain.

3 **ACTIVITY: Measuring Angles**

Work with a partner.

a. How many angles are formed by the intersecting roads? Number the angles.

b. **CHOOSE TOOLS** Measure each angle formed by the intersecting roads. What do you notice?

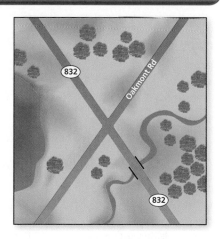

What Is Your Answer?

4. **IN YOUR OWN WORDS** What can you conclude about the angles formed by two intersecting lines?

5. Draw two acute angles that are adjacent.

Practice Use what you learned about angles and intersecting lines to complete Exercises 3 and 4 on page 274.

 **Key Ideas**

Adjacent Angles

Words Two angles are **adjacent angles** when they share a common side and have the same vertex.

Examples

∠1 and ∠2 are adjacent.

∠2 and ∠4 are not adjacent.

Vertical Angles

Words Two angles are **vertical angles** when they are opposite angles formed by the intersection of two lines. Vertical angles are **congruent angles**, meaning they have the same measure.

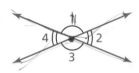

Examples

∠1 and ∠3 are vertical angles.

∠2 and ∠4 are vertical angles.

EXAMPLE 1 Naming Angles

Use the figure shown.

a. Name a pair of adjacent angles.

∠ABC and ∠ABF share a common side and have the same vertex B.

⋮ So, ∠ABC and ∠ABF are adjacent angles.

b. Name a pair of vertical angles.

∠ABF and ∠CBD are opposite angles formed by the intersection of two lines.

⋮ So, ∠ABF and ∠CBD are vertical angles.

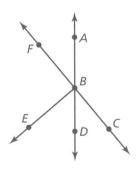

⬤ On Your Own

Now You're Ready
Exercises 5 and 6

Name two pairs of adjacent angles and two pairs of vertical angles in the figure.

1.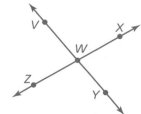

2.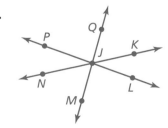

EXAMPLE **2** **Using Adjacent and Vertical Angles**

Tell whether the angles are *adjacent* or *vertical*. Then find the value of *x*.

a.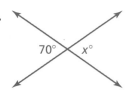

The angles are vertical angles. Because vertical angles are congruent, the angles have the same measure.

⁝• So, the value of *x* is 70.

> **Remember**
>
> You can add angle measures. When two or more adjacent angles form a larger angle, the sum of the measures of the smaller angles is equal to the measure of the larger angle.

b.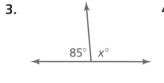

The angles are adjacent angles. Because the angles make up a right angle, the sum of their measures is 90°.

$$(x + 4) + 31 = 90 \qquad \text{Write equation.}$$
$$x + 35 = 90 \qquad \text{Combine like terms.}$$
$$x = 55 \qquad \text{Subtract 35 from each side.}$$

⁝• So, the value of *x* is 55.

EXAMPLE **3** **Constructing Angles**

Draw a pair of vertical angles with a measure of 40°.

Step 1: Use a protractor to draw a 40° angle.

Step 2: Use a straightedge to extend the sides to form two intersecting lines.

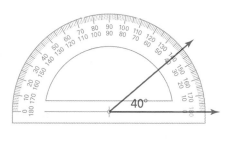

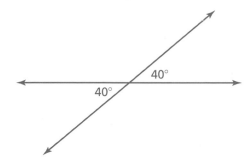

On Your Own

Now You're Ready
Exercises 8–17

Tell whether the angles are *adjacent* or *vertical*. Then find the value of *x*.

3.

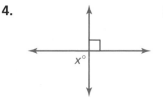

4.

5.

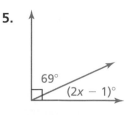

6. Draw a pair of vertical angles with a measure of 75°.

Vocabulary and Concept Check

1. **VOCABULARY** When two lines intersect, how many pairs of vertical angles are formed? How many pairs of adjacent angles are formed?

2. **REASONING** Identify the congruent angles in the figure. Explain your reasoning.

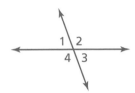

Practice and Problem Solving

Use the figure at the right.

3. Measure each angle formed by the intersecting lines.

4. Name two angles that are adjacent to ∠ABC.

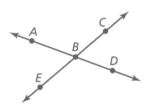

Name two pairs of adjacent angles and two pairs of vertical angles in the figure.

5.

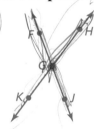

6.

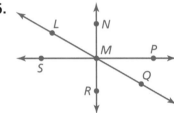

7. **ERROR ANALYSIS** Describe and correct the error in naming a pair of vertical angles.

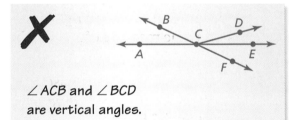

∠ACB and ∠BCD
are vertical angles.

Tell whether the angles are *adjacent* or *vertical*. Then find the value of *x*.

8.

$x°$
$35°$

9.

$x°$
$128°$

10.

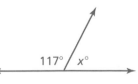

$117°$ $x°$

11.

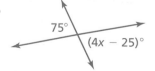

$75°$
$(4x - 25)°$

12.

$4x°$
$2x°$

13.

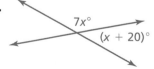

$7x°$
$(x + 20)°$

Draw a pair of vertical angles with the given measure.

③ 14. 25°   **15.** 85°   **16.** 110°   **17.** 135°

18. IRON CROSS The iron cross is a skiing trick in which the tips of the skis are crossed while the skier is airborne. Find the value of x in the iron cross shown.

127°

$(2x + 41)°$

19. OPEN-ENDED Draw a pair of adjacent angles with the given description.

    a. Both angles are acute.

    b. One angle is acute, and one is obtuse.

    c. The sum of the angle measures is 135°.

20. PRECISION Explain two procedures that you can use to draw adjacent angles with given measures.

Determine whether the statement is *always, sometimes,* or *never* true.

21. When the measure of ∠1 is 70°, the measure of ∠3 is 110°.

22. When the measure of ∠4 is 120°, the measure of ∠1 is 60°.

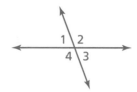

23. ∠2 and ∠3 are congruent.

24. The measure of ∠1 plus the measure of ∠2 equals the measure of ∠3 plus the measure of ∠4.

25. REASONING Draw a figure in which ∠1 and ∠2 are acute vertical angles, ∠3 is a right angle adjacent to ∠2, and the sum of the measure of ∠1 and the measure of ∠4 is 180°.

26. Structure For safety reasons, a ladder should make a 15° angle with a wall. Is the ladder shown leaning at a safe angle? Explain.

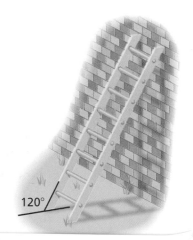

120°

Fair Game Review What you learned in previous grades & lessons

Solve the inequality. Graph the solution. *(Section 4.3)*

27. $-6n > 54$ **28.** $-\dfrac{1}{2}x \le 17$ **29.** $-1.6 < \dfrac{m}{-2.5}$

30. MULTIPLE CHOICE What is the slope of the line that passes through the points (2, 3) and (6, 8)? *(Section 5.5)*

  Ⓐ $\dfrac{4}{5}$ **Ⓑ** $\dfrac{5}{4}$ **Ⓒ** $\dfrac{4}{3}$ **Ⓓ** $\dfrac{3}{2}$

Complementary and Supplementary Angles

Essential Question
How can you classify two angles as complementary or supplementary?

1 ACTIVITY: Complementary and Supplementary Angles

Work with a partner.

a. The graph represents the measures of *complementary angles*. Use the graph to complete the table.

x		20°		30°	45°		75°
y	80°		65°	60°		40°	

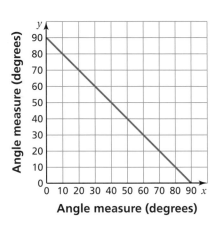

b. How do you know when two angles are complementary? Explain.

c. The graph represents the measures of *supplementary angles*. Use the graph to complete the table.

x	20°		60°	90°		140°	
y		150°		90°	50°		30°

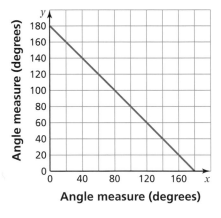

d. How do you know when two angles are supplementary? Explain.

2 ACTIVITY: Exploring Rules About Angles

Work with a partner. Copy and complete each sentence with *always*, *sometimes*, or *never*.

a. If x and y are complementary angles, then both x and y are _____ acute.

b. If x and y are supplementary angles, then x is _____ acute.

c. If x is a right angle, then x is _____ acute.

d. If x and y are complementary angles, then x and y are _____ adjacent.

e. If x and y are supplementary angles, then x and y are _____ vertical.

COMMON CORE

Geometry

In this lesson, you will
• classify pairs of angles as complementary, supplementary, or neither.
• find angle measures using complementary and supplementary angles.

Learning Standard
7.G.5

ACTIVITY: Classifying Pairs of Angles

Work with a partner. Tell whether the two angles shown on the clocks are *complementary*, *supplementary*, or *neither*. Explain your reasoning.

a.

b.

c.

d.

4 **ACTIVITY: Identifying Angles**

Work with a partner. Use a protractor and the figure shown.

a. Name four pairs of complementary angles and four pairs of supplementary angles.

Math Practice 3

Use Definitions
How can you use the definitions of *complementary*, *supplementary*, and *vertical angles* to answer the questions?

b. Name two pairs of vertical angles.

What Is Your Answer?

5. **IN YOUR OWN WORDS** How can you classify two angles as complementary or supplementary? Give examples of each type.

Practice ➤ Use what you learned about complementary and supplementary angles to complete Exercises 3–5 on page 280.

Check It Out
Lesson Tutorials
BigIdeasMath $\checkmark$com

Key Vocabulary 🔊
complementary
  angles, *p. 278*
supplementary
  angles, *p. 278*

 **Key Ideas**

Complementary Angles

Words Two angles are **complementary angles** when the sum of their
measures is 90°.

Examples

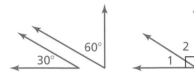

∠1 and ∠2 are
complementary angles.

Supplementary Angles

Words Two angles are **supplementary angles** when the sum of their
measures is 180°.

Examples

∠3 and ∠4 are
supplementary angles.

EXAMPLE 1 **Classifying Pairs of Angles**

Tell whether the angles are *complementary*, *supplementary*, or *neither*.

a.   70° 110°

$70° + 110° = 180°$

∴ So, the angles are supplementary.

b.   49° 41°

$41° + 49° = 90°$

∴ So, the angles are complementary.

c.   128° 62°

$128° + 62° = 190°$

∴ So, the angles are *neither* complementary
nor supplementary.

🔴 **On Your Own**

Now You're Ready
Exercises 6–11

Tell whether the angles are *complementary*, *supplementary*, or *neither*.

1.   26° 64°

2.   136° 44°

3.   70° 19°

EXAMPLE 2 **Using Complementary and Supplementary Angles**

**Tell whether the angles are *complementary* or *supplementary*.
Then find the value of x.**

a.

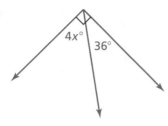

The two angles make up a right angle.
So, the angles are complementary angles,
and the sum of their measures is 90°.

$$4x + 36 = 90 \quad \text{Write equation.}$$
$$4x = 54 \quad \text{Subtract 36 from each side.}$$
$$x = 13.5 \quad \text{Divide each side by 4.}$$

b.

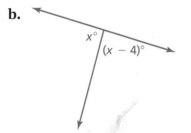

The two angles make up a straight angle.
So, the angles are supplementary angles,
and the sum of their measures is 180°.

$$x + (x - 4) = 180 \quad \text{Write equation.}$$
$$2x - 4 = 180 \quad \text{Combine like terms.}$$
$$2x = 184 \quad \text{Add 4 to each side.}$$
$$x = 92 \quad \text{Divide each side by 2.}$$

EXAMPLE 3 **Constructing Angles**

**Draw a pair of adjacent supplementary angles so that one angle
has a measure of 60°.**

Step 1: Use a protractor to
draw a 60° angle.

Step 2: Extend one of the sides
to form a line.

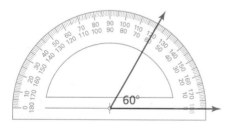

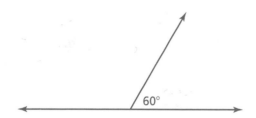

● **On Your Own**

Now You're Ready
Exercises 12–14
and 17–20

**Tell whether the angles are *complementary* or *supplementary*.
Then find the value of x.**

4.

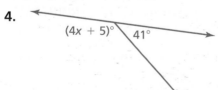

5.

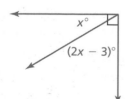

6. Draw a pair of adjacent supplementary angles so that
one angle has a measure of 15°.

 Check It Out
Help with Homework
BigIdeasMath ✓com

✓ Vocabulary and Concept Check

1. **VOCABULARY** Explain how complementary angles and supplementary angles are different.

2. **REASONING** Can adjacent angles be supplementary? complementary? neither? Explain.

Practice and Problem Solving

Tell whether the statement is *always*, *sometimes*, or *never* true. Explain.

3. If x and y are supplementary angles, then x is obtuse.

4. If x and y are right angles, then x and y are supplementary angles.

5. If x and y are complementary angles, then y is a right angle.

Tell whether the angles are *complementary*, *supplementary*, or *neither*.

① 6.

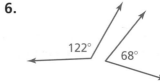

7.

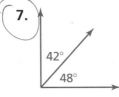

8.

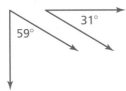

9.

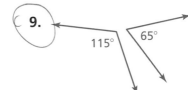

10.

11.

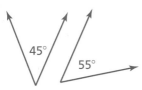

Tell whether the angles are *complementary* or *supplementary*. Then find the value of x.

② 12.

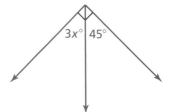

13.

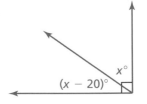

14.

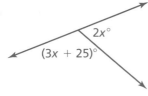

15. **INTERSECTION** What are the measures of the other three angles formed by the intersection?

16. **TRIBUTARY** A tributary joins a river at an angle. Find the value of x.

Draw a pair of adjacent supplementary angles so that one angle has the given measure.

3 **17.** 20° **18.** 35° **19.** 80° **20.** 130°

21. PRECISION Explain two procedures that you can use to draw two adjacent complementary angles. Then draw a pair of adjacent complementary angles so that one angle has a measure of 30°.

22. OPEN-ENDED Give an example of an angle that can be a supplementary angle but cannot be a complementary angle. Explain.

23. VANISHING POINT The vanishing point of the picture is represented by point *B*.

   a. The measure of ∠*ABD* is 6.2 times greater than the measure of ∠*CBD*. Find the measure of ∠*CBD*.

   b. ∠*FBE* and ∠*EBD* are congruent. Find the measure of ∠*FBE*.

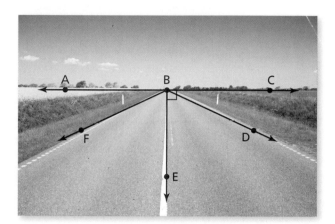

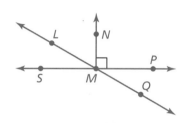

24. LOGIC Your friend says that ∠*LMN* and ∠*PMQ* are complementary angles. Is she correct? Explain.

25. RATIO The measures of two complementary angles have a ratio of 3 : 2. What is the measure of the larger angle?

26. REASONING Two angles are vertical angles. What are their measures if they are also complementary angles? supplementary angles?

27. **Problem Solving** Find the values of *x* and *y*.

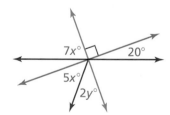

Fair Game Review What you learned in previous grades & lessons

Solve the equation. Check your solution. *(Section 3.3)*

28. $x + 7 = -8$

29. $\dfrac{1}{3} = n + \dfrac{3}{4}$

30. $-12.7 = y - 3.4$

31. MULTIPLE CHOICE Which decimal is equal to 3.7%? *(Section 6.1)*

 A 0.0037 **B** 0.037 **C** 0.37 **D** 3.7

Essential Question How can you construct triangles?

1 ACTIVITY: Constructing Triangles Using Side Lengths

Work with a partner. Cut different-colored straws to the lengths shown. Then construct a triangle with the specified straws if possible. Compare your results with those of others in your class.

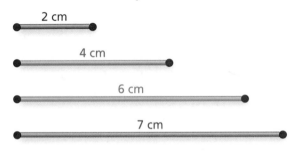

2 cm

4 cm

6 cm

7 cm

  a. blue, green, purple   **b.** red, green, purple

  c. red, blue, purple   **d.** red, blue, green

2 ACTIVITY: Using Technology to Draw Triangles (Side Lengths)

Work with a partner. Use geometry software to draw a triangle with the two given side lengths. What is the length of the third side of your triangle? Compare your results with those of others in your class.

  a. 4 units, 7 units

COMMON CORE

Geometry

In this lesson, you will
- construct triangles with given angle measures.
- construct triangles with given side lengths.

Learning Standard
7.G.2

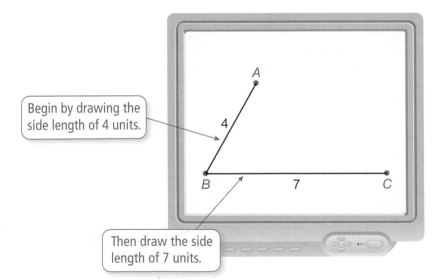

Begin by drawing the side length of 4 units.

Then draw the side length of 7 units.

  b. 3 units, 5 units   **c.** 2 units, 8 units   **d.** 1 unit, 1 unit

3 **ACTIVITY: Constructing Triangles Using Angle Measures**

Work with a partner. Two angle measures of a triangle are given. Draw the triangle. What is the measure of the third angle? Compare your results with those of others in your class.

a. 40°, 70°

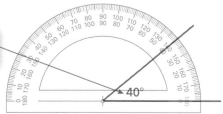

Begin by drawing the angle measure of 40°.

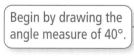

40°

b. 60°, 75°   **c.** 90°, 30°   **d.** 100°, 40°

4 **ACTIVITY: Using Technology to Draw Triangles (Angle Measures)**

Math Practice **5**

Recognize Usefulness of Tools

What are some advantages and disadvantages of using geometry software to draw a triangle?

Work with a partner. Use geometry software to draw a triangle with the two given angle measures. What is the measure of the third angle? Compare your results with those of others in your class.

a. 45°, 55°

b. 50°, 40°

c. 110°, 35°

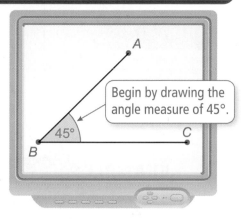

Begin by drawing the angle measure of 45°.

What Is Your Answer?

5. IN YOUR OWN WORDS How can you construct triangles?

6. REASONING Complete the table below for each set of side lengths in Activity 2. Write a rule that compares the sum of any two side lengths to the third side length.

Side Length			
Sum of Other Two Side Lengths			

7. REASONING Use a table to organize the angle measures of each triangle you formed in Activity 3. Include the sum of the angle measures. Then describe the pattern in the table and write a conclusion based on the pattern.

Practice

Use what you learned about constructing triangles to complete Exercises 3–5 on page 286.

Check It Out
Lesson Tutorials
BigIdeasMath com

You can use side lengths and angle measures to classify triangles.

 Key Ideas

Classifying Triangles Using Angles

acute triangle	*obtuse* triangle	*right* triangle	*equiangular* triangle
all acute angles	1 obtuse angle	1 right angle	3 congruent angles

Classifying Triangles Using Sides

Congruent sides have the same length.

scalene triangle	*isosceles* triangle	*equilateral* triangle
no congruent sides	at least 2 congruent sides	3 congruent sides

Reading

Red arcs indicate congruent angles. Red tick marks indicate congruent sides.

EXAMPLE **1** **Classifying Triangles**

Classify each triangle.

a.

The triangle has one obtuse angle and no congruent sides.

⋮ So, the triangle is an obtuse scalene triangle.

b.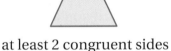

The triangle has all acute angles and two congruent sides.

⋮ So, the triangle is an acute isosceles triangle.

 **On Your Own**

Now You're Ready
Exercises 6–11

Classify the triangle.

1.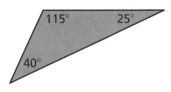

2.

Multi-Language Glossary at BigIdeasMath.com

EXAMPLE 2 **Constructing a Triangle Using Angle Measures**

Draw a triangle with angle measures of 30°, 60°, and 90°. Then classify the triangle.

Step 1: Use a protractor to draw the 30° angle.

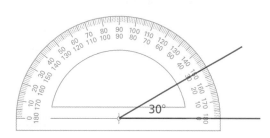

Step 2: Use a protractor to draw the 60° angle.

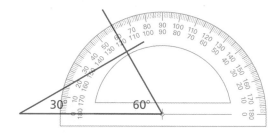

Step 3: The protractor shows that the measure of the remaining angle is 90°.

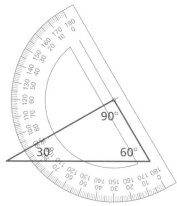

Study Tip

After drawing the first two angles, make sure you check the remaining angle.

∴ The triangle is a right scalene triangle.

EXAMPLE 3 **Constructing a Triangle Using Side Lengths**

Draw a triangle with a 3-centimeter side and a 4-centimeter side that meet at a 20° angle. Then classify the triangle.

Step 1: Use a protractor to draw a 20° angle.

Step 2: Use a ruler to mark 3 centimeters on one ray and 4 centimeters on the other ray.

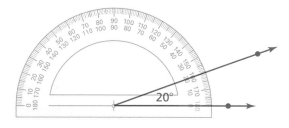

Step 3: Draw the third side to form the triangle.

∴ The triangle is an obtuse scalene triangle.

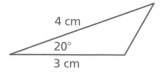

● **On Your Own**

Now You're Ready
Exercises 14–19

3. Draw a triangle with angle measures of 45°, 45°, and 90°. Then classify the triangle.

4. Draw a triangle with a 1-inch side and a 2-inch side that meet at a 60° angle. Then classify the triangle.

✓ Vocabulary and Concept Check

1. **WRITING** How can you classify triangles using angles? using sides?

2. **DIFFERENT WORDS, SAME QUESTION** Which is different? Find "both" answers.

Construct an equilateral triangle.	Construct a triangle with 3 congruent sides.
Construct an equiangular triangle.	Construct a triangle with no congruent sides.

Practice and Problem Solving

Construct a triangle with the given description.

3. side lengths: 4 cm, 6 cm   4. side lengths: 5 cm, 12 cm   5. angles: 65°, 55°

Classify the triangle.

① 6.
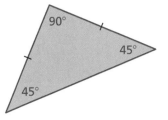
90° 45° 45°

7.
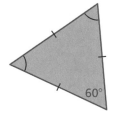
60°

8.
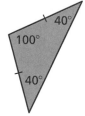
40° 100° 40°

9.
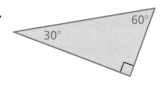
60° 30°

10.
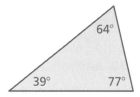
64° 39° 77°

11.
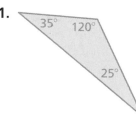
35° 120° 25°

12. **ERROR ANALYSIS** Describe and correct the error in classifying the triangle.

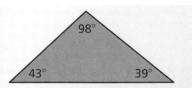
98° 43° 39°

✗ The triangle is acute and scalene because it has two acute angles and no congruent sides.

70° 40° 70°

13. **MOSAIC TILE** A mosaic is a pattern or picture made of small pieces of colored material. Classify the yellow triangle used in the mosaic.

Draw a triangle with the given angle measures. Then classify the triangle.

2 **14.** 15°, 75°, 90°   **15.** 20°, 60°, 100°   **16.** 30°, 30°, 120°

Draw a triangle with the given description.

3 **17.** a triangle with a 2-inch side and a 3-inch side that meet at a 40° angle

18. a triangle with a 45° angle connected to a 60° angle by an 8-centimeter side

19. an acute scalene triangle

20. **LOGIC** You are constructing a triangle. You draw the first angle, as shown. Your friend says that you must be constructing an acute triangle. Is your friend correct? Explain your reasoning.

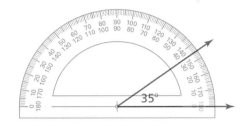

Determine whether you can construct *many, one,* or *no* triangle(s) with the given description. Explain your reasoning.

21. a triangle with angle measures of 50°, 70°, and 100°

22. a triangle with one angle measure of 60° and one 4-centimeter side

23. a scalene triangle with a 3-centimeter side and a 7-centimeter side

24. an isosceles triangle with two 4-inch sides that meet at an 80° angle

25. an isosceles triangle with two 2-inch sides and one 5-inch side

26. a right triangle with three congruent sides

27. **Critical Thinking** Consider the three isosceles triangles.

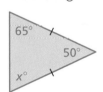
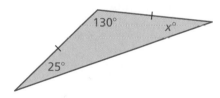

    a. Find the value of *x* for each triangle.

    b. What do you notice about the angle measures of each triangle?

    c. Write a rule about the angle measures of an isosceles triangle.

Fair Game Review *What you learned in previous grades & lessons*

Tell whether *x* and *y* show direct variation. Explain your reasoning. If so, find the constant of proportionality. *(Section 5.6)*

28. $x = 2y$ **29.** $y - x = 6$ **30.** $xy = 5$

31. **MULTIPLE CHOICE** A savings account earns 6% simple interest per year. The principal is $800. What is the balance after 18 months? *(Section 6.7)*

    Ⓐ $864   **Ⓑ** $872   **Ⓒ** $1664   **Ⓓ** $7200

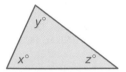

Check It Out
Lesson Tutorials
BigIdeasMath.com

 Key Idea

Sum of the Angle Measures of a Triangle

Words The sum of the angle measures of a triangle is 180°.

Algebra $x + y + z = 180$

EXAMPLE **1** **Finding Angle Measures**

Find each value of x. Then classify each triangle.

a.

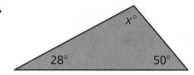

$x + 28 + 50 = 180$

$x + 78 = 180$

$x = 102$

⋮• The value of x is 102. The triangle has one obtuse angle and no congruent sides. So, it is an obtuse scalene triangle.

b.

$x + 45 + 90 = 180$

$x + 135 = 180$

$x = 45$

⋮• The value of x is 45. The triangle has a right angle and two congruent sides. So, it is a right isosceles triangle.

Practice

Find the value of x. Then classify the triangle.

1.

2.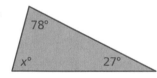

3.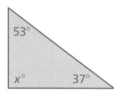

4.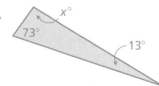

5.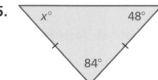

6.

Tell whether a triangle can have the given angle measures. If not, change the first angle measure so that the angle measures form a triangle.

7. $76.2°, 81.7°, 22.1°$

8. $115.1°, 47.5°, 93°$

9. $5\frac{2}{3}°, 64\frac{1}{3}°, 87°$

10. $31\frac{3}{4}°, 53\frac{1}{2}°, 94\frac{3}{4}°$

EXAMPLE 2 Finding Angle Measures

Find each value of x. Then classify each triangle.

a. Flag of Jamaica

$$x + x + 128 = 180$$
$$2x + 128 = 180$$
$$2x = 52$$
$$x = 26$$

∴ The value of x is 26. The triangle has one obtuse angle and two congruent sides. So, it is an obtuse isosceles triangle.

b. Flag of Cuba

$$x + x + 60 = 180$$
$$2x + 60 = 180$$
$$2x = 120$$
$$x = 60$$

∴ The value of x is 60. All three angles are congruent. So, it is an equilateral and equiangular triangle.

Math Practice 1

Analyze Givens
What information is given in the problem? How can you use this information to answer the question?

Practice

Find the value of x. Then classify the triangle.

11.

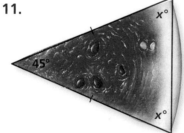

12.

13.

14.

15.

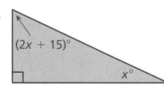

16. REASONING Explain why all triangles have at least two acute angles.

17. CARDS One method of stacking cards is shown.

a. Find the value of x.

b. Describe how to stack the cards with different angles. Is the value of x limited? If so, what are the limitations? Explain your reasoning.

You can use an **example and non-example chart** to list examples and non-examples of a vocabulary word or item. Here is an example and non-example chart for complementary angles.

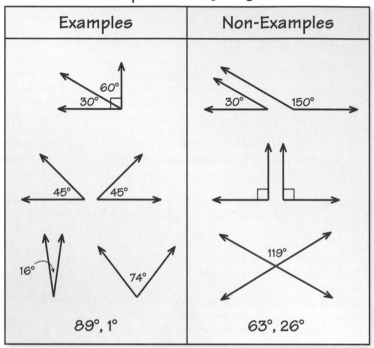

On Your Own

Make example and non-example charts to help you study these topics.

1. adjacent angles

2. vertical angles

3. supplementary angles

After you complete this chapter, make example and non-example charts for the following topics.

4. quadrilaterals

5. scale factor

"What do you think of my example & non-example chart for popular cat toys?"

Name two pairs of adjacent angles and two pairs of vertical angles in the figure.   *(Section 7.1)*

1.

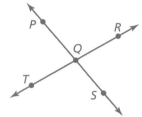

2.

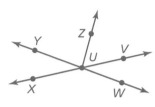

Tell whether the angles are *adjacent* or *vertical*. Then find the value of *x*.   *(Section 7.1)*

3.

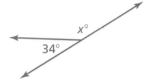

4.

5.

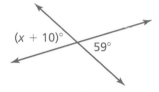

Tell whether the angles are *complementary* or *supplementary*. Then find the value of *x*. *(Section 7.2)*

6.

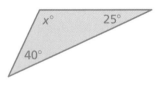

7.

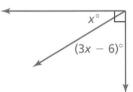

Draw a triangle with the given description.   *(Section 7.3)*

8. a triangle with angle measures of 35°, 65°, and 80°

9. a triangle with a 5-centimeter side and a 7-centimeter side that meet at a 70° angle

10. an obtuse scalene triangle

Find the value of *x*. Then classify the triangle.   *(Section 7.3)*

11.

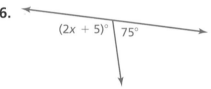

12.

13.

14. **RAILROAD CROSSING** Describe two ways to find the measure of ∠2.   *(Section 7.1 and Section 7.2)*

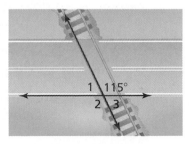

Essential Question How can you classify quadrilaterals?

Quad means *four* and *lateral* means *side*. So, *quadrilateral* means a polygon with *four sides*.

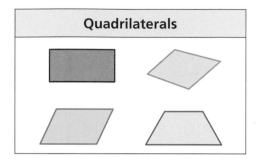

Quadrilaterals

1 ACTIVITY: Using Descriptions to Form Quadrilaterals

Work with a partner. Use a geoboard to form a quadrilateral that fits the given description. Record your results on geoboard dot paper.

a. Form a quadrilateral with exactly one pair of parallel sides.

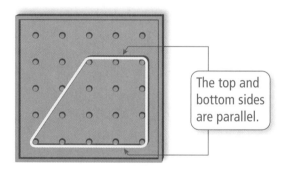

The top and bottom sides are parallel.

b. Form a quadrilateral with four congruent sides and four right angles.

c. Form a quadrilateral with four right angles that is *not* a square.

d. Form a quadrilateral with four congruent sides that is *not* a square.

e. Form a quadrilateral with two pairs of congruent adjacent sides and whose opposite sides are *not* congruent.

f. Form a quadrilateral with congruent and parallel opposite sides that is *not* a rectangle.

2 ACTIVITY: Naming Quadrilaterals

Work with a partner. Match the names *square, rectangle, rhombus, parallelogram, trapezoid,* and *kite* with your 6 drawings in Activity 1.

COMMON CORE

Geometry

In this lesson, you will

- understand that the sum of the angle measures of any quadrilateral is 360°.
- find missing angle measures in quadrilaterals.
- construct quadrilaterals.

Learning Standard 7.G.2

ACTIVITY: Forming Quadrilaterals

Work with a partner. Form each quadrilateral on your geoboard. Then move
only one **vertex to create the new type of quadrilateral. Record your results**
on geoboard dot paper.

a. Trapezoid ⟹ Kite

b. Kite ⟹ Rhombus (*not* a square)

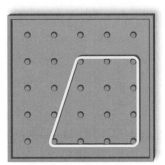

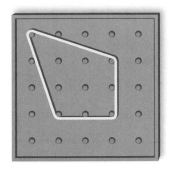

4 ACTIVITY: Using Technology to Draw Quadrilaterals

Math Practice 5

Use Technology to Explore

How does geometry software help you learn about the characteristics of a quadrilateral?

Work with a partner. Use geometry software to draw a quadrilateral that fits
the given description.

a. a square with a side length of 3 units

b. a rectangle with a width of 2 units and a length of 5 units

c. a parallelogram with side lengths of 6 units and 1 unit

d. a rhombus with a side length of 4 units

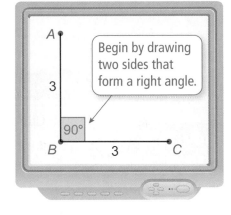

Begin by drawing two sides that form a right angle.

What Is Your Answer?

5. **REASONING** Measure the angles of each quadrilateral you formed in Activity 1. Record your results in a table. Include the sum of the angle measures. Then describe the pattern in the table and write a conclusion based on the pattern.

6. **IN YOUR OWN WORDS** How can you classify quadrilaterals? Explain using properties of sides and angles.

Practice

Use what you learned about quadrilaterals to complete Exercises 4–6 on page 296.

7.4 Lesson

Check It Out
Lesson Tutorials
BigIdeasMath ✓com

Key Vocabulary 🔊
kite, *p. 294*

A quadrilateral is a polygon with four sides. The diagram shows properties of different types of quadrilaterals and how they are related. When identifying a quadrilateral, use the name that is most specific.

Reading

Red arrows indicate parallel sides.

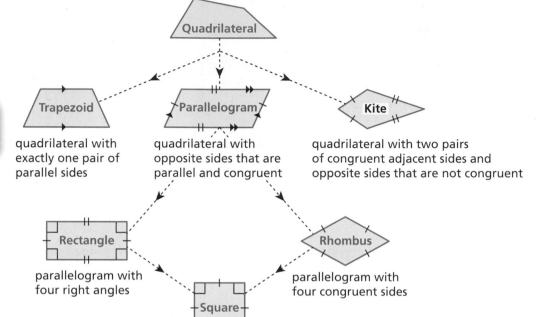

Quadrilateral

Trapezoid
quadrilateral with exactly one pair of parallel sides

Parallelogram
quadrilateral with opposite sides that are parallel and congruent

Kite
quadrilateral with two pairs of congruent adjacent sides and opposite sides that are not congruent

Rectangle
parallelogram with four right angles

Rhombus
parallelogram with four congruent sides

Square
parallelogram with four congruent sides and four right angles

---

EXAMPLE ① **Classifying Quadrilaterals**

Study Tip

In Example 1(a), the square is also a parallelogram, a rectangle, and a rhombus. Square is the most specific name.

Classify the quadrilateral.

a.

The quadrilateral has four congruent sides and four right angles.

⁙· So, the quadrilateral is a square.

b.

The quadrilateral has two pairs of congruent adjacent sides and opposite sides that are not congruent.

⁙· So, the quadrilateral is a kite.

On Your Own

Now You're Ready
Exercises 4–9

Classify the quadrilateral.

1.

2.

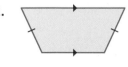

3.

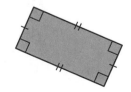

Key Idea

Sum of the Angle Measures of a Quadrilateral

Words The sum of the angle measures of a quadrilateral is 360°.

Algebra $w + x + y + z = 360$

EXAMPLE 2 **Finding an Angle Measure of a Quadrilateral**

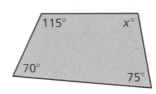

Find the value of x.

$$70 + 75 + 115 + x = \boxed{360} \qquad \text{Write an equation.}$$

$$260 + x = 360 \qquad \text{Combine like terms.}$$

$$\underline{-\,260} \qquad \underline{-\,260} \qquad \text{Subtraction Property of Equality}$$

$$x = 100 \qquad \text{Simplify.}$$

∴ The value of x is 100.

EXAMPLE 3 **Constructing a Quadrilateral**

Draw a parallelogram with a 60° angle and a 120° angle.

Step 1: Draw a line.

Step 2: Draw a 60° angle and a 120° angle that each have one side on the line.

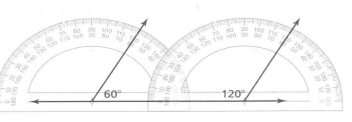

Step 3: Draw the remaining side. Make sure that both pairs of opposite sides are parallel and congruent.

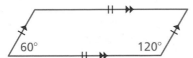

● **On Your Own**

Find the value of x.

4.

5.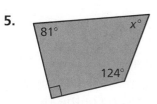

6. Draw a right trapezoid whose parallel sides have lengths of 3 centimeters and 5 centimeters.

Vocabulary and Concept Check

1. **VOCABULARY** Which statements are true?

   a. All squares are rectangles.   b. All squares are parallelograms.

   c. All rectangles are parallelograms.   d. All squares are rhombuses.

   e. All rhombuses are parallelograms.

2. **REASONING** Name two types of quadrilaterals with four right angles.

3. **WHICH ONE DOESN'T BELONG?** Which type of quadrilateral
   does *not* belong with the other three? Explain your reasoning.

rectangle	parallelogram	square	kite

Practice and Problem Solving

Classify the quadrilateral.

① 4.    5.    6.

7.   8.   9.

Find the value of x.

② 10.
65° 115°
115° x°

11.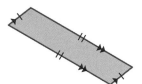
128° x°
82°
40°

12.
x° 52°

13. **KITE MAKING** What is the
    measure of the angle at
    the tail end of the kite?

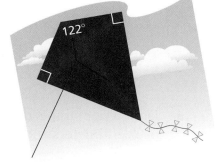
122°

Draw a quadrilateral with the given description.

③ 14. a trapezoid with a pair of congruent, nonparallel sides

15. a rhombus with 3-centimeter sides and two 100° angles

16. a parallelogram with a 45° angle and a 135° angle

17. a parallelogram with a 75° angle and a 4-centimeter side

Copy and complete using *always*, *sometimes*, or *never*.

18. A square is __?__ a rectangle.

19. A square is __?__ a rhombus.

20. A rhombus is __?__ a square.

21. A parallelogram is __?__ a trapezoid.

22. A trapezoid is __?__ a kite.

23. A rhombus is __?__ a rectangle.

24. DOOR The dashed line shows how you cut the bottom of a rectangular door so it opens more easily.

a. Identify the new shape of the door. Explain.

b. What is the new angle at the bottom left side of the door? Explain.

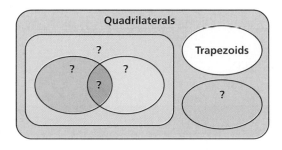

25. VENN DIAGRAM The diagram shows that some quadrilaterals are trapezoids, and all trapezoids are quadrilaterals. Copy the diagram. Fill in the names of the types of quadrilaterals to show their relationships.

26. Structure Consider the parallelogram.

a. Find the values of *x* and *y*.

b. Make a conjecture about opposite angles in a parallelogram.

c. In polygons, consecutive interior angles share a common side. Make a conjecture about consecutive interior angles in a parallelogram.

 Fair Game Review What you learned in previous grades & lessons

Write the ratio as a fraction in simplest form. *(Section 5.1)*

27. 3 turnovers : 12 assists

28. 18 girls to 27 boys

29. 42 pens : 35 pencils

30. MULTIPLE CHOICE Computer sales decreased from 40 to 32. What is the percent of decrease? *(Section 6.5)*

    Ⓐ 8%   Ⓑ 20%   Ⓒ 25%   Ⓓ 80%

7.5 Scale Drawings

Essential Question How can you enlarge or reduce a drawing proportionally?

---

1 **ACTIVITY: Comparing Measurements**

Work with a partner. The diagram shows a food court at a shopping mall. Each centimeter in the diagram represents 40 meters.

a. Find the length and the width of the drawing of the food court.

length: ▮ cm   width: ▮ cm

b. Find the actual length and width of the food court. Explain how you found your answers.

length: ▮ m   width: ▮ m

c. Find the ratios $\dfrac{\text{drawing length}}{\text{actual length}}$ and $\dfrac{\text{drawing width}}{\text{actual width}}$. What do you notice?

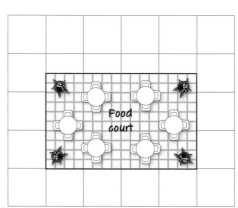

---

2 **ACTIVITY: Recreating a Drawing**

Work with a partner. Draw the food court in Activity 1 on the grid paper so that each centimeter represents 20 meters.

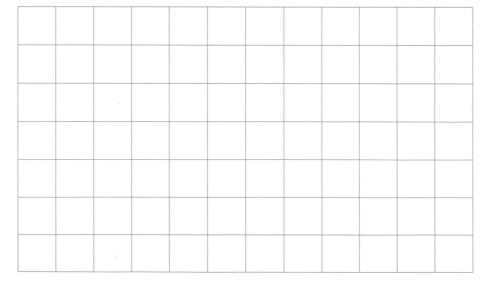

COMMON CORE

Geometry

In this lesson, you will

- use scale drawings to find actual distances.
- find scale factors.
- use scale drawings to find actual perimeters and areas.
- recreate scale drawings at a different scale.

Learning Standard
7.G.1

a. What happens to the size of the drawing?

b. Find the length and the width of your drawing. Compare these dimensions to the dimensions of the original drawing in Activity 1.

3   ACTIVITY: Comparing Measurements

Work with a partner. The diagram shows a sketch of a painting.
Each unit in the sketch represents 8 inches.

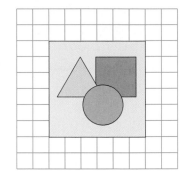

a. Find the length and the width of the sketch.

   length: ▢ units   width: ▢ units

b. Find the actual length and width of the painting.
   Explain how you found your answers.

   length: ▢ in.   width: ▢ in.

c. Find the ratios $\dfrac{\text{sketch length}}{\text{actual length}}$ and $\dfrac{\text{sketch width}}{\text{actual width}}$.
   What do you notice?

4   ACTIVITY: Recreating a Drawing

Math Practice 6

Specify Units
How do you know whether to use feet or units for each measurement?

Work with a partner. Let each unit in the grid
paper represent 2 feet. Now sketch the painting
in Activity 3 onto the grid paper.

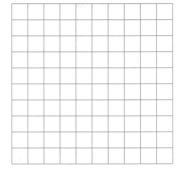

a. What happens to the size of the sketch?

b. Find the length and the width of your sketch.
   Compare these dimensions to the dimensions
   of the original sketch in Activity 3.

What Is Your Answer?

5. **IN YOUR OWN WORDS** How can you enlarge or reduce a drawing
   proportionally?

6. Complete the table for both the food court and the painting.

	Actual Object	Original Drawing	Your Drawing
Perimeter			
Area			

   Compare the measurements in each table. What conclusions can you make?

7. **RESEARCH** Look at some maps in your school library or on the
   Internet. Make a list of the different scales used on the maps.

8. When you view a map on the Internet, how does the scale change
   when you zoom out? How does the scale change when you zoom in?

Practice

Use what you learned about enlarging or reducing drawings
to complete Exercises 4–7 on page 303.

Check It Out
Lesson Tutorials
BigIdeasMath ✓com

Key Vocabulary ◀》
scale drawing, *p. 300*
scale model, *p. 300*
scale, *p. 300*
scale factor, *p. 301*

Key Ideas

Scale Drawings and Models

A **scale drawing** is a proportional, two-dimensional drawing of an object.
A **scale model** is a proportional, three-dimensional model of an object.

Scale

The measurements in scale drawings and models are proportional to
the measurements of the actual object. The **scale** gives the ratio
that compares the measurements of the drawing or model with
the actual measurements.

Study Tip

Scales are written
so that the drawing
distance comes first in
the ratio.

$$\frac{1 \text{ in.}}{10 \text{ mi}}$$ ← drawing distance
← actual distance

$$1 \text{ in.} : 10 \text{ mi}$$
↑ drawing   ↑ actual

EXAMPLE ① **Finding an Actual Distance**

What is the actual distance *d* between Cadillac and Detroit?

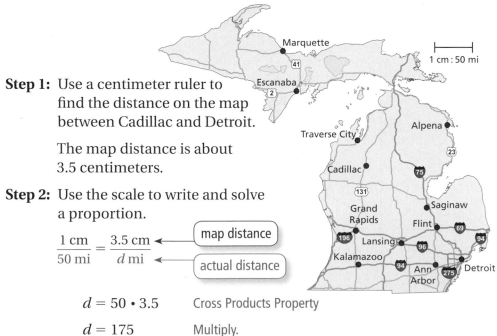

1 cm : 50 mi

Step 1: Use a centimeter ruler to
find the distance on the map
between Cadillac and Detroit.

The map distance is about
3.5 centimeters.

Step 2: Use the scale to write and solve
a proportion.

$$\frac{1 \text{ cm}}{50 \text{ mi}} = \frac{3.5 \text{ cm}}{d \text{ mi}}$$ ← map distance
← actual distance

$d = 50 \cdot 3.5$ Cross Products Property

$d = 175$ Multiply.

∴ So, the distance between Cadillac and Detroit is about 175 miles.

On Your Own

Now You're Ready
Exercises 8–11

1. What is the actual distance between Traverse City and Marquette?

◀》 Multi-Language Glossary at BigIdeasMath✓com

EXAMPLE 2 Finding a Distance in a Model

The liquid outer core of Earth is 2300 kilometers thick. A scale model of the layers of Earth has a scale of 1 in. : 500 km. How thick is the liquid outer core of the model?

(A) 0.2 in.   (B) 4.6 in.   (C) 0.2 km   (D) 4.6 km

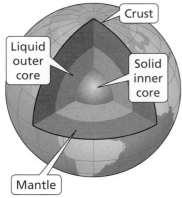

$$\frac{1 \text{ in.}}{500 \text{ km}} = \frac{x \text{ in.}}{2300 \text{ km}}$$ ← model thickness
← actual thickness

$$\frac{1 \text{ in.}}{500 \text{ km}} \cdot 2300 \text{ km} = \frac{x \text{ in.}}{2300 \text{ km}} \cdot 2300 \text{ km}$$ Multiplication Property of Equality

$$4.6 = x$$ Simplify.

So, the liquid outer core of the model is 4.6 inches thick. The correct answer is (B).

On Your Own

2. The mantle of Earth is 2900 kilometers thick. How thick is the mantle of the model?

A scale can be written without units when the units are the same. A scale without units is called a **scale factor**.

EXAMPLE 3 Finding a Scale Factor

A scale model of the Sergeant Floyd Monument is 10 inches tall. The actual monument is 100 feet tall.

a. What is the scale of the model?

$$\frac{\text{model height}}{\text{actual height}} = \frac{10 \text{ in.}}{100 \text{ ft}} = \frac{1 \text{ in.}}{10 \text{ ft}}$$

The scale is 1 in. : 10 ft.

b. What is the scale factor of the model?

Write the scale with the same units. Use the fact that 1 ft = 12 in.

$$\text{scale factor} = \frac{1 \text{ in.}}{10 \text{ ft}} = \frac{1 \text{ in.}}{120 \text{ in.}} = \frac{1}{120}$$

The scale factor is 1 : 120.

On Your Own

Now You're Ready
Exercises 12–16

3. A drawing has a scale of 1 mm : 20 cm. What is the scale factor of the drawing?

EXAMPLE 4 **Finding an Actual Perimeter and Area**

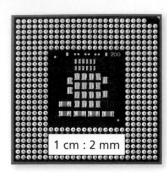

1 cm : 2 mm

The scale drawing of a computer chip helps you see the individual components on the chip.

a. Find the perimeter and the area of the computer chip in the scale drawing.

When measured using a centimeter ruler, the scale drawing of the computer chip has a side length of 4 centimeters.

∴ So, the perimeter of the computer chip in the scale drawing is 4(4) = 16 centimeters, and the area is 4^2 = 16 square centimeters.

b. Find the actual perimeter and area of the computer chip.

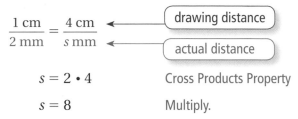

$$\frac{1 \text{ cm}}{2 \text{ mm}} = \frac{4 \text{ cm}}{s \text{ mm}}$$ ← drawing distance
    ← actual distance

$s = 2 \cdot 4$ Cross Products Property

$s = 8$ Multiply.

The side length of the actual computer chip is 8 millimeters.

∴ So, the actual perimeter of the computer chip is 4(8) = 32 millimeters, and the actual area is 8^2 = 64 square millimeters.

c. Compare the ratios $\dfrac{\text{drawing perimeter}}{\text{actual perimeter}}$ and $\dfrac{\text{drawing area}}{\text{actual area}}$ to the scale factor.

Use the fact that 1 cm = 10 mm.

$$\text{scale factor} = \frac{1 \text{ cm}}{2 \text{ mm}} = \frac{10 \text{ mm}}{2 \text{ mm}} = \frac{5}{1}$$

$$\frac{\text{drawing perimeter}}{\text{actual perimeter}} = \frac{16 \text{ cm}}{32 \text{ mm}} = \frac{1 \text{ cm}}{2 \text{ mm}} = \frac{5}{1}$$

$$\frac{\text{drawing area}}{\text{actual area}} = \frac{16 \text{ cm}^2}{64 \text{ mm}^2} = \frac{1 \text{ cm}^2}{4 \text{ mm}^2} = \left(\frac{1 \text{ cm}}{2 \text{ mm}}\right)^2 = \left(\frac{5}{1}\right)^2$$

∴ So, the ratio of the perimeters is equal to the scale factor, and the ratio of the areas is equal to the square of the scale factor.

Study Tip

The ratios tell you that the perimeter of the drawing is 5 times the actual perimeter, and the area of the drawing is 5^2 = 25 times the actual area.

● **On Your Own**

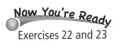
Now You're Ready
Exercises 22 and 23

4. **WHAT IF?** The scale of the drawing of the computer chip is 1 cm : 3 mm. How do the answers in parts (a)–(c) change? Justify your answer.

 ## Vocabulary and Concept Check

1. **VOCABULARY** Compare and contrast the terms *scale* and *scale factor*.

2. **CRITICAL THINKING** The scale of a drawing is 2 cm : 1 mm. Is the scale drawing *larger* or *smaller* than the actual object? Explain.

3. **REASONING** How would you find the scale factor of a drawing that shows a length of 4 inches when the actual object is 8 feet long?

 ## Practice and Problem Solving

Use the drawing and a centimeter ruler. Each centimeter in the drawing represents 5 feet.

4. What is the actual length of the flower garden?

5. What are the actual dimensions of the rose bed?

6. What are the actual perimeters of the perennial beds?

7. The area of the tulip bed is what percent of the area of the rose bed?

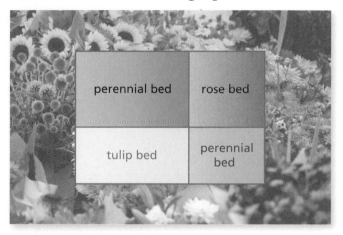

Use the map in Example 1 to find the actual distance between the cities.

① 8. Kalamazoo and Ann Arbor

9. Lansing and Flint

10. Grand Rapids and Escanaba

11. Saginaw and Alpena

Find the missing dimension. Use the scale factor 1 : 12.

	Item	Model	Actual
② ③ 12.	Mattress	Length: 6.25 in.	Length: ___ in.
13.	Corvette	Length: ___ in.	Length: 15 ft
14.	Water tower	Depth: 32 cm	Depth: ___ m
15.	Wingspan	Width: 5.4 ft	Width: ___ yd
16.	Football helmet	Diameter: ___ mm	Diameter: 21 cm

17. **ERROR ANALYSIS** A scale is 1 cm : 20 m. Describe and correct the error in finding the actual distance that corresponds to 5 centimeters.

$$\times \quad \frac{1\,cm}{20\,m} = \frac{x\,m}{5\,cm}$$

$$x = 0.25\,m$$

Use a centimeter ruler to measure the segment shown. Find the scale of the drawing.

18. |—— 120 m ——|

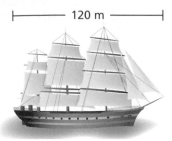

19.

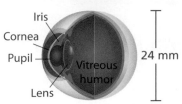

Iris
Cornea
Pupil
Vitreous humor
Lens
24 mm

20. REASONING You know the length and the width of a scale model. What additional information do you need to know to find the scale of the model?

21. OPEN-ENDED You are in charge of creating a billboard advertisement with the dimensions shown.

   a. Choose a product. Then design the billboard using words and a picture.

   b. What is the scale factor of your design?

16 ft

8 ft

YOUR AD HERE

④ 22. CENTRAL PARK Central Park is a rectangular park in New York City.

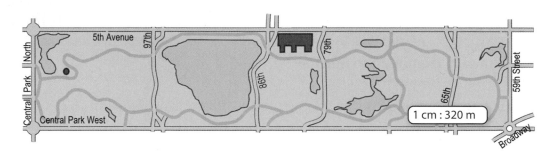

5th Avenue
97th
79th
86th
65th
59th Street
North
Central Park
Central Park West
Broadway
1 cm : 320 m

   a. Find the perimeter and the area of Central Park in the scale drawing.

   b. Find the actual perimeter and area of Central Park.

23. ICON You are designing an icon for a mobile app.

   a. Find the perimeter and the area of the icon in the scale drawing.

   b. Find the actual perimeter and area of the icon.

1 cm : 2.5 mm

24. CRITICAL THINKING Use the results of Exercises 22 and 23 to make a conjecture about the relationship between the scale factor of a drawing and the ratios $\dfrac{\text{drawing perimeter}}{\text{actual perimeter}}$ and $\dfrac{\text{drawing area}}{\text{actual area}}$.

Recreate the scale drawing so that it has a scale of 1 cm : 4 m.

25.

1 cm : 8 m

26.

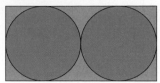

1 cm : 2 m

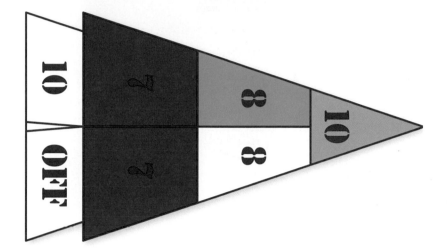

The shuffleboard diagram has a scale of 1 cm : 1 ft. Find the actual area of the region.

27. red region

28. blue region

29. green region

30. **BLUEPRINT** In a blueprint, each square has a side length of $\frac{1}{4}$ inch.

   a. Ceramic tile costs $5 per square foot. How much would it cost to tile the bathroom?

   b. Carpet costs $18 per square yard. How much would it cost to carpet the bedroom and living room?

   c. Which has a greater unit cost, the tile or the carpet? Explain.

Reduced Drawing of Blueprint

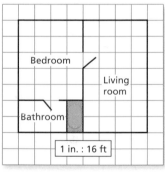

Bedroom
Living room
Bathroom

1 in. : 16 ft

31. **Modeling** You are making a scale model of the solar system. The radius of Earth is 6378 kilometers. The radius of the Sun is 695,500 kilometers. Is it reasonable to choose a baseball as a model of Earth? Explain your reasoning.

Fair Game Review *What you learned in previous grades & lessons*

Plot and label the ordered pair in a coordinate plane. *(Skills Review Handbook)*

32. $A(-4, 3)$ 33. $B(2, -6)$ 34. $C(5, 1)$ 35. $D(-3, -7)$

36. **MULTIPLE CHOICE** Which set of numbers is ordered from least to greatest? *(Section 6.2)*

   Ⓐ $\frac{7}{20}$, 32%, 0.45   Ⓑ 17%, 0.21, $\frac{3}{25}$ Ⓒ 0.88, $\frac{7}{8}$, 93%   Ⓓ 57%, $\frac{11}{16}$, 5.7

Classify the quadrilateral.   *(Section 7.4)*

1.

2.

Find the value of x.   *(Section 7.4)*

3.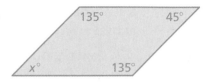

4.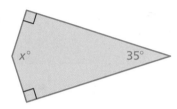

Draw a quadrilateral with the given description.   *(Section 7.4)*

5. a rhombus with 2-centimeter sides and two 50° angles

6. a parallelogram with a 65° angle and a 5-centimeter side

Find the missing dimension. Use the scale factor 1 : 20.   *(Section 7.5)*

	Item	Model	Actual
7.	Basketball player	Height: in.	Height: 90 in.
8.	Dinosaur	Length: 3.75 ft	Length: ft

9. **SHED** The side of the storage shed is in the shape of a trapezoid. Find the value of *x*.   *(Section 7.4)*

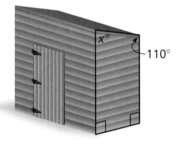

10. **DOLPHIN** A dolphin in an aquarium is 12 feet long. A scale model of the dolphin is $3\frac{1}{2}$ inches long. What is the scale factor of the model?   *(Section 7.5)*

11. **SOCCER** A scale drawing of a soccer field is shown. The actual soccer field is 300 feet long.
(Section 7.5)

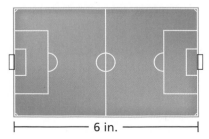

    a. What is the scale of the drawing?

    b. What is the scale factor of the drawing?

Check It Out
Vocabulary Help
BigIdeasMath ✓com

Review Key Vocabulary

adjacent angles, *p. 272*
vertical angles, *p. 272*
congruent angles, *p. 272*
complementary angles,
   p. 278

supplementary angles,
   p. 278
congruent sides, *p. 284*
kite, *p. 294*
scale drawing, *p. 300*

scale model, *p. 300*
scale, *p. 300*
scale factor, *p. 301*

Review Examples and Exercises

7.1 Adjacent and Vertical Angles   *(pp. 270–275)*

Tell whether the angles are *adjacent* or *vertical*. Then find the value of *x*.

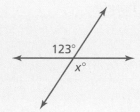

The angles are vertical angles. Because vertical angles are congruent, the angles have the same measure.

⁖ So, the value of *x* is 123.

Exercises

Tell whether the angles are *adjacent* or *vertical*. Then find the value of *x*.

1.

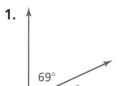

2.

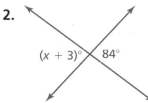

7.2 Complementary and Supplementary Angles   *(pp. 276–281)*

Tell whether the angles are *complementary* or *supplementary*. Then find the value of *x*.

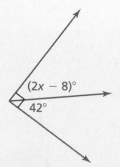

The two angles make up a right angle. So, the angles are complementary angles, and the sum of their measures is 90°.

$(2x - 8) + 42 = 90$	Write equation.
$2x + 34 = 90$	Combine like terms.
$2x = 56$	Subtract 34 from each side.
$x = 28$	Divide each side by 2.

⁖ So, the value of *x* is 28.

Exercises

Tell whether the angles are *complementary* or *supplementary*. Then find the value of *x*.

3.

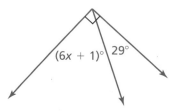

4.

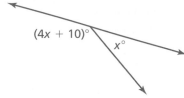

7.3
Triangles *(pp. 282–289)*

Draw a triangle with a 3.5-centimeter side and a 4-centimeter side that meet at a 25° angle. Then classify the triangle.

Step 1: Use a protractor to draw a 25° angle.

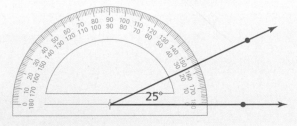

Step 2: Use a ruler to mark 3.5 centimeters on one ray and 4 centimeters on the other ray.

Step 3: Draw the third side to form the triangle.

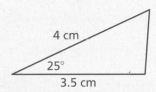

∴ The triangle is an obtuse scalene triangle.

Exercises

Draw a triangle with the given description.

5. a triangle with angle measures of 40°, 50°, and 90°

6. a triangle with a 3-inch side and a 4-inch side that meet at a 30° angle

Find the value of *x*. Then classify the triangle.

7.

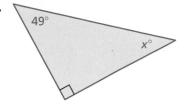

8.

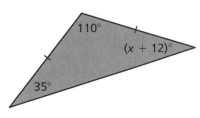

7.4 Quadrilaterals *(pp. 292–297)*

Draw a parallelogram with a 50° angle and a 130° angle.

Step 1: Draw a line.

Step 2: Draw a 50° angle and a 130° angle that each have one side on the line.

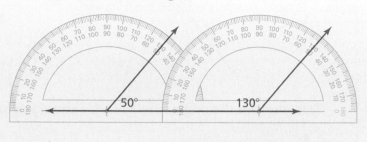

Step 3: Draw the remaining side. Make sure that both pairs of opposite sides are parallel and congruent.

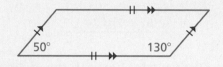

Exercises

Find the value of *x*.

9.

128°   *x*°

10.

80° *x*°

95°   38°

11. Draw a rhombus with 5-centimeter sides and two 120° angles.

7.5 Scale Drawings *(pp. 298–305)*

A lighthouse is 160 feet tall. A scale model of the lighthouse has a scale of 1 in. : 8 ft. How tall is the model of the lighthouse?

$$\frac{1 \text{ in.}}{8 \text{ ft}} = \frac{x \text{ in.}}{160 \text{ ft}}$$ ← model height

← actual height

$$\frac{1 \text{ in.}}{8 \text{ ft}} \cdot 160 \text{ ft} = \frac{x \text{ in.}}{160 \text{ ft}} \cdot 160 \text{ ft}$$ Multiplication Property of Equality

$$20 = x$$ Simplify.

∴ So, the model of the lighthouse is 20 inches tall.

Exercises

Use a centimeter ruler to measure the segment shown. Find the scale of the drawing.

12. |———— 30 in. ————|

13. |— 7.5 in. —|

Check It Out
Test Practice
BigIdeasMath V.com

Tell whether the angles are *adjacent* or *vertical*. Then find the value of *x*.

1.
113°
x°

2.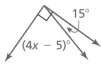
(*x* + 6)°
56°

Tell whether the angles are *complementary* or *supplementary*. Then find the value of *x*.

3.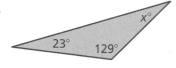
(8*x* + 2)° / 74°

4.
15°
(4*x* − 5)°

Draw a triangle with the given angle measures. Then classify the triangle.

5. 10°, 80°, 90°

6. 30°, 40°, 110°

Draw a triangle with the given description.

7. a triangle with a 5-inch side and a 6-inch side that meet at a 50° angle

8. a right isosceles triangle

Find the value of *x*. Then classify the triangle.

9.
x°
23° 129°

10.
x°
68° *x*°

11.
x°
x° *x*°

Find the value of *x*.

12.
x°

13.
95° *x*°
95°

14.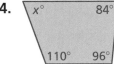
x° 84°
110° 96°

Draw a quadrilateral with the given description.

15. a rhombus with 6-centimeter sides and two 80° angles

16. a parallelogram with a 20° angle and a 160° angle

17. FISH Use a centimeter ruler to measure the fish. Find the scale factor of the drawing.

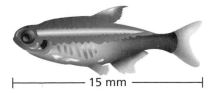
15 mm

18. CAD An engineer is using computer-aided design (CAD) software to design a component for a space shuttle. The scale of the drawing is 1 cm : 60 in. The actual length of the component is 12.5 feet. What is the length of the component in the drawing?

1. The number of calories you burn by playing basketball is proportional to the number of minutes you play. Which of the following is a valid interpretation of the graph below? *(7.RP.2d)*

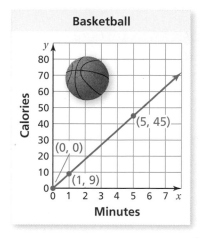

Basketball

(5, 45)
(0, 0)
(1, 9)

   A. The unit rate is $\frac{1}{9}$ calorie per minute.

   B. You burn 5 calories by playing basketball for 45 minutes.

   C. You do not burn any calories if you do not play basketball for at least 1 minute.

   D. You burn an additional 9 calories for each minute of basketball you play.

2. A lighting store is holding a clearance sale. The store is offering discounts on all the lamps it sells. As the sale progresses, the store will increase the percent of discount it is offering.

   You want to buy a lamp that has an original price of $40. You will buy the lamp when its price is marked down to $10. What percent discount will you have received? *(7.RP.3)*

3. What is the value of the expression below? *(7.NS.1c)*

$$2 - 6 - (-9)$$

   F. -13 **H.** 5

   G. -5 **I.** 13

4. What is the solution to the proportion below?   *(7.RP.2c)*

$$\frac{8}{12} = \frac{x}{18}$$

5. Which graph represents the inequality below?   *(7.EE.4b)*

$$-5 - 6x \le -23$$

A.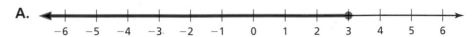

B.

C.

D.

6. You are building a scale model of a park that is planned for a city. The model uses the scale below.

$$1 \text{ centimeter} = 2 \text{ meters}$$

The park will have a rectangular reflecting pool with a length of 20 meters and a width of 12 meters. In your scale model, what will be the area of the reflecting pool?   *(7.G.1)*

F. 60 cm^2 **H.** 480 cm^2

G. 120 cm^2 **I.** 960 cm^2

7. The quantities x and y are proportional. What is the missing value in the table?   *(7.RP.2a)*

x	y
$\frac{5}{7}$	10
$\frac{9}{7}$	18
$\frac{15}{7}$	30
4	

A. 38   **C.** 46

B. 42   **D.** 56

8. ∠1 and ∠2 form a straight angle. ∠1 has a measure of 28°. What is the measure of ∠2? *(7.G.5)*

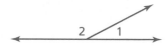

   F. 62°

   G. 118°

   H. 152°

   I. 208°

9. Brett solved the equation in the box below. *(7.EE.4a)*

$$\frac{c}{5} - (-15) = -35$$

$$\frac{c}{5} + 15 = -35$$

$$\frac{c}{5} + 15 - 15 = -35 - 15$$

$$\frac{c}{5} = -50$$

$$\frac{c}{5} = \frac{-50}{5}$$

$$c = -10$$

   What should Brett do to correct the error that he made?

   A. Subtract 15 from −35 to get −20.

   B. Rewrite $\frac{c}{5} - (-15)$ as $\frac{c}{5} - 15$.

   C. Multiply each side of the equation by 5 to get $c = -250$.

   D. Multiply each side of the equation by −5 to get $c = 250$.

10. A map of the state where Donna lives has the scale shown below. *(7.G.1)*

$$\frac{1}{2} \text{ inch} = 10 \text{ miles}$$

    Part A Donna measured the distance between her town and the state capital on the map. Her measurement was $4\frac{1}{2}$ inches. Based on Donna's measurement, what is the actual distance, in miles, between her town and the state capital? Show your work and explain your reasoning.

    Part B Donna wants to mark her favorite campsite on the map. She knows that the campsite is 65 miles north of her town. What distance on the map, in inches, represents an actual distance of 65 miles? Show your work and explain your reasoning.

8 Circles and Area

$\pi \approx 3.141592\ldots$

"Think of any number between 1 and 9."

"Okay, now add 4 to the number, multiply by 3, subtract 12, and divide by your original number."

Can I start over?

"You end up with 3, don't you?"

"What do you get when you divide the circumference of a jack-o-lantern by its diameter?"

I love math!

"Pumpkin pi, HE HE HE."

What You Learned Before

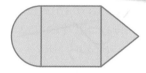

"The area of the circle is pi *r* squared. The area of the triangle is one-half *bh*."

Classifying Figures (4.G.2)

Identify the basic shapes in the figure.

Example 1

∴ Rectangle, right triangle

Example 2

∴ Semicircle, square, and triangle

Try It Yourself
Identify the basic shapes in the figure.

1.

2.

3.

4.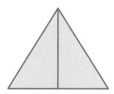

5.

6.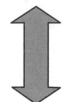

Squaring Numbers and Using Order of Operations (6.EE.1)

Example 3 Evaluate 4^2.

$$4^2 = 4 \cdot 4 = 16$$

4^2 means to multiply 4 by itself.

Example 4 Evaluate $3 \cdot 6^2$.

$$3 \cdot 6^2 = 3 \cdot (6 \cdot 6) = 3 \cdot 36 = 108$$

Use order of operations. Evaluate the exponent, and then multiply.

Try It Yourself
Evaluate the expression.

7. 5^2

8. 12^2

9. $3 \cdot 2^2$

10. $4 \cdot 7^2$

11. $3(1 + 8)^2$

12. $2(3 + 7)^2 - 3 \cdot 4$

Essential Question How can you find the circumference of a circle?

Archimedes was a Greek mathematician, physicist, engineer, and astronomer.

Archimedes discovered that in any circle the ratio of circumference to diameter is always the same. Archimedes called this ratio pi, or π (a letter from the Greek alphabet).

$$\pi = \frac{\text{circumference}}{\text{diameter}}$$

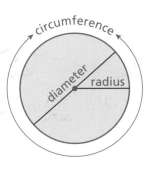

In Activities 1 and 2, you will use the same strategy Archimedes used to approximate π.

1 ACTIVITY: Approximating Pi

Work with a partner. Copy the table. Record your results in the table.

- **Measure the perimeter of the large square in millimeters.**

- **Measure the diameter of the circle in millimeters.**

- **Measure the perimeter of the small square in millimeters.**

- **Calculate the ratios of the two perimeters to the diameter.**

- **The average of these two ratios is an approximation of π.**

Large Square
Small Square

COMMON CORE

Geometry
In this lesson, you will
- describe a circle in terms of radius and diameter.
- understand the concept of pi.
- find circumferences of circles and perimeters of semicircles.

Learning Standard
7.G.4

Sides	Large Perimeter	Diameter of Circle	Small Perimeter	$\dfrac{\text{Large Perimeter}}{\text{Diameter}}$	$\dfrac{\text{Small Perimeter}}{\text{Diameter}}$	Average of Ratios
4						
6						
8						
10						

A page from *Sir Cumference and the First Round Table* by Cindy Neuschwander

2 **ACTIVITY: Approximating Pi**

Math Practice 3

Make Conjectures

How can you use the results of the activity to find an approximation of pi?

Continue your approximation of pi. Complete the table from Activity 1 using a hexagon (6 sides), an octagon (8 sides), and a decagon (10 sides).

a.
Large Hexagon   Small Hexagon

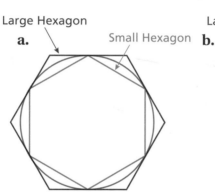

b.
Large Octagon   Small Octagon

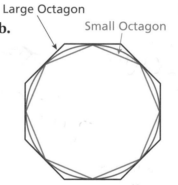

c.
Large Decagon   Small Decagon

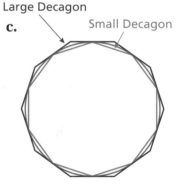

d. From the table, what can you conclude about the value of π? Explain your reasoning.

e. Archimedes calculated the value of π using polygons with 96 sides. Do you think his calculations were more or less accurate than yours?

What Is Your Answer?

3. **IN YOUR OWN WORDS** Now that you know an approximation for pi, explain how you can use it to find the circumference of a circle. Write a formula for the circumference C of a circle whose diameter is d.

4. **CONSTRUCTION** Use a compass to draw three circles. Use your formula from Question 3 to find the circumference of each circle.

Practice ➤ Use what you learned about circles and circumference to complete Exercises 9–11 on page 321.

Check It Out
Lesson Tutorials
BigIdeasMath ✓com

Key Vocabulary 🔊
circle, *p. 318*
center, *p. 318*
radius, *p. 318*
diameter, *p. 318*
circumference, *p. 319*
pi, *p. 319*
semicircle, *p. 320*

A **circle** is the set of all points in a plane that are the same distance from a point called the **center**.

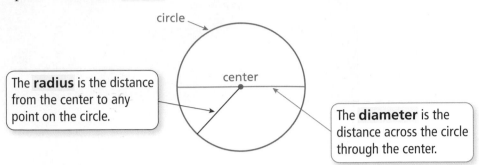

circle

The **radius** is the distance from the center to any point on the circle.

center

The **diameter** is the distance across the circle through the center.

🔑 Key Idea

Radius and Diameter

Words The diameter d of a circle is twice the radius r. The radius r of a circle is one-half the diameter d.

Algebra **Diameter:** $d = 2r$ **Radius:** $r = \dfrac{d}{2}$

EXAMPLE **1** **Finding a Radius and a Diameter**

a. **The diameter of a circle is 20 feet. Find the radius.**

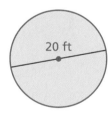

20 ft

b. **The radius of a circle is 7 meters. Find the diameter.**

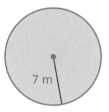

7 m

$r = \dfrac{d}{2}$ Radius of a circle

$ = \dfrac{20}{2}$ Substitute 20 for d.

$ = 10$ Divide.

∴ The radius is 10 feet.

$d = 2r$ Diameter of a circle

$ = 2(7)$ Substitute 7 for r.

$ = 14$ Multiply.

∴ The diameter is 14 meters.

⬤ On Your Own

Now You're Ready
Exercises 3–8

1. The diameter of a circle is 16 centimeters. Find the radius.

2. The radius of a circle is 9 yards. Find the diameter.

The distance around a circle is called the **circumference**. The ratio $\frac{circumference}{diameter}$ is the same for *every* circle and is represented by the Greek letter π, called **pi**. The value of π can be approximated as 3.14 or $\frac{22}{7}$.

Study Tip

When the radius or diameter is a multiple of 7, it is easier to use $\frac{22}{7}$ as the estimate of π.

 **Key Idea**

Circumference of a Circle

Words The circumference C of a circle is equal to π times the diameter d or π times twice the radius r.

Algebra $C = \pi d$ or $C = 2\pi r$

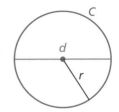

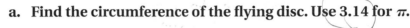

EXAMPLE 2 **Finding Circumferences of Circles**

a. **Find the circumference of the flying disc. Use 3.14 for π.**

$C = 2\pi r$ Write formula for circumference.

$\approx 2 \cdot 3.14 \cdot 5$ Substitute 3.14 for π and 5 for r.

$= 31.4$ Multiply.

∴ The circumference is about 31.4 inches.

b. **Find the circumference of the watch face. Use $\frac{22}{7}$ for π.**

$C = \pi d$ Write formula for circumference.

$\approx \frac{22}{7} \cdot 28$ Substitute $\frac{22}{7}$ for π and 28 for d.

$= 88$ Multiply.

∴ The circumference is about 88 millimeters.

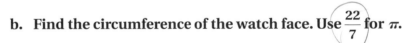 **On Your Own**

Now You're Ready
Exercises 9–11

Find the circumference of the object. Use 3.14 or $\frac{22}{7}$ for π.

3.

2 cm

4.

14 ft

5.

9 in.

EXAMPLE **3** **Estimating a Diameter**

C = 31.4 in.

The circumference of the roll of caution tape decreases 10.5 inches after a construction worker uses some of the tape. Which is the best estimate of the diameter of the roll after the decrease?

(A) 5 inches   (B) 7 inches   (C) 10 inches   (D) 12 inches

After the decrease, the circumference of the roll is
31.4 − 10.5 = 20.9 inches.

$C = \pi d$	Write formula for circumference.
$20.9 \approx 3.14 \cdot d$	Substitute 20.9 for C and 3.14 for π.
$21 \approx 3d$	Round 20.9 up to 21. Round 3.14 down to 3.
$7 = d$	Divide each side by 3.

∴ The correct answer is (B).

● **On Your Own**

6. **WHAT IF?** The circumference of the roll of tape decreases 5.25 inches. Estimate the diameter of the roll after the decrease.

EXAMPLE **4** **Finding the Perimeter of a Semicircular Region**

A semicircle is one-half of a circle. Find the perimeter of the semicircular region.

The straight side is 6 meters long. The distance around the curved part is one-half the circumference of a circle with a diameter of 6 meters.

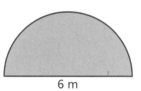

6 m

$\dfrac{C}{2} = \dfrac{\pi d}{2}$	Divide the circumference by 2.
$\approx \dfrac{3.14 \cdot 6}{2}$	Substitute 3.14 for π and 6 for d.
$= 9.42$	Simplify.

∴ So, the perimeter is about 6 + 9.42 = 15.42 meters.

● **On Your Own**

Now You're Ready
Exercises 15 and 16

Find the perimeter of the semicircular region.

7.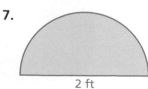

2 ft

8. 7 cm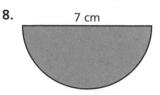

9.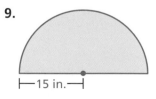

├─15 in.─┤

 Vocabulary and Concept Check

1. **VOCABULARY** What is the relationship between the radius and the diameter of a circle?

2. **WHICH ONE DOESN'T BELONG?** Which phrase does *not* belong with the other three? Explain your reasoning.

> the distance around a circle π times twice the radius

> π times the diameter   the distance from the center to any point on the circle

 Practice and Problem Solving

Find the radius of the button.

① **3.**

5 cm

4.

28 mm

5.

$3\frac{1}{2}$ in.

Find the diameter of the object.

6.

6 cm

7.

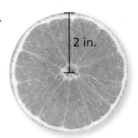
2 in.

8.

0.8 ft

Find the circumference of the pizza. Use 3.14 or $\frac{22}{7}$ for π.

② **9.**

10 in.

10.

7 in.

11.

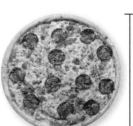
18 in.

12. **CHOOSE TOOLS** Choose a real-life circular object. Explain why you might need to know its circumference. Then find the circumference.

13. **SINKHOLE** A circular sinkhole has a circumference of 75.36 meters. A week later, it has a circumference of 150.42 meters.

   a. Estimate the diameter of the sinkhole each week.

   b. How many times greater is the diameter of the sinkhole now compared to the previous week?

14. **REASONING** Consider the circles *A*, *B*, *C*, and *D*.

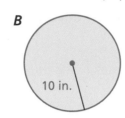
A 8 ft

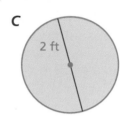
B 10 in.

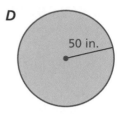
C 2 ft

D 50 in.

   a. Without calculating, which circle has the greatest circumference?

   b. Without calculating, which circle has the least circumference?

Find the perimeter of the window.

④ 15.

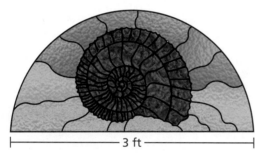
3 ft

16.

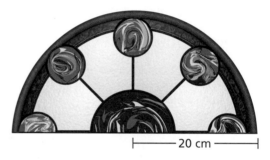
20 cm

Find the circumferences of both circles.

17.

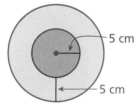
5 cm
5 cm

18.

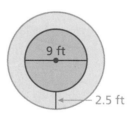
9 ft
2.5 ft

19.

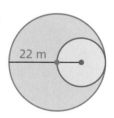
22 m

20. **STRUCTURE** Because the ratio $\dfrac{\text{circumference}}{\text{diameter}}$ is the same for every circle, is the ratio $\dfrac{\text{circumference}}{\text{radius}}$ the same for every circle? Explain.

21. **WIRE** A wire is bent to form four semicircles. How long is the wire?

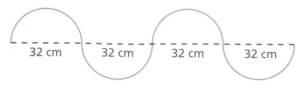
32 cm   32 cm   32 cm   32 cm

22. **CRITICAL THINKING** Explain how to draw a circle with a circumference of π^2 inches. Then draw the circle.

23. AROUND THE WORLD "Lines" of latitude on Earth are actually circles. The Tropic of Cancer is the northernmost line of latitude at which the Sun appears directly overhead at noon. The Tropic of Cancer has a radius of 5854 kilometers.

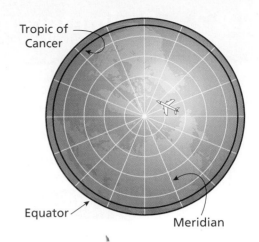

To qualify for an around-the-world speed record, a pilot must cover a distance no less than the circumference of the Tropic of Cancer, cross all meridians, and land on the same airfield where he started.

   a. What is the minimum distance that a pilot must fly to qualify for an around-the-world speed record?

   b. RESEARCH Estimate the time it would take for a pilot to qualify for the speed record.

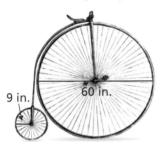

24. PROBLEM SOLVING Bicycles in the late 1800s looked very different than they do today.

   a. How many rotations does each tire make after traveling 600 feet? Round your answers to the nearest whole number.

   b. Would you rather ride a bicycle made with two large wheels or two small wheels? Explain.

25. **Logic** The length of the minute hand is 150% of the length of the hour hand.

   a. What distance will the tip of the minute hand move in 45 minutes? Explain how you found your answer.

   b. In 1 hour, how much farther does the tip of the minute hand move than the tip of the hour hand? Explain how you found your answer.

 Fair Game Review *What you learned in previous grades & lessons*

Find the perimeter of the polygon. *(Skills Review Handbook)*

26.
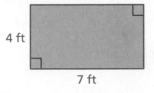
4 ft
7 ft

27.
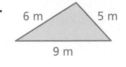
6 m   5 m
9 m

28.
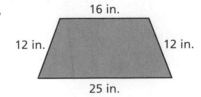
16 in.
12 in.   12 in.
25 in.

29. MULTIPLE CHOICE What is the median of the data set? *(Skills Review Handbook)*

    12, 25, 16, 9, 5, 22, 27, 20

    (A) 7   (B) 16   (C) 17   (D) 18

Essential Question How can you find the perimeter of a composite figure?

1 ACTIVITY: Finding a Pattern

Work with a partner. Describe the pattern of the perimeters. Use your pattern to find the perimeter of the tenth figure in the sequence. (Each small square has a perimeter of 4.)

a.

b.

c.

2 ACTIVITY: Combining Figures

Work with a partner.

a. A rancher is constructing a rectangular corral and a trapezoidal corral, as shown. How much fencing does the rancher need to construct both corrals?

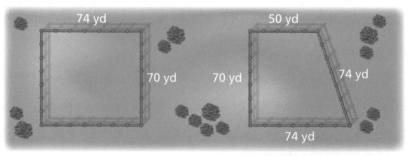

COMMON CORE

Geometry

In this lesson, you will

• find perimeters of composite figures.

Applying Standard 7.G.4

b. Another rancher is constructing one corral by combining the two corrals above, as shown. Does this rancher need more or less fencing? Explain your reasoning.

c. How can the rancher in part (b) combine the two corrals to use even less fencing?

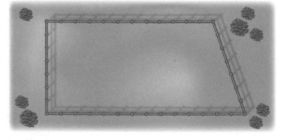

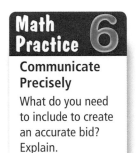

3 ACTIVITY: Submitting a Bid

Work with a partner. You want to bid on a tiling contract. You will be supplying and installing the brown tile that borders the swimming pool. In the figure, each grid square represents 1 square foot.

- **Your cost for the tile is $4 per linear foot.**
- **It takes about 15 minutes to prepare, install, and clean each foot of tile.**

a. How many brown tiles do you need for the border?

b. Write a bid for how much you will charge to supply and install the tile. Include what you want to charge as an hourly wage. Estimate what you think your profit will be.

Math Practice 6

Communicate Precisely

What do you need to include to create an accurate bid? Explain.

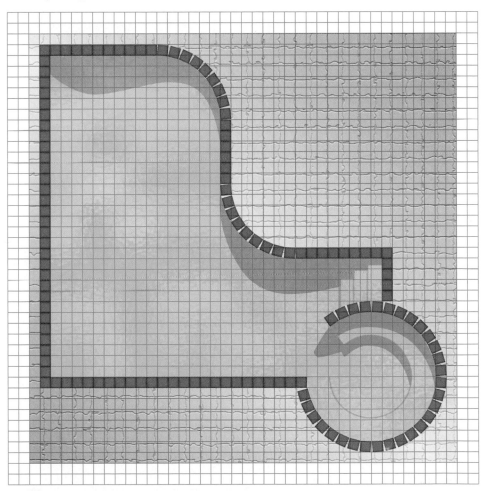

What Is Your Answer?

4. IN YOUR OWN WORDS How can you find the perimeter of a composite figure? Use a semicircle, a triangle, and a parallelogram to draw a composite figure. Label the dimensions. Find the perimeter of the figure.

Practice ▶ Use what you learned about perimeters of composite figures to complete Exercises 3–5 on page 328.

Key Vocabulary
composite figure,
   p. 326

A **composite figure** is made up of triangles, squares, rectangles, semicircles, and other two-dimensional figures. Here are two examples.

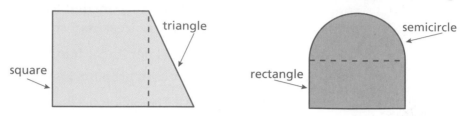

To find the perimeter of a composite figure, find the distance around the figure.

EXAMPLE **1** **Estimating a Perimeter Using Grid Paper**

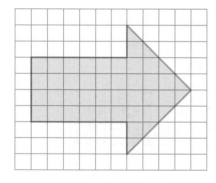

Estimate the perimeter of the arrow.

Count the number of grid square lengths around the arrow. There are 20.

Count the number of diagonal lengths around the arrow. There are 8.

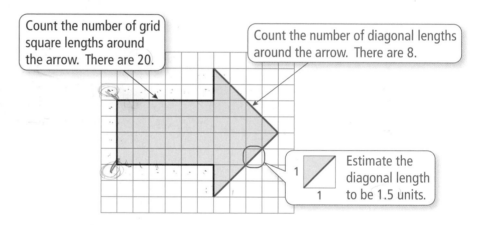

Estimate the diagonal length to be 1.5 units.

Length of 20 grid square lengths: $20 \times 1 = 20$ units

Length of 8 diagonal lengths: $8 \times 1.5 = 12$ units

So, the perimeter is about $20 + 12 = 32$ units.

On Your Own

Now You're Ready
Exercises 3–8

Estimate the perimeter of the figure.

1.

2.

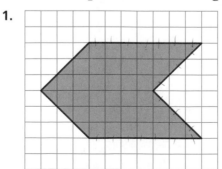

Multi-Language Glossary at BigIdeasMath✓com

EXAMPLE 2 **Finding a Perimeter**

The figure is made up of a semicircle and a triangle. Find the perimeter.

The distance around the triangular part of the figure is $6 + 8 = 14$ feet.

The distance around the semicircle is one-half the circumference of a circle with a diameter of 10 feet.

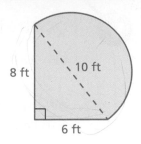

8 ft   10 ft   6 ft

$$\frac{C}{2} = \frac{\pi d}{2} \qquad \text{Divide the circumference by 2.}$$

$$\approx \frac{3.14 \cdot 10}{2} \qquad \text{Substitute 3.14 for } \pi \text{ and 10 for } d.$$

$$= 15.7 \qquad \text{Simplify.}$$

⋮• So, the perimeter is about $14 + 15.7 = 29.7$ feet.

EXAMPLE 3 **Finding a Perimeter**

The running track is made up of a rectangle and two semicircles. Find the perimeter.

The semicircular ends of the track form a circle with a radius of 32 meters. Find its circumference.

$$C = 2\pi r \qquad \text{Write formula for circumference.}$$

$$\approx 2 \cdot 3.14 \cdot 32 \qquad \text{Substitute 3.14 for } \pi \text{ and 32 for } r.$$

$$= 200.96 \qquad \text{Multiply.}$$

⋮• So, the perimeter is about $100 + 100 + 200.96 = 400.96$ meters.

[track figure on left]

32 m

100 m

Now You're Ready
Exercises 9–11

● **On Your Own**

3. The figure is made up of a semicircle and a triangle. Find the perimeter.

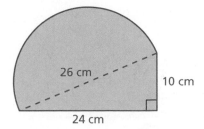

26 cm   10 cm   24 cm

4. The figure is made up of a square and two semicircles. Find the perimeter.

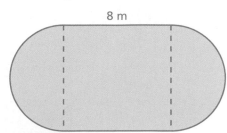

8 m

Check It Out
Help with Homework
BigIdeasMath ✓com

Vocabulary and Concept Check

1. **REASONING** Is the perimeter of the composite figure equal to the sum of the perimeters of the individual figures? Explain.

2. **OPEN-ENDED** Draw a composite figure formed by a parallelogram and a trapezoid.

Practice and Problem Solving

Estimate the perimeter of the figure.

3.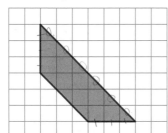

4.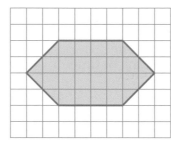

5.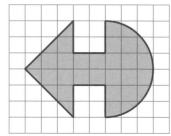

6.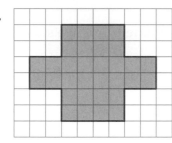

7.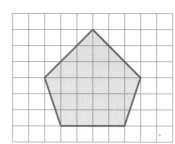

8.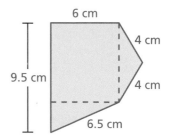

Find the perimeter of the figure.

9.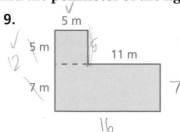
5 m
5 m
11 m
7 m

10.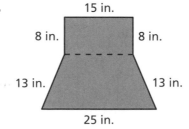
15 in.
8 in.   8 in.
13 in.   13 in.
25 in.

11.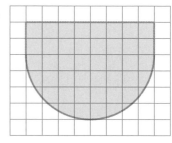
6 cm
4 cm
9.5 cm
4 cm
6.5 cm

12. **ERROR ANALYSIS** Describe and correct the error in finding the perimeter of the figure.

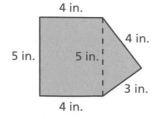
4 in.
4 in.
5 in.   5 in.
3 in.
4 in.

Perimeter = 4 + 3 + 4 + 5 + 4 + 5
= 25 in.

Find the perimeter of the figure.

13.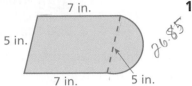

7 in.

5 in.

7 in.   5 in.

26.85

14.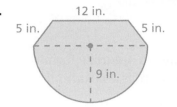

12 in.

5 in.   5 in.

9 in.

15.

3 ft   3 ft

3 ft

3 ft   3 ft

|— 12 ft —|

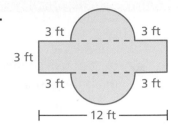

240 ft   285 ft

450 ft   450 ft

450 ft

16. PASTURE A farmer wants to fence a section of land for a horse pasture. Fencing costs $27 per yard. How much will it cost to fence the pasture?

17. BASEBALL You run around the perimeter of the baseball field at a rate of 9 feet per second. How long does it take you to run around the baseball field?

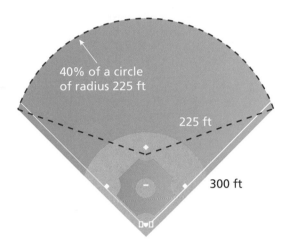

40% of a circle of radius 225 ft

225 ft

300 ft

18. TRACK In Example 3, the running track has six lanes. Explain why the starting points for the six runners are staggered. Draw a diagram as part of your explanation.

19. **Critical Thinking** How can you add a figure to a composite figure without increasing its perimeter? Draw a diagram to support your answer.

Fair Game Review *What you learned in previous grades & lessons*

Evaluate the expression. *(Skills Review Handbook)*

20. $2.15(3)^2$

21. $4.37(8)^2$

22. $3.14(7)^2$

23. $8.2(5)^2$

24. MULTIPLE CHOICE Which expression is equivalent to $(5y + 4) - 2(7 - 2y)$? *(Section 3.2)*

(A) $y - 10$ (B) $9y + 18$ (C) $3y - 10$ (D) $9y - 10$

You can use a **word magnet** to organize formulas or phrases that are associated with a vocabulary word or term. Here is an example of a word magnet for circle.

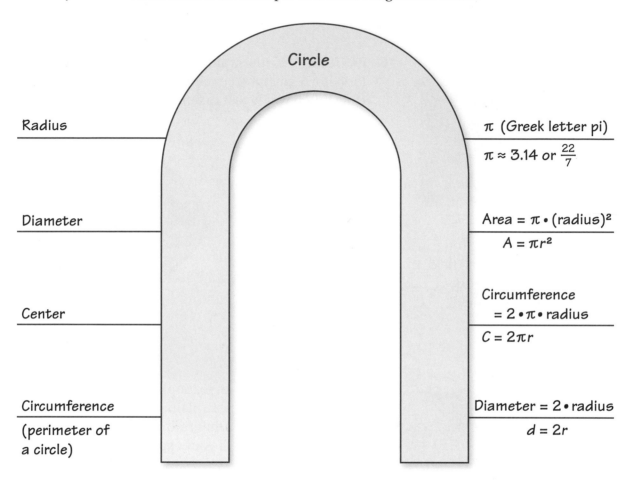

Circle

Radius — π (Greek letter pi)

$\pi \approx 3.14$ or $\frac{22}{7}$

Diameter — Area $= \pi \cdot (\text{radius})^2$

$A = \pi r^2$

Center — Circumference
$= 2 \cdot \pi \cdot \text{radius}$

$C = 2\pi r$

Circumference (perimeter of a circle) — Diameter $= 2 \cdot \text{radius}$

$d = 2r$

On Your Own

Make word magnets to help you study these topics.

1. semicircle

2. composite figure

3. perimeter

After you complete this chapter, make word magnets for the following topics.

4. area of a circle

5. area of a composite figure

"I'm trying to make a word magnet for happiness, but I can only think of two words."

1. The diameter of a circle is 36 centimeters. Find the radius.   *(Section 8.1)*

2. The radius of a circle is 11 inches. Find the diameter.   *(Section 8.1)*

Estimate the perimeter of the figure.   *(Section 8.2)*

3.
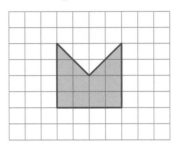

4.
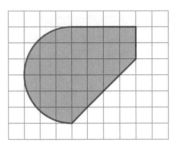

5.
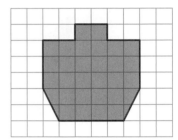

Find the circumference of the circle. Use 3.14 or $\frac{22}{7}$ for π.   *(Section 8.1)*

6.
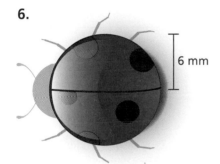
6 mm

7.

1.5 ft

8.
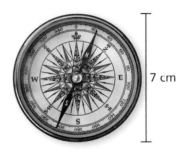
7 cm

Find the perimeter of the figure.   *(Section 8.1 and Section 8.2)*

9.
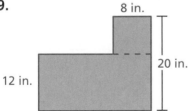
8 in.
20 in.
12 in.
24 in.

10.
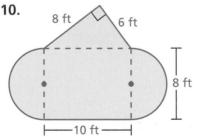
8 ft
6 ft
8 ft
10 ft

11.
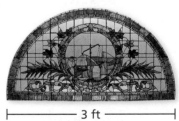
3 ft

12. **BUTTON** What is the circumference of a circular button with a diameter of 8 millimeters?   *(Section 8.1)*

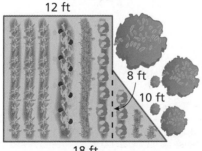
12 ft
14 ft
8 ft
10 ft
18 ft

13. **GARDEN** You want to fence part of a yard to make a vegetable garden. How many feet of fencing do you need to surround the garden?   *(Section 8.2)*

14. **BAKING** A baker is using two circular pans. The larger pan has a diameter of 12 inches. The smaller pan has a diameter of 7 inches. How much greater is the circumference of the larger pan than that of the smaller pan?   *(Section 8.1)*

Essential Question How can you find the area of a circle?

1 ACTIVITY: Estimating the Area of a Circle

Work with a partner. Each square in the grid is 1 unit by 1 unit.

a. Find the area of the large 10-by-10 square.

b. Copy and complete the table.

Region	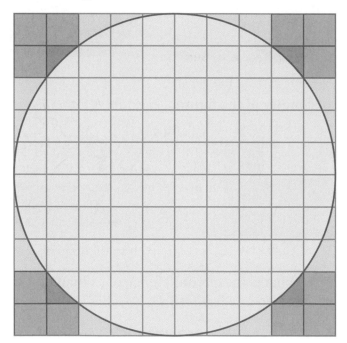		
Area (square units)			

c. Use your results to estimate the area of the circle. Explain your reasoning.

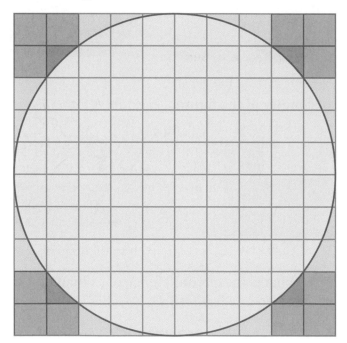

COMMON CORE

Geometry

In this lesson, you will
- find areas of circles and semicircles.

Learning Standard
7.G.4

d. Fill in the blanks. Explain your reasoning.

$$\text{Area of large square} = \boxed{} \cdot 5^2 \text{ square units}$$

$$\text{Area of circle} \approx \boxed{} \cdot 5^2 \text{ square units}$$

e. What dimension of the circle does 5 represent? What can you conclude?

Work with a partner.

a. Draw a circle. Label the radius as *r*.

b. Divide the circle into 24 equal sections.

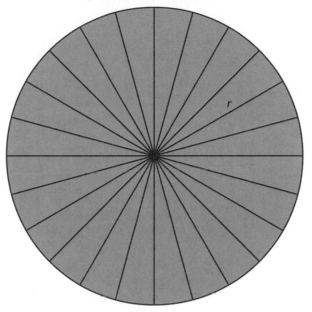

c. Cut the sections apart. Then arrange them to approximate a parallelogram.

Math Practice 1

Interpret a Solution

What does the area of the parallelogram represent? Explain.

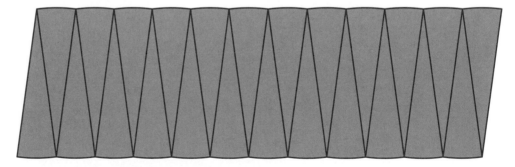

d. What is the approximate height and base of the parallelogram?

e. Find the area of the parallelogram. What can you conclude?

What Is Your Answer?

3. **IN YOUR OWN WORDS** How can you find the area of a circle?

4. Write a formula for the area of a circle with radius *r*. Find an object that is circular. Use your formula to find the area.

Practice ➤ Use what you learned about areas of circles to complete Exercises 3–5 on page 336.

 **Key Idea**

Area of a Circle

Words   The area A of a circle is the product of π and the square of the radius.

Algebra $A = \pi r^2$

EXAMPLE ① **Finding Areas of Circles**

7 cm

a. **Find the area of the circle. Use $\dfrac{22}{7}$ for π.**

Estimate $3 \times 7^2 \approx 3 \times 50 = 150$

$$A = \pi r^2 \qquad \text{Write formula for area.}$$

$$\approx \frac{22}{7} \cdot 7^2 \qquad \text{Substitute } \frac{22}{7} \text{ for } \pi \text{ and 7 for } r.$$

$$= \frac{22}{\overset{}{\underset{1}{\cancel{7}}}} \cdot \overset{7}{\cancel{49}} \qquad \text{Evaluate } 7^2. \text{ Divide out the common factor.}$$

$$= 154 \qquad \text{Multiply.}$$

∴ The area is about 154 square centimeters.

Reasonable? $154 \approx 150$ ✔

b. **Find the area of the circle. Use 3.14 for π.**

26 in.

The radius is $26 \div 2 = 13$ inches.

Estimate $3 \times 13^2 \approx 3 \times 170 = 510$

$$A = \pi r^2 \qquad \text{Write formula for area.}$$

$$\approx 3.14 \cdot 13^2 \qquad \text{Substitute 3.14 for } \pi \text{ and 13 for } r.$$

$$= 3.14 \cdot 169 \qquad \text{Evaluate } 13^2.$$

$$= 530.66 \qquad \text{Multiply.}$$

∴ The area is about 530.66 square inches.

Reasonable? $530.66 \approx 510$ ✔

● **On Your Own**

Now You're Ready
Exercises 3–10

1. Find the area of a circle with a radius of 6 feet. Use 3.14 for π.

2. Find the area of a circle with a diameter of 28 meters. Use $\dfrac{22}{7}$ for π.

EXAMPLE 2 **Describing a Distance**

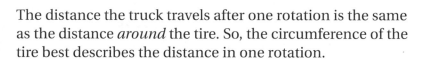

You want to find the distance the monster truck travels when the tires make one 360-degree rotation. Which best describes this distance?

Ⓐ the radius of the tire   Ⓑ the diameter of the tire

Ⓒ the circumference of the tire   Ⓓ the area of the tire

The distance the truck travels after one rotation is the same as the distance *around* the tire. So, the circumference of the tire best describes the distance in one rotation.

∴ The correct answer is Ⓒ.

● **On Your Own**

3. You want to find the height of one of the tires. Which measurement would best describe the height?

EXAMPLE 3 **Finding the Area of a Semicircle**

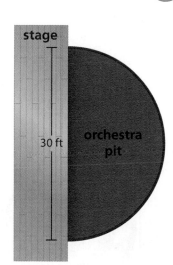

Find the area of the semicircular orchestra pit.

The area of the orchestra pit is one-half the area of a circle with a diameter of 30 feet.

The radius of the circle is 30 ÷ 2 = 15 feet.

$$\frac{A}{2} = \frac{\pi r^2}{2}$$ Divide the area by 2.

$$\approx \frac{3.14 \cdot 15^2}{2}$$ Substitute 3.14 for π and 15 for r.

$$= \frac{3.14 \cdot 225}{2}$$ Evaluate 15^2.

$$= 353.25$$ Simplify.

∴ So, the area of the orchestra pit is about 353.25 square feet.

● **On Your Own**

Now You're Ready
Exercises 13–15

Find the area of the semicircle.

4.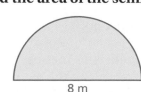
8 m

5. 5 yd
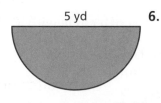

6. 11 cm

 ## Vocabulary and Concept Check

1. **VOCABULARY** Explain how to find the area of a circle given its diameter.

2. **DIFFERENT WORDS, SAME QUESTION** Which is different? Find "both" answers.

> What is the area of a circle with a diameter of 1 m?

> What is the area of a circle with a diameter of 100 cm?

> What is the area of a circle with a radius of 100 cm?

> What is the area of a circle with a radius of 500 mm?

 ## Practice and Problem Solving

Find the area of the circle. Use 3.14 or $\frac{22}{7}$ for π.

3.  9 mm

4.  14 cm

5.  10 in.

6.  3 in.

7.  2 cm

8.  1.5 ft

9. Find the area of a circle with a diameter of 56 millimeters.

10. Find the area of a circle with a radius of 5 feet.

11. **TORTILLA** The diameter of a flour tortilla is 12 inches. What is the area?

12. **LIGHTHOUSE** The Hillsboro Inlet Lighthouse lights up how much more area than the Jupiter Inlet Lighthouse?

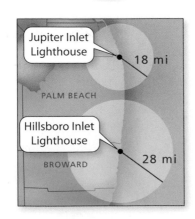
Jupiter Inlet Lighthouse — 18 mi
PALM BEACH
Hillsboro Inlet Lighthouse — 28 mi
BROWARD

Find the area of the semicircle.

③ 13.

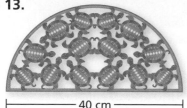

|— 40 cm —|

14.

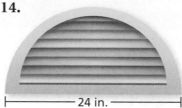

|— 24 in. —|

15.

|— 2 ft —|

16. REPEATED REASONING Consider five circles with radii of 1, 2, 4, 8, and 16 inches.

  a. Copy and complete the table. Write your answers in terms of π.

  b. Compare the areas and circumferences. What happens to the circumference of a circle when you double the radius? What happens to the area?

  c. What happens when you triple the radius?

Radius	Circumference	Area
1	2π in.	π in.^{2}
2		
4		
8		
16		

20 ft

17. DOG A dog is leashed to the corner of a house. How much running area does the dog have? Explain how you found your answer.

18. CRITICAL THINKING Is the area of a semicircle with a diameter of *x greater than*, *less than*, or *equal to* the area of a circle with a diameter of $\frac{1}{2}x$? Explain.

Reasoning Find the area of the shaded region. Explain how you found your answer.

19.

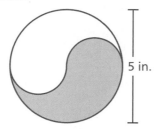

5 in.

20.

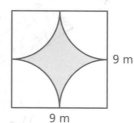

9 m

9 m

21.

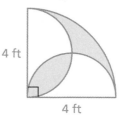

4 ft

4 ft

Fair Game Review *What you learned in previous grades & lessons*

Evaluate the expression. *(Skills Review Handbook)*

22. $\frac{1}{2}(7)(4) + 6(5)$

23. $\frac{1}{2} \cdot 8^2 + 3(7)$

24. $12(6) + \frac{1}{4} \cdot 2^2$

25. MULTIPLE CHOICE What is the product of $-8\frac{1}{3}$ and $3\frac{2}{5}$? *(Section 2.4)*

  Ⓐ $-28\frac{1}{3}$ **Ⓑ** $-24\frac{2}{15}$ **Ⓒ** $24\frac{2}{15}$ **Ⓓ** $28\frac{1}{3}$

Essential Question How can you find the area of a composite figure?

Work with a partner.

a. Choose a state. On grid paper, draw a larger outline of the state.

b. Use your drawing to estimate the area (in square miles) of the state.

c. Which state areas are easy to find? Which are difficult? Why?

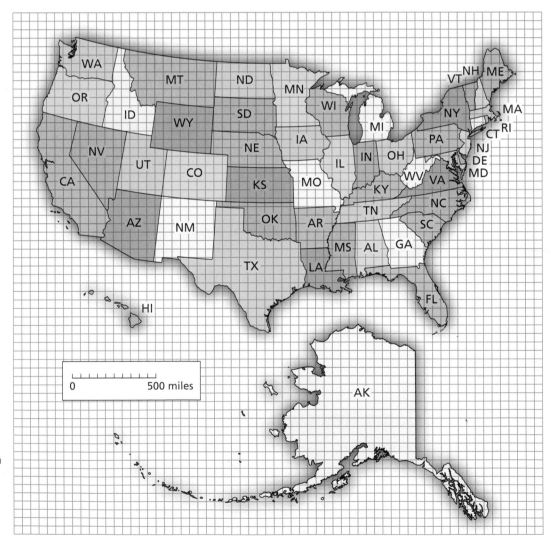

Geometry

In this lesson, you will

• find areas of composite figures by separating them into familiar figures.

• solve real-life problems.

Learning Standard
7.G.6

2 ACTIVITY: Estimating Areas

Work with a partner. The completed puzzle has an area of 150 square centimeters.

a. Estimate the area of each puzzle piece.

b. Check your work by adding the six areas. Why is this a check?

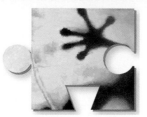
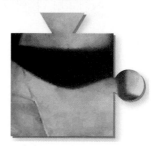
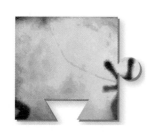

3 ACTIVITY: Filling a Square with Circles

Work with a partner. Which pattern fills more of the square with circles? Explain.

a.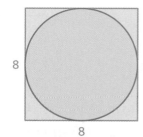

b.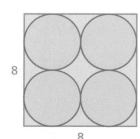

c.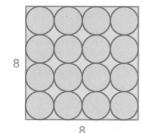

d.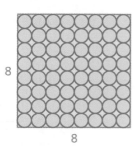

What Is Your Answer?

4. **IN YOUR OWN WORDS** How can you find the area of a composite figure?

5. Summarize the area formulas for all the basic figures you have studied. Draw a single composite figure that has each type of basic figure. Label the dimensions and find the total area.

Practice Use what you learned about areas of composite figures to complete Exercises 3–5 on page 342.

To find the area of a composite figure, separate it into figures with areas you know how to find. Then find the sum of the areas of those figures.

EXAMPLE 1 **Finding an Area Using Grid Paper**

Find the area of the yellow figure.

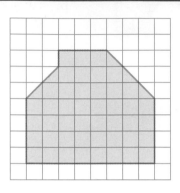

> Count the number of squares that lie entirely in the figure. There are 45.

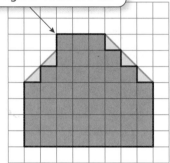

> Count the number of half squares in the figure. There are 5.

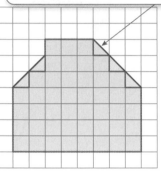

The area of a half square is $1 \div 2 = 0.5$ square unit.

Area of 45 squares: $45 \times 1 = 45$ square units

Area of 5 half squares: $5 \times 0.5 = 2.5$ square units

So, the area is $45 + 2.5 = 47.5$ square units.

On Your Own

Now You're Ready
Exercises 3–8

Find the area of the shaded figure.

1.

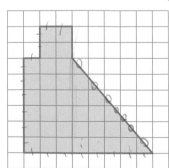

2.

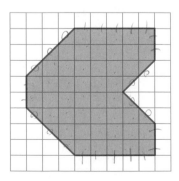

EXAMPLE **2** **Finding an Area**

Find the area of the portion of the basketball court shown.

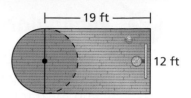

The figure is made up of a rectangle and a semicircle. Find the area of each figure.

Area of Rectangle

$$A = \ell w$$

$$= 19(12)$$

$$= 228$$

Area of Semicircle

$$A = \frac{\pi r^2}{2}$$

$$\approx \frac{3.14 \cdot 6^2}{2}$$

$$= 56.52$$

> The semicircle has a radius of $\frac{12}{2} = 6$ feet.

So, the area is about $228 + 56.52 = 284.52$ square feet.

EXAMPLE **3** **Finding an Area**

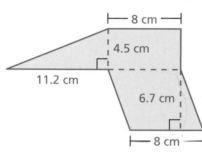

Find the area of the figure.

The figure is made up of a triangle, a rectangle, and a parallelogram. Find the area of each figure.

Area of Triangle

$$A = \frac{1}{2}bh$$

$$= \frac{1}{2}(11.2)(4.5)$$

$$= 25.2$$

Area of Rectangle

$$A = \ell w$$

$$= 8(4.5)$$

$$= 36$$

Area of Parallelogram

$$A = bh$$

$$= 8(6.7)$$

$$= 53.6$$

So, the area is $25.2 + 36 + 53.6 = 114.8$ square centimeters.

On Your Own

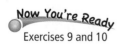

Exercises 9 and 10

Find the area of the figure.

3.

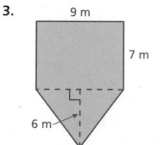

4.

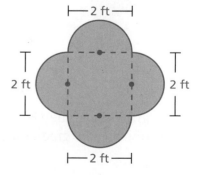

 Vocabulary and Concept Check

1. **REASONING** Describe two different ways to find the area of the figure. Name the types of figures you used and the dimensions of each.

2. **REASONING** Draw a trapezoid. Explain how you can think of the trapezoid as a composite figure to find its area.

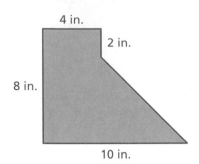

 Practice and Problem Solving

Find the area of the figure.

① 3.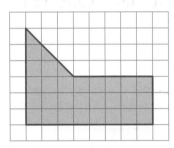

4.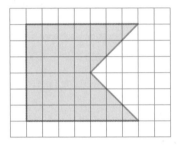

5.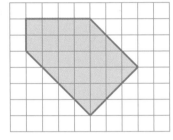

6.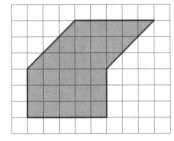

7.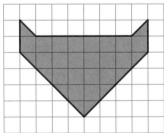

8.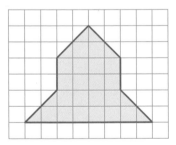

Find the area of the figure.

② ③ 9.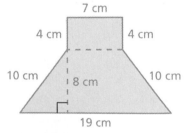

10.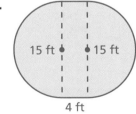

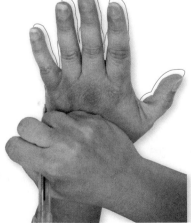

11. **OPEN-ENDED** Trace your hand and your foot on grid paper. Then estimate the area of each. Which one has the greater area?

Find the area of the figure.

12.

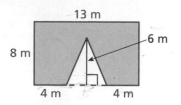

13 m
6 m
8 m
4 m   4 m

13.

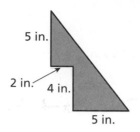

5 in.
2 in.
4 in.
5 in.

14.

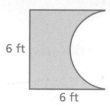

6 ft
6 ft

15. STRUCTURE The figure is made up of a square and a rectangle. Find the area of the shaded region.

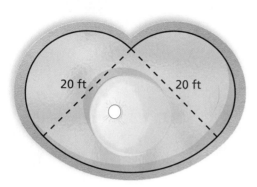

7 m
16 m
3 m

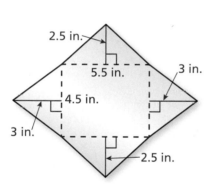

20 ft   20 ft

16. FOUNTAIN The fountain is made up of two semicircles and a quarter circle. Find the perimeter and the area of the fountain.

17. ★*Critical Thinking* You are deciding on two different designs for envelopes.

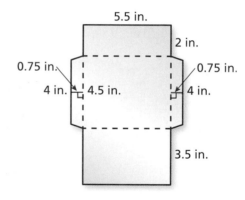

2.5 in.
5.5 in.
3 in.
4.5 in.
3 in.
2.5 in.

5.5 in.
2 in.
0.75 in.
0.75 in.
4 in.   4.5 in.   4 in.
3.5 in.

a. Which design has the greater area?

b. You make 500 envelopes using the design with the greater area. Using the same amount of paper, how many more envelopes can you make with the other design?

Fair Game Review What you learned in previous grades & lessons

Write the phrase as an expression. *(Skills Review Handbook)*

18. 12 less than a number x

19. a number y divided by 6

20. a number b increased by 3

21. the product of 7 and a number w

22. MULTIPLE CHOICE What number is 0.02% of 50? *(Section 6.4)*

Ⓐ 0.01   Ⓑ 0.1   Ⓒ 1   Ⓓ 100

Check It Out
Progress Check
BigIdeasMath ✓.com

Find the area of the figure. *(Section 8.4)*

1.
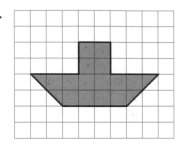

2.
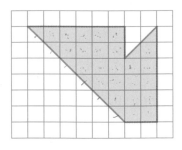

3.
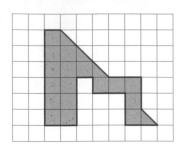

Find the area of the circle. Use 3.14 or $\frac{22}{7}$ for π. *(Section 8.3)*

4.

12 in.

5.

6 cm

6.
$3\frac{1}{2}$ in.

Find the area of the figure. *(Section 8.4)*

7.
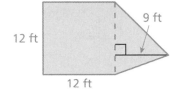
9 ft
12 ft
12 ft

8.
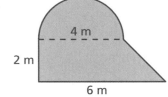
4 m
2 m
6 m

9.
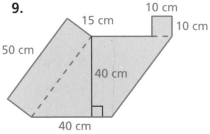
10 cm
15 cm
10 cm
50 cm
40 cm
40 cm

10. **POT HOLDER** A knitted pot holder is shaped like a circle. Its radius is 3.5 inches. What is its area? *(Section 8.3)*

11. **CARD** The heart-shaped card is made up of a square and two semicircles. What is the area of the card? *(Section 8.4)*

LOVE
8 cm   8 cm

├── 14 ft ──┤

12. **DESK** A desktop is shaped like a semicircle with a diameter of 28 inches. What is the area of the desktop? *(Section 8.3)*

13. **RUG** The circular rug is placed on a square floor. The rug touches all four walls. How much of the floor space is *not* covered by the rug? *(Section 8.4)*

Review Key Vocabulary

circle, *p. 318*

center, *p. 318*

radius, *p. 318*

diameter, *p. 318*

circumference, *p. 319*

pi, *p. 319*

semicircle, *p. 320*

composite figure, *p. 326*

Review Examples and Exercises

8.1 Circles and Circumference *(pp. 316–323)*

Find the circumference of the circle. Use 3.14 for π.

The radius is 4 millimeters.

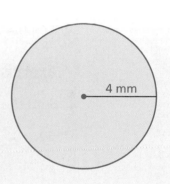

4 mm

$C = 2\pi r$	Write formula for circumference.
$\approx 2 \cdot 3.14 \cdot 4$	Substitute 3.14 for π and 4 for r.
$= 25.12$	Multiply.

⋮⋅ The circumference is about 25.12 millimeters.

Exercises

Find the radius of the circle with the given diameter.

1. 8 inches

2. 60 millimeters

3. 100 meters

4. 3 yards

Find the diameter of the circle with the given radius.

5. 20 feet

6. 5 meters

7. 1 inch

8. 25 millimeters

Find the circumference of the circle. Use 3.14 or $\dfrac{22}{7}$ for π.

9.

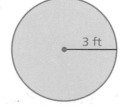

3 ft

10.

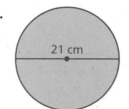

21 cm

11.

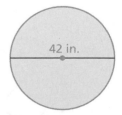

42 in.

8.2 Perimeters of Composite Figures (pp. 324–329)

The figure is made up of a semicircle and a square. Find the perimeter.

The distance around the square part is $6 + 6 + 6 = 18$ meters. The distance around the semicircle is one-half the circumference of a circle with $d = 6$ meters.

$$\frac{C}{2} = \frac{\pi d}{2}$$ Divide the circumference by 2.

$$\approx \frac{3.14 \cdot 6}{2}$$ Substitute 3.14 for π and 6 for d.

$$= 9.42$$ Simplify.

6 m

So, the perimeter is about $18 + 9.42 = 27.42$ meters.

Exercises

Find the perimeter of the figure.

12.

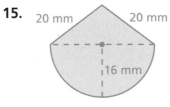

5 in.
4 in.
3 in.
5 in.
9 in.

13.

9 ft
9 ft
9 ft
9 ft
30 ft

14.

13 cm   15 cm
10 cm
10 cm   14 cm
10 cm

15.

20 mm   20 mm
16 mm

16.

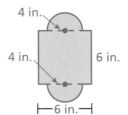

4 in.
4 in.   6 in.
6 in.

17.

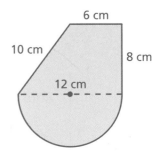

6 cm
10 cm   8 cm
12 cm

8.3 Areas of Circles (pp. 332–337)

Find the area of the circle. Use 3.14 for π.

$$A = \pi r^2$$ Write formula for area.

$$\approx 3.14 \cdot 20^2$$ Substitute 3.14 for π and 20 for r.

$$= 1256$$ Multiply.

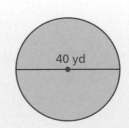

40 yd

The area is about 1256 square yards.

Exercises

Find the area of the circle. Use 3.14 or $\frac{22}{7}$ for π.

18.
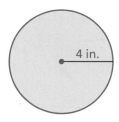
4 in.

19.
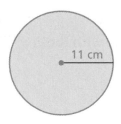
11 cm

20.

42 mm

8.4 **Areas of Composite Figures** *(pp. 338–343)*

Find the area of the figure.

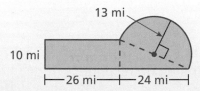
13 mi
10 mi
├─26 mi─┼─24 mi─┤

The figure is made up of a rectangle, a triangle and a semicircle. Find the area of each figure.

Area of Rectangle	*Area of Triangle*	*Area of Semicircle*

$A = \ell w$

$= 26(10)$

$= 260$

$A = \frac{1}{2}bh$

$= \frac{1}{2}(10)(24)$

$= 120$

$A = \frac{\pi r^2}{2}$

$\approx \frac{3.14 \cdot 13^2}{2}$

$= 265.33$

So, the area is about $260 + 120 + 265.33 = 645.33$ square miles.

Exercises

Find the area of the figure.

21.
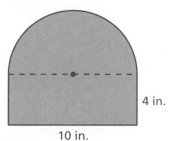
4 in.
10 in.

22.
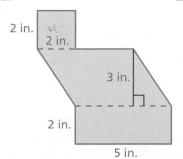
2 in.
2 in.
3 in.
2 in.
5 in.

23.
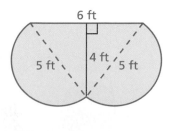
6 ft
5 ft   4 ft   5 ft

Check It Out
Test Practice
BigIdeasMath ✓com

Find the radius of the circle with the given diameter.

1. 10 inches

2. 5 yards

Find the diameter of the circle with the given radius.

3. 34 feet

4. 19 meters

Find the circumference and the area of the circle. Use 3.14 or $\frac{22}{7}$ for π.

5.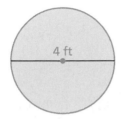
4 ft

6.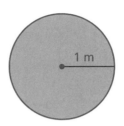
1 m

7.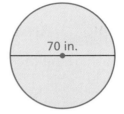
70 in.

8. Estimate the perimeter of the figure. Then find the area.

Find the perimeter and the area of the figure. Use 3.14 or $\frac{22}{7}$ for π.

9.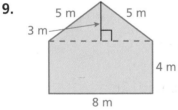
5 m 5 m
3 m
4 m
8 m

10.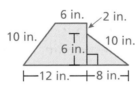
6 in. 2 in.
10 in. 10 in.
6 in.
12 in. 8 in.

11.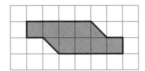
7 m 14 m 7 m
14 m

12. **MUSEUM** A museum plans to rope off the perimeter of the L-shaped exhibit. How much rope does it need?

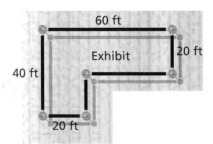
60 ft
Exhibit
20 ft
40 ft
20 ft

13. **ANIMAL PEN** You unfold chicken wire to make a circular pen with a diameter of 2.9 meters. How many meters of chicken wire do you need?

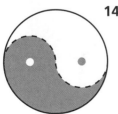

14. **YIN AND YANG** In the Chinese symbol for yin and yang, the dashed curve shows two semicircles formed by the curve separating the yin (dark) and the yang (light). Is the circumference of the entire yin and yang symbol *less than*, *greater than*, or *equal to* the perimeter of the yin?

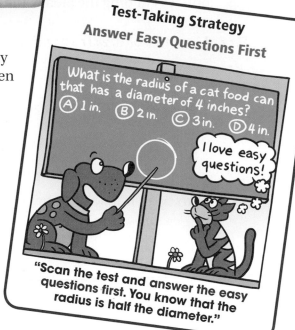

Test-Taking Strategy
Answer Easy Questions First

What is the radius of a cat food can that has a diameter of 4 inches?
(A) 1 in. (B) 2 in. (C) 3 in. (D) 4 in.

I love easy questions!

"Scan the test and answer the easy questions first. You know that the radius is half the diameter."

1. To make 6 servings of soup, you need 5 cups of chicken broth. You want to know how many servings you can make with 2 quarts of chicken broth. Which proportion should you use? *(7.RP.2c)*

A. $\dfrac{6}{5} = \dfrac{2}{x}$

B. $\dfrac{6}{5} = \dfrac{x}{2}$

C. $\dfrac{6}{5} = \dfrac{x}{8}$

D. $\dfrac{5}{6} = \dfrac{x}{8}$

2. What is the value of x? *(7.G.5)*

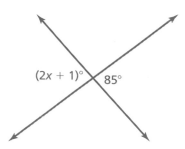

$(2x + 1)°$ $85°$

3. Your mathematics teacher described an equation in words. Her description is in the box below.

> "5 less than the product of 7 and an unknown number is equal to 42."

Which equation matches your mathematics teacher's description? *(7.EE.4a)*

F. $(5 - 7)n = 42$

G. $(7 - 5)n = 42$

H. $5 - 7n = 42$

I. $7n - 5 = 42$

4. What is the area of the circle below? $\left(\text{Use } \dfrac{22}{7} \text{ for } \pi.\right)$ *(7.G.4)*

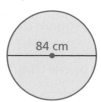

84 cm

A. 132 cm^2

B. 264 cm^2

C. 5544 cm^2

D. $22{,}176 \text{ cm}^2$

5. John was finding the area of the figure below.

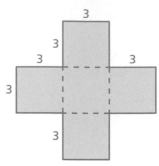

John's work is in the box below.

area of horizontal rectangle

A = 3 × (3 + 3 + 3)

= 3 × 9

= 27 square units

area of vertical rectangle

A = (3 + 3 + 3) × 3

= 9 × 3

= 27 square units

total area of figure

A = 27 + 27

= 54 square units

What should John do to correct the error that he made?   *(7.G.6)*

F. Add the area of the center square to the 54 square units.

G. Find the area of one square and multiply this number by 4.

H. Subtract the area of the center square from the 54 square units.

I. Subtract 54 from the area of a large square that is 9 units on each side.

6. Which value of x makes the equation below true?   *(7.EE.4a)*

$$5x - 3 = 11$$

A. 1.6

C. 40

B. 2.8

D. 70

7. What is the perimeter of the figure below? (Use 3.14 for π.)   *(7.G.4)*

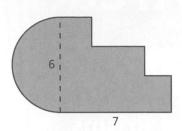

8. Which inequality has 5 in its solution set?   *(7.EE.4b)*

F. $5 - 2x \geq 3$

G. $3x - 4 \geq 8$

H. $8 - 3x > -7$

I. $4 - 2x < -6$

9. Four jewelry stores are selling an identical pair of earrings.

- Store A: original price of $75; 20% off during sale
- Store B: original price of $100; 35% off during sale
- Store C: original price of $70; 10% off during sale
- Store D: original price of $95; 30% off during sale

Which store has the least sale price for the pair of earrings?   *(7.RP.3)*

A. Store A

B. Store B

C. Store C

D. Store D

10. A lawn sprinkler sprays water onto part of a circular region, as shown below.   *(7.G.4)*

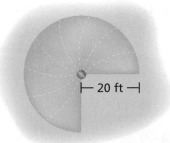

Part A   What is the area, in square feet, of the region that the sprinkler sprays with water? Show your work and explain your reasoning. (Use 3.14 for π.)

Part B   What is the perimeter, in feet, of the region that the sprinkler sprays with water? Show your work and explain your reasoning. (Use 3.14 for π.)

9 Surface Area and Volume

"I was thinking that I want the Pagodal roof instead of the Swiss chalet roof for my new doghouse."

"Because PAGODAL rearranges to spell 'A DOG PAL.'"

"Take a deep breath and hold it."

"Now, do you feel like your surface area or your volume is increasing more?"

What You Learned Before

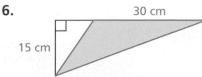

"Descartes, how would you like it if I could double the height of your cat food can?"

● Finding Areas of Squares and Rectangles (4.MD.3, 5.NF.4b)

Example 1 Find the area of the rectangle.

3 mm

7 mm

$$\text{Area} = \ell w \qquad \text{Write formula for area.}$$
$$= 7(3) \qquad \text{Substitute 7 for } \ell \text{ and 3 for } w.$$
$$= 21 \qquad \text{Multiply.}$$

∴ The area of the rectangle is 21 square millimeters.

Try It Yourself

Find the area of the square or rectangle.

1.

9 m

11 m

2. 4.2 ft

8.5 ft

3.
$\frac{2}{3}$ in.

$\frac{2}{3}$ in.

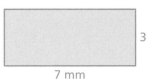

● Finding Areas of Triangles (6.G.1)

Example 2 Find the area of the triangle.

$$A = \frac{1}{2}bh \qquad \text{Write formula.}$$

$$= \frac{1}{2}(6)(7) \qquad \text{Substitute 6 for } b \text{ and 7 for } h.$$

$$= \frac{1}{2}(42) \qquad \text{Multiply 6 and 7.}$$

$$= 21 \qquad \text{Multiply } \frac{1}{2} \text{ and 42.}$$

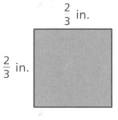

7 in.

6 in.

∴ The area of the triangle is 21 square inches.

Try It Yourself

Find the area of the triangle.

4.
6 ft

13 ft

5.
14 m

20 m

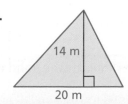

6.
30 cm

15 cm

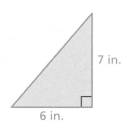

9.1 Surface Areas of Prisms

Essential Question How can you find the surface area of a prism?

1 ACTIVITY: Surface Area of a Rectangular Prism

Work with a partner. Copy the net for a rectangular prism. Label each side as h, w, or ℓ. Then use your drawing to write a formula for the surface area of a rectangular prism.

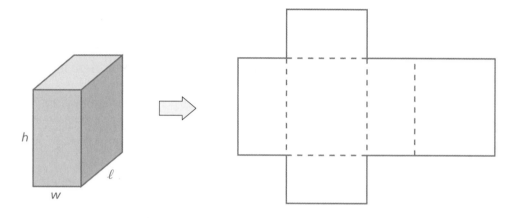

2 ACTIVITY: Surface Area of a Triangular Prism

Work with a partner.

a. Find the surface area of the solid shown by the net. Copy the net, cut it out, and fold it to form a solid. Identify the solid.

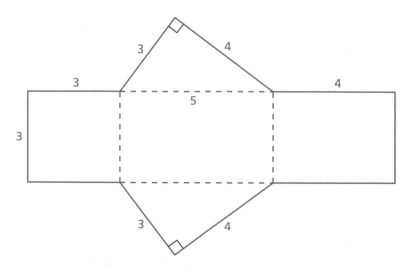

COMMON CORE

Geometry

In this lesson, you will

- use two-dimensional nets to represent three-dimensional solids.
- find surface areas of rectangular and triangular prisms.
- solve real-life problems.

Learning Standard
7.G.6

b. Which of the surfaces of the solid are bases? Why?

3 ACTIVITY: Forming Rectangular Prisms

Math Practice 3

Construct Arguments

What method did you use to find the surface area of the rectangular prism? Explain.

Work with a partner.

- **Use 24 one-inch cubes to form a rectangular prism that has the given dimensions.**

- **Draw each prism.**

- **Find the surface area of each prism.**

a. $4 \times 3 \times 2$ *Drawing*   *Surface Area*

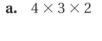

$\boxed{}$ in.^{2}

b. $1 \times 1 \times 24$ **c.** $1 \times 2 \times 12$ **d.** $1 \times 3 \times 8$

e. $1 \times 4 \times 6$ **f.** $2 \times 2 \times 6$ **g.** $2 \times 4 \times 3$

What Is Your Answer?

4. Use your formula from Activity 1 to verify your results in Activity 3.

5. **IN YOUR OWN WORDS** How can you find the surface area of a prism?

6. **REASONING** When comparing ice blocks with the same volume, the ice with the greater surface area will melt faster. Which will melt faster, the bigger block or the three smaller blocks? Explain your reasoning.

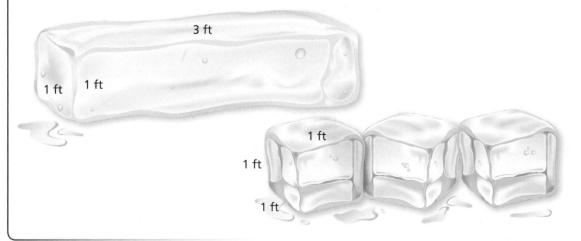

Practice

Use what you learned about the surface areas of rectangular prisms to complete Exercises 4–6 on page 359.

Key Vocabulary
lateral surface area,
p. 358

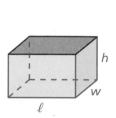 **Key Idea**

Surface Area of a Rectangular Prism

Words The surface area S of a rectangular prism is the sum of the areas of the bases and the lateral faces.

Algebra $S = 2\ell w + 2\ell h + 2wh$

Areas of bases ↑

Areas of lateral faces ↑ ↑

EXAMPLE **1** **Finding the Surface Area of a Rectangular Prism**

Find the surface area of the prism.

Draw a net.

$$S = 2\ell w + 2\ell h + 2wh$$
$$= 2(3)(5) + 2(3)(6) + 2(5)(6)$$
$$= 30 + 36 + 60$$
$$= 126$$

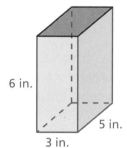
6 in.
3 in.
5 in.

3 in.
5 in.
5 in.   5 in. 3 in.
6 in.

∴ The surface area is 126 square inches.

 On Your Own

Now You're Ready
Exercises 7–9

Find the surface area of the prism.

1.

4 ft
3 ft
2 ft

2.
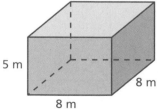
5 m
8 m
8 m

 **Key Idea**

Surface Area of a Prism

The surface area S of any prism is the sum of the areas of the bases and the lateral faces.

$$S = \text{areas of bases} + \text{areas of lateral faces}$$

EXAMPLE ② **Finding the Surface Area of a Triangular Prism**

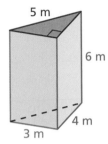

Find the surface area of the prism.

Draw a net.

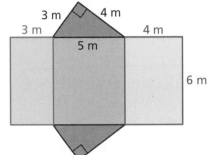

 **Remember**

The area A of a triangle with base b and height h is $A = \frac{1}{2}bh$.

Area of a Base

Red base: $\frac{1}{2} \cdot 3 \cdot 4 = 6$

Areas of Lateral Faces

Green lateral face: $3 \cdot 6 = 18$

Purple lateral face: $5 \cdot 6 = 30$

Blue lateral face: $4 \cdot 6 = 24$

Add the areas of the bases and the lateral faces.

$S = \text{areas of bases} + \text{areas of lateral faces}$

$\quad = \underbrace{6 + 6} + 18 + 30 + 24$

> There are two identical bases. Count the area twice.

$\quad = 84$

∴ The surface area is 84 square meters.

● **On Your Own**

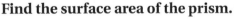
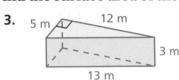 *Now You're Ready*
Exercises 10–12

Find the surface area of the prism.

3.

4.

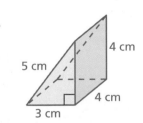

When all the edges of a rectangular prism have the same length s, the rectangular prism is a cube. The formula for the surface area of a cube is

$$S = 6s^2.$$ Formula for surface area of a cube

EXAMPLE ③   **Finding the Surface Area of a Cube**

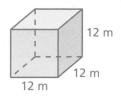

Find the surface area of the cube.

$S = 6s^2$ Write formula for surface area of a cube.

$\quad = 6(12)^2$ Substitute 12 for s.

$\quad = 864$ Simplify.

∴ The surface area of the cube is 864 square meters.

The **lateral surface area** of a prism is the sum of the areas of the lateral faces.

EXAMPLE ④   **Real-Life Application**

The outsides of purple traps are coated with glue to catch emerald ash borers. You make your own trap in the shape of a rectangular prism with an open top and bottom. What is the surface area that you need to coat with glue?

Find the lateral surface area.

20 in.

10 in.

12 in.

$S = 2\ell h + 2wh$ ← Do not include the areas of the bases in the formula.

$\quad = 2(12)(20) + 2(10)(20)$ Substitute.

$\quad = 480 + 400$ Multiply.

$\quad = 880$ Add.

∴ So, you need to coat 880 square inches with glue.

● **On Your Own**

Now You're Ready
Exercises 13–15

5. Which prism has the greater surface area?

9 cm

9 cm

9 cm

7 cm

15 cm

5 cm

6. **WHAT IF?** In Example 4, both the length and the width of your trap are 12 inches. What is the surface area that you need to coat with glue?

9.1 Exercises

Vocabulary and Concept Check

1. **VOCABULARY** Describe two ways to find the surface area of a rectangular prism.

2. **WRITING** Compare and contrast a rectangular prism to a cube.

3. **DIFFERENT WORDS, SAME QUESTION** Which is different? Find "both" answers.

| Find the surface area of the prism. | Find the area of the bases of the prism. |
| Find the area of the net of the prism. | Find the sum of the areas of the bases and the lateral faces of the prism. |

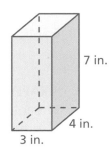

7 in.
4 in.
3 in.

Practice and Problem Solving

Use one-inch cubes to form a rectangular prism that has the given dimensions. Then find the surface area of the prism.

4. $1 \times 2 \times 3$

5. $3 \times 4 \times 1$

6. $2 \times 3 \times 2$

Find the surface area of the prism.

 7.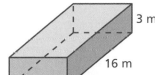

3 m
16 m
6 m

8.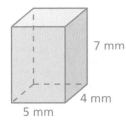

7 mm
4 mm
5 mm

9.

3 yd
5 yd
$1\frac{1}{5}$ yd

10.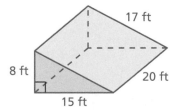

17 ft
8 ft
20 ft
15 ft

11.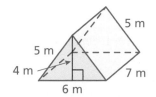

5 m
5 m
4 m
6 m
7 m

12.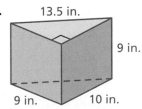

13.5 in.
9 in.
9 in.
10 in.

13.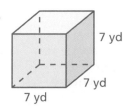

7 yd
7 yd
7 yd

14.

0.5 cm
0.5 cm
0.5 cm

15.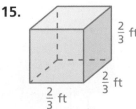

$\frac{2}{3}$ ft
$\frac{2}{3}$ ft
$\frac{2}{3}$ ft

16. ERROR ANALYSIS Describe and correct the error in finding the surface area of the prism.

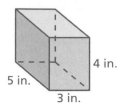

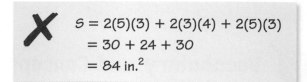

$$S = 2(5)(3) + 2(3)(4) + 2(5)(3)$$
$$= 30 + 24 + 30$$
$$= 84 \text{ in.}^2$$

17. GAME Find the surface area of the tin game case.

10 in.   10 in.

8.7 in.

3 in.   10 in.

18. WRAPPING PAPER A cube-shaped gift is 11 centimeters long. What is the least amount of wrapping paper you need to wrap the gift?

3 in.

9 in.

13 in.

19. FROSTING One can of frosting covers about 280 square inches. Is one can of frosting enough to frost the cake? Explain.

Find the surface area of the prism.

20.

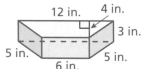

12 in.   4 in.

3 in.

5 in.   5 in.

6 in.

21.

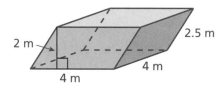

2 m

2.5 m

4 m

4 m

22. OPEN-ENDED Draw and label a rectangular prism that has a surface area of 158 square yards.

23. LABEL A label that wraps around a box of golf balls covers 75% of its lateral surface area. What is the value of x?

3 in.

SUPER
Golf Balls

SUPER
Golf Balls
because
YOU are a
super golfer

2 in.

2 in.

x in.

h

10 cm

10 cm

24. BREAD Fifty percent of the surface area of the bread is crust. What is the height h?

Compare the dimensions of the prisms. How many times greater is the surface area of the red prism than the surface area of the blue prism?

25.

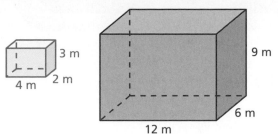

3 m
4 m
2 m

9 m
6 m
12 m

26.

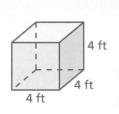

4 ft
4 ft
4 ft

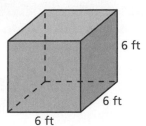

6 ft
6 ft
6 ft

27. STRUCTURE You are painting the prize pedestals shown (including the bottoms). You need 0.5 pint of paint to paint the red pedestal.

a. The side lengths of the green pedestal are one-half the side lengths of the red pedestal. How much paint do you need to paint the green pedestal?

b. The side lengths of the blue pedestal are triple the side lengths of the green pedestal. How much paint do you need to paint the blue pedestal?

c. Compare the ratio of paint amounts to the ratio of side lengths for the green and red pedestals. Repeat for the green and blue pedestals. What do you notice?

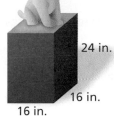

24 in.
16 in.
16 in.

28. **Number Sense** A keychain-sized Rubik's Cube® is made up of small cubes. Each small cube has a surface area of 1.5 square inches.

a. What is the side length of each small cube?

b. What is the surface area of the entire Rubik's Cube®?

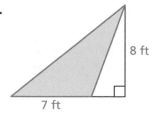

Fair Game Review *What you learned in previous grades & lessons*

Find the area of the triangle *(Skills Review Handbook)*

29.

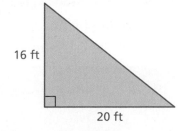

16 ft
20 ft

30.

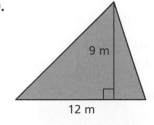

9 m
12 m

31.

8 ft
7 ft

32. MULTIPLE CHOICE What is the circumference of the basketball? Use 3.14 for π. *(Section 8.1)*

9 in.

(A) 14.13 in.   **(B)** 28.26 in.   **(C)** 56.52 in.   **(D)** 254.34 in.

Essential Question How can you find the surface area of a pyramid?

Even though many well-known pyramids have square bases, the base of a pyramid can be any polygon.

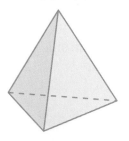

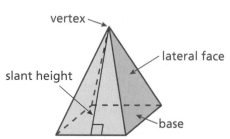

vertex

lateral face

slant height

base

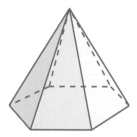

Triangular Base   **Square Base**   **Hexagonal Base**

1 ACTIVITY: Making a Scale Model

Work with a partner. Each pyramid has a square base.

● **Draw a net for a scale model of one of the pyramids. Describe your scale.**

● **Cut out the net and fold it to form a pyramid.**

● **Find the lateral surface area of the real-life pyramid.**

a. Cheops Pyramid in Egypt

Side = 230 m, Slant height ≈ 186 m

b. Muttart Conservatory in Edmonton

Side = 26 m, Slant height ≈ 27 m

c. Louvre Pyramid in Paris

Side = 35 m, Slant height ≈ 28 m

d. Pyramid of Caius Cestius in Rome

Side = 22 m, Slant height ≈ 29 m

COMMON CORE

Geometry

In this lesson, you will
● find surface areas of regular pyramids.
● solve real-life problems.
Learning Standard
7.G.6

2 ACTIVITY: Estimation

Work with a partner. There are many different types of gemstone cuts. Here is one called a brilliant cut.

Top View *Side View* *Bottom View*

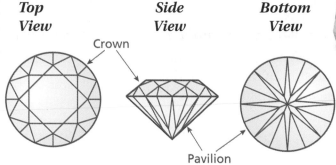

Crown

Pavilion

The size and shape of the pavilion can be approximated by an octagonal pyramid.

a. What does *octagonal* mean?

b. Draw a net for the pyramid.

c. Find the lateral surface area of the pyramid.

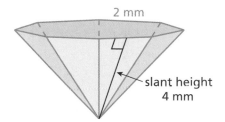

2 mm

slant height 4 mm

3 ACTIVITY: Comparing Surface Areas

Work with a partner. Both pyramids have the same side lengths of the base and the same slant heights.

a. **REASONING** Without calculating, which pyramid has the greater surface area? Explain.

b. Verify your answer to part (a) by finding the surface area of each pyramid.

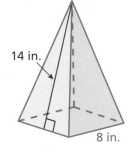

14 in.

8 in.

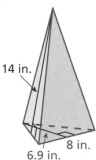

14 in.

8 in.

6.9 in.

What Is Your Answer?

4. **IN YOUR OWN WORDS** How can you find the surface area of a pyramid? Draw a diagram with your explanation.

Practice

Use what you learned about the surface area of a pyramid to complete Exercises 4–6 on page 366.

Check It Out
Lesson Tutorials
BigIdeasMath com

A **regular pyramid** is a pyramid whose base is a regular polygon. The lateral faces are triangles. The height of each triangle is the **slant height** of the pyramid.

Key Vocabulary
regular pyramid, p. 364
slant height, p. 364

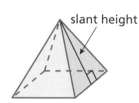 **Key Idea**

Remember

In a regular polygon, all the sides are congruent and all the angles are congruent.

Surface Area of a Pyramid

The surface area S of a pyramid is the sum of the areas of the base and the lateral faces.

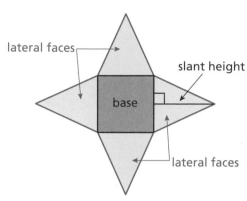

S = area of base + areas of lateral faces

EXAMPLE 1 **Finding the Surface Area of a Square Pyramid**

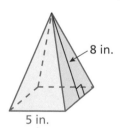

Find the surface area of the regular pyramid.

Draw a net.

Area of Base	**Area of a Lateral Face**
$5 \cdot 5 = 25$	$\dfrac{1}{2} \cdot 5 \cdot 8 = 20$

Find the sum of the areas of the base and the lateral faces.

$$S = \text{area of base} + \text{areas of lateral faces}$$
$$= 25 + 20 + 20 + 20 + 20$$
$$= 105$$

There are 4 identical lateral faces. Count the area 4 times.

∴ The surface area is 105 square inches.

On Your Own

1. What is the surface area of a square pyramid with a base side length of 9 centimeters and a slant height of 7 centimeters?

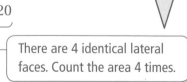 Multi-Language Glossary at BigIdeasMath.com

EXAMPLE 2 **Finding the Surface Area of a Triangular Pyramid**

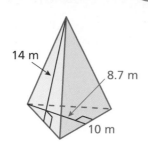

Find the surface area of the regular pyramid.

Draw a net.

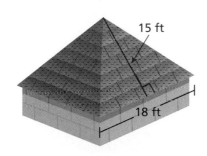

Area of Base	***Area of a Lateral Face***
$\frac{1}{2} \cdot 10 \cdot 8.7 = 43.5$	$\frac{1}{2} \cdot 10 \cdot 14 = 70$

Find the sum of the areas of the base and the lateral faces.

$$S = \text{area of base} + \text{areas of lateral faces}$$

$$= 43.5 + \underbrace{70 + 70 + 70}$$

$$= 253.5$$

> There are 3 identical lateral faces. Count the area 3 times.

∴ The surface area is 253.5 square meters.

EXAMPLE 3 **Real-Life Application**

A roof is shaped like a square pyramid. One bundle of shingles covers 25 square feet. How many bundles should you buy to cover the roof?

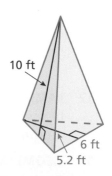

The base of the roof does not need shingles. So, find the sum of the areas of the lateral faces of the pyramid.

Area of a Lateral Face

$$\frac{1}{2} \cdot 18 \cdot 15 = 135$$

There are four identical lateral faces. So, the lateral surface area is

$$135 + 135 + 135 + 135 = 540.$$

Because one bundle of shingles covers 25 square feet, it will take $540 \div 25 = 21.6$ bundles to cover the roof.

∴ So, you should buy 22 bundles of shingles.

On Your Own

Now You're Ready
Exercises 7–12

2. What is the surface area of the regular pyramid at the right?

3. WHAT IF? In Example 3, one bundle of shingles covers 32 square feet. How many bundles should you buy to cover the roof?

✓ Vocabulary and Concept Check

1. **VOCABULARY** Can a pyramid have rectangles as lateral faces? Explain.

2. **CRITICAL THINKING** Why is it helpful to know the slant height of a pyramid to find its surface area?

3. **WHICH ONE DOESN'T BELONG?** Which description of the solid does *not* belong with the other three? Explain your answer.

square pyramid	regular pyramid
rectangular pyramid	triangular pyramid

5 m
5 m

Practice and Problem Solving

Use the net to find the surface area of the regular pyramid.

4.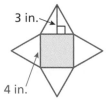
3 in.
4 in.

5.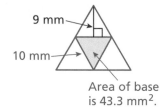
9 mm
10 mm
Area of base is 43.3 mm².

6.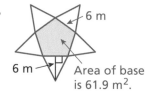
6 m
6 m
Area of base is 61.9 m².

In Exercises 7–11, find the surface area of the regular pyramid.

① ② 7.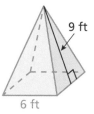
9 ft
6 ft

8.
6 cm
4 cm

9.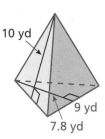
10 yd
9 yd
7.8 yd

10.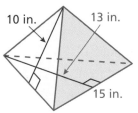
10 in.
13 in.
15 in.

11.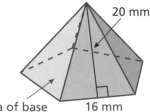
20 mm
Area of base is 440.4 mm².
16 mm

10 in.

③ 12. **LAMPSHADE** The base of the lampshade is a regular hexagon with a side length of 8 inches. Estimate the amount of glass needed to make the lampshade.

13. **GEOMETRY** The surface area of a square pyramid is 85 square meters. The base length is 5 meters. What is the slant height?

Find the surface area of the composite solid.

14.

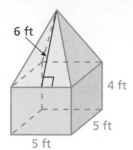

6 ft
4 ft
5 ft
5 ft

15.

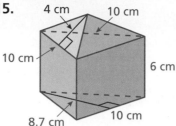

4 cm
10 cm
10 cm
6 cm
8.7 cm
10 cm

16.

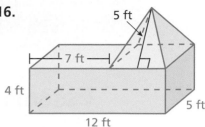

5 ft
7 ft
4 ft
5 ft
12 ft

17. PROBLEM SOLVING You are making an umbrella that is shaped like a regular octagonal pyramid.

5 ft

4 ft

a. Estimate the amount of fabric that you need to make the umbrella.

b. The fabric comes in rolls that are 72 inches wide. You don't want to cut the fabric "on the bias." Find out what this means. Then draw a diagram of how you can cut the fabric most efficiently.

c. How much fabric is wasted?

18. REASONING The *height* of a pyramid is the perpendicular distance between the base and the top of the pyramid. Which is greater, the height of a pyramid or the slant height? Explain your reasoning.

pyramid height

19. TETRAHEDRON A tetrahedron is a triangular pyramid whose four faces are identical equilateral triangles. The total lateral surface area is 93 square centimeters. Find the surface area of the tetrahedron.

20. Reasoning Is the total area of the lateral faces of a pyramid *greater than*, *less than*, or *equal to* the area of the base? Explain.

Fair Game Review What you learned in previous grades & lessons

Find the area and the circumference of the circle. Use 3.14 for π.
(Section 8.1 and Section 8.3)

21.

12

22.

8

23.

27

24. MULTIPLE CHOICE The distance between bases on a youth baseball field is proportional to the distance between bases on a professional baseball field. The ratio of the youth distance to the professional distance is 2 : 3. Bases on a youth baseball field are 60 feet apart. What is the distance between bases on a professional baseball field? *(Section 5.4)*

    (A) 40 ft   (B) 90 ft   (C) 120 ft   (D) 180 ft

Essential Question How can you find the surface area of a cylinder?

A *cylinder* is a solid that has two parallel, identical circular bases.

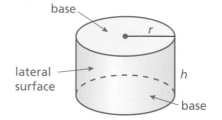

1 ACTIVITY: Finding Area

Work with a partner. Use a cardboard cylinder.

- **Talk about how you can find the area of the outside of the roll.**

- **Estimate the area using the methods you discussed.**

- **Use the roll and the scissors to find the actual area of the cardboard.**

- **Compare the actual area to your estimates.**

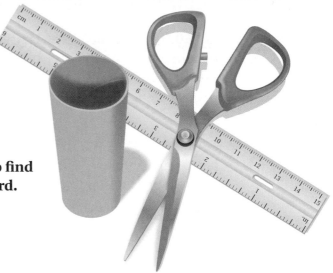

2 ACTIVITY: Finding Surface Area

Work with a partner.

- **Make a net for the can. Name the shapes in the net.**

- **Find the surface area of the can.**

- **How are the dimensions of the rectangle related to the dimensions of the can?**

COMMON CORE

Geometry

In this lesson, you will
- find surface areas of cylinders.

Applying Standard
7.G.4

3 ACTIVITY: Estimation

Work with a partner. From memory, estimate the dimensions of the real-life item in inches. Then use the dimensions to estimate the surface area of the item in square inches.

a.

b.

c.

d.

What Is Your Answer?

4. **IN YOUR OWN WORDS** How can you find the surface area of a cylinder? Give an example with your description. Include a drawing of the cylinder.

5. To eight decimal places, $\pi \approx 3.14159265$. Which of the following is closest to π?

  a. 3.14
  b. $\frac{22}{7}$
  c. $\frac{355}{113}$

"To approximate $\pi \approx 3.141593$, I simply remember 1, 1, 3, 3, 5, 5."

"Then I compute $\frac{355}{113} \approx 3.141593$."

Practice

Use what you learned about the surface area of a cylinder to complete Exercises 3–5 on page 372.

Check It Out
Lesson Tutorials
BigIdeasMath ✓com

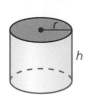 Key Idea

Surface Area of a Cylinder

Words The surface area S of a cylinder is the sum of the areas of the bases and the lateral surface.

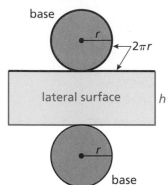

Remember

Pi can be approximated as 3.14 or $\frac{22}{7}$.

Algebra $S = 2\pi r^2 + 2\pi rh$

Areas of bases

Area of lateral surface

EXAMPLE **1** **Finding the Surface Area of a Cylinder**

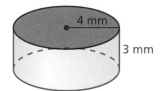

Find the surface area of the cylinder. Round your answer to the nearest tenth.

Draw a net.

$$S = 2\pi r^2 + 2\pi rh$$

$$= 2\pi(4)^2 + 2\pi(4)(3)$$

$$= 32\pi + 24\pi$$

$$= 56\pi$$

$$\approx 175.8$$

∴ The surface area is about 175.8 square millimeters.

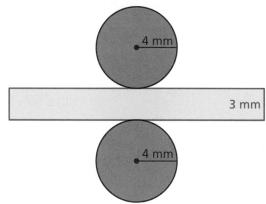

● **On Your Own**

Now You're Ready
Exercises 6–8

Find the surface area of the cylinder. Round your answer to the nearest tenth.

1.

2.

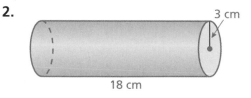

How much paper is used for the label on the can of peas?

Find the lateral surface area of the cylinder.

1 in.

2 in.

SWEET PEAS

NET WT. 8 OZ.

$S = 2\pi rh$ ← *Do not include the areas of the bases in the formula.*

$= 2\pi(1)(2)$ Substitute.

$= 4\pi \approx 12.56$ Multiply.

∴ About 12.56 square inches of paper is used for the label.

You earn \$0.01 for recycling the can in Example 2. How much can you expect to earn for recycling the tomato can? Assume that the recycle value is proportional to the surface area.

2 in.

Defiggio's

WHOLE PEELED

TOMATOES

5.5 in.

No Added Salt

NET WT. 28 OZ. (1 LB. 12 OZ.)

Find the surface area of each can.

Tomatoes

$S = 2\pi r^2 + 2\pi rh$

$= 2\pi(2)^2 + 2\pi(2)(5.5)$

$= 8\pi + 22\pi$

$= 30\pi$

Peas

$S = 2\pi r^2 + 2\pi rh$

$= 2\pi(1)^2 + 2\pi(1)(2)$

$= 2\pi + 4\pi$

$= 6\pi$

Use a proportion to find the recycle value x of the tomato can.

$$\frac{30\pi \text{ in.}^2}{x} = \frac{6\pi \text{ in.}^2}{\$0.01}$$ ← *surface area* ← *recycle value*

$30\pi \cdot 0.01 = x \cdot 6\pi$ Cross Products Property

$5 \cdot 0.01 = x$ Divide each side by 6π.

$0.05 = x$ Simplify.

∴ You can expect to earn \$0.05 for recycling the tomato can.

● **On Your Own**

Now You're Ready
Exercises 9–11

3. **WHAT IF?** In Example 3, the height of the can of peas is doubled.

     a. Does the amount of paper used in the label double?

     b. Does the recycle value double? Explain.

 Vocabulary and Concept Check

1. **CRITICAL THINKING** Which part of the formula $S = 2\pi r^2 + 2\pi rh$ represents the lateral surface area of a cylinder?

2. **CRITICAL THINKING** You are given the height and the circumference of the base of a cylinder. Describe how to find the surface area of the entire cylinder.

 Practice and Problem Solving

Make a net for the cylinder. Then find the surface area of the cylinder. Round your answer to the nearest tenth.

3.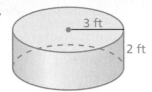
3 ft
2 ft

4.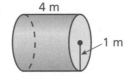
4 m
1 m

5.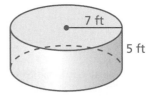
7 ft
5 ft

Find the surface area of the cylinder. Round your answer to the nearest tenth.

① 6.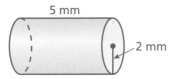
5 mm
2 mm

7.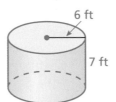
6 ft
7 ft

8.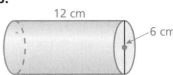
12 cm
6 cm

Find the lateral surface area of the cylinder. Round your answer to the nearest tenth.

② 9.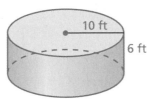
10 ft
6 ft

10.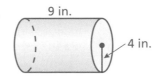
9 in.
4 in.

11.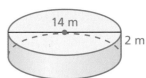
14 m
2 m

12. **ERROR ANALYSIS** Describe and correct the error in finding the surface area of the cylinder.

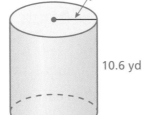
5 yd
10.6 yd

✗ $S = \pi r^2 + 2\pi rh$
$= \pi(5)^2 + 2\pi(5)(10.6)$
$= 25\pi + 106\pi$
$= 131\pi \approx 411.3 \text{ yd}^2$

13. **TANKER** The truck's tank is a stainless steel cylinder. Find the surface area of the tank.

50 ft
radius = 4 ft

14. OTTOMAN What percent of the surface area of the ottoman is green (not including the bottom)?

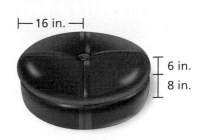
├— 16 in. —┤
6 in.
8 in.

15. REASONING You make two cylinders using 8.5-by-11-inch pieces of paper. One has a height of 8.5 inches, and the other has a height of 11 inches. Without calculating, compare the surface areas of the cylinders.

10 cm
3.5 cm
24.5 cm
5.5 cm

16. INSTRUMENT A *ganza* is a percussion instrument used in samba music.

  a. Find the surface area of each of the two labeled ganzas.

  b. The weight of the smaller ganza is 1.1 pounds. Assume that the surface area is proportional to the weight. What is the weight of the larger ganza?

17. BRIE CHEESE The cut wedge represents one-eighth of the cheese.

  a. Find the surface area of the cheese before it is cut.

  b. Find the surface area of the remaining cheese after the wedge is removed. Did the surface area increase, decrease, or remain the same?

├— 3 in. —┤
1 in.

18. *Repeated Reasoning* A cylinder has radius *r* and height *h*.

  a. How many times greater is the surface area of a cylinder when both dimensions are multiplied by a factor of 2? 3? 5? 10?

  b. Describe the pattern in part (a). How many times greater is the surface area of a cylinder when both dimensions are multiplied by a factor of 20?

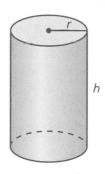
r
h

 Fair Game Review What you learned in previous grades & lessons

Find the area. *(Skills Review Handbook)*

19.
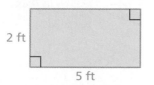
2 ft
5 ft

20.
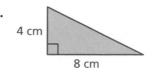
4 cm
8 cm

21.
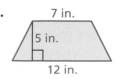
7 in.
5 in.
12 in.

22. MULTIPLE CHOICE 40% of what number is 80? *(Section 6.4)*

  (A) 32   (B) 48   (C) 200   (D) 320

Check It Out
Graphic Organizer
BigIdeasMath ✓com

You can use an **information frame** to help you organize and remember concepts. Here is an example of an information frame for surface areas of rectangular prisms.

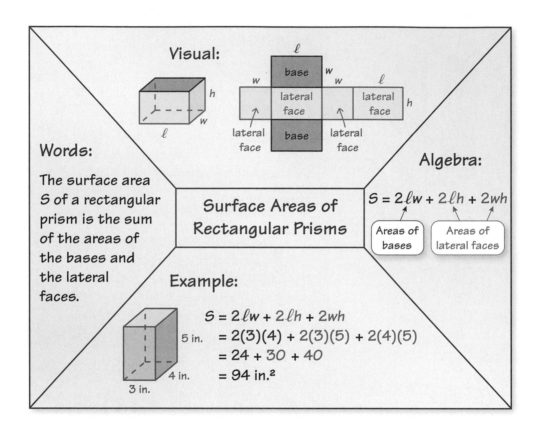

Visual:

base

lateral face

lateral face   base   lateral face

lateral face

Words:

The surface area S of a rectangular prism is the sum of the areas of the bases and the lateral faces.

Surface Areas of Rectangular Prisms

Algebra:

$S = 2\ell w + 2\ell h + 2wh$

Areas of bases

Areas of lateral faces

Example:

$S = 2\ell w + 2\ell h + 2wh$
$= 2(3)(4) + 2(3)(5) + 2(4)(5)$
$= 24 + 30 + 40$
$= 94$ in.²

5 in.
4 in.
3 in.

On Your Own

Make information frames to help you study the topics.

1. surface areas of prisms

2. surface areas of pyramids

3. surface areas of cylinders

After you complete this chapter, make information frames for the following topics.

4. volumes of prisms

5. volumes of pyramids

Lizards

Birds   ?   Mice

Squirrels

Is this one of those "what the cat dragged home" jokes?

"I'm having trouble thinking of a good title for my information frame."

Find the surface area of the prism. *(Section 9.1)*

1.

3 cm   4 cm

10 cm

5 cm

2.

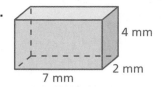

4 mm

2 mm

7 mm

Find the surface area of the regular pyramid. *(Section 9.2)*

3.

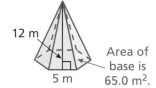

12 m

Area of base is 65.0 m².

5 m

4.

6 cm

2 cm

Find the surface area of the cylinder. Round your answer to the nearest tenth. *(Section 9.3)*

5.

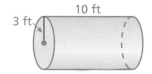

10 ft

3 ft

6.

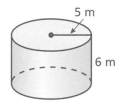

5 m

6 m

Find the lateral surface area of the cylinder. Round your answer to the nearest tenth. *(Section 9.3)*

7.

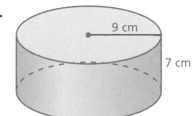

9 cm

7 cm

8.

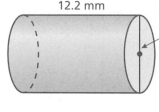

12.2 mm

8 mm

9. SKYLIGHT You are making a skylight that has 12 triangular pieces of glass and a slant height of 3 feet. Each triangular piece has a base of 1 foot. *(Section 9.2)*

   a. How much glass will you need to make the skylight?

   b. Can you cut the 12 glass triangles from a sheet of glass that is 4 feet by 8 feet? If so, draw a diagram showing how this can be done.

10. MAILING TUBE What is the least amount of material needed to make the mailing tube? *(Section 9.3)*

3 ft

3 in.

11. WOODEN CHEST All the faces of the wooden chest will be painted except for the bottom. Find the area to be painted, in *square inches*. *(Section 9.1)*

4 ft

4 ft   4 ft

Essential Question How can you find the volume of a prism?

1 **ACTIVITY: Pearls in a Treasure Chest**

Work with a partner. A treasure chest is filled with valuable pearls.
Each pearl is about 1 centimeter in diameter and is worth about $80.

Use the diagrams below to describe two
ways that you can estimate the number
of pearls in the treasure chest.

a.

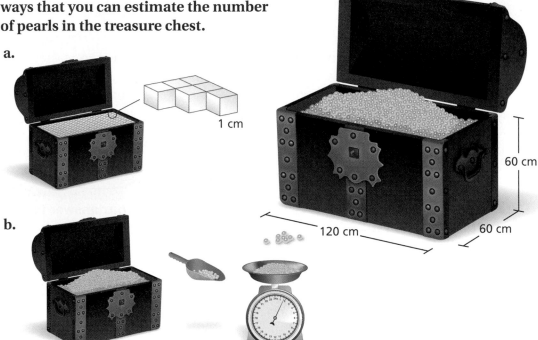

1 cm

60 cm

120 cm

60 cm

b.

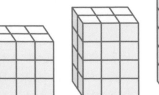

c. Use the method in part (a) to estimate the value of the pearls in the chest.

COMMON CORE

Geometry

In this lesson, you will
• find volumes of prisms.
• solve real-life problems.
Learning Standard
7.G.6

2 **ACTIVITY: Finding a Formula for Volume**

Work with a partner. You know that the formula for the volume of a
rectangular prism is $V = \ell wh$.

a. Write a formula that gives the volume in terms of the area of the
base B and the height h.

b. Use both formulas to find the volume of
each prism. Do both formulas
give you the same
volume?

ACTIVITY: Finding a Formula for Volume

Work with a partner. Use the concept in Activity 2 to find a formula that gives the volume of any prism.

Triangular Prism

Rectangular Prism

Pentagonal Prism

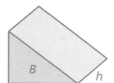

Triangular Prism

Hexagonal Prism

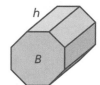

Octagonal Prism

4 ACTIVITY: Using a Formula

Work with a partner. A ream of paper has 500 sheets.

a. Does a single sheet of paper have a volume? Why or why not?

b. If so, explain how you can find the volume of a single sheet of paper.

What Is Your Answer?

5. **IN YOUR OWN WORDS** How can you find the volume of a prism?

6. **STRUCTURE** Draw a prism that has a trapezoid as its base. Use your formula to find the volume of the prism.

Use what you learned about the volumes of prisms to complete Exercises 4–6 on page 380.

Check It Out
Lesson Tutorials
BigIdeasMath com

The *volume* of a three-dimensional figure is a measure of the amount of space that it occupies. Volume is measured in cubic units.

Key Idea

Volume of a Prism

Words The volume V of a prism is the product of the area of the base and the height of the prism.

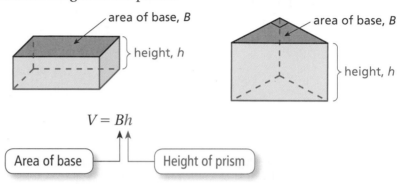

Remember

The volume V of a cube with an edge length of s is $V = s^3$.

Algebra $V = Bh$

Area of base → ↑↑ ← Height of prism

EXAMPLE **1** **Finding the Volume of a Prism**

Study Tip

The area of the base of a rectangular prism is the product of the length ℓ and the width w.

You can use $V = \ell wh$ to find the volume of a rectangular prism.

Find the volume of the prism.

$V = Bh$	Write formula for volume.
$= 6(8) \cdot 15$	Substitute.
$= 48 \cdot 15$	Simplify.
$= 720$	Multiply.

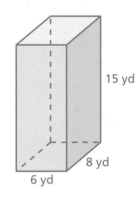

15 yd

8 yd

6 yd

∴ The volume is 720 cubic yards.

EXAMPLE **2** **Finding the Volume of a Prism**

Find the volume of the prism.

$V = Bh$	Write formula for volume.
$= \dfrac{1}{2}(5.5)(2) \cdot 4$	Substitute.
$= 5.5 \cdot 4$	Simplify.
$= 22$	Multiply.

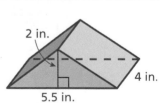

2 in.

4 in.

5.5 in.

∴ The volume is 22 cubic inches.

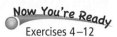

● **On Your Own**

Now You're Ready
Exercises 4–12

Find the volume of the prism.

1.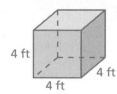

4 ft

4 ft

4 ft

2.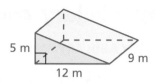

5 m

12 m

9 m

EXAMPLE 3 Real-Life Application

A movie theater designs two bags to hold 96 cubic inches of popcorn. (a) Find the height of each bag. (b) Which bag should the theater choose to reduce the amount of paper needed? Explain.

Bag A

3 in.

4 in.

h

Bag B

4 in.

4 in.

h

a. Find the height of each bag.

Bag A	**Bag B**
$V = Bh$	$V = Bh$
$96 = 4(3)(h)$	$96 = 4(4)(h)$
$96 = 12h$	$96 = 16h$
$8 = h$	$6 = h$

⠇∙ The height is 8 inches. ⠇∙ The height is 6 inches.

b. To determine the amount of paper needed, find the surface area of each bag. Do not include the top base.

Bag A	**Bag B**
$S = \ell w + 2\ell h + 2wh$	$S = \ell w + 2\ell h + 2wh$
$= 4(3) + 2(4)(8) + 2(3)(8)$	$= 4(4) + 2(4)(6) + 2(4)(6)$
$= 12 + 64 + 48$	$= 16 + 48 + 48$
$= 124 \text{ in.}^2$	$= 112 \text{ in.}^2$

⠇∙ The surface area of Bag B is less than the surface area of Bag A. So, the theater should choose Bag B.

● **On Your Own**

3. You design Bag C that has a volume of 96 cubic inches. Should the theater in Example 3 choose your bag? Explain.

Bag C

h

4 in.

4.8 in.

Vocabulary and Concept Check

1. **VOCABULARY** What types of units are used to describe volume?

2. **VOCABULARY** Explain how to find the volume of a prism.

3. **CRITICAL THINKING** How are volume and surface area different?

Practice and Problem Solving

Find the volume of the prism.

 4.

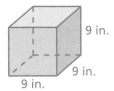
9 in.
9 in.
9 in.

5.

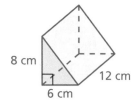
8 cm
12 cm
6 cm

6.

$8\frac{1}{2}$ m
7 m
4 m

7.

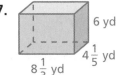
6 yd
$4\frac{1}{5}$ yd
$8\frac{1}{3}$ yd

8.

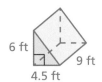
6 ft
9 ft
4.5 ft

9.

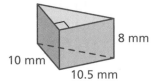
8 mm
10 mm
10.5 mm

10.

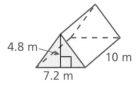
4.8 m
10 m
7.2 m

11.

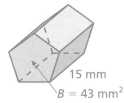
15 mm
$B = 43$ mm^2

12.

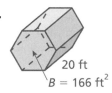
20 ft
$B = 166$ ft^2

13. **ERROR ANALYSIS** Describe and correct the error in finding the volume of the triangular prism.

7 cm
10 cm
5 cm

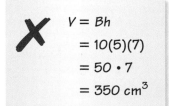
$\times$
$V = Bh$
$= 10(5)(7)$
$= 50 \cdot 7$
$= 350$ cm^3

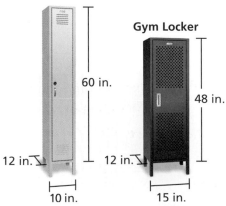

School Locker

Gym Locker

60 in.

48 in.

12 in.

12 in.

10 in.

15 in.

14. **LOCKER** Each locker is shaped like a rectangular prism. Which has more storage space? Explain.

15. **CEREAL BOX** A cereal box is 9 inches by 2.5 inches by 10 inches. What is the volume of the box?

Find the volume of the prism.

16.

12 in.

12 in.   10 in.

17.

24 ft

30 ft

20 ft

18. LOGIC Two prisms have the same volume. Do they *always*, *sometimes*, or *never* have the same surface area? Explain.

19. CUBIC UNITS How many cubic inches are in a cubic foot? Use a sketch to explain your reasoning.

20. CAPACITY As a gift, you fill the calendar with packets of chocolate candy. Each packet has a volume of 2 cubic inches. Find the maximum number of packets you can fit inside the calendar.

6 in.

8 in.   4 in.

21. PRECISION Two liters of water are poured into an empty vase shaped like an octagonal prism. The base area is 100 square centimeters. What is the height of the water? $(1 \text{ L} = 1000 \text{ cm}^3)$

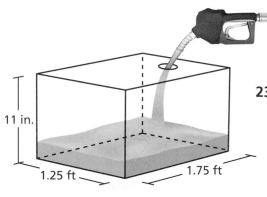

11 in.

1.25 ft   1.75 ft

22. GAS TANK The gas tank is 20% full. Use the current price of regular gasoline in your community to find the cost to fill the tank. $(1 \text{ gal} = 231 \text{ in.}^3)$

23. OPEN-ENDED You visit an aquarium. One of the tanks at the aquarium holds 450 gallons of water. Draw a diagram to show one possible set of dimensions of the tank. $(1 \text{ gal} = 231 \text{ in.}^3)$

24. *Critical Thinking* How many times greater is the volume of a triangular prism when one of its dimensions is doubled? when all three dimensions are doubled?

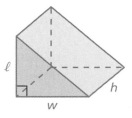

ℓ h w

 Fair Game Review *What you learned in previous grades & lessons*

Find the selling price. *(Section 6.6)*

25. Cost to store: $75
Markup: 20%

26. Cost to store: $90
Markup: 60%

27. Cost to store: $130
Markup: 85%

28. MULTIPLE CHOICE What is the approximate surface area of a cylinder with a radius of 3 inches and a height of 10 inches? *(Section 9.3)*

Ⓐ 30 in.^{2} **Ⓑ** 87 in.^{2} **Ⓒ** 217 in.^{2} **Ⓓ** 245 in.^{2}

9.5 Volumes of Pyramids

Essential Question How can you find the volume of a pyramid?

1 ACTIVITY: Finding a Formula Experimentally

Work with a partner.

- Draw the two nets on cardboard and cut them out.

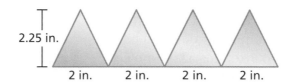

2.25 in.

2 in.   2 in.   2 in.   2 in.

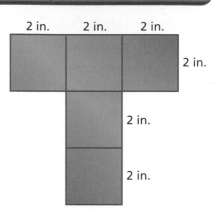

2 in.   2 in.   2 in.

2 in.

2 in.

2 in.

- Fold and tape the nets to form an open square box and an open pyramid.

- Both figures should have the same size square base and the same height.

- Fill the pyramid with pebbles. Then pour the pebbles into the box. Repeat this until the box is full. How many pyramids does it take to fill the box?

- Use your result to find a formula for the volume of a pyramid.

2 ACTIVITY: Comparing Volumes

Work with a partner. You are an archaeologist studying two ancient pyramids. What factors would affect how long it took to build each pyramid? Given similar conditions, which pyramid took longer to build? Explain your reasoning.

COMMON CORE

Geometry

In this lesson, you will
- find volumes of pyramids.
- solve real-life problems.

Learning Standard
7.G.6

The Sun Pyramid in Mexico
Height: about 246 ft
Base: about 738 ft by 738 ft

Cheops Pyramid in Egypt
Height: about 480 ft
Base: about 755 ft by 755 ft

3 ACTIVITY: Finding and Using a Pattern

Math Practice 7

Look for Patterns

As the height and the base lengths increase, how does this pattern affect the volume? Explain.

Work with a partner.

- **Find the volumes of the pyramids.**
- **Organize your results in a table.**
- **Describe the pattern.**
- **Use your pattern to find the volume of a pyramid with a base length and a height of 20.**

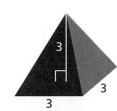
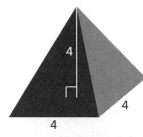
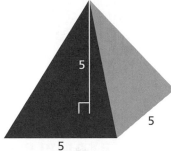

4 ACTIVITY: Breaking a Prism into Pyramids

Work with a partner. The rectangular prism can be cut to form three pyramids. Show that the sum of the volumes of the three pyramids is equal to the volume of the prism.

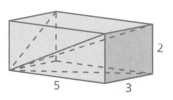

a.

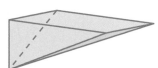

b.

c.

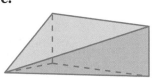

What Is Your Answer?

5. **IN YOUR OWN WORDS** How can you find the volume of a pyramid?

6. **STRUCTURE** Write a general formula for the volume of a pyramid.

Practice

Use what you learned about the volumes of pyramids to complete Exercises 4–6 on page 386.

 **Key Idea**

Volume of a Pyramid

Study Tip

The *height* of a pyramid is the perpendicular distance from the base to the vertex.

Words The volume V of a pyramid is one-third the product of the area of the base and the height of the pyramid.

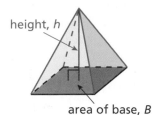
height, *h*

area of base, *B*

Area of base

Algebra $V = \dfrac{1}{3}Bh$

Height of pyramid

EXAMPLE 1 **Finding the Volume of a Pyramid**

Find the volume of the pyramid.

$V = \dfrac{1}{3}Bh$
Write formula for volume.

$= \dfrac{1}{3}(48)(9)$
Substitute.

$= 144$
Multiply.

9 mm

$B = 48$ mm²

⋮⋱ The volume is 144 cubic millimeters.

EXAMPLE 2 **Finding the Volume of a Pyramid**

Find the volume of the pyramid.

Study Tip

The area of the base of a rectangular pyramid is the product of the length ℓ and the width w.

You can use $V = \dfrac{1}{3}\ell wh$ to find the volume of a rectangular pyramid.

a.

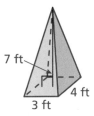
7 ft
4 ft
3 ft

$V = \dfrac{1}{3}Bh$

$= \dfrac{1}{3}(4)(3)(7)$

$= 28$

⋮⋱ The volume is 28 cubic feet.

b.

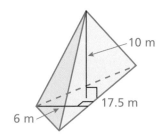
10 m
17.5 m
6 m

$V = \dfrac{1}{3}Bh$

$= \dfrac{1}{3}\left(\dfrac{1}{2}\right)(17.5)(6)(10)$

$= 175$

⋮⋱ The volume is 175 cubic meters.

On Your Own

Now You're Ready
Exercises 4–11

Find the volume of the pyramid.

1.

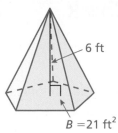

6 ft

$B = 21$ ft^2

2.

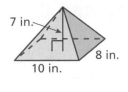

7 in.

8 in.

10 in.

3.

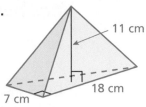

11 cm

7 cm

18 cm

EXAMPLE ③ **Real-Life Application**

a. **The volume of sunscreen in Bottle B is about how many times the volume in Bottle A?**

b. **Which is the better buy?**

a. Use the formula for the volume of a pyramid to estimate the amount of sunscreen in each bottle.

Bottle A
$9.96

6 in.

1 in.

2 in.

Bottle B
$14.40

4 in.

1.5 in.

3 in.

Bottle A

$$V = \frac{1}{3}Bh$$

$$= \frac{1}{3}(2)(1)(6)$$

$$= 4 \text{ in.}^3$$

Bottle B

$$V = \frac{1}{3}Bh$$

$$= \frac{1}{3}(3)(1.5)(4)$$

$$= 6 \text{ in.}^3$$

∴ So, the volume of sunscreen in Bottle B is about $\frac{6}{4} = 1.5$ times the volume in Bottle A.

b. Find the unit cost for each bottle.

Bottle A

$$\frac{\text{cost}}{\text{volume}} = \frac{\$9.96}{4 \text{ in.}^3}$$

$$= \frac{\$2.49}{1 \text{ in.}^3}$$

Bottle B

$$\frac{\text{cost}}{\text{volume}} = \frac{\$14.40}{6 \text{ in.}^3}$$

$$= \frac{\$2.40}{1 \text{ in.}^3}$$

∴ The unit cost of Bottle B is less than the unit cost of Bottle A. So, Bottle B is the better buy.

On Your Own

Now You're Ready
Exercise 16

4. Bottle C is on sale for $13.20. Is Bottle C a better buy than Bottle B in Example 3? Explain.

Bottle C

3 in.

2 in.

3 in.

✓ Vocabulary and Concept Check

1. **WRITING** How is the formula for the volume of a pyramid different from the formula for the volume of a prism?

2. **OPEN-ENDED** Describe a real-life situation that involves finding the volume of a pyramid.

3. **REASONING** A triangular pyramid and a triangular prism have the same base and height. The volume of the prism is how many times the volume of the pyramid?

Practice and Problem Solving

Find the volume of the pyramid.

① ② **4.**

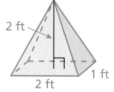

2 ft
1 ft
2 ft

5.

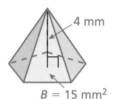

4 mm
$B = 15$ mm²

6.

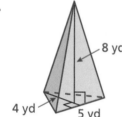

8 yd
4 yd
5 yd

7.

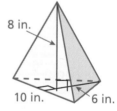

8 in.
10 in.
6 in.

8.

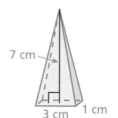

7 cm
3 cm
1 cm

9.

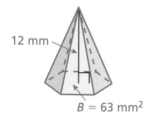

12 mm
$B = 63$ mm²

10.

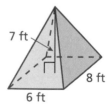

7 ft
8 ft
6 ft

11.

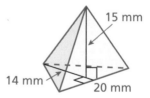

15 mm
14 mm
20 mm

12. **PARACHUTE** In 1483, Leonardo da Vinci designed a parachute. It is believed that this was the first parachute ever designed. In a notebook, he wrote, "If a man is provided with a length of gummed linen cloth with a length of 12 yards on each side and 12 yards high, he can jump from any great height whatsoever without injury." Find the volume of air inside Leonardo's parachute.

Not drawn to scale

Find the volume of the composite solid.

13.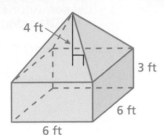
4 ft

3 ft

6 ft

6 ft

14.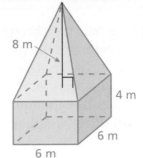
8 m

4 m

6 m

6 m

15.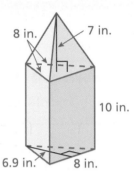
8 in. 7 in.

10 in.

6.9 in. 8 in.

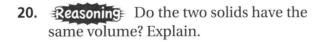
6 in.

$B = 30$ in.^{2}

Spire A

8 in.

$B = 24$ in.^{2}

Spire B

③ 16. **SPIRE** Which sand-castle spire has a greater volume? How much more sand do you need to make the spire with the greater volume?

17. **PAPERWEIGHT** How much glass is needed to manufacture 1000 paperweights? Explain your reasoning.

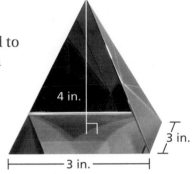
4 in.

3 in.

3 in.

Paperweight

18. **PROBLEM SOLVING** Use the photo of the tepee.

   a. What is the shape of the base? How can you tell?

   b. The tepee's height is about 10 feet. Estimate the volume of the tepee.

19. **OPEN-ENDED** A pyramid has a volume of 40 cubic feet and a height of 6 feet. Find one possible set of dimensions of the rectangular base.

20. **Reasoning** Do the two solids have the same volume? Explain.

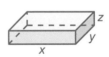
z
y
x

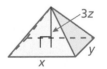
3z
y
x

Fair Game Review What you learned in previous grades & lessons

For the given angle measure, find the measure of a supplementary angle and the measure of a complementary angle, if possible. *(Section 7.2)*

21. $27°$

22. $82°$

23. $120°$

24. **MULTIPLE CHOICE** The circumference of a circle is 44 inches. Which estimate is closest to the area of the circle? *(Section 8.3)*

   Ⓐ 7 in.^{2}

   Ⓑ 14 in.^{2}

   Ⓒ 154 in.^{2}

   Ⓓ 484 in.^{2}

Check It Out
Lesson Tutorials
BigIdeasMath ✓com

Consider a plane "slicing" through a solid. The intersection of the plane and the solid is a two-dimensional shape called a **cross section**. For example, the diagram shows that the intersection of the plane and the rectangular prism is a rectangle.

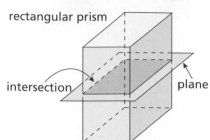

rectangular prism

intersection

plane

EXAMPLE 1 Describing the Intersection of a Plane and a Solid

Describe the intersection of the plane and the solid.

COMMON CORE

Geometry

In this extension, you will
• describe the intersections of planes and solids.

Learning Standard
7.G.3

a.
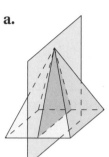

b.
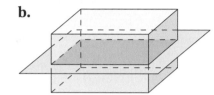

c.
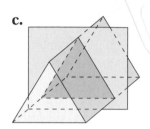

a. The intersection is a triangle.

b. The intersection is a rectangle.

c. The intersection is a triangle.

● Practice

Describe the intersection of the plane and the solid.

1.
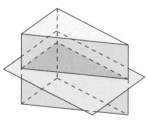

2.
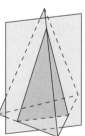

3.
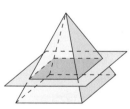

4.
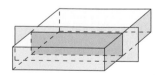

5.
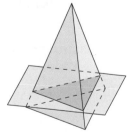

6.
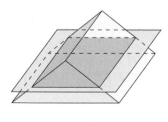

7. REASONING A plane that intersects a prism is parallel to the bases of the prism. Describe the intersection of the plane and the prism.

🔊 Multi-Language Glossary at BigIdeasMath ✓com

Example 1 shows how a plane intersects a polyhedron. Now consider the intersection of a plane and a solid having a curved surface, such as a cylinder or cone. As shown, a *cone* is a solid that has one circular base and one vertex.

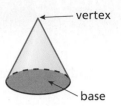
vertex
base

EXAMPLE 2 Describing the Intersection of a Plane and a Solid

Describe the intersection of the plane and the solid.

a.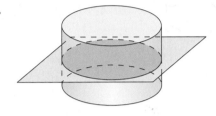

b.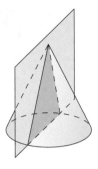

a. The intersection is a circle.

b. The intersection is a triangle.

Practice

Describe the intersection of the plane and the solid.

8.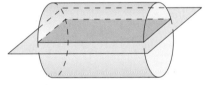

9.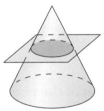

10.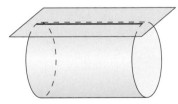

11.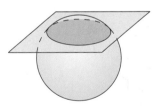

Describe the shape that is formed by the cut made in the food shown.

12.

13.

14.

15. **REASONING** Explain how a plane can be parallel to the base of a cone and intersect the cone at exactly one point.

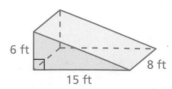

Check It Out
Progress Check
BigIdeasMath ✓com

Find the volume of the prism. *(Section 9.4)*

1.

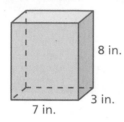

8 in.

3 in.

7 in.

2.

6 ft

8 ft

15 ft

3.

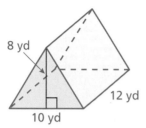

8 yd

12 yd

10 yd

4.

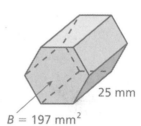

25 mm

$B = 197 \text{ mm}^2$

Find the volume of the solid. Round your answer to the nearest tenth if necessary. *(Section 9.5)*

5.

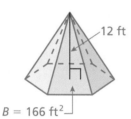

12 ft

$B = 166 \text{ ft}^2$

6.

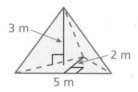

3 m

2 m

5 m

Describe the intersection of the plane and the solid. *(Section 9.5)*

7.

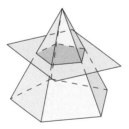

8.

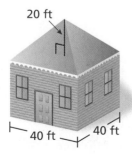

20 ft

40 ft

40 ft

9. ROOF A pyramid hip roof is a good choice for a house in a hurricane area. What is the volume of the roof to the nearest tenth? *(Section 9.5)*

10. CUBIC UNITS How many cubic feet are in a cubic yard? Use a sketch to explain your reasoning. *(Section 9.4)*

Review Key Vocabulary

lateral surface area, *p. 358*   slant height, *p. 364*
regular pyramid, *p. 364*   cross section, *p. 388*

Review Examples and Exercises

9.1 Surface Areas of Prisms (pp. 354–361)

Find the surface area of the prism.

Draw a net.

$$S = 2\ell w + 2\ell h + 2wh$$
$$= 2(6)(4) + 2(6)(5) + 2(4)(5)$$
$$= 48 + 60 + 40$$
$$= 148$$

∴ The surface area is 148 square feet.

Exercises

Find the surface area of the prism.

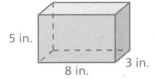

1. 5 in. 8 in. 3 in.

2. 17 cm 15 cm 8 cm 7 cm

3. 3 m 4 m 8 m 5 m

9.2 Surface Areas of Pyramids (pp. 362–367)

Find the surface area of the regular pyramid.

Draw a net.

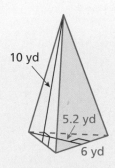

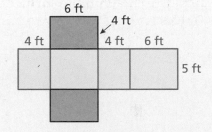

Area of Base	**Area of a Lateral Face**
$\frac{1}{2} \cdot 6 \cdot 5.2 = 15.6$	$\frac{1}{2} \cdot 6 \cdot 10 = 30$

Find the sum of the areas of the base and all three lateral faces.

$$S = 15.6 + \underbrace{30 + 30 + 30}$$
$$= 105.6$$

There are 3 identical lateral faces. Count the area 3 times.

∴ The surface area is 105.6 square yards.

Exercises

Find the surface area of the regular pyramid.

4.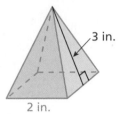
3 in.

2 in.

5.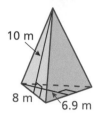
10 m

8 m
6.9 m

6.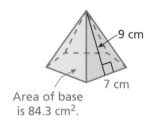
9 cm

7 cm

Area of base is 84.3 cm².

9.3 **Surface Areas of Cylinders** *(pp. 368–373)*

Find the surface area of the cylinder. Round your answer to the nearest tenth.

Draw a net.

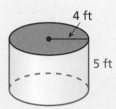
4 ft

5 ft

$S = 2\pi r^2 + 2\pi rh$

$= 2\pi(4)^2 + 2\pi(4)(5)$

$= 32\pi + 40\pi$

$= 72\pi \approx 226.1$

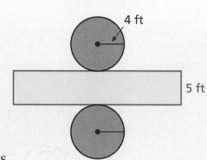
4 ft

5 ft

✦ The surface area is about 226.1 square millimeters.

Exercises

Find the surface area of the cylinder. Round your answer to the nearest tenth.

7.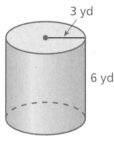
3 yd

6 yd

8. 0.8 cm

6 cm

9. ORANGES Find the lateral surface area of the can of mandarin oranges.

4 cm

11 cm

9.4 **Volumes of Prisms** *(pp. 376–381)*

Find the volume of the prism.

$V = Bh$ Write formula for volume.

$= \dfrac{1}{2}(7)(3) \cdot 5$ Substitute.

$= 52.5$ Multiply.

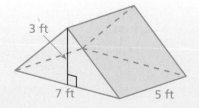
3 ft

7 ft   5 ft

✦ The volume is 52.5 cubic feet.

Exercises

Find the volume of the prism.

10.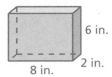
6 in.
2 in.
8 in.

11.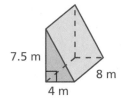
7.5 m
8 m
4 m

12.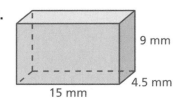
9 mm
4.5 mm
15 mm

9.5 **Volumes of Pyramids** *(pp. 382–389)*

a. **Find the volume of the pyramid.**

$$V = \frac{1}{3}Bh \qquad \text{Write formula for volume.}$$

$$= \frac{1}{3}(6)(5)(10) \qquad \text{Substitute.}$$

$$= 100 \qquad \text{Multiply.}$$

∴ The volume is 100 cubic yards.

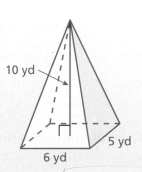
10 yd
5 yd
6 yd

b. **Describe the intersection of the plane and the solid.**

i.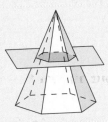

The intersection is a hexagon.

ii.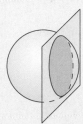

The intersection is a circle.

Exercises

Find the volume of the pyramid.

13.
20 ft
17 ft 15 ft

14.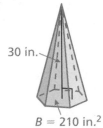
30 in.
$B = 210 \text{ in.}^2$

15.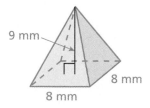
9 mm
8 mm
8 mm

Describe the intersection of the plane and the solid.

16.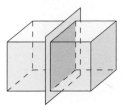

17.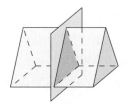

Find the surface area of the prism or regular pyramid.

1.
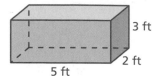
3 ft
2 ft
5 ft

2.
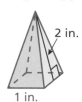
2 in.
1 in.

3.

15 m
11 m
9.5 m

Find the surface area of the cylinder. Round your answer to the nearest tenth.

4.
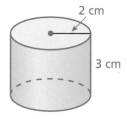
2 cm
3 cm

5.
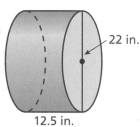
22 in.
12.5 in.

Find the volume of the solid.

6.
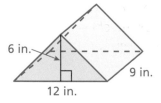
6 in.
9 in.
12 in.

7.
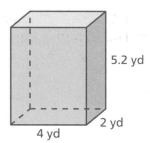
5.2 yd
2 yd
4 yd

8.
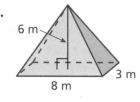
6 m
8 m
3 m

9. **SKATEBOARD RAMP** A quart of paint covers 80 square feet. How many quarts should you buy to paint the ramp with two coats? (Assume you will not paint the bottom of the ramp.)

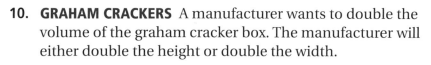
15.2 ft
6 ft
19.5 ft
14 ft

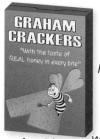
GRAHAM CRACKERS
"With the taste of REAL honey in every bite"
h = 9 in.
ℓ = 6 in.   w = 2 in.

10. **GRAHAM CRACKERS** A manufacturer wants to double the volume of the graham cracker box. The manufacturer will either double the height or double the width.

   a. Which option uses less cardboard? Justify your answer.

   b. What is the volume of the new graham cracker box?

11. **SOUP** The label on the can of soup covers about 354.2 square centimeters. What is the height of the can? Round your answer to the nearest whole number.

4.7 cm
TOMATO SOUP

1. A gift box and its dimensions are shown below.

2 in.

8 in.   4 in.

What is the least amount of wrapping paper that you could have used to wrap the box? *(7.G.6)*

A. 20 in.²   C. 64 in.²

B. 56 in.²   D. 112 in.²

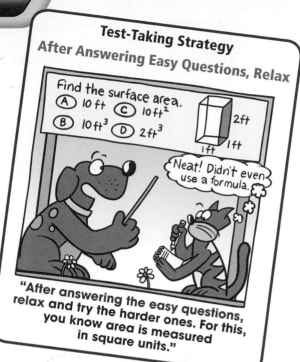

Test-Taking Strategy
After Answering Easy Questions, Relax

Find the surface area.
(A) 10 ft   (C) 10 ft²
(B) 10 ft³   (D) 2 ft³

2ft
1 ft   1 ft

Neat! Didn't even use a formula.

"After answering the easy questions, relax and try the harder ones. For this, you know area is measured in square units."

2. A student scored 600 the first time she took the mathematics portion of her college entrance exam. The next time she took the exam, she scored 660. Her second score represents what percent increase over her first score? *(7.RP.3)*

F. 9.1%   H. 39.6%

G. 10%   I. 60%

3. Raj was solving the proportion in the box below.

$$\frac{3}{8} = \frac{x-3}{24}$$

$$3 \cdot 24 = (x - 3) \cdot 8$$

$$72 = x - 24$$

$$96 = x$$

What should Raj do to correct the error that he made? *(7.RP.2c)*

A. Set the product of the numerators equal to the product of the denominators.

B. Distribute 8 to get $8x - 24$.

C. Add 3 to each side to get $\frac{3}{8} + 3 = \frac{x}{24}$.

D. Divide both sides by 24 to get $\frac{3}{8} \div 24 = x - 3$.

4. A line contains the two points plotted in the coordinate plane below.

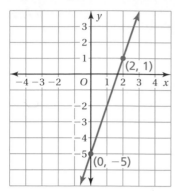

What is the slope of the line?   *(7.RP.2b)*

F. $\dfrac{1}{3}$

H. 3

G. 2

I. 6

5. James is getting ready for wrestling season. As part of his preparation, he plans to lose 5% of his body weight. James currently weighs 160 pounds. How much will he weigh, in pounds, after he loses 5% of his weight?   *(7.RP.3)*

6. How much material is needed to make the popcorn container?   *(7.G.4)*

4 in.

9.5 in.

A. 76π in.^{2}

C. 92π in.^{2}

B. 84π in.^{2}

D. 108π in.^{2}

7. To make 10 servings of soup you need 4 cups of broth. You want to know how many servings you can make with 8 pints of broth. Which proportion should you use?   *(7.RP.2c)*

F. $\dfrac{10}{4} = \dfrac{x}{8}$

H. $\dfrac{10}{4} = \dfrac{8}{x}$

G. $\dfrac{4}{10} = \dfrac{x}{16}$

I. $\dfrac{10}{4} = \dfrac{x}{16}$

8. A rectangular prism and its dimensions are shown below.

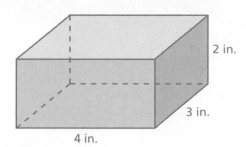

2 in.

3 in.

4 in.

What is the volume, in cubic inches, of a rectangular prism whose dimensions are three times greater? *(7.G.6)*

9. What is the value of *x*? *(7.G.5)*

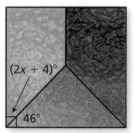

(2x + 4)°

46°

A. 20   **C.** 44

B. 43   **D.** 65

10. Which of the following could be the angle measures of a triangle? *(7.G.5)*

F. 60°, 50°, 20°   **H.** 30°, 60°, 90°

G. 40°, 80°, 90°   **I.** 0°, 90°, 90°

11. The table below shows the costs of buying matinee movie tickets. *(7.RP.2b)*

Matinee Tickets, x	2	3	4	5
Cost, y	$9	$13.50	$18	$22.50

Part A Graph the data.

Part B Find and interpret the slope of the line through the points.

Part C How much does it cost to buy 8 matinee movie tickets?

10 Probability and Statistics

"If there are 7 cats in a sack and I draw one at random,..."

"... what is the probability that I will draw you?"

"I'm just about finished making my two number cubes."

"Now, here's how the game works. You toss the two cubes."

"If the sum is even, I win. If it's odd, you win."

What You Learned Before

M = Mouse
B = Dog Biscuit

"Why do we always have to use this spinner?"

"Let's spin to decide what we will have for lunch."

● Writing Ratios (6.RP.1)

Example 1 There are 32 football players and 16 cheerleaders at your school. Write the ratio of cheerleaders to football players.

cheerleaders → $\dfrac{16}{32} = \dfrac{1}{2}$ Write in simplest form.

football players →

⋮· So, the ratio of cheerleaders to football players is $\dfrac{1}{2}$.

Example 2

a. Write the ratio of girls to boys in Classroom A.

$\dfrac{\text{Girls in Classroom A}}{\text{Boys in Classroom A}} = \dfrac{11}{14}$

	Boys	Girls
Classroom A	14	11
Classroom B	12	8

⋮· So, the ratio of girls to boys in Classroom A is $\dfrac{11}{14}$.

b. Write the ratio of boys in Classroom B to the total number of students in both classes.

$\dfrac{\text{Boys in Classroom B}}{\text{Total number of students}} = \dfrac{12}{14 + 11 + 12 + 8} = \dfrac{12}{45} = \dfrac{4}{15}$ Write in simplest form.

⋮· So, the ratio of boys in Classroom B to the total number of students is $\dfrac{4}{15}$.

Try It Yourself
Write the ratio in simplest form.

1. baseballs to footballs

2. footballs to total pieces of equipment

3. sneakers to ballet slippers

4. sneakers to total number of shoes

5. green beads to blue beads

6. red beads : green beads

7. green beads : total number of beads

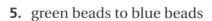

Essential Question

In an experiment, how can you determine the number of possible results?

An *experiment* is an investigation or a procedure that has varying results. Flipping a coin, rolling a number cube, and spinning a spinner are all examples of experiments.

1 ACTIVITY: Conducting Experiments

Work with a partner.

a. You flip a dime.

There are ☐ possible results.

Out of 20 flips, you think you will flip heads ☐ times.

Flip a dime 20 times. Tally your results in a table. How close was your guess?

b. You spin the spinner shown.

There are ☐ possible results.

Out of 20 spins, you think you will spin orange ☐ times.

Spin the spinner 20 times. Tally your results in a table. How close was your guess?

c. You spin the spinner shown.

There are ☐ possible results.

Out of 20 spins, you think you will spin a 4 ☐ times.

Spin the spinner 20 times. Tally your results in a table. How close was your guess?

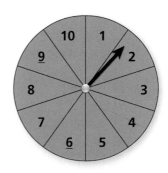

COMMON CORE

Probability and Statistics

In this lesson, you will

- identify and count the outcomes of experiments.

Preparing for Standard 7.SP.5

2 ACTIVITY: Comparing Different Results

Work with a partner. Use the spinner in Activity 1(c).

a. Do you have a better chance of spinning an even number or a multiple of 4? Explain your reasoning.

b. Do you have a better chance of spinning an even number or an odd number? Explain your reasoning.

3 ACTIVITY: Rock Paper Scissors

Work with a partner.

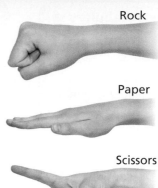

Rock

a. Play Rock Paper Scissors 30 times. Tally your results in the table.

b. How many possible results are there?

Paper

c. Of the possible results, in how many ways can Player A win? Player B win? the players tie?

d. Does one of the players have a better chance of winning than the other player? Explain your reasoning.

Scissors

Math Practice 1

Interpret a Solution

How do your results compare to the possible results? Explain.

Rock *breaks* scissors.
Paper *covers* rock.
Scissors *cut* paper.

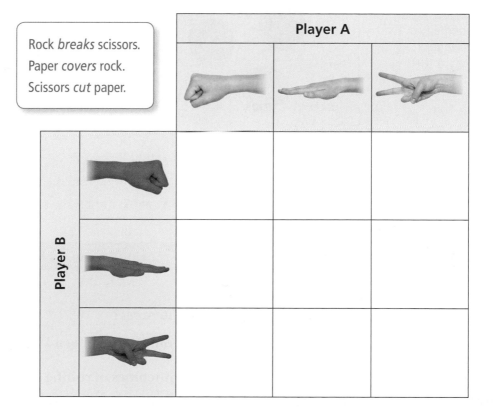

What Is Your Answer?

4. IN YOUR OWN WORDS In an experiment, how can you determine the number of possible results?

Use what you learned about experiments to complete Exercises 3 and 4 on page 404.

Check It Out
Lesson Tutorials
BigIdeasMath.com

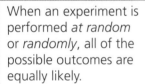

Key Vocabulary
experiment, *p. 402*
outcomes, *p. 402*
event, *p. 402*
favorable outcomes,
   p. 402

 **Key Ideas**

Outcomes and Events

An **experiment** is an investigation or a procedure that has varying results. The possible results of an experiment are called **outcomes**. A collection of one or more outcomes is an **event**. The outcomes of a specific event are called **favorable outcomes**.

For example, randomly selecting a marble from a group of marbles is an experiment. Each marble in the group is an outcome. Selecting a green marble from the group is an event.

Reading

When an experiment is performed *at random* or *randomly*, all of the possible outcomes are equally likely.

Possible outcomes

Event: Choosing a green marble
Number of favorable outcomes: 2

EXAMPLE 1 **Identifying Outcomes**

You roll the number cube.

a. What are the possible outcomes?

⁚⁚· The six possible outcomes are rolling a 1, 2, 3, 4, 5, and 6.

b. What are the favorable outcomes of rolling an even number?

even	*not* even
2, 4, 6	1, 3, 5

⁚⁚· The favorable outcomes of the event are rolling a 2, 4, and 6.

c. What are the favorable outcomes of rolling a number greater than 5?

greater than 5	*not* greater than 5
6	1, 2, 3, 4, 5

⁚⁚· The favorable outcome of the event is rolling a 6.

◀) Multi-Language Glossary at BigIdeasMath.com

On Your Own

Now You're Ready
Exercises 5–11

1. You randomly choose a letter from a hat that contains the letters A through K.

    a. What are the possible outcomes?

    b. What are the favorable outcomes of choosing a vowel?

EXAMPLE 2 **Counting Outcomes**

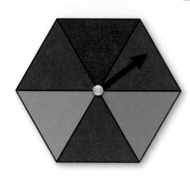

You spin the spinner.

  a. **How many possible outcomes are there?**

    The spinner has 6 sections. So, there are 6 possible outcomes.

  b. **In how many ways can spinning red occur?**

    The spinner has 3 red sections. So, spinning red can occur in 3 ways.

  c. **In how many ways can spinning *not* purple occur? What are the favorable outcomes of spinning *not* purple?**

    The spinner has 5 sections that are *not* purple. So, spinning *not* purple can occur in 5 ways.

purple	*not* purple
purple	red, red, red, green, blue

    The favorable outcomes of the event are red, red, red, green, and blue.

On Your Own

Now You're Ready
Exercises 12–17

2. You randomly choose a marble.

    a. How many possible outcomes are there?

    b. In how many ways can choosing blue occur?

    c. In how many ways can choosing *not* yellow occur? What are the favorable outcomes of choosing *not* yellow?

Check It Out
Help with Homework
BigIdeasMath.com

 ## Vocabulary and Concept Check

1. **VOCABULARY** Is rolling an even number on a number cube an *outcome* or an *event*? Explain.

2. **WRITING** Describe how an outcome and a favorable outcome are different.

 ## Practice and Problem Solving

You spin the spinner shown.

3. How many possible results are there?

4. Of the possible results, in how many ways can you spin an even number? an odd number?

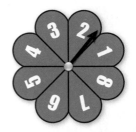

① 5. **TILES** What are the possible outcomes of randomly choosing one of the tiles shown?

You randomly choose one of the tiles shown above. Find the favorable outcomes of the event.

6. Choosing a 6

7. Choosing an odd number

8. Choosing a number greater than 5

9. Choosing an odd number less than 5

10. Choosing a number less than 3

11. Choosing a number divisible by 3

You randomly choose one marble from the bag. (a) Find the number of ways the event can occur. (b) Find the favorable outcomes of the event.

② 12. Choosing blue

13. Choosing green

14. Choosing purple

15. Choosing yellow

16. Choosing *not* red

17. Choosing *not* blue

18. **ERROR ANALYSIS** Describe and correct the error in finding the number of ways that choosing *not* purple can occur.

purple	*not* purple
purple	red, blue, green, yellow

Choosing *not* purple can occur in 4 ways.

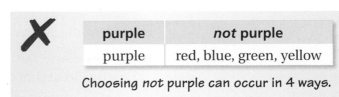

19. COINS You have 10 coins in your pocket. Five are Susan B. Anthony dollars, two are Kennedy half-dollars, and three are presidential dollars. You randomly choose a coin. In how many ways can choosing *not* a presidential dollar occur?

Susan B. Anthony dollar

Kennedy half-dollar

Presidential dollar

Spinner A

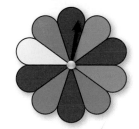

Tell whether the statement is *true* or *false*. If it is false, change the italicized word to make the statement true.

20. Spinning blue and spinning *green* have the same number of favorable outcomes on Spinner A.

21. Spinning blue has one *more* favorable outcome than spinning green on Spinner B.

22. There are *three* possible outcomes of spinning Spinner A.

23. Spinning *red* can occur in four ways on Spinner B.

24. Spinning not green can occur in *three* ways on Spinner B.

Spinner B

25. MUSIC A bargain bin contains classical and rock CDs. There are 60 CDs in the bin. Choosing a rock CD and *not* choosing a rock CD have the same number of favorable outcomes. How many rock CDs are in the bin?

26. Precision You randomly choose one of the cards and set it aside. Then you randomly choose a second card. Describe how the number of possible outcomes changes after the first card is chosen.

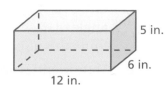

Fair Game Review *What you learned in previous grades & lessons*

Solve the proportion. *(Section 5.4)*

27. $\dfrac{x}{10} = \dfrac{1}{5}$

28. $\dfrac{60}{n} = \dfrac{20}{7}$

29. $\dfrac{1}{3} = \dfrac{w}{36}$

30. $\dfrac{25}{17} = \dfrac{100}{b}$

31. MULTIPLE CHOICE What is the surface area of the rectangular prism? *(Section 9.1)*

(A) 162 in.^{2}

(B) 264 in.^{2}

(C) 324 in.^{2}

(D) 360 in.^{2}

5 in.

6 in.

12 in.

Essential Question How can you describe the likelihood of an event?

1 ACTIVITY: Black-and-White Spinner Game

Work with a partner. You work for a game company. You need to create a game that uses the spinner below.

a. Write rules for a game that uses the spinner. Then play it.

b. After playing the game, do you want to revise the rules? Explain.

COMMON CORE

Probability and Statistics

In this lesson, you will

- understand the concept of probability and the relationship between probability and likelihood.
- find probabilities of events.

Learning Standards
7.SP.5
7.SP.7a

c. **CHOOSE TOOLS** Using the center of the spinner as the vertex, measure the angle of each pie-shaped section. Is each section the same size? How do you think this affects the likelihood of spinning a given number?

d. Your friend is about to spin the spinner and wants to know how likely it is to spin a 3. How would you describe the likelihood of this event to your friend?

Work with a partner. For each spinner, do the following.

- Measure the angle of each pie-shaped section.
- Tell whether you are more likely to spin a particular number. Explain your reasoning.
- Tell whether your rules from Activity 1 make sense for these spinners. Explain your reasoning.

a.   b.

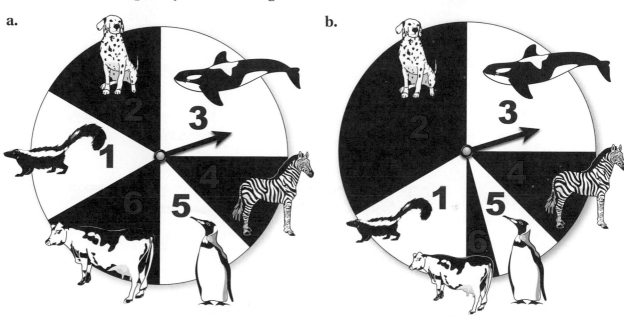

3 **ACTIVITY: Is This Game Fair?**

Math Practice 3

Use Prior Results

How can you use the results of the previous activities to determine whether the game is fair?

Work with a partner. Apply the following rules to each spinner in Activities 1 and 2. Is the game fair? Why or why not? If not, who has the better chance of winning?

- Take turns spinning the spinner.
- If you spin an odd number, Player 1 wins.
- If you spin an even number, Player 2 wins.

What Is Your Answer?

4. **IN YOUR OWN WORDS** How can you describe the likelihood of an event?

5. Describe the likelihood of spinning an 8 in Activity 1.

6. Describe a career in which it is important to know the likelihood of an event.

Practice Use what you learned about the likelihood of an event to complete Exercises 4 and 5 on page 410.

Check It Out
Lesson Tutorials
BigIdeasMath.com

Key Vocabulary 🔊
probability, *p. 408*

 **Key Idea**

Probability

The **probability** of an event is a number that measures the likelihood that the event will occur. Probabilities are between 0 and 1, including 0 and 1. The diagram relates likelihoods (above the diagram) and probabilities (below the diagram).

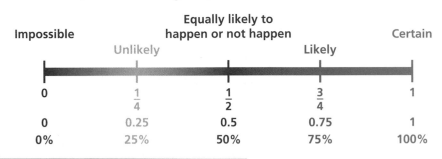

Study Tip

Probabilities can be written as fractions, decimals, or percents.

EXAMPLE ① **Describing the Likelihood of an Event**

80% chance

There is an 80% chance of thunderstorms tomorrow. Describe the likelihood of the event.

The probability of thunderstorms tomorrow is 80%.

⋮∙ Because 80% is close to 75%, it is *likely* that there will be thunderstorms tomorrow.

On Your Own

Now You're Ready
Exercises 6–9

Describe the likelihood of the event given its probability.

1. The probability that you land a jump on a snowboard is $\frac{1}{2}$.

2. There is a 100% chance that the temperature will be less than 120°F tomorrow.

 **Key Idea**

Finding the Probability of an Event

When all possible outcomes are equally likely, the probability of an event is the ratio of the number of favorable outcomes to the number of possible outcomes. The probability of an event is written as P(event).

$$P(\text{event}) = \frac{\text{number of favorable outcomes}}{\text{number of possible outcomes}}$$

🔊 Multi-Language Glossary at BigIdeasMath.com

EXAMPLE 2 Finding a Probability

You roll the number cube. What is the probability of rolling an odd number?

$$P(\text{event}) = \frac{\text{number of favorable outcomes}}{\text{number of possible outcomes}}$$

$$P(\text{odd}) = \frac{3}{6} \quad \longleftarrow \boxed{\text{There are 3 odd numbers (1, 3, and 5).}}$$
$$\quad\quad\quad\quad\ \longleftarrow \boxed{\text{There is a total of 6 numbers.}}$$

$$\quad\quad = \frac{1}{2} \quad\quad \text{Simplify.}$$

∴ The probability of rolling an odd number is $\frac{1}{2}$, or 50%.

EXAMPLE 3 Using a Probability

The probability that you randomly draw a short straw from a group of 40 straws is $\frac{3}{20}$. How many are short straws?

    Ⓐ 4   Ⓑ 6

    Ⓒ 15   Ⓓ 34

$$P(\text{short}) = \frac{\text{number of short straws}}{\text{total number of straws}}$$

$$\frac{3}{20} = \frac{n}{40} \quad\quad \text{Substitute. Let } n \text{ be the number of short straws.}$$

$$6 = n \quad\quad \text{Solve for } n.$$

There are 6 short straws.

∴ So, the correct answer is Ⓑ.

On Your Own

Exercises 11–15

3. In Example 2, what is the probability of rolling a number greater than 2?

4. In Example 2, what is the probability of rolling a 7?

5. The probability that you randomly draw a short straw from a group of 75 straws is $\frac{1}{15}$. How many are short straws?

 Vocabulary and Concept Check

1. **VOCABULARY** Explain how to find the probability of an event.

2. **REASONING** Can the probability of an event be 1.5? Explain.

3. **OPEN-ENDED** Give a real-life example of an event that is impossible. Give a real-life example of an event that is certain.

 Practice and Problem Solving

You are playing a game using the spinners shown.

4. You want to move down. On which spinner are you more likely to spin "Down"? Explain.

5. You want to move forward. Which spinner would you spin? Explain.

Spinner A   Spinner B

Describe the likelihood of the event given its probability.

① 6. Your soccer team wins $\frac{3}{4}$ of the time.

7. There is a 0% chance that you will grow 12 more feet.

8. The probability that the sun rises tomorrow is 1.

9. It rains on $\frac{1}{5}$ of the days in July.

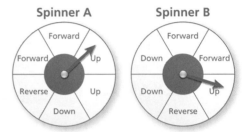

10. **VIOLIN** You have a 50% chance of playing the correct note on a violin. Describe the likelihood of playing the correct note.

You randomly choose one shirt from the shelves. Find the probability of the event.

② 11. Choosing a red shirt

12. Choosing a green shirt

13. *Not* choosing a white shirt

14. *Not* choosing a black shirt

15. Choosing an orange shirt

16. **ERROR ANALYSIS** Describe and correct the error in finding the probability of *not* choosing a blue shirt from the shelves above.

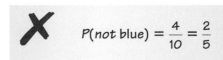
$P(not\ blue) = \frac{4}{10} = \frac{2}{5}$

17. CONTEST The rules of a contest say that there is a 5% chance of winning a prize. Four hundred people enter the contest. Predict how many people will win a prize.

18. RUBBER DUCKS At a carnival, the probability that you choose a winning rubber duck from 25 ducks is 0.24.

    a. How many are *not* winning ducks?

    b. Describe the likelihood of *not* choosing a winning duck.

19. DODECAHEDRON A dodecahedron has twelve sides numbered 1 through 12. Find the probability and describe the likelihood of each event.

    a. Rolling a number less than 9

    b. Rolling a multiple of 3

    c. Rolling a number greater than 6

A Punnett square is a grid used to show possible gene combinations for the offspring of two parents. In the Punnett square shown, a boy is represented by *XY*. A girl is represented by *XX*.

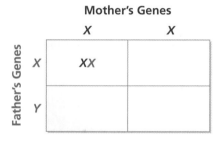

20. Complete the Punnett square.

21. Explain why the probability of two parents having a boy or having a girl is equally likely.

22. **Critical Thinking** Two parents each have the gene combination *Cs*. The gene *C* is for curly hair. The gene *s* is for straight hair.

    a. Make a Punnett square for the two parents. When all outcomes are equally likely, what is the probability of a child having the gene combination *CC*?

    b. Any gene combination that includes a *C* results in curly hair. When all outcomes are equally likely, what is the probability of a child having curly hair?

 Fair Game Review What you learned in previous grades & lessons

Solve the inequality. Graph the solution. *(Section 4.2 and Section 4.3)*

23. $x + 5 < 9$ **24.** $b - 2 \geq -7$ **25.** $1 > -\dfrac{w}{3}$ **26.** $6 \leq -2g$

27. MULTIPLE CHOICE Find the value of *x*. *(Section 7.4)*

    Ⓐ 85   **Ⓑ** 90

    Ⓒ 93   **Ⓓ** 102

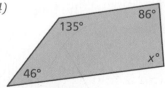

Essential Question How can you use relative frequencies to find probabilities?

When you conduct an experiment, the **relative frequency** of an event is the fraction or percent of the time that the event occurs.

$$\text{relative frequency} = \frac{\text{number of times the event occurs}}{\text{total number of times you conduct the experiment}}$$

1 ACTIVITY: Finding Relative Frequencies

Work with a partner.

a. Flip a quarter 20 times and record your results. Then complete the table. Are the relative frequencies the same as the probability of flipping heads or tails? Explain.

	Flipping Heads	Flipping Tails
Relative Frequency		

b. Compare your results with those of other students in your class. Are the relative frequencies the same? If not, why do you think they differ?

c. Combine all of the results in your class. Then complete the table again. Did the relative frequencies change? What do you notice? Explain.

d. Suppose everyone in your school conducts this experiment and you combine the results. How do you think the relative frequencies will change?

2 ACTIVITY: Using Relative Frequencies

COMMON CORE

Probability and Statistics
In this lesson, you will
- find relative frequencies.
- use experimental probabilities to make predictions.
- use theoretical probabilities to find quantities.
- compare experimental and theoretical probabilities.
Learning Standards
7.SP.5
7.SP.6
7.SP.7a
7.SP.7b

Work with a partner. You have a bag of colored chips. You randomly select a chip from the bag and replace it. The table shows the number of times you select each color.

Red	Blue	Green	Yellow
24	12	15	9

a. There are 20 chips in the bag. Can you use the table to find the exact number of each color in the bag? Explain.

b. You randomly select a chip from the bag and replace it. You do this 50 times, then 100 times, and you calculate the relative frequencies after each experiment. Which experiment do you think gives a better approximation of the exact number of each color in the bag? Explain.

Work with a partner. You toss a thumbtack onto a table. There are two ways the thumbtack can land.

Point up   On its side

a. Your friend says that because there are two outcomes, the probability of the thumbtack landing point up must be $\frac{1}{2}$. Do you think this conclusion is true? Explain.

b. Toss a thumbtack onto a table 50 times and record your results. In a *uniform probability model*, each outcome is equally likely to occur. Do you think this experiment represents a uniform probability model? Explain.

Use the relative frequencies to complete the following.

$$P(\text{point up}) = \quad\quad\quad\quad P(\text{on its side}) = $$

What Is Your Answer?

4. **IN YOUR OWN WORDS** How can you use relative frequencies to find probabilities? Give an example.

5. Your friend rolls a number cube 500 times. How many times do you think your friend will roll an odd number? Explain your reasoning.

6. In Activity 2, your friend says, "There are no orange-colored chips in the bag." Do you think this conclusion is true? Explain.

7. Give an example of an experiment that represents a uniform probability model.

8. Tell whether you can use each spinner to represent a uniform probability model. Explain your reasoning.

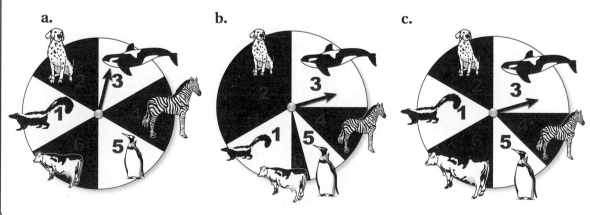

a.   b.   c.

Practice

Use what you learned about relative frequencies to complete Exercises 6 and 7 on page 417.

Check It Out
Lesson Tutorials
BigIdeasMath ✓com

Key Vocabulary 🔊
relative frequency, *p. 412*
experimental probability, *p. 414*
theoretical probability, *p. 415*

🔑 Key Idea

Experimental Probability

Probability that is based on repeated trials of an experiment is called **experimental probability**.

$$P(\text{event}) = \frac{\text{number of times the event occurs}}{\text{total number of trials}}$$

EXAMPLE 1 Finding an Experimental Probability

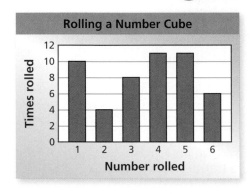

The bar graph shows the results of rolling a number cube 50 times. What is the experimental probability of rolling an odd number?

The bar graph shows 10 ones, 8 threes, and 11 fives. So, an odd number was rolled $10 + 8 + 11 = 29$ times in a total of 50 rolls.

$$P(\text{event}) = \frac{\text{number of times the event occurs}}{\text{total number of trials}}$$

$$P(\text{odd}) = \frac{29}{50}$$

An odd number was rolled 29 times.

There was a total of 50 rolls.

∴ The experimental probability is $\frac{29}{50}$, 0.58, or 58%.

EXAMPLE 2 Making a Prediction

It rains 2 out of the last 12 days in March. If this trend continues, how many rainy days would you expect in April?

Find the experimental probability of a rainy day.

$$P(\text{event}) = \frac{\text{number of times the event occurs}}{\text{total number of trials}}$$

$$P(\text{rain}) = \frac{2}{12} = \frac{1}{6}$$

It rains 2 days.

There is a total of 12 days.

"April showers bring May flowers." Old Proverb, 1557

To make a prediction, multiply the probability of a rainy day by the number of days in April.

$$\frac{1}{6} \cdot 30 = 5$$

∴ So, you can predict that there will be 5 rainy days in April.

🔊 Multi-Language Glossary at BigIdeasMath ✓com

On Your Own

1. In Example 1, what is the experimental probability of rolling an even number?

2. At a clothing company, an inspector finds 5 defective pairs of jeans in a shipment of 200. If this trend continues, about how many pairs of jeans would you expect to be defective in a shipment of 5000?

🔑 Key Idea

Theoretical Probability

When all possible outcomes are equally likely, the **theoretical probability** of an event is the ratio of the number of favorable outcomes to the number of possible outcomes.

$$P(\text{event}) = \frac{\text{number of favorable outcomes}}{\text{number of possible outcomes}}$$

EXAMPLE 3 **Finding a Theoretical Probability**

You randomly choose one of the letters shown. What is the theoretical probability of choosing a vowel?

$$P(\text{event}) = \frac{\text{number of favorable outcomes}}{\text{number of possible outcomes}}$$

$$P(\text{vowel}) = \frac{3}{7} \longleftarrow \boxed{\text{There are 3 vowels.}}$$
$$\phantom{P(\text{vowel}) = \frac{3}{7}} \longleftarrow \boxed{\text{There is a total of 7 letters.}}$$

∴ The probability of choosing a vowel is $\frac{3}{7}$, or about 43%.

EXAMPLE 4 **Using a Theoretical Probability**

The theoretical probability of winning a bobblehead when spinning a prize wheel is $\frac{1}{6}$. The wheel has 3 bobblehead sections. How many sections are on the wheel?

$$P(\text{bobblehead}) = \frac{\text{number of bobblehead sections}}{\text{total number of sections}}$$

$$\frac{1}{6} = \frac{3}{s} \qquad \text{Substitute. Let } s \text{ be the total number of sections.}$$

$$s = 18 \qquad \text{Cross Products Property}$$

∴ So, there are 18 sections on the wheel.

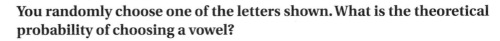

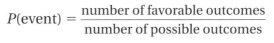

On Your Own

Now You're Ready
Exercises 15–23

3. In Example 3, what is the theoretical probability of choosing an X?

4. The theoretical probability of spinning an odd number on a spinner is 0.6. The spinner has 10 sections. How many sections have odd numbers?

5. The prize wheel in Example 4 was spun 540 times at a baseball game. About how many bobbleheads would you expect were won?

EXAMPLE **5** **Comparing Experimental and Theoretical Probability**

The bar graph shows the results of rolling a number cube 300 times.

a. What is the experimental probability of rolling an odd number?

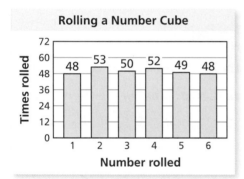

The bar graph shows 48 ones, 50 threes, and 49 fives. So, an odd number was rolled $48 + 50 + 49 = 147$ times in a total of 300 rolls.

$$P(\text{event}) = \frac{\text{number of times the event occurs}}{\text{total number of trials}}$$

$$P(\text{odd}) = \frac{147}{300}$$

An odd number was rolled 147 times.

There was a total of 300 rolls.

$$= \frac{49}{100}, \text{ or } 49\%$$

b. How does the experimental probability compare with the theoretical probability of rolling an odd number?

In Section 10.2, Example 2, you found that the theoretical probability of rolling an odd number is 50%. The experimental probability, 49%, is close to the theoretical probability.

c. Compare the experimental probability in part (a) to the experimental probability in Example 1.

As the number of trials increased from 50 to 300, the experimental probability decreased from 58% to 49%. So, it became closer to the theoretical probability of 50%.

On Your Own

Now You're Ready
Exercises 25–27

6. Use the bar graph in Example 5 to find the experimental probability of rolling a number greater than 1. Compare the experimental probability to the theoretical probability of rolling a number greater than 1.

Vocabulary and Concept Check

1. **VOCABULARY** Describe how to find the experimental probability of an event.

2. **REASONING** You flip a coin 10 times and find the experimental probability of flipping tails to be 0.7. Does this seem reasonable? Explain.

3. **VOCABULARY** An event has a theoretical probability of 0.5. What does this mean?

4. **OPEN-ENDED** Describe an event that has a theoretical probability of $\frac{1}{4}$.

5. **LOGIC** A pollster surveys randomly selected individuals about an upcoming election. Do you think the pollster will use experimental probability or theoretical probability to make predictions? Explain.

Practice and Problem Solving

Use the bar graph to find the relative frequency of the event.

6. Spinning a 6

7. Spinning an even number

Use the bar graph to find the experimental probability of the event.

8. Spinning a number less than 3

9. *Not* spinning a 1

10. Spinning a 1 or a 3

11. Spinning a 7

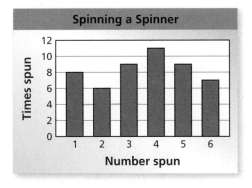

12. **EGGS** You check 20 cartons of eggs. Three of the cartons have at least one cracked egg. What is the experimental probability that a carton of eggs has at least one cracked egg?

13. **BOARD GAME** There are 105 lettered tiles in a board game. You choose the tiles shown. How many of the 105 tiles would you expect to be vowels?

14. **CARDS** You have a package of 20 assorted thank-you cards. You pick the four cards shown. How many of the 20 cards would you expect to have flowers on them?

Use the spinner to find the theoretical probability of the event.

③ 15. Spinning red

16. Spinning a 1

17. Spinning an odd number

18. Spinning a multiple of 2

19. Spinning a number less than 7

20. Spinning a 9

21. LETTERS Each letter of the alphabet is printed on an index card. What is the theoretical probability of randomly choosing any letter except Z?

④ 22. GAME SHOW On a game show, a contestant randomly chooses a chip from a bag that contains numbers and strikes. The theoretical probability of choosing a strike is $\frac{3}{10}$. The bag contains 9 strikes. How many chips are in the bag?

23. MUSIC The theoretical probability that a pop song plays on your MP3 player is 0.45. There are 80 songs on your MP3 player. How many of the songs are pop songs?

24. MODELING There are 16 females and 20 males in a class.

a. What is the theoretical probability that a randomly chosen student is female?

b. One week later, there are 45 students in the class. The theoretical probability that a randomly chosen student is a female is the same as last week. How many males joined the class?

The bar graph shows the results of spinning the spinner 200 times. Compare the theoretical and experimental probabilities of the event.

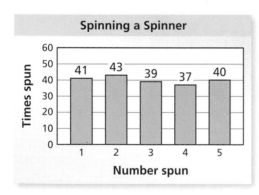

⑤ 25. Spinning a 4

26. Spinning a 3

27. Spinning a number greater than 4

28. Should you use *theoretical* or *experimental* probability to predict the number of times you will spin a 3 in 10,000 spins?

29. NUMBER SENSE The table at the right shows the results of flipping two coins 12 times each.

HH	HT	TH	TT
2	6	3	1

a. What is the experimental probability of flipping two tails? Using this probability, how many times can you expect to flip two tails in 600 trials?

HH	HT	TH	TT
23	29	26	22

b. The table at the left shows the results of flipping the same two coins 100 times each. What is the experimental probability of flipping two tails? Using this probability, how many times can you expect to flip two tails in 600 trials?

c. Why is it important to use a large number of trials when using experimental probability to predict results?

You roll a pair of number cubes 60 times. You record your results in the bar graph shown.

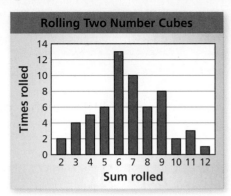

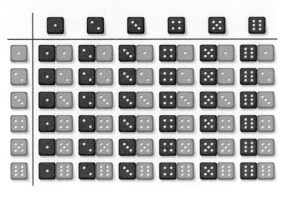

30. Use the bar graph to find the experimental probability of rolling each sum. Is each sum equally likely? Explain. If not, which is most likely?

31. Use the table to find the theoretical probability of rolling each sum. Is each sum equally likely? Explain. If not, which is most likely?

32. PROBABILITIES Compare the probabilities you found in Exercises 30 and 31.

33. REASONING Consider the results of Exercises 30 and 31.

    a. Which sum would you expect to be most likely after 500 trials? 1000 trials? 10,000 trials?

    b. Explain how experimental probability is related to theoretical probability as the number of trials increases.

34. **Project** When you toss a paper cup into the air, there are three ways for the cup to land: *open-end up*, *open-end down*, or *on its side*.

    a. Toss a paper cup 100 times and record your results. Do the outcomes for tossing the cup appear to be equally likely? Explain.

    b. What is the probability of the cup landing open-end up? open-end down? on its side?

    c. Use your results to predict the number of times the cup lands on its side in 1000 tosses.

    d. Suppose you tape a quarter to the bottom of the cup. Do you think the cup will be *more likely* or *less likely* to land open-end up? Justify your answer.

 Fair Game Review *What you learned in previous grades & lessons*

Find the annual interest rate. *(Section 6.7)*

35. $I = \$16$, $P = \$200$, $t = 2$ years

36. $I = \$26.25$, $P = \$500$, $t = 18$ months

37. MULTIPLE CHOICE The volume of a prism is 9 cubic yards. What is its volume in cubic feet? *(Section 9.4)*

    A 3 ft^3 **B** 27 ft^3 **C** 81 ft^3 **D** 243 ft^3

10.4 Compound Events

Essential Question How can you find the number of possible outcomes of one or more events?

1 ACTIVITY: Comparing Combination Locks

Work with a partner. You are buying a combination lock. You have three choices.

a. This lock has 3 wheels. Each wheel is numbered from 0 to 9.

The least three-digit combination possible is [   ] .

The greatest three-digit combination possible is [   ] .

How many possible combinations are there?

b. Use the lock in part (a).

There are [   ] possible outcomes for the first wheel.

There are [   ] possible outcomes for the second wheel.

There are [   ] possible outcomes for the third wheel.

How can you use multiplication to determine the number of possible combinations?

c. This lock is numbered from 0 to 39. Each combination uses three numbers in a right, left, right pattern. How many possible combinations are there?

COMMON CORE

Probability and Statistics

In this lesson, you will

• use tree diagrams, tables, or a formula to find the number of possible outcomes.
• find probabilities of compound events.

Learning Standards
7.SP.8a
7.SP.8b

d. This lock has 4 wheels.

Wheel 1: 0–9
Wheel 2: A–J
Wheel 3: K–T
Wheel 4: 0–9

How many possible combinations are there?

e. For which lock is it most difficult to guess the combination? Why?

2 ACTIVITY: Comparing Password Security

Work with a partner. Which password requirement is most secure? Explain your reasoning. Include the number of different passwords that are possible for each requirement.

a. The password must have four digits.

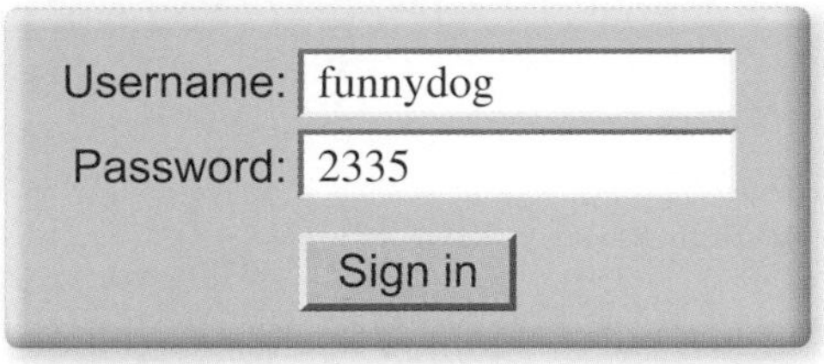

Username: funnydog
Password: 2335
Sign in

b. The password must have five digits.

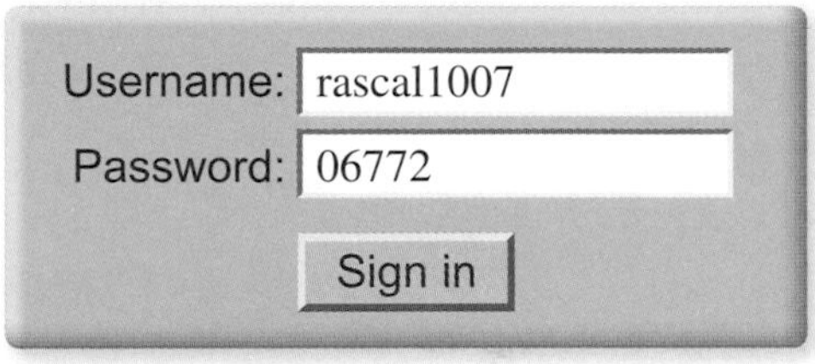

Username: rascal1007
Password: 06772
Sign in

c. The password must have six letters.

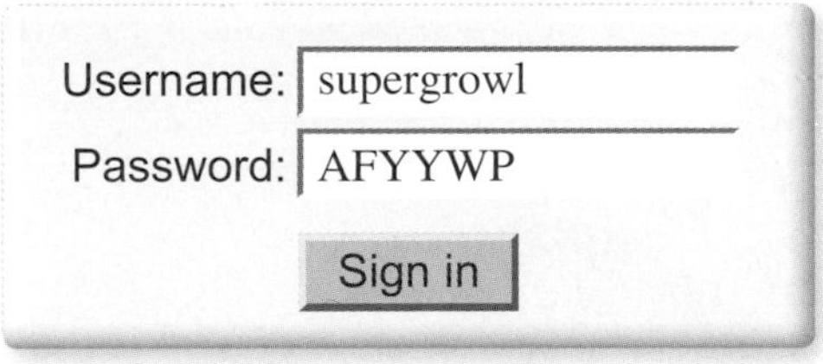

Username: supergrowl
Password: AFYYWP
Sign in

d. The password must have eight digits or letters.

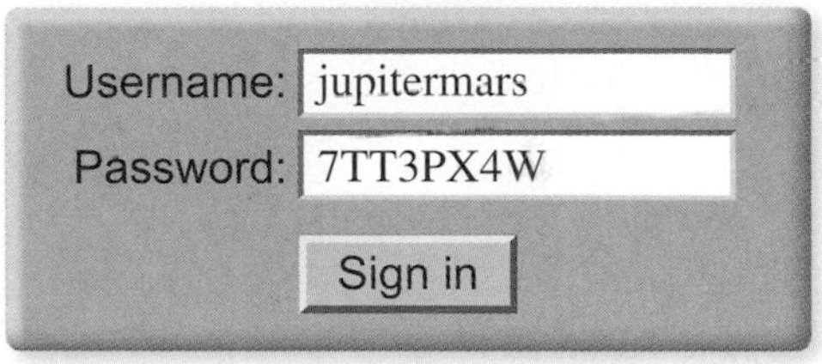

Username: jupitermars
Password: 7TT3PX4W
Sign in

Math Practice 7

View as Components

What is the number of possible outcomes for each character of the password? Explain.

What Is Your Answer?

3. **IN YOUR OWN WORDS** How can you find the number of possible outcomes of one or more events?

4. **SECURITY** A hacker uses a software program to guess the passwords in Activity 2. The program checks 600 passwords per minute. What is the greatest amount of time it will take the program to guess each of the four types of passwords?

Practice Use what you learned about the total number of possible outcomes of one or more events to complete Exercise 5 on page 425.

Key Vocabulary 🔊
sample space, *p. 422*
Fundamental
   Counting Principle,
   p. 422
compound event,
   p. 424

The set of all possible outcomes of one or more events is called the **sample space**.

You can use tables and tree diagrams to find the sample space of two or more events.

EXAMPLE **1** **Finding a Sample Space**

You randomly choose a crust and style of pizza. Find the sample space. How many different pizzas are possible?

Crust

• Thin Crust
• Stuffed Crust

Style

• Hawaiian
• Mexican
• Pepperoni
• Veggie

Use a tree diagram to find the sample space.

Crust	*Style*	*Outcome*
Thin	Hawaiian	Thin Crust Hawaiian
	Mexican	Thin Crust Mexican
	Pepperoni	Thin Crust Pepperoni
	Veggie	Thin Crust Veggie
Stuffed	Hawaiian	Stuffed Crust Hawaiian
	Mexican	Stuffed Crust Mexican
	Pepperoni	Stuffed Crust Pepperoni
	Veggie	Stuffed Crust Veggie

⋮ There are 8 different outcomes in the sample space.
   So, there are 8 different pizzas possible.

On Your Own

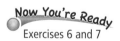
Now You're Ready
Exercises 6 and 7

1. **WHAT IF?** The pizza shop adds a deep dish crust. Find the sample space. How many pizzas are possible?

Another way to find the total number of possible outcomes is to use the **Fundamental Counting Principle**.

Key Idea

Study Tip

The Fundamental Counting Principle can be extended to more than two events.

Fundamental Counting Principle

An event M has m possible outcomes. An event N has n possible outcomes. The total number of outcomes of event M followed by event N is $m \times n$.

EXAMPLE 2 **Finding the Total Number of Possible Outcomes**

Find the total number of possible outcomes of rolling a number cube and flipping a coin.

Method 1: Use a table to find the sample space. Let H = heads and T = tails.

	1	2	3	4	5	6
🪙	1H	2H	3H	4H	5H	6H
🪙	1T	2T	3T	4T	5T	6T

∴ There are 12 possible outcomes.

Method 2: Use the Fundamental Counting Principle. Identify the number of possible outcomes of each event.

    Event 1: Rolling a number cube has 6 possible outcomes.

    Event 2: Flipping a coin has 2 possible outcomes.

        $6 \times 2 = 12$ Fundamental Counting Principle

∴ There are 12 possible outcomes.

EXAMPLE 3 **Finding the Total Number of Possible Outcomes**

How many different outfits can you make from the T-shirts, jeans, and shoes in the closet?

Use the Fundamental Counting Principle. Identify the number of possible outcomes for each event.

    Event 1: Choosing a T-shirt has 7 possible outcomes.

    Event 2: Choosing jeans has 4 possible outcomes.

    Event 3: Choosing shoes has 3 possible outcomes.

        $7 \times 4 \times 3 = 84$ Fundamental Counting Principle

∴ So, you can make 84 different outfits.

On Your Own

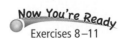

Now You're Ready
Exercises 8–11

2. Find the total number of possible outcomes of spinning the spinner and choosing a number from 1 to 5.

3. How many different outfits can you make from 4 T-shirts, 5 pairs of jeans, and 5 pairs of shoes?

A **compound event** consists of two or more events. As with a single event, the probability of a compound event is the ratio of the number of favorable outcomes to the number of possible outcomes.

EXAMPLE 4 Finding the Probability of a Compound Event

In Example 2, what is the probability of rolling a number greater than 4 and flipping tails?

There are two favorable outcomes in the sample space for rolling a number greater than 4 and flipping tails: 5T and 6T.

$$P(\text{event}) = \frac{\text{number of favorable outcomes}}{\text{number of possible outcomes}}$$

$$P(\text{greater than 4 and tails}) = \frac{2}{12} \qquad \text{Substitute.}$$

$$= \frac{1}{6} \qquad \text{Simplify.}$$

∴ The probability is $\frac{1}{6}$, or $16\frac{2}{3}\%$.

EXAMPLE 5 Finding the Probability of a Compound Event

You flip three nickels. What is the probability of flipping two heads and one tails?

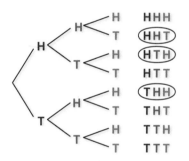

Use a tree diagram to find the sample space. Let H = heads and T = tails.

There are three favorable outcomes in the sample space for flipping two heads and one tails: HHT, HTH, and THH.

$$P(\text{event}) = \frac{\text{number of favorable outcomes}}{\text{number of possible outcomes}}$$

$$P(\text{2 heads and 1 tails}) = \frac{3}{8} \qquad \text{Substitute.}$$

∴ The probability is $\frac{3}{8}$, or 37.5%.

On Your Own

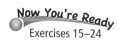
Now You're Ready
Exercises 15–24

4. In Example 2, what is the probability of rolling at most 4 and flipping heads?

5. In Example 5, what is the probability of flipping at least two tails?

6. You roll two number cubes. What is the probability of rolling double threes?

7. In Example 1, what is the probability of choosing a stuffed crust Hawaiian pizza?

10.4 Exercises

Vocabulary and Concept Check

1. **VOCABULARY** What is the sample space of an event? How can you find the sample space of two or more events?

2. **WRITING** Explain how to use the Fundamental Counting Principle.

3. **WRITING** Describe two ways to find the total number of possible outcomes of spinning the spinner and rolling the number cube.

4. **OPEN-ENDED** Give a real-life example of a compound event.

Practice and Problem Solving

5. **COMBINATIONS** The lock is numbered from 0 to 49. Each combination uses three numbers in a right, left, right pattern. Find the total number of possible combinations for the lock.

Use a tree diagram to find the sample space and the total number of possible outcomes.

6.

Birthday Party	
Event	Miniature golf, Laser tag, Roller skating
Time	1:00 P.M.–3:00 P.M., 6:00 P.M.–8:00 P.M.

7.

New School Mascot	
Type	Lion, Bear, Hawk, Dragon
Style	Realistic, Cartoon

Use the Fundamental Counting Principle to find the total number of possible outcomes.

8.

Beverage	
Size	Small, Medium, Large
Flavor	Root beer, Cola, Diet cola, Iced tea, Lemonade, Water, Coffee

9.

MP3 Player	
Memory	2 GB, 4 GB, 8 GB, 16 GB
Color	Silver, Green, Blue, Pink, Black

10.

Clown	
Suit	Dots, Stripes, Checkers board
Wig	One color, Multicolor
Talent	Balloon animals, Juggling, Unicycle, Magic

11.

Meal	
Appetizer	Nachos, Soup, Spinach dip, Salad, Fruit
Entrée	Chicken, Beef, Spaghetti, Fish
Dessert	Cake, Cookies, Ice cream

12. **NOTE CARDS** A store sells three types of note cards. There are three sizes of each type. Show two ways to find the total number of note cards the store sells.

13. **ERROR ANALYSIS** A true-false quiz has five questions. Describe and correct the error in using the Fundamental Counting Principle to find the total number of ways that you can answer the quiz.

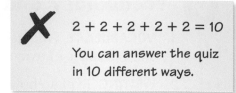

✗ 2 + 2 + 2 + 2 + 2 = 10

You can answer the quiz in 10 different ways.

14. **CHOOSE TOOLS** You randomly choose one of the marbles. Without replacing the first marble, you choose a second marble.

   a. Name two ways you can find the total number of possible outcomes.

   b. Find the total number of possible outcomes.

You spin the spinner and flip a coin. Find the probability of the compound event.

④ **15.** Spinning a 1 and flipping heads

  16. Spinning an even number and flipping heads

  17. Spinning a number less than 3 and flipping tails

  18. Spinning a 6 and flipping tails

  19. *Not* spinning a 5 and flipping heads

  20. Spinning a prime number and *not* flipping heads

You spin the spinner, flip a coin, then spin the spinner again. Find the probability of the compound event.

⑤ **21.** Spinning blue, flipping heads, then spinning a 1

  22. Spinning an odd number, flipping heads, then spinning yellow

  23. Spinning an even number, flipping tails, then spinning an odd number

  24. *Not* spinning red, flipping tails, then *not* spinning an even number

25. **TAKING A TEST** You randomly guess the answers to two questions on a multiple-choice test. Each question has three choices: A, B, and C.

  a. What is the probability that you guess the correct answers to both questions?

  b. Suppose you can eliminate one of the choices for each question. How does this change the probability that your guesses are correct?

26. **PASSWORD** You forget the last two digits of your password for a website.

    a. What is the probability that you randomly choose the correct digits?

    b. Suppose you remember that both digits are even. How does this change the probability that your choices are correct?

27. **COMBINATION LOCK** The combination lock has 3 wheels, each numbered from 0 to 9.

    a. What is the probability that someone randomly guesses the correct combination in one attempt?

    b. Explain how to find the probability that someone randomly guesses the correct combination in five attempts.

28. **TRAINS** Your model train has one engine and eight train cars. Find the total number of ways you can arrange the train. (The engine must be first.)

29. **REPEATED REASONING** You have been assigned a 9-digit identification number.

    a. Why should you use the Fundamental Counting Principle instead of a tree diagram to find the total number of possible identification numbers?

    b. How many identification numbers are possible?

    c. **RESEARCH** Use the Internet to find out why the possible number of Social Security numbers is not the same as your answer to part (b).

30. **Problem Solving** From a group of 5 candidates, a committee of 3 people is selected. In how many different ways can the committee be selected?

 Fair Game Review *What you learned in previous grades & lessons*

Name two pairs of adjacent angles and two pairs of vertical angles in the figure. *(Section 7.1)*

31.
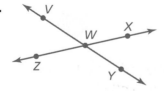

32.
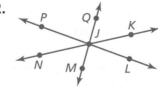

33. **MULTIPLE CHOICE** A drawing has a scale of 1 cm : 1 m. What is the scale factor of the drawing? *(Section 7.5)*

    A 1 : 1   **B** 1 : 100   **C** 10 : 1   **D** 100 : 1

Essential Question What is the difference between dependent and independent events?

1   ACTIVITY: Drawing Marbles from a Bag (With Replacement)

Work with a partner. You have three marbles in a bag. There are two green marbles and one purple marble. Randomly draw a marble from the bag. Then put the marble back in the bag and draw a second marble.

a. Complete the tree diagram. Let G = green and P = purple. Find the probability that both marbles are green.

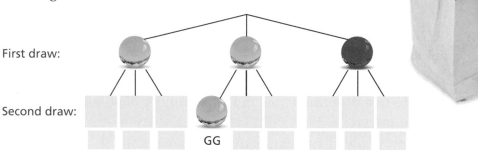

First draw:

Second draw:

GG

b. Does the probability of getting a green marble on the second draw *depend* on the color of the first marble? Explain.

2   ACTIVITY: Drawing Marbles from a Bag (Without Replacement)

Work with a partner. Using the same marbles from Activity 1, randomly draw two marbles from the bag.

a. Complete the tree diagram. Let G = green and P = purple. Find the probability that both marbles are green.

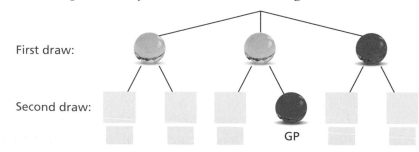

First draw:

Second draw:

GP

Is this event more likely than the event in Activity 1? Explain.

b. Does the probability of getting a green marble on the second draw *depend* on the color of the first marble? Explain.

COMMON CORE

Probability and Statistics

In this lesson, you will

- identify independent and dependent events.
- use formulas to find probabilities of independent and dependent events.

Applying Standards
7.SP.8a
7.SP.8b

Work with a partner. Conduct two experiments.

a. In the first experiment, randomly draw one marble from the bag. Put it back. Draw a second marble. Repeat this 36 times. Record each result. Make a bar graph of your results.

b. In the second experiment, randomly draw two marbles from the bag 36 times. Record each result. Make a bar graph of your results.

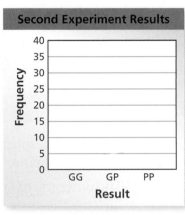

Math Practice 3

Use Definitions

In what other mathematical context have you seen the terms *independent* and *dependent*? How does knowing these definitions help you answer the questions in part (d)?

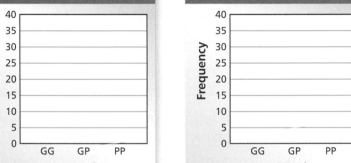

First Experiment Results

Frequency: 40, 35, 30, 25, 20, 15, 10, 5, 0

Result: GG, GP, PP

Second Experiment Results

Frequency: 40, 35, 30, 25, 20, 15, 10, 5, 0

Result: GG, GP, PP

c. For each experiment, estimate the probability of drawing two green marbles.

d. Which experiment do you think represents *dependent events*? Which represents *independent events*? Explain your reasoning.

What Is Your Answer?

4. IN YOUR OWN WORDS What is the difference between dependent and independent events? Describe a real-life example of each.

In Questions 5–7, tell whether the events are *independent* or *dependent*. Explain your reasoning.

5. You roll a 5 on a number cube and spin blue on a spinner.

6. Your teacher chooses one student to lead a group, and then chooses another student to lead another group.

7. You spin red on one spinner and green on another spinner.

8. In Activities 1 and 2, what is the probability of drawing a green marble on the first draw? on the second draw? How do you think you can use these two probabilities to find the probability of drawing two green marbles?

Practice Use what you learned about independent and dependent events to complete Exercises 3 and 4 on page 433.

10.5 Lesson

Key Vocabulary 🔊
independent events, p. 430
dependent events, p. 431

Compound events may be *independent events* or *dependent events*. Events are **independent events** if the occurrence of one event *does not* affect the likelihood that the other event(s) will occur.

🔑 Key Idea

Probability of Independent Events

Words The probability of two or more independent events is the product of the probabilities of the events.

Symbols $P(A \text{ and } B) = P(A) \cdot P(B)$

$P(A \text{ and } B \text{ and } C) = P(A) \cdot P(B) \cdot P(C)$

EXAMPLE 1 Finding the Probability of Independent Events

You spin the spinner and flip the coin. What is the probability of spinning a prime number and flipping tails?

The outcome of spinning the spinner does not affect the outcome of flipping the coin. So, the events are independent.

$P(\text{prime}) = \dfrac{3}{5}$ ← There are 3 prime numbers (2, 3, and 5).
← There is a total of 5 numbers.

$P(\text{tails}) = \dfrac{1}{2}$ ← There is 1 tails side.
← There is a total of 2 sides.

Use the formula for the probability of independent events.

$P(A \text{ and } B) = P(A) \cdot P(B)$

$P(\text{prime and tails}) = P(\text{prime}) \cdot P(\text{tails})$

$= \dfrac{3}{5} \cdot \dfrac{1}{2}$ Substitute.

$= \dfrac{3}{10}$ Multiply.

∴ The probability of spinning a prime number and flipping tails is $\dfrac{3}{10}$, or 30%.

⬤ On Your Own

Now You're Ready
Exercises 5–8

1. What is the probability of spinning a multiple of 2 and flipping heads?

🔊 Multi-Language Glossary at BigIdeasMath✓com

Events are **dependent events** if the occurrence of one event *does* affect the likelihood that the other event(s) will occur.

 **Key Idea**

Probability of Dependent Events

Words The probability of two dependent events *A* and *B* is the probability of *A* times the probability of *B* after *A* occurs.

Symbols $P(A \text{ and } B) = P(A) \cdot P(B \text{ after } A)$

EXAMPLE ② **Finding the Probability of Dependent Events**

People are randomly chosen to be game show contestants from an audience of 100 people. You are with 5 of your relatives and 6 other friends. What is the probability that one of your relatives is chosen first, and then one of your friends is chosen second?

Choosing an audience member changes the number of audience members left. So, the events are dependent.

> There are 5 relatives.

$$P(\text{relative}) = \frac{5}{100} = \frac{1}{20}$$

> There is a total of 100 audience members.

> There are 6 friends.

$$P(\text{friend}) = \frac{6}{99} = \frac{2}{33}$$

> There is a total of 99 audience members left.

Use the formula for the probability of dependent events.

$$P(A \text{ and } B) = P(A) \cdot P(B \text{ after } A)$$

$$P(\text{relative and friend}) = P(\text{relative}) \cdot P(\text{friend after relative})$$

$$= \frac{1}{20} \cdot \frac{2}{33} \qquad \text{Substitute.}$$

$$= \frac{1}{330} \qquad \text{Simplify.}$$

∴ The probability is $\frac{1}{330}$, or about 0.3%.

● **On Your Own**

Now You're Ready
Exercises 9–12

2. What is the probability that you, your relatives, and your friends are *not* chosen to be either of the first two contestants?

EXAMPLE ③ **Finding the Probability of a Compound Event**

A student randomly guesses the answer for each of the multiple-choice questions. What is the probability of answering all three questions correctly?

> **1.** In what year did the United States gain independence from Britain?
>   **A.** 1492   **B.** 1776   **C.** 1788   **D.** 1795   **E.** 2000
>
> **2.** Which amendment to the Constitution grants citizenship to all persons born in the United States and guarantees them equal protection under the law?
>   **A.** 1st   **B.** 5th   **C.** 12th   **D.** 13th   **E.** 14th
>
> **3.** In what year did the Boston Tea Party occur?
>   **A.** 1607   **B.** 1773   **C.** 1776   **D.** 1780   **E.** 1812

Choosing the answer for one question does not affect the choice for the other questions. So, the events are independent.

Method 1: Use the formula for the probability of independent events.

P(#1 and #2 and #3 correct) = P(#1 correct) • P(#2 correct) • P(#3 correct)

$$= \frac{1}{5} \cdot \frac{1}{5} \cdot \frac{1}{5} \qquad \text{Substitute.}$$

$$= \frac{1}{125} \qquad \text{Multiply.}$$

∴ The probability of answering all three questions correctly is $\frac{1}{125}$, or 0.8%.

Method 2: Use the Fundamental Counting Principle.

There are 5 choices for each question, so there are $5 \cdot 5 \cdot 5 = 125$ possible outcomes. There is only 1 way to answer all three questions correctly.

$$P(\#1 \text{ and } \#2 \text{ and } \#3 \text{ correct}) = \frac{1}{125}$$

∴ The probability of answering all three questions correctly is $\frac{1}{125}$, or 0.8%.

● **On Your Own**

Now You're Ready
Exercises 18–22

3. The student can eliminate Choice A for all three questions. What is the probability of answering all three questions correctly? Compare this probability with the probability in Example 3. What do you notice?

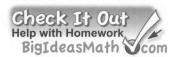

Check It Out
Help with Homework
BigIdeasMath √com

 Vocabulary and Concept Check

1. **DIFFERENT WORDS, SAME QUESTION** You randomly choose one of the chips. Without replacing the first chip, you choose a second chip. Which question is different? Find "both" answers.

> What is the probability of choosing a 1 and then a blue chip?

> What is the probability of choosing a 1 and then an even number?

> What is the probability of choosing a green chip and then a chip that is *not* red?

> What is the probability of choosing a number less than 2 and then an even number?

2. **WRITING** How do you find the probability of two events *A* and *B* when *A* and *B* are independent? dependent?

 Practice and Problem Solving

Tell whether the events are *independent* or *dependent*. Explain.

3. You roll a 4 on a number cube. Then you roll an even number on a different number cube.

4. You randomly draw a lane number for a 100-meter race. Then your friend randomly draws a lane number for the same race.

You spin the spinner and flip a coin. Find the probability of the compound event.

① 5. Spinning a 3 and flipping heads

6. Spinning an even number and flipping tails

7. Spinning a number greater than 1 and flipping tails

8. *Not* spinning a 2 and flipping heads

You randomly choose one of the tiles. Without replacing the first tile, you choose a second tile. Find the probability of the compound event.

② 9. Choosing a 5 and then a 6

10. Choosing an odd number and then a 20

11. Choosing a number less than 7 and then a multiple of 4

12. Choosing two even numbers

13. **ERROR ANALYSIS** Describe and correct the error in finding the probability.

> You randomly choose one of the marbles. Without replacing the first marble, you choose a second marble. What is the probability of choosing red and then green?
>
> $P(\text{red and green}) = \dfrac{1}{4} \cdot \dfrac{1}{4} = \dfrac{1}{16}$

14. **LOGIC** A bag contains three marbles. Does the tree diagram show the outcomes for *independent* or *dependent* events? Explain.

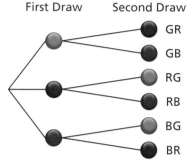

15. **EARRINGS** A jewelry box contains two gold hoop earrings and two silver hoop earrings. You randomly choose two earrings. What is the probability that both are silver hoop earrings?

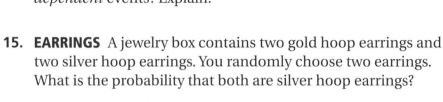

16. **HIKING** You are hiking to a ranger station. There is one correct path. You come to a fork and randomly take the path on the left. You come to another fork and randomly take the path on the right. What is the probability that you are still on the correct path?

17. **CARNIVAL** At a carnival game, you randomly throw two darts at the board and break two balloons. What is the probability that both of the balloons you break are purple?

You spin the spinner, flip a coin, then spin the spinner again. Find the probability of the compound event.

(3) 18. Spinning a 4, flipping heads, then spinning a 7

19. Spinning an odd number, flipping heads, then spinning a 3

20. Spinning an even number, flipping tails, then spinning an odd number

21. *Not* spinning a 5, flipping heads, then spinning a 1

22. Spinning an odd number, *not* flipping heads, then *not* spinning a 6

23. LANGUAGES There are 16 students in your Spanish class. Your teacher randomly chooses one student at a time to take a verbal exam. What is the probability that you are *not* one of the first four students chosen?

24. SHOES Twenty percent of the shoes in a factory are black. One shoe is chosen and replaced. A second shoe is chosen and replaced. Then a third shoe is chosen. What is the probability that *none* of the shoes are black?

25. PROBLEM SOLVING Your teacher divides your class into two groups, and then randomly chooses a leader for each group. The probability that you are chosen to be a leader is $\frac{1}{12}$. The probability that both you and your best friend are chosen is $\frac{1}{132}$.

   a. Is your best friend in your group? Explain.

   b. What is the probability that your best friend is chosen as a group leader?

   c. How many students are in the class?

26. **Structure** After ruling out some of the answer choices, you randomly guess the answer for each of the story questions below.

> **1.** Who was the oldest?
>   **A.** Ned   **B.** Yvonne   **C.** Sun Li   **D.** Angel   **E.** Dusty
>
> **2.** What city was Stacey from?
>   **A.** Raleigh   **B.** New York   **C.** Roanoke   **D.** Dallas   **E.** San Diego

   a. How can the probability of getting both answers correct be 25%?

   b. How can the probability of getting both answers correct be $8\frac{1}{3}\%$?

Fair Game Review *What you learned in previous grades & lessons*

Draw a triangle with the given angle measures. Then classify the triangle. *(Section 7.3)*

27. 30°, 60°, 90°   **28.** 20°, 50°, 110°   **29.** 50°, 50°, 80°

30. MULTIPLE CHOICE Which set of numbers is in order from least to greatest? *(Section 6.2)*

   Ⓐ $\frac{2}{3}$, 0.6, 67%   **Ⓑ** 44.5%, $\frac{4}{9}$, $0.4\overline{6}$

   Ⓒ 0.269, 27%, $\frac{3}{11}$ **Ⓓ** $2\frac{1}{7}$, 214%, $2.\overline{14}$

Key Vocabulary 🔊
simulation, *p. 436*

A **simulation** is an experiment that is designed to reproduce the conditions of a situation or process. Simulations allow you to study situations that are impractical to create in real life.

EXAMPLE 1 — Simulating Outcomes That Are Equally Likely

HTH	HTT
HTT	HTH
HTT	TTT
(HHH)	HTT
HTT	TTT
HTT	HTH
HTH	(HHH)
HTT	HTT
TTT	HTH
HTH	HTT

A couple plans on having three children. The gender of each child is equally likely. (a) Design a simulation involving 20 trials that you can use to model the genders of the children. (b) Use your simulation to find the experimental probability that all three children are boys.

a. Choose an experiment that has two equally likely outcomes for each event (gender), such as tossing three coins. Let heads (H) represent a boy and tails (T) represent a girl.

b. To find the experimental probability, you need repeated trials of the simulation. The table shows 20 trials.

$$P(\text{three boys}) = \frac{2}{20} = \frac{1}{10}$$

HHH occurred 2 times.

There is a total of 20 trials.

∴ The experimental probability is $\frac{1}{10}$, 0.1, or 10%.

EXAMPLE 2 — Simulating Outcomes That Are Not Equally Likely

Study Tip

In Example 2, the digits 1 through 6 represent 60% of the possible digits (0 through 9) in the tens place. Likewise, the digits 1 and 2 represent 20% of the possible digits in the ones place.

There is a 60% chance of rain on Monday and a 20% chance of rain on Tuesday. Design and use a simulation involving 50 randomly generated numbers to find the experimental probability that it will rain on both days.

Use the random number generator on a graphing calculator. Randomly generate 50 numbers from 0 to 99. The table below shows the results.

Let the digits 1 through 6 in the tens place represent rain on Monday. Let digits 1 and 2 in the ones place represent rain on Tuesday. Any number that meets these criteria represents rain on both days.

```
randInt(0,99,50)
 {52 66 73 68 75...
```

(52)	66	73	68	75	28	35	47	48	2
16	68	49	3	77	35	92	78	6	6
58	18	89	39	24	80	(32)	(41)	77	(21)
(32)	40	96	59	86	1	(12)	0	94	73
40	71	28	(61)	1	24	37	25	3	25

$$P(\text{rain both days}) = \frac{7}{50}$$

7 numbers meet the criteria.

There is a total of 50 trials.

∴ The experimental probability is $\frac{7}{50}$, 0.14, or 14%.

EXAMPLE 3 Using a Spreadsheet to Simulate Outcomes

COMMON CORE

Probability and Statistics

In this extension, you will

- use simulations to find experimental probabilities.

Learning Standard
7.SP.8c

Each school year, there is a 50% chance that weather causes one or more days of school to be canceled. Design and use a simulation involving 50 randomly generated numbers to find the experimental probability that weather will cause school to be canceled in at least three of the next four school years.

Use a random number table in a spreadsheet. Randomly generate 50 four-digit whole numbers. The spreadsheet below shows the results.

Let the digits 1 through 5 represent school years with a cancellation. The numbers in the spreadsheet that contain at least three digits from 1 through 5 represent four school years in which at least three of the years have a cancellation.

Study Tip

To create a four-digit random number table in a spreadsheet, follow these steps.

1. Highlight the group of cells to use for your table.
2. Format the cells to display four-digit whole numbers.
3. Enter the formula RAND()*10000 into each cell.

	A	B	C	D	E	F
1	7584	3974	8614	2500	4629	
2	3762	3805	2725	7320	6487	
3	3024	1554	2708	1126	9395	
4	4547	6220	9497	7530	3036	
5	1719	0662	1814	6218	2766	
6	7938	9551	8552	4321	8043	
7	6951	0578	5560	0740	4479	
8	4714	4511	5115	6952	5609	
9	0797	3022	9067	2193	6553	
10	3300	5454	5351	6319	0387	
11						

$$P\left(\begin{matrix}\text{cancellation in at least three}\\\text{of the next four school years}\end{matrix}\right) = \frac{17}{50}$$

> 17 numbers contain at least three digits from 1 to 5.

> There is a total of 50 trials.

The experimental probability is $\frac{17}{50}$, 0.34, or 34%.

Practice

1. **QUIZ** You randomly guess the answers to four true-false questions. (a) Design a simulation that you can use to model the answers. (b) Use your simulation to find the experimental probability that you answer all four questions correctly.

2. **BASEBALL** A baseball team wins 70% of its games. Assuming this trend continues, design and use a simulation to find the experimental probability that the team wins the next three games.

3. **WHAT IF?** In Example 3, there is a 40% chance that weather causes one or more days of school to be canceled each school year. Find the experimental probability that weather will cause school to be canceled in at least three of the next four school years.

4. **REASONING** In Examples 1–3 and Exercises 1–3, try to find the theoretical probability of the event. What do you think happens to the experimental probability when you increase the number of trials in the simulation?

You can use a **notetaking organizer** to write notes, vocabulary, and questions about a topic. Here is an example of a notetaking organizer for probability.

Write important vocabulary or formulas in this space.

If $P(\text{event}) = 0$, the event is *impossible*.

If $P(\text{event}) = 0.25$, the event is *unlikely*.

If $P(\text{event}) = 0.5$, the event is *equally likely to happen or not happen*.

If $P(\text{event}) = 0.75$, the event is *likely*.

If $P(\text{event}) = 1$, the event is *certain*.

Probability

A number that measures the likelihood that an event will occur

Can be written as a fraction, decimal, or percent

Always between 0 and 1, inclusive

Write your notes about the topic in this space.

Write your questions about the topic in this space.

How do you find the probability of two or more events?

On Your Own

Make notetaking organizers to help you study these topics.

1. experimental probability

2. theoretical probability

3. Fundamental Counting Principle

4. independent events

5. dependent events

After you complete this chapter, make notetaking organizers for the following topics.

6. sample

7. population

Formulas:
Newton = beagle

Notes:
Likes dog biscuits

Questions: What is my greatest accomplishment?

I hope it's better than your 200-page essay on bacon.

"I am using a notetaking organizer to plan my autobiography."

You randomly choose one butterfly. Find the number of ways the event can occur. *(Section 10.1)*

1. Choosing red

2. Choosing brown

3. Choosing *not* blue

6 Green
3 White
4 Red
2 Blue
5 Yellow

You randomly choose one paper clip from the jar. Find the probability of the event. *(Section 10.2)*

4. Choosing a green paper clip

5. Choosing a yellow paper clip

6. *Not* choosing a yellow paper clip

7. Choosing a purple paper clip

Use the bar graph to find the experimental probability of the event. *(Section 10.3)*

8. Rolling a 4

9. Rolling a multiple of 3

10. Rolling a 2 or a 3

11. Rolling a number less than 7

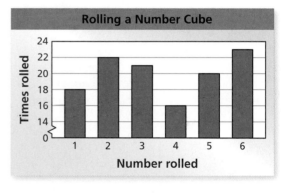

Use the Fundamental Counting Principle to find the total number of possible outcomes. *(Section 10.4)*

12.

Calculator	
Type	Basic display, Scientific, Graphing, Financial
Color	Black, White, Silver

13.

Vacation	
Destination	Florida, Italy, Mexico, England
Length	1 week, 2 weeks

14. **BLACK PENS** You randomly choose one of the pens shown. What is the theoretical probability of choosing a black pen? *(Section 10.3)*

15. **BLUE PENS** You randomly choose one of the five pens shown. Your friend randomly chooses one of the remaining pens. What is the probability that you and your friend both choose a blue pen? *(Section 10.5)*

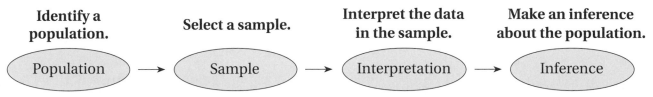

Essential Question How can you determine whether a sample accurately represents a population?

A **population** is an entire group of people or objects. A **sample** is a part of the population. You can use a sample to make an *inference*, or conclusion, about a population.

Identify a population.		Select a sample.		Interpret the data in the sample.		Make an inference about the population.
(Population)	→	(Sample)	→	(Interpretation)	→	(Inference)

1 ACTIVITY: Identifying Populations and Samples

Work with a partner. Identify the population and the sample.

a.

The students
in a school

The students in
a math class

b.

The grizzly bears with
GPS collars in a park

The grizzly bears
in a park

c.

150 quarters

All quarters
in circulation

d.

All books in a library

10 fiction books
in a library

COMMON CORE

Probability and Statistics

In this lesson, you will
- determine when samples are representative of populations.
- use data from random samples to make predictions about populations.

Learning Standards
7.SP.1
7.SP.2

2 ACTIVITY: Identifying Random Samples

Work with a partner. When a sample is selected at random, each member of the population is equally likely to be selected. You want to know the favorite extracurricular activity of students at your school. Determine whether each method will result in a random sample. Explain your reasoning.

a. You ask members of the school band.

b. You publish a survey in the school newspaper.

c. You ask every eighth student who enters the school in the morning.

d. You ask students in your class.

There are many different ways to select a sample from a population. To make valid inferences about a population, you must choose a random sample very carefully so that it accurately represents the population.

3 ACTIVITY: Identifying Representative Samples

Work with a partner. A new power plant is being built outside a town. In each situation below, residents of the town are asked how they feel about the new power plant. Determine whether each conclusion is valid. Explain your reasoning.

a. A local radio show takes calls from 500 residents. The table shows the results. The radio station concludes that most of the residents of the town oppose the new power plant.

New Power Plant	
For	70
Against	425
Don't know	5

Math Practice 2

Understand Quantities

Can the size of a sample affect the validity of a conclusion about a population?

New Power Plant

b. A news reporter randomly surveys 2 residents outside a supermarket. The graph shows the results. The reporter concludes that the residents of the town are evenly divided on the new power plant.

c. You randomly survey 250 residents at a shopping mall. The table shows the results. You conclude that there are about twice as many residents of the town against the new power plant than for the new power plant.

New Power Plant	
For	32%
Against	62%
Don't know	6%

What Is Your Answer?

4. IN YOUR OWN WORDS How can you determine whether a sample accurately represents a population?

5. RESEARCH Choose a topic that you would like to ask people's opinions about, and then write a survey question. How would you choose people to survey so that your sample is random? How many people would you survey? Conduct your survey and display your results. Would you change any part of your survey to make it more accurate? Explain.

6. Does increasing the size of a sample necessarily make the sample representative of a population? Give an example to support your explanation.

Practice

Use what you learned about populations and samples to complete Exercises 3 and 4 on page 444.

Key Vocabulary
population, *p. 440*
sample, *p. 440*
unbiased sample, *p. 442*
biased sample, *p. 442*

An **unbiased sample** is representative of a population. It is selected at random and is large enough to provide accurate data.

A **biased sample** is not representative of a population. One or more parts of the population are favored over others.

EXAMPLE **1** **Identifying an Unbiased Sample**

You want to estimate the number of students in a high school who ride the school bus. Which sample is unbiased?

   A   4 students in the hallway

   B   all students in the marching band

   C   50 seniors at random

   D   100 students at random during lunch

Choice A is not large enough to provide accurate data.

Choice B is not selected at random.

Choice C is not representative of the population because seniors are more likely to drive to school than other students.

Choice D is representative of the population, selected at random, and large enough to provide accurate data.

    So, the correct answer is **D**.

On Your Own

Now You're Ready
Exercises 5–7

1. **WHAT IF?** You want to estimate the number of seniors in a high school who ride the school bus. Which sample is unbiased? Explain.

2. You want to estimate the number of eighth-grade students in your school who consider it relaxing to listen to music. You randomly survey 15 members of the band. Your friend surveys every fifth student whose name appears on an alphabetical list of eighth graders. Which sample is unbiased? Explain.

The results of an unbiased sample are proportional to the results of the population. So, you can use unbiased samples to make predictions about the population.

Biased samples are not representative of the population. So, you should not use them to make predictions about the population because the predictions may not be valid.

EXAMPLE 2 **Determining Whether Conclusions Are Valid**

You want to know how the residents of your town feel about adding a new stop sign. Determine whether each conclusion is valid.

a. You survey the 20 residents who live closest to the new sign. Fifteen support the sign, and five do not. So, you conclude that 75% of the residents of your town support the new sign.

The sample is not representative of the population because residents who live close to the sign are more likely to support it.

⋮ So, the sample is biased, and the conclusion is not valid.

b. You survey 100 residents at random. Forty support the new sign, and sixty do not. So, you conclude that 40% of the residents of your town support the new sign.

The sample is representative of the population, selected at random, and large enough to provide accurate data.

⋮ So, the sample is unbiased, and the conclusion is valid.

EXAMPLE 3 **Making Predictions**

Movies per Week

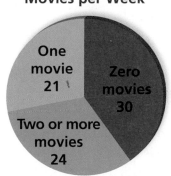

You ask 75 randomly chosen students how many movies they watch each week. There are 1200 students in the school. Predict the number n of students in the school who watch one movie each week.

The sample is representative of the population, selected at random, and large enough to provide accurate data. So, the sample is unbiased, and you can use it to make a prediction about the population.

Write and solve a proportion to find n.

Sample	**Population**

$$\frac{\text{students in survey (one movie)}}{\text{number of students in survey}} = \frac{\text{students in school (one movie)}}{\text{number of students in school}}$$

$$\frac{21}{75} = \frac{n}{1200} \qquad \text{Substitute.}$$

$$336 = n \qquad \text{Solve for } n.$$

⋮ So, about 336 students in the school watch one movie each week.

On Your Own

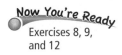
Exercises 8, 9, and 12

3. In Example 2, each of 25 randomly chosen firefighters supports the new sign. So, you conclude that 100% of the residents of your town support the new sign. Is the conclusion valid? Explain.

4. In Example 3, predict the number of students in the school who watch two or more movies each week.

10.6 Exercises

✓ Vocabulary and Concept Check

1. **VOCABULARY** Why would you survey a sample instead of a population?

2. **CRITICAL THINKING** What should you consider when conducting a survey?

Practice and Problem Solving

Identify the population and the sample.

3.

Residents of New Jersey   Residents of Ocean County

4.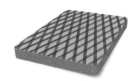

4 cards   All cards in a deck

Determine whether the sample is *biased* or *unbiased*. Explain.

① 5. You want to estimate the number of students in your school who play a musical instrument. You survey the first 15 students who arrive at a band class.

6. You want to estimate the number of books students in your school read over the summer. You survey every fourth student who enters the school.

7. You want to estimate the number of people in a town who think that a park needs to be remodeled. You survey every 10th person who enters the park.

Determine whether the conclusion is valid. Explain.

② 8. You want to determine the number of students in your school who have visited a science museum. You survey 50 students at random. Twenty have visited a science museum, and thirty have not. So, you conclude that 40% of the students in your school have visited a science museum.

9. You want to know how the residents of your town feel about building a new baseball stadium. You randomly survey 100 people who enter the current stadium. Eighty support building a new stadium, and twenty do not. So, you conclude that 80% of the residents of your town support building a new baseball stadium.

Which sample is better for making a prediction? Explain.

10.

Predict the number of students in a school who like gym class.	
Sample A	A random sample of 8 students from the yearbook
Sample B	A random sample of 80 students from the yearbook

11.

Predict the number of defective pencils produced per day.	
Sample A	A random sample of 500 pencils from 20 machines
Sample B	A random sample of 500 pencils from 1 machine

③ **12. FOOD** You ask 125 randomly chosen students to name their favorite food. There are 1500 students in the school. Predict the number of students in the school whose favorite food is pizza.

Favorite Food	
Pizza	58
Hamburger	36
Pasta	14
Other	17

Determine whether you would survey the population or a sample. Explain.

13. You want to know the average height of seventh graders in the United States.

14. You want to know the favorite types of music of students in your homeroom.

15. You want to know the number of students in your state who have summer jobs.

Theater Ticket Sales	
Adults	**Students**
522	210

16. THEATER You survey 72 randomly chosen students about whether they are going to attend the school play. Twelve say yes. Predict the number of students who attend the school.

17. CRITICAL THINKING Explain why 200 people with email addresses may not be a random sample. When might it be a random sample?

18. LOGIC A person surveys residents of a town to determine whether a skateboarding ban should be overturned.

   a. Describe how the person could conduct the survey so that the sample is biased toward overturning the ban.

   b. Describe how the person could conduct the survey so that the sample is biased toward keeping the ban.

19. ⚡Reasoning⚡ A guidance counselor surveys a random sample of 60 out of 900 high school students. Using the survey results, the counselor predicts that approximately 720 students plan to attend college. Do you agree with her prediction? Explain.

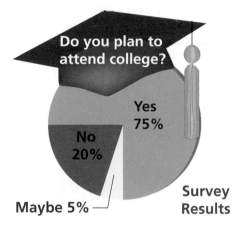

Do you plan to attend college?

Yes 75%

No 20%

Maybe 5%

Survey Results

Fair Game Review What you learned in previous grades & lessons

Write and solve a proportion to answer the question. *(Section 6.3)*

20. What percent of 60 is 18?

21. 70% of what number is 98?

22. 30 is 15% of what number?

23. What number is 0.6% of 500?

24. MULTIPLE CHOICE What is the volume of the pyramid? *(Section 9.5)*

   Ⓐ 40 cm³
   Ⓑ 50 cm³
   Ⓒ 100 cm³
   Ⓓ 120 cm³

5 cm

6 cm

4 cm

You have already used unbiased samples to make inferences about a population. In some cases, making an inference about a population from only one sample is not as precise as using multiple samples.

1 ACTIVITY: Using Multiple Random Samples

Work with a partner. You and a group of friends want to know how many students in your school listen to pop music. There are 840 students in your school. Each person in the group randomly surveys 20 students.

Step 1: The table shows your results. Make an inference about the number of students in your school who prefer pop music.

Favorite Type of Music			
Country	Pop	Rock	Rap
4	10	5	1

Step 2: The table shows Kevin's results. Use these results to make another inference about the number of students in your school who prefer pop music.

Favorite Type of Music			
Country	Pop	Rock	Rap
2	13	4	1

Compare the results of Steps 1 and 2.

COMMON CORE

Probability and Statistics

In this extension, you will
- use multiple samples to make predictions about populations.

Learning Standard
7.SP.2

Step 3: The table shows the results of three other friends. Use these results to make three more inferences about the number of students in your school who prefer pop music.

	Favorite Type of Music			
	Country	Pop	Rock	Rap
Steve	3	8	7	2
Laura	5	10	4	1
Ming	5	9	3	3

Step 4: Describe the variation of the five inferences. Which one would you use to describe the number of students in your school who prefer pop music? Explain your reasoning.

Step 5: Show how you can use all five samples to make an inference.

Practice

1. **PACKING PEANUTS** Work with a partner. Mark 24 packing peanuts with either a red or a black marker. Put the peanuts into a paper bag. Trade bags with other students in the class.

   a. Generate a sample by choosing a peanut from your bag six times, replacing the peanut each time. Record the number of times you choose each color. Repeat this process to generate four more samples. Organize the results in a table.

   b. Use each sample to make an inference about the number of red peanuts in the bag. Then describe the variation of the five inferences. Make inferences about the numbers of red and black peanuts in the bag based on all the samples.

   c. Take the peanuts out of the bag. How do your inferences compare to the population? Do you think you can make a more accurate prediction? If so, explain how.

Hours Worked Each Week
1: 6, 8, 6, 6, 7, 4, 10, 8, 7, 8
2: 10, 4, 4, 6, 8, 6, 7, 12, 8, 8
3: 10, 9, 8, 6, 5, 8, 6, 6, 9, 10
4: 4, 8, 4, 4, 5, 4, 4, 6, 5, 6
5: 6, 8, 8, 6, 12, 4, 10, 8, 6, 12
6: 10, 10, 8, 9, 16, 8, 7, 12, 16, 14
7: 4, 5, 6, 6, 4, 5, 6, 6, 4, 4
8: 16, 20, 8, 12, 10, 8, 8, 14, 16, 8

Work with a partner. You want to know the mean number of hours students with part-time jobs work each week. You go to 8 different schools. At each school, you randomly survey 10 students with part-time jobs. Your results are shown at the left.

Step 1: Find the mean of each sample.

Step 2: Make a box-and-whisker plot of the sample means.

Step 3: Use the box-and-whisker plot to estimate the actual mean number of hours students with part-time jobs work each week.

How does your estimate compare to the mean of the entire data set?

Work with a partner. Another way to generate multiple samples of data is to use a simulation. Suppose 70% of all seventh graders watch reality shows on television.

Step 1: Design a simulation involving 50 packing peanuts by marking 70% of the peanuts with a certain color. Put the peanuts into a paper bag.

Step 2: Simulate choosing a sample of 30 students by choosing peanuts from the bag, replacing the peanut each time. Record the results. Repeat this process to generate eight more samples. How much variation do you expect among the samples? Explain.

Step 3: Display your results.

Practice

2. **SPORTS DRINKS** You want to know whether student-athletes prefer water or sports drinks during games. You go to 10 different schools. At each school, you randomly survey 10 student-athletes. The percents of student-athletes who prefer water are shown.

   60%   70%   60%   50%   80%   70%   30%   70%   80%   40%

   a. Make a box-and-whisker plot of the data.

   b. Use the box-and-whisker plot to estimate the actual percent of student-athletes who prefer water. How does your estimate compare to the mean of the data?

3. **PART-TIME JOBS** Repeat Activity 2 using the medians of the samples.

4. **TELEVISION** In Activity 3, how do the percents in your samples compare to the given percent of seventh graders who watch reality shows on television?

5. **REASONING** Why is it better to make inferences about a population based on multiple samples instead of only one sample? What additional information do you gain by taking multiple random samples? Explain.

10.7 Comparing Populations

Essential Question How can you compare data sets that represent two populations?

1 ACTIVITY: Comparing Two Data Distributions

Work with a partner. You want to compare the shoe sizes of male students in two classes. You collect the data shown in the table.

Male Students in Eighth-Grade Class															
7	9	8	$7\frac{1}{2}$	$8\frac{1}{2}$	10	6	$6\frac{1}{2}$	8	8	$8\frac{1}{2}$	9	11	$7\frac{1}{2}$	$8\frac{1}{2}$	
Male Students in Sixth-Grade Class															
6	$5\frac{1}{2}$	6	$6\frac{1}{2}$	$7\frac{1}{2}$	$8\frac{1}{2}$	7	$5\frac{1}{2}$	5	$5\frac{1}{2}$	$6\frac{1}{2}$	7	$4\frac{1}{2}$	6	6	

a. How can you display both data sets so that you can visually compare the measures of center and variation? Make the data display you chose.

b. Describe the shape of each distribution.

c. Complete the table.

	Mean	Median	Mode	Range	Interquartile Range (IQR)	Mean Absolute Deviation (MAD)
Male Students in Eighth-Grade Class						
Male Students in Sixth-Grade Class						

d. Compare the measures of center for the data sets.

e. Compare the measures of variation for the data sets. Does one data set show more variation than the other? Explain.

f. Do the distributions overlap? How can you tell using the data display you chose in part (a)?

g. The double box-and-whisker plot below shows the shoe sizes of the members of two girls basketball teams. Can you conclude that at least one girl from each team has the same shoe size? Can you conclude that at least one girl from the Bobcats has a larger shoe size than one of the girls from the Tigers? Explain your reasoning.

COMMON CORE

Probability and Statistics

In this lesson, you will
- use measures of center and variation to compare populations.
- use random samples to compare populations.

Learning Standards
7.SP.3
7.SP.4

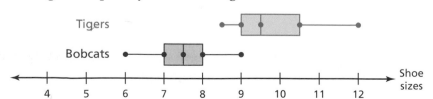

Work with a partner. Compare the shapes of the distributions. Do the two data sets overlap? Explain. If so, use measures of center and the least and the greatest values to describe the overlap between the two data sets.

a.

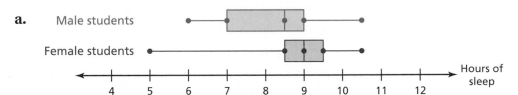

b. **Heights (inches)**

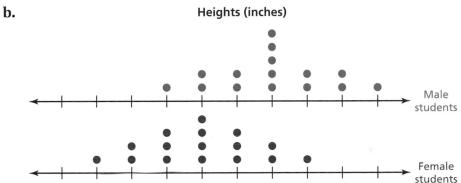

Math
Practice 5

Recognize Usefulness of Tools

How is each type of data display useful? Which do you prefer? Explain.

c. **Ages of People in Two Exercise Classes**

10:00 A.M. Class							8:00 P.M. Class
						1	8 9
						2	1 2 2 7 9 9
						3	0 3 4 5 7
9	7 3 2 2 2					4	0
	7 5 4 3 1					5	
		7 0 0				6	
			0			7	

Key: 1 | 8 = 18

What Is Your Answer?

3. IN YOUR OWN WORDS   How can you compare data sets that represent two populations?

Use what you learned about comparing data sets to complete Exercise 3 on page 452.

Check It Out
Lesson Tutorials
BigIdeasMath.com

Recall that you use the mean and the mean absolute deviation (MAD) to describe symmetric distributions of data. You use the median and the interquartile range (IQR) to describe skewed distributions of data.

To compare two populations, use the mean and the MAD when both distributions are symmetric. Use the median and the IQR when either one or both distributions are skewed.

EXAMPLE 1 **Comparing Populations**

The double dot plot shows the time that each candidate in a debate spent answering each of 15 questions.

> **Study Tip**
>
> You can more easily see the visual overlap of dot plots that are aligned vertically.

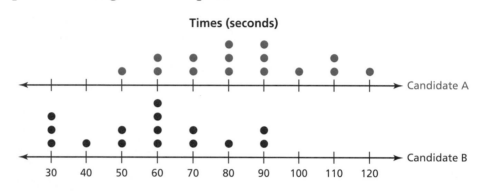

Times (seconds)

Candidate A

Candidate B

30 40 50 60 70 80 90 100 110 120

a. **Compare the populations using measures of center and variation.**

Both distributions are approximately symmetric, so use the mean and the MAD.

Candidate A	Candidate B
Mean $= \dfrac{1260}{15} = 84$	Mean $= \dfrac{870}{15} = 58$
MAD $= \dfrac{244}{15} \approx 16$	MAD $= \dfrac{236}{15} \approx 16$

> **Study Tip**
>
> When two populations have similar variabilities, the value in part (b) describes the visual overlap between the data. In general, the greater the value, the less the overlap.

∴ So, the variation in the times was about the same, but Candidate A had a greater mean time.

b. **Express the difference in the measures of center as a multiple of the measure of variation.**

$$\frac{\text{mean for Candidate A} - \text{mean for Candidate B}}{\text{MAD}} = \frac{26}{16} \approx 1.6$$

∴ So, the difference in the means is about 1.6 times the MAD.

On Your Own

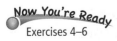
Now You're Ready
Exercises 4–6

1. **WHAT IF?** Each value in the dot plot for Candidate A increases by 30 seconds. How does this affect the answers in Example 1? Explain.

You do not need to have all the data from two populations to make comparisons. You can use random samples to make comparisons.

EXAMPLE 2 **Using Random Samples to Compare Populations**

You want to compare the costs of speeding tickets in two states.

a. **The double box-and-whisker plot shows a random sample of 10 speeding tickets issued in two states. Compare the samples using measures of center and variation. Can you use this to make a valid comparison about speeding tickets in the two states? Explain.**

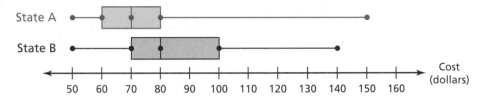

Both distributions are skewed right, so use the median and the IQR.

⋮⋰ The median and the IQR for State A, 70 and 20, are less than the median and the IQR for State B, 80 and 30. However, the sample size is too small and the variability is too great to conclude that speeding tickets generally cost more in State B.

b. **The double box-and-whisker plot shows the medians of 100 random samples of 10 speeding tickets for each state. Compare the variability of the sample medians to the variability of the sample costs in part (a).**

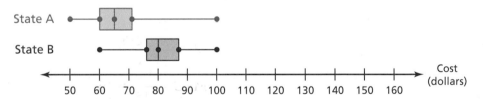

The IQR of the sample medians for each state is about 10.

⋮⋰ So, the sample medians vary much less than the sample costs.

c. **Make a conclusion about the costs of speeding tickets in the two states.**

The sample medians show less variability. Most of the sample medians for State B are greater than the sample medians for State A.

⋮⋰ So, speeding tickets generally cost more in State B than in State A.

⬤ **On Your Own**

Now You're Ready
Exercise 8

2. **WHAT IF?** A random sample of 8 speeding tickets issued in State C has a median of $120. Can you conclude that a speeding ticket in State C costs more than in States A and B? Explain.

 Vocabulary and Concept Check

1. **REASONING** When comparing two populations, when should you use the mean and the MAD? the median and the IQR?

2. **WRITING** Two data sets have similar variabilities. Suppose the measures of center of the data sets differ by 4 times the measure of variation. Describe the visual overlap of the data.

 Practice and Problem Solving

3. **SNAKES** The tables show the lengths of two types of snakes at an animal store.

Garter Snake Lengths (inches)					
26	30	22	15	21	24
28	32	24	25	18	35

Water Snake Lengths (inches)					
34	25	24	35	40	32
41	27	37	32	21	30

a. Find the mean, median, mode, range, interquartile range, and mean absolute deviation for each data set.

b. Compare the data sets.

4. **HOCKEY** The double box-and-whisker plot shows the goals scored per game by two hockey teams during a 20-game season.

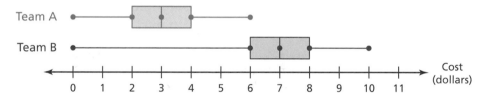

a. Compare the populations using measures of center and variation.

b. Express the difference in the measures of center as a multiple of the measure of variation.

5. **TEST SCORES** The dot plots show the test scores for two classes taught by the same teacher.

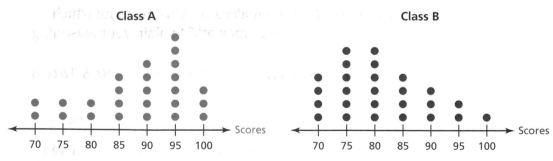

a. Compare the populations using measures of center and variation.

b. Express the difference in the measures of center as a multiple of each measure of variation.

6. ATTENDANCE The tables show the attendances at volleyball games and basketball games at a school during the year.

Volleyball Game Attendance						
112	95	84	106	62	68	53
75	88	93	127	98	117	60
49	54	85	74	88	132	

Basketball Game Attendance						
202	190	173	155	169	188	195
176	141	152	181	198	214	179
163	186	184	207	219	228	

a. Compare the populations using measures of center and variation.

b. Express the difference in the measures of center as a multiple of each measure of variation.

7. NUMBER SENSE Compare the answers to Exercises 4(b), 5(b), and 6(b). Which value is the greatest? What does this mean?

8. MAGAZINES You want to compare the number of words per sentence in a sports magazine to the number of words per sentence in a political magazine.

a. The data represent random samples of 10 sentences in each magazine. Compare the samples using measures of center and variation. Can you use this to make a valid comparison about the magazines? Explain.

Sports magazine: 9, 21, 15, 14, 25, 26, 9, 19, 22, 30

Political magazine: 31, 22, 17, 5, 23, 15, 10, 20, 20, 17

b. The double box-and-whisker plot shows the means of 200 random samples of 20 sentences. Compare the variability of the sample means to the variability of the sample numbers of words in part (a).

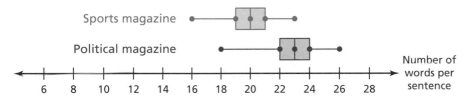

c. Make a conclusion about the numbers of words per sentence in the magazines.

9. **Project** You want to compare the average amounts of time students in sixth, seventh, and eighth grade spend on homework each week.

a. Design an experiment involving random sampling that can help you make a comparison.

b. Perform the experiment. Can you make a conclusion about which students spend the most time on homework? Explain your reasoning.

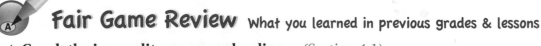

Fair Game Review What you learned in previous grades & lessons

Graph the inequality on a number line. *(Section 4.1)*

10. $x > 5$ **11.** $b \leq -3$ **12.** $n < -1.6$ **13.** $p \geq 2.5$

14. MULTIPLE CHOICE The number of students in the marching band increased from 100 to 125. What is the percent of increase? *(Section 6.5)*

Ⓐ 20% Ⓑ 25% Ⓒ 80% Ⓓ 500%

Check It Out
Progress Check
BigIdeasMath ✓com

1. Which sample is better for making a prediction? Explain.   *(Section 10.6)*

Predict the number of students in your school who play at least one sport.	
Sample A	A random sample of 10 students from the school student roster
Sample B	A random sample of 80 students from the school student roster

2. **GYMNASIUM** You want to estimate the number of students in your school who think the gymnasium should be remodeled. You survey 12 students on the basketball team. Determine whether the sample is *biased* or *unbiased*. Explain.   *(Section 10.6)*

3. **TOWN COUNCIL** You want to know how the residents of your town feel about a recent town council decision. You survey 100 residents at random. Sixty-five support the decision, and thirty-five do not. So, you conclude that 65% of the residents of your town support the decision. Determine whether the conclusion is valid. Explain.   *(Section 10.6)*

4. **FIELD TRIP** Of 60 randomly chosen students surveyed, 16 chose the aquarium as their favorite field trip. There are 720 students in the school. Predict the number of students in the school who would choose the aquarium as their favorite field trip.   *(Section 10.6)*

5. **FOOTBALL** The double box-and-whisker plot shows the points scored per game by two football teams during the regular season.   *(Section 10.7)*

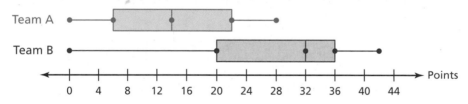

   a. Compare the populations using measures of center and variation.

   b. Express the difference in the measures of center as a multiple of the measure of variation.

6. **SUMMER CAMP** The dot plots show the ages of campers at two summer camps. *(Section 10.7)*

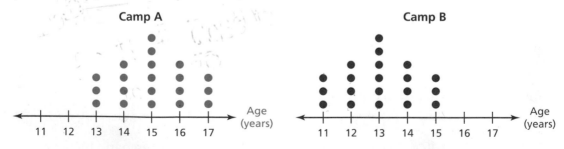

   a. Compare the populations using measures of center and variation.

   b. Express the difference in the measures of center as a multiple of the measure of variation.

Check It Out
Vocabulary Help
BigIdeasMath ✓.com

Review Key Vocabulary

experiment, *p. 402*
outcomes, *p. 402*
event, *p. 402*
favorable outcomes, *p. 402*
probability, *p. 408*
relative frequency, *p. 412*

experimental probability, *p. 414*
theoretical probability, *p. 415*
sample space, *p. 422*
Fundamental Counting Principle, *p. 422*
compound event, *p. 424*

independent events, *p. 430*
dependent events, *p. 431*
simulation, *p. 436*
population, *p. 440*
sample, *p. 440*
unbiased sample, *p. 442*
biased sample, *p. 442*

Review Examples and Exercises

 Outcomes and Events *(pp. 400–405)*

You randomly choose one toy race car.

a. In how many ways can choosing a green car occur?

b. In how many ways can choosing a car that is *not* green occur? What are the favorable outcomes of choosing a car that is *not* green?

a. There are 5 green cars. So, choosing a green car can occur in 5 ways.

b. There are 2 cars that are *not* green. So, choosing a car that is *not* green can occur in 2 ways.

green	*not* green
green, green, green, green, green	blue, red

∴ The favorable outcomes of the event are blue and red.

Exercises

You spin the spinner. (a) Find the number of ways the event can occur. (b) Find the favorable outcomes of the event.

1. Spinning a 1

2. Spinning a 3

3. Spinning an odd number

4. Spinning an even number

5. Spinning a number greater than 0

6. Spinning a number less than 3

10.2 **Probability** *(pp. 406–411)*

You flip a coin. What is the probability of flipping tails?

$$P(\text{event}) = \frac{\text{number of favorable outcomes}}{\text{number of possible outcomes}}$$

$P(\text{tails}) = \dfrac{1}{2}$ ← There is 1 tails.

← There is a total of 2 sides.

∴ The probability of flipping tails is $\dfrac{1}{2}$, or 50%.

Exercises

7. You roll a number cube. Find the probability of rolling an even number.

10.3 **Experimental and Theoretical Probability** *(pp. 412–419)*

a. The bar graph shows the results of spinning the spinner 70 times. What is the experimental probability of spinning a 2?

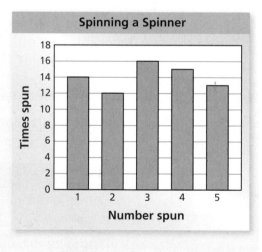

Spinning a Spinner

The bar graph shows 12 twos. So, the spinner landed on two 12 times in a total of 70 spins.

$$P(\text{event}) = \frac{\text{number of times the event occurs}}{\text{total number of trials}}$$

Two was landed on 12 times.

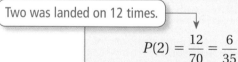

$P(2) = \dfrac{12}{70} = \dfrac{6}{35}$

There was a total of 70 spins.

∴ The experimental probability is $\dfrac{6}{35}$, or about 17%.

b. The theoretical probability of choosing a purple grape from a bag is $\dfrac{2}{9}$. There are 8 purple grapes in the bag. How many grapes are in the bag?

$$P(\text{purple}) = \frac{\text{number of purple grapes}}{\text{total number of grapes}}$$

$\dfrac{2}{9} = \dfrac{8}{g}$ Substitute. Let g be the total number of grapes.

$g = 36$ Solve for g.

∴ So, there are 36 grapes in the bag.

Exercises

Use the bar graph on page 456 to find the experimental probability of the event.

8. Spinning a 3

9. Spinning an odd number

10. *Not* spinning a 5

11. Spinning a number greater than 3

Use the spinner to find the theoretical probability of the event.

12. Spinning blue

13. Spinning a 1

14. Spinning an even number

15. Spinning a 4

16. The theoretical probability of spinning an even number on a spinner is $\frac{2}{3}$. The spinner has 8 even-numbered sections. How many sections are on the spinner?

10.4 **Compound Events** *(pp. 420–427)*

a. **How many different home theater systems can you make from 6 DVD players, 8 TVs, and 3 brands of speakers?**

$$6 \times 8 \times 3 = 144 \qquad \text{Fundamental Counting Principle}$$

∴ So, you can make 144 different home theater systems.

b. **You flip two pennies. What is the probability of flipping two heads?**

Use a tree diagram to find the probability. Let H = heads and T = tails.

There is one favorable outcome in the sample space for flipping two heads: HH.

$$P(\text{event}) = \frac{\text{number of favorable outcomes}}{\text{number of possible outcomes}}$$

$$P(2 \text{ heads}) = \frac{1}{4} \qquad \text{Substitute.}$$

∴ The probability is $\frac{1}{4}$, or 25%.

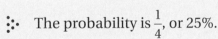

Exercises

17. You have 6 bracelets and 15 necklaces. Find the number of ways you can wear one bracelet and one necklace.

18. You flip two coins and roll a number cube. What is the probability of flipping two tails and rolling an even number?

10.5 **Independent and Dependent Events** *(pp. 428–437)*

You randomly choose one of the tiles and flip the coin. What is the probability of choosing a vowel and flipping heads?

Choosing one of the tiles does not affect the outcome of flipping the coin. So, the events are independent.

$P(\text{vowel}) = \dfrac{2}{7}$ ← There are 2 vowels (A and E).

← There is a total of 7 tiles.

$P(\text{tails}) = \dfrac{1}{2}$ ← There is 1 tails side.

← There is a total of 2 sides.

Use the formula for the probability of independent events.

$P(A \text{ and } B) = P(A) \cdot P(B)$

$= \dfrac{2}{7} \cdot \dfrac{1}{2} = \dfrac{1}{7}$

The probability of choosing a vowel and flipping heads is $\dfrac{1}{7}$, or about 14%.

Exercises

You randomly choose one of the tiles above and flip the coin. Find the probability of the compound event.

19. Choosing a blue tile and flipping tails

20. Choosing the letter G and flipping tails

You randomly choose one of the tiles above. Without replacing the first tile, you randomly choose a second tile. Find the probability of the compound event.

21. Choosing a green tile and then a blue tile

22. Choosing a red tile and then a vowel

10.6 **Samples and Populations** *(pp. 440–447)*

You want to estimate the number of students in your school whose favorite subject is math. You survey every third student who leaves the school. Determine whether the sample is *biased* or *unbiased*.

The sample is representative of the population, selected at random, and large enough to provide accurate data.

So, the sample is unbiased.

Exercises

23. You want to estimate the number of students in your school whose favorite subject is biology. You survey the first 10 students who arrive at biology club. Determine whether the sample is *biased* or *unbiased*. Explain.

10.7 **Comparing Populations** *(pp. 448–453)*

The double box-and-whisker plot shows the test scores for two French classes taught by the same teacher.

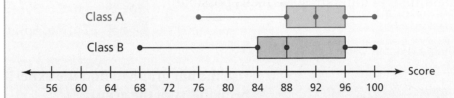

a. **Compare the populations using measures of center and variation.**

Both distributions are skewed left, so use the median and the IQR.

⋮ The median for Class A, 92, is greater than the median for Class B, 88. The IQR for Class B, 12, is greater than the IQR for Class A, 8. The scores in Class A are generally greater and have less variability than the scores in Class B.

b. **Express the difference in the measures of center as a multiple of each measure of variation.**

$$\frac{\text{median for Class A} - \text{median for Class B}}{\text{IQR for Class A}} = \frac{4}{8} = 0.5$$

$$\frac{\text{median for Class A} - \text{median for Class B}}{\text{IQR for Class B}} = \frac{4}{12} = 0.3$$

⋮ So, the difference in the medians is about 0.3 to 0.5 times the IQR.

Exercises

24. **SPANISH TEST** The double box-and-whisker plot shows the test scores of two Spanish classes taught by the same teacher.

a. Compare the populations using measures of center and variation.

b. Express the difference in the measures of center as a multiple of each measure of variation.

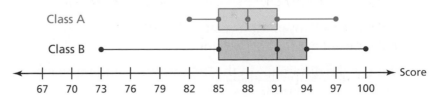

Check It Out
Test Practice
BigIdeasMath com

You randomly choose one game piece. (a) Find the number of ways the event can occur. (b) Find the favorable outcomes of the event.

1. Choosing green

2. Choosing *not* yellow

3. Use the Fundamental Counting Principle to find the total number of different sunscreens possible.

Sunscreen	
SPF	10, 15, 30, 45, 50
Type	Lotion, Spray, Gel

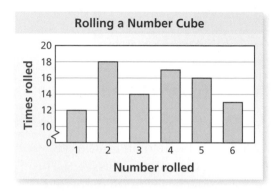

Rolling a Number Cube

Use the bar graph to find the experimental probability of the event.

4. Rolling a 1 or a 2

5. Rolling an odd number

6. *Not* rolling a 5

Use the spinner to find the theoretical probability of the event(s).

7. Spinning an even number

8. Spinning a 1 and then a 2

Knight Queen King Bishop Rook Pawn

You randomly choose one chess piece. Without replacing the first piece, you randomly choose a second piece. Find the probability of choosing the first piece, then the second piece.

9. Bishop and bishop   10. King and queen

11. **LUNCH** You want to estimate the number of students in your school who prefer to bring a lunch from home rather than buy one at school. You survey five students who are standing in the lunch line. Determine whether the sample is *biased* or *unbiased*. Explain.

12. **AGES** The double box-and-whisker plot shows the ages of the viewers of two television shows in a small town.

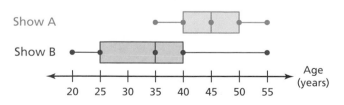

Show A

Show B

Age (years)

a. Compare the populations using measures of center and variation.

b. Express the difference in the measures of center as a multiple of each measure of variation.

1. A school athletic director asked each athletic team member to name his or her favorite professional sports team. The results are below:

- D.C. United: 3
- Florida Panthers: 8
- Jacksonville Jaguars: 26
- Jacksonville Sharks: 7
- Miami Dolphins: 22
- Miami Heat: 15
- Miami Marlins: 20
- Minnesota Lynx: 4
- New York Knicks: 5
- Orlando Magic: 18
- Tampa Bay Buccaneers: 17
- Tampa Bay Lightning: 12
- Tampa Bay Rays: 28
- Other: 6

Test-Taking Strategy
Use Intelligent Guessing

What's the probability of drawing 1 hyena out of a bag with 2 hyenas and 3 mice?
Ⓐ -10% Ⓑ 40% Ⓒ 60% Ⓓ 500%

40% < 60% I'm hoping 40%.

"You know it can't be -10% or 500%. So, you can intelligently guess between 40% and 60%."

One athletic team member is picked at random. What is the likelihood that this team member's favorite professional sports team is *not* located in Florida? *(7.SP.5)*

A. certain

B. likely, but not certain

C. unlikely, but not impossible

D. impossible

2. Each student in your class voted for his or her favorite day of the week. Their votes are shown below:

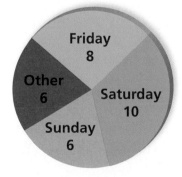

Favorite Day of the Week

Friday 8

Other 6

Saturday 10

Sunday 6

A student from your class is picked at random. What is the probability that this student's favorite day of the week is Sunday? *(7.SP.7b)*

3. How far, in millimeters, will the tip of the hour hand of the clock travel in 2 hours? (Use $\frac{22}{7}$ for π.)   *(7.G.4)*

84 mm

F. 44 mm   H. 264 mm

G. 88 mm   I. 528 mm

4. Nathaniel solved the proportion in the box below.

$$\frac{16}{40} = \frac{p}{27}$$

$$16 \cdot p = 40 \cdot 27$$

$$16p = 1080$$

$$\frac{16p}{16} = \frac{1080}{16}$$

$$p = 67.5$$

What should Nathaniel do to correct the error that he made?   *(7.RP.2c)*

A. Add 40 to 16 and 27 to p.

B. Subtract 16 from 40 and 27 from p.

C. Multiply 16 by 27 and p by 40.

D. Divide 16 by 27 and p by 40.

5. A North American hockey rink contains 5 face-off circles. Each of these circles has a radius of 15 feet. What is the total area, in square feet, of all the face-off circles? (Use 3.14 for π.)   *(7.G.4)*

F. 706.5 ft^2 H. 3532.5 ft^2

G. 2826 ft^2 I. 14,130 ft^2

6. A spinner is divided into eight congruent sections, as shown below.

You spin the spinner twice. What is the probability that the arrow will stop in a yellow section both times? *(7.SP.8a)*

7. What is the surface area, in square inches, of the square pyramid? *(7.G.6)*

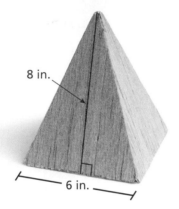

8 in.

6 in.

A. 24 in.²

C. 132 in.²

B. 96 in.²

D. 228 in.²

8. The value of one of Kevin's baseball cards was $6.00 when he first got it. The value of this card is now $15.00. What is the percent increase in the value of the card? *(7.RP.3)*

F. 40%

H. 150%

G. 90%

I. 250%

9. You roll a number cube twice. You want to roll two even numbers. *(7.SP.8a)*

Part A   Determine whether the events are independent or dependent.

Part B   Find the number of favorable outcomes and the number of possible outcomes of each roll.

Part C   Find the probability of rolling two even numbers. Explain your reasoning.

11 Transformations

"Just 2 more minutes. I'm almost done with my 'cat tessellation' painting."

"If you hold perfectly still..."

"...each frame becomes a horizontal..."

"...translation of the previous frame..."

What You Learned Before

"Did you know that when you look at yourself in the mirror, your left and right get switched?"

Does that mean that my mirror image is better at music than I am?

● Reflecting Points (6.NS.6b)

Example 1   **Reflect (3, −4) in the x-axis.**

Plot (3, −4).

To reflect (3, −4) in the x-axis, use the same x-coordinate, 3, and take the opposite of the y-coordinate. The opposite of −4 is 4.

∴ So, the reflection of (3, −4) in the x-axis is (3, 4).

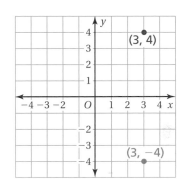

Try It Yourself

Reflect the point in (a) the x-axis and (b) the y-axis.

1. (7, 3)
2. (−4, 6)
3. (5, −5)
4. (−8, −3)

5. (0, 1)
6. (−5, 0)
7. (4, −6.5)
8. $\left(-3\frac{1}{2}, -4\right)$

● Drawing a Polygon in a Coordinate Plane (6.G.3)

Example 2   **The vertices of a quadrilateral are A(1, 5), B(2, 9), C(6, 8), and D(8, 1). Draw the quadrilateral in a coordinate plane.**

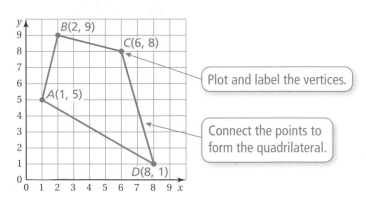

Plot and label the vertices.

Connect the points to form the quadrilateral.

Try It Yourself

Draw the polygon with the given vertices in a coordinate plane.

9. J(1, 1), K(5, 6), M(9, 3)

10. Q(2, 3), R(2, 8), S(7, 8), T(7, 3)

Essential Question How can you identify congruent triangles?

Two figures are congruent when they have the same size and the same shape.

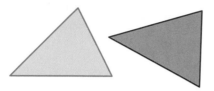

Congruent

Same size *and* shape

Not Congruent

Same shape, but not same size

1 ACTIVITY: Identifying Congruent Triangles

Work with a partner.

- **Which of the geoboard triangles below are congruent to the geoboard triangle at the right?**
- **Form each triangle on a geoboard.**
- **Measure each side with a ruler. Record your results in a table.**
- **Write a conclusion about the side lengths of triangles that are congruent.**

a.

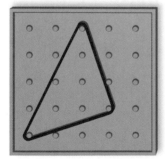

b.

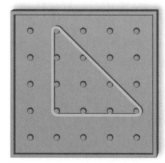

c.

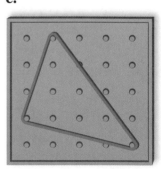

d.

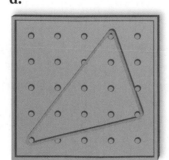

e.

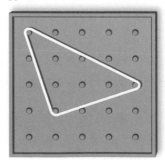

f.

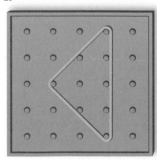

COMMON CORE

Geometry

In this lesson, you will

- name corresponding angles and corresponding sides of congruent figures.
- identify congruent figures.

Preparing for Standard 8.G.2

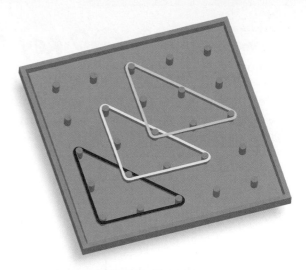

Math Practice 5

Recognize Usefulness of Tools

What are some advantages and disadvantages of using a geoboard to construct congruent triangles?

The geoboard at the right shows three congruent triangles.

2 ACTIVITY: Forming Congruent Triangles

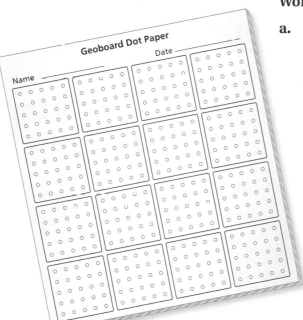

Work with a partner.

a. Form the yellow triangle in Activity 1 on your geoboard. Record the triangle on geoboard dot paper.

b. Move each vertex of the triangle one peg to the right. Is the new triangle congruent to the original triangle? How can you tell?

c. On a 5-by-5 geoboard, make as many different triangles as possible, each of which is congruent to the yellow triangle in Activity 1. Record each triangle on geoboard dot paper.

What Is Your Answer?

3. **IN YOUR OWN WORDS** How can you identify congruent triangles? Use the conclusion you wrote in Activity 1 as part of your answer.

4. Can you form a triangle on your geoboard whose side lengths are 3, 4, and 5 units? If so, draw such a triangle on geoboard dot paper.

Practice Use what you learned about congruent triangles to complete Exercises 4 and 5 on page 470.

Key Vocabulary
congruent figures,
  p. 468
corresponding angles,
  p. 468
corresponding sides,
  p. 468

Key Idea

Congruent Figures

Figures that have the same size and the same shape are called **congruent figures**. The triangles below are congruent.

Matching angles are called **corresponding angles**.

Matching sides are called **corresponding sides**.

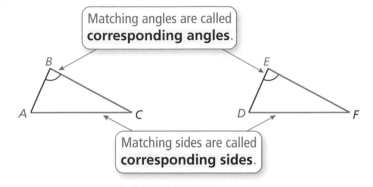

EXAMPLE 1 Naming Corresponding Parts

The figures are congruent. Name the corresponding angles and the corresponding sides.

Corresponding Angles	*Corresponding Sides*
∠A and ∠W	Side AB and Side WX
∠B and ∠X	Side BC and Side XY
∠C and ∠Y	Side CD and Side YZ
∠D and ∠Z	Side AD and Side WZ

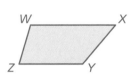

On Your Own

Exercises 6 and 7

1. The figures are congruent. Name the corresponding angles and the corresponding sides.

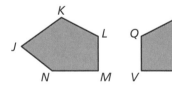

Key Idea

Identifying Congruent Figures

Two figures are congruent when corresponding angles and corresponding sides are congruent.

Triangle *ABC* is congruent to Triangle *DEF*.

$$\triangle ABC \cong \triangle DEF$$

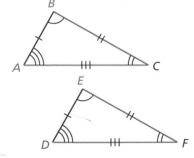

Reading

The symbol ≅ means *is congruent to.*

EXAMPLE 2 — Identifying Congruent Figures

Which square is congruent to Square A?

Square A

8
8   8
8

Square B

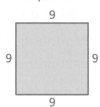
9
9   9
9

Square C

8   8
8   8

Each square has four right angles. So, corresponding angles are congruent. Check to see if corresponding sides are congruent.

Square A and Square B

Each side length of Square A is 8, and each side length of Square B is 9. So, corresponding sides are not congruent.

Square A and Square C

Each side length of Square A and Square C is 8. So, corresponding sides are congruent.

∴ So, Square C is congruent to Square A.

EXAMPLE 3 — Using Congruent Figures

Trapezoids *ABCD* and *JKLM* are congruent.

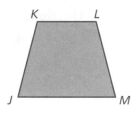

a. **What is the length of side *JM*?**

   Side *JM* corresponds to side *AD*.

   ∴ So, the length of side *JM* is 10 feet.

b. **What is the perimeter of *JKLM*?**

   The perimeter of *ABCD* is $10 + 8 + 6 + 8 = 32$ feet. Because the trapezoids are congruent, their corresponding sides are congruent.

   ∴ So, the perimeter of *JKLM* is also 32 feet.

On Your Own

Now You're Ready
Exercises 8, 9, and 12

2. Which square in Example 2 is congruent to Square D?

3. In Example 3, which angle of *JKLM* corresponds to ∠*C*? What is the length of side *KJ*?

Square D

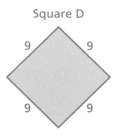
9   9
9   9

Vocabulary and Concept Check

1. **VOCABULARY** △*ABC* is congruent to △*DEF.*

   a. Identify the corresponding angles.
   b. Identify the corresponding sides.

2. **VOCABULARY** Explain how you can tell that two figures are congruent.

3. **WHICH ONE DOESN'T BELONG?** Which one does *not* belong with the other three? Explain your reasoning.

   | ∠*R* | ∠*U* | ∠*V* | ∠*Q* |

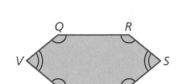
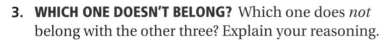

Practice and Problem Solving

Tell whether the triangles are *congruent* or *not congruent*.

4.

5.

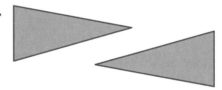

The figures are congruent. Name the corresponding angles and the corresponding sides.

① 6.

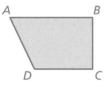

7.

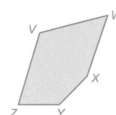

Tell whether the two figures are congruent. Explain your reasoning.

② 8.

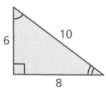

9.

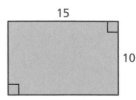

10. **PUZZLE** Describe the relationship between the unfinished puzzle and the missing piece.

11. ERROR ANALYSIS Describe and correct the error in telling whether the two figures are congruent.

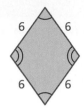

 Both figures have four sides, and the corresponding side lengths are equal. So, they are congruent.

12. HOUSES The fronts of the houses are identical.

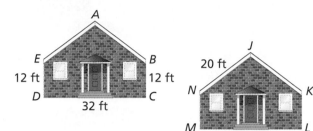

a. What is the length of side *LM*?

b. Which angle of *JKLMN* corresponds to ∠*D*?

c. Side *AB* is congruent to side *AE*. What is the length of side *AB*?

d. What is the perimeter of *ABCDE*?

13. REASONING Here are two ways to draw *one* line to divide a rectangle into two congruent figures. Draw three other ways.

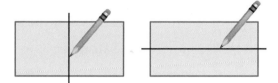

14. CRITICAL THINKING Are the areas of two congruent figures equal? Explain. Draw a diagram to support your answer.

15. **True or False?** The trapezoids are congruent. Determine whether the statement is *true* or *false*. Explain your reasoning.

a. Side *AB* is congruent to side *YZ*.

b. ∠*A* is congruent to ∠*X*.

c. ∠*A* corresponds to ∠*X*.

d. The sum of the angle measures of *ABCD* is 360°.

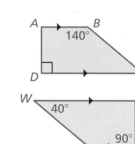

 Fair Game Review *What you learned in previous grades & lessons*

Plot and label the ordered pair in a coordinate plane. *(Skills Review Handbook)*

16. *A*(5, 3)   **17.** *B*(4, −1)   **18.** *C*(−2, 6)   **19.** *D*(−4, −2)

20. MULTIPLE CHOICE You have 2 quarters and 5 dimes in your pocket. Write the ratio of quarters to the total number of coins. *(Skills Review Handbook)*

Ⓐ $\frac{2}{5}$ Ⓑ 2 : 7   Ⓒ 5 to 7   Ⓓ $\frac{7}{2}$

Essential Question How can you arrange tiles to make a tessellation?

The Meaning of a Word ● Translate

When you **translate** a tile, you slide it from one place to another.

When tiles cover a floor with no empty spaces, the collection of tiles is called a *tessellation*.

1 ACTIVITY: Describing Tessellations

Work with a partner. Can you make the tessellation by translating single tiles that are all of the same shape and design? If so, show how.

a. Sample:

Tile Pattern   Single Tiles

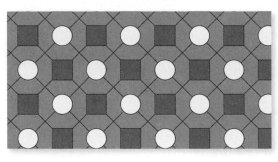
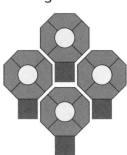

b.

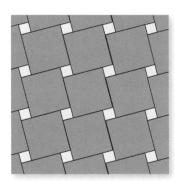

c.

COMMON CORE

Geometry

In this lesson, you will
- identify translations.
- translate figures in the coordinate plane.

Learning Standards
8.G.1
8.G.2
8.G.3

2 ACTIVITY: Tessellations and Basic Shapes

Work with a partner.

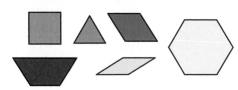

a. Which pattern blocks can you use to make a tessellation? For each one that works, draw the tessellation.

b. Can you make the tessellation by translating? Or do you have to rotate or flip the pattern blocks?

3 ACTIVITY: Designing Tessellations

Work with a partner. Design your own tessellation. Use one of the basic shapes from Activity 2.

Sample:

Step 1: Start with a square.

Step 2: Cut a design out of one side.

Step 3: Tape it to the other side to make your pattern.

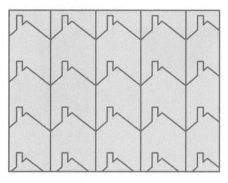 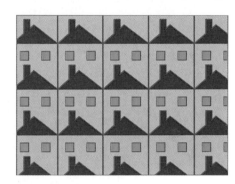

Step 4: Translate the pattern to make your tessellation.

Step 5: Color the tessellation.

4 ACTIVITY: Translating in the Coordinate Plane

Work with a partner.

Math Practice 3

Justify Conclusions

What information do you need to conclude that two figures are congruent?

a. Draw a rectangle in a coordinate plane. Find the dimensions of the rectangle.

b. Move each vertex 3 units right and 4 units up. Draw the new figure. List the vertices.

c. Compare the dimensions and the angle measures of the new figure to those of the original rectangle.

d. Are the opposite sides of the new figure still parallel? Explain.

e. Can you conclude that the two figures are congruent? Explain.

f. Compare your results with those of other students in your class. Do you think the results are true for any type of figure?

What Is Your Answer?

5. **IN YOUR OWN WORDS** How can you arrange tiles to make a tessellation? Give an example.

6. **PRECISION** Explain why any parallelogram can be translated to make a tessellation.

Practice

Use what you learned about translations to complete Exercises 4–6 on page 476.

Check It Out
Lesson Tutorials
BigIdeasMath.com

Key Vocabulary
transformation,
    p. 474
image, p. 474
translation, p. 474

A **transformation** changes a figure into another figure. The new figure is called the **image**.

A **translation** is a transformation in which a figure *slides* but does not turn. Every point of the figure moves the same distance and in the same direction.

Slide

EXAMPLE **1** **Identifying a Translation**

Tell whether the blue figure is a translation of the red figure.

a.

b.

The red figure *slides* to form the blue figure.

⋮⋮• So, the blue figure is a translation of the red figure.

The red figure *turns* to form the blue figure.

⋮⋮• So, the blue figure is *not* a translation of the red figure.

On Your Own

Now You're Ready
Exercises 4–9

Tell whether the blue figure is a translation of the red figure. Explain.

1.   **2.** 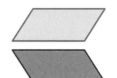   **3.**

Key Idea

Reading

A' is read "*A* prime."
Use *prime* symbols
when naming
an image.

$A \longrightarrow A'$
$B \longrightarrow B'$
$C \longrightarrow C'$

Translations in the Coordinate Plane

Words   To translate a figure *a* units horizontally and *b* units vertically in a coordinate plane, add *a* to the *x*-coordinates and *b* to the *y*-coordinates of the vertices.

Positive values of *a* and *b* represent translations up and right. Negative values of *a* and *b* represent translations down and left.

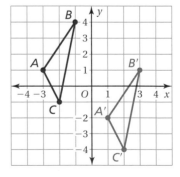

Algebra $(x, y) \longrightarrow (x + a, y + b)$

In a translation, the original figure and its image are congruent.

EXAMPLE **2**   **Translating a Figure in the Coordinate Plane**

Translate the red triangle 3 units right and 3 units down. What are the coordinates of the image?

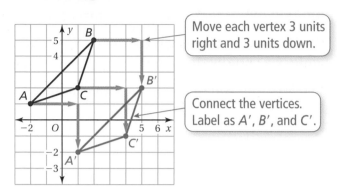

Move each vertex 3 units right and 3 units down.

Connect the vertices. Label as *A′*, *B′*, and *C′*.

∴ The coordinates of the image are $A'(1, -2)$, $B'(5, 2)$, and $C'(4, -1)$.

On Your Own

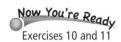

Exercises 10 and 11

4.   WHAT IF? The red triangle is translated 4 units left and 2 units up. What are the coordinates of the image?

EXAMPLE **3**   **Translating a Figure Using Coordinates**

The vertices of a square are $A(1, -2)$, $B(3, -2)$, $C(3, -4)$, and $D(1, -4)$. Draw the figure and its image after a translation 4 units left and 6 units up.

Add −4 to each *x*-coordinate. So, subtract 4 from each *x*-coordinate.

Add 6 to each *y*-coordinate.

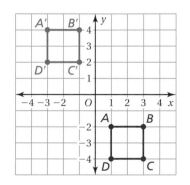

Vertices of *ABCD*	$(x - 4, y + 6)$	Vertices of *A′B′C′D′*
$A(1, -2)$	$(1 - 4, -2 + 6)$	$A'(-3, 4)$
$B(3, -2)$	$(3 - 4, -2 + 6)$	$B'(-1, 4)$
$C(3, -4)$	$(3 - 4, -4 + 6)$	$C'(-1, 2)$
$D(1, -4)$	$(1 - 4, -4 + 6)$	$D'(-3, 2)$

∴ The figure and its image are shown at the above right.

On Your Own

Exercises 12–15

5.   The vertices of a triangle are $A(-2, -2)$, $B(0, 2)$, and $C(3, 0)$. Draw the figure and its image after a translation 1 unit left and 2 units up.

Check It Out
Help with Homework
BigIdeasMath ✓com

 Vocabulary and Concept Check

1. **VOCABULARY** Which figure is the image?

2. **VOCABULARY** How do you translate a figure in a coordinate plane?

3. **WRITING** Can you translate the letters in the word TOKYO to form the word KYOTO? Explain.

Slide

A

B

 Practice and Problem Solving

Tell whether the blue figure is a translation of the red figure.

① 4.
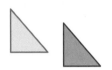

5.
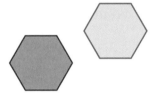

6.
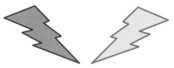

7.
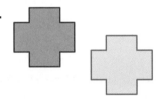

8.
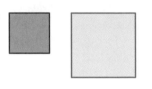

9.
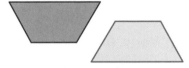

② 10. Translate the triangle 4 units right and 3 units down. What are the coordinates of the image?

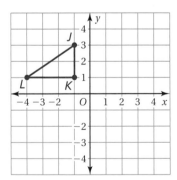

11. Translate the figure 2 units left and 4 units down. What are the coordinates of the image?

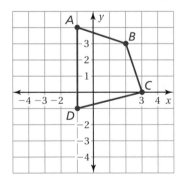

The vertices of a triangle are $L(0, 1)$, $M(1, -2)$, and $N(-2, 1)$. Draw the figure and its image after the translation.

③ 12. 1 unit left and 6 units up

13. 5 units right

14. $(x + 2, y + 3)$

15. $(x - 3, y - 4)$

16. **ICONS** You can click and drag an icon on a computer screen. Is this an example of a translation? Explain.

HURRY!

Describe the translation of the point to its image.

17. $(3, -2) \longrightarrow (1, 0)$

18. $(-8, -4) \longrightarrow (-3, 5)$

Describe the translation from the red figure to the blue figure.

19.
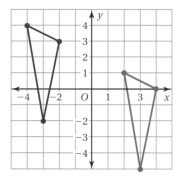

20.
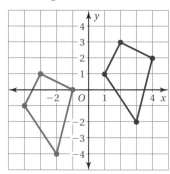

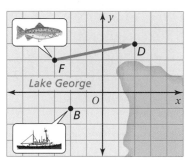

21. FISHING A school of fish translates from point F to point D.

   a. Describe the translation of the school of fish.

   b. Can the fishing boat make the same translation? Explain.

   c. Describe a translation the fishing boat could make to get to point D.

22. REASONING The vertices of a triangle are $A(0, -3)$, $B(2, -1)$, and $C(3, -3)$. You translate the triangle 5 units right and 2 units down. Then you translate the image 3 units left and 8 units down. Is the original triangle congruent to the final image? If so, give two ways to show that they are congruent.

23. **Problem Solving** In chess, a knight can move only in an L-shaped pattern:

- *two* vertical squares, then *one* horizontal square;
- *two* horizontal squares, then *one* vertical square;
- *one* vertical square, then *two* horizontal squares; or
- *one* horizontal square, then *two* vertical squares.

Write a series of translations to move the knight from g8 to g5.

Tell whether you can fold the figure in half so that one side matches the other.
(Skills Review Handbook)

24.

25.

26.

27.

28. MULTIPLE CHOICE You put $550 in an account that earns 4.4% simple interest per year. How much interest do you earn in 6 months? *(Section 6.7)*

   (A) $1.21   **(B)** $12.10   **(C)** $121.00   **(D)** $145.20

Essential Question

How can you use reflections to classify a frieze pattern?

The Meaning of a Word ● Reflection

When you look at a mountain by a lake, you can see the **reflection**, or mirror image, of the mountain in the lake.

If you fold the photo on its axis, the mountain and its reflection will align.

Actual mountain

Axis

Reflection of mountain

Frieze

A *frieze* is a horizontal band that runs at the top of a building. A frieze is often decorated with a design that repeats.

● All frieze patterns are translations of themselves.
● Some frieze patterns are reflections of themselves.

1 ACTIVITY: Frieze Patterns and Reflections

Work with a partner. Consider the frieze pattern shown.

COMMON CORE

Geometry
In this lesson, you will
● identify reflections.
● reflect figures in the *x*-axis or the *y*-axis of the coordinate plane.
Learning Standards
8.G.1
8.G.2
8.G.3

a. Is the frieze pattern a reflection of itself when folded horizontally? Explain.

b. Is the frieze pattern a reflection of itself when folded vertically? Explain.

2 ACTIVITY: Frieze Patterns and Reflections

Work with a partner. Is the frieze pattern a reflection of itself when folded *horizontally*, *vertically*, or *neither*?

a.

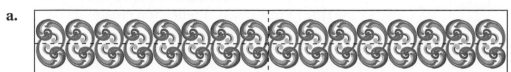

b.

3 ACTIVITY: Reflecting in the Coordinate Plane

Work with a partner.

a. Draw a rectangle in Quadrant I of a coordinate plane. Find the dimensions of the rectangle.

b. Copy the axes and the rectangle onto a piece of transparent paper.

Flip the transparent paper once so that the rectangle is in Quadrant IV. Then align the origin and the axes with the coordinate plane.

Draw the new figure in the coordinate plane. List the vertices.

c. Compare the dimensions and the angle measures of the new figure to those of the original rectangle.

d. Are the opposite sides of the new figure still parallel? Explain.

e. Can you conclude that the two figures are congruent? Explain.

f. Flip the transparent paper so that the original rectangle is in Quadrant II. Draw the new figure in the coordinate plane. List the vertices. Then repeat parts (c) – (e).

g. Compare your results with those of other students in your class. Do you think the results are true for any type of figure?

Math Practice 7

Look for Patterns

What do you notice about the vertices of the original figure and the image? How does this help you determine whether the figures are congruent?

What Is Your Answer?

4. IN YOUR OWN WORDS How can you use reflections to classify a frieze pattern?

Practice ➤ Use what you learned about reflections to complete Exercises 4–6 on page 482.

Check It Out
Lesson Tutorials
BigIdeasMath ⍀com

Key Vocabulary ◀))
reflection, *p. 480*
line of reflection,
   p. 480

A **reflection**, or *flip*, is a transformation in which a figure is reflected in a line called the **line of reflection**. A reflection creates a mirror image of the original figure.

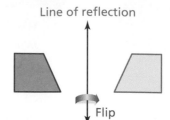
Line of reflection

Flip

EXAMPLE ① **Identifying a Reflection**

Tell whether the blue figure is a reflection of the red figure.

a.

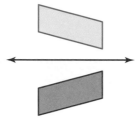

The red figure can be *flipped* to form the blue figure.

∴ So, the blue figure is a reflection of the red figure.

b.

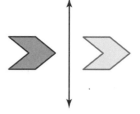

If the red figure were *flipped*, it would point to the left.

∴ So, the blue figure is *not* a reflection of the red figure.

⬤ **On Your Own**

Now You're Ready
Exercises 4–9

Tell whether the blue figure is a reflection of the red figure. Explain.

1.

2.

3.

 Key Idea

Reflections in the Coordinate Plane

Words To reflect a figure in the *x*-axis, take the opposite of the *y*-coordinate.

To reflect a figure in the *y*-axis, take the opposite of the *x*-coordinate.

Algebra Reflection in *x*-axis: $(x, y) \rightarrow (x, -y)$
Reflection in *y*-axis: $(x, y) \rightarrow (-x, y)$

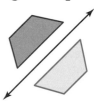

In a reflection, the original figure and its image are congruent.

◀)) Multi-Language Glossary at BigIdeasMath ⍀com

EXAMPLE 2 **Reflecting a Figure in the x-axis**

The vertices of a triangle are $A(-1, 1)$, $B(-1, 3)$, and $C(6, 3)$. Draw the figure and its reflection in the x-axis. What are the coordinates of the image?

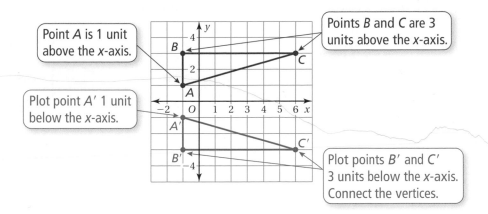

Point A is 1 unit above the x-axis.

Points B and C are 3 units above the x-axis.

Plot point A′ 1 unit below the x-axis.

Plot points B′ and C′ 3 units below the x-axis. Connect the vertices.

⁝ The coordinates of the image are $A'(-1, -1)$, $B'(-1, -3)$, and $C'(6, -3)$.

EXAMPLE 3 **Reflecting a Figure in the y-axis**

The vertices of a quadrilateral are $P(-2, 5)$, $Q(-1, -1)$, $R(-4, 2)$, and $S(-4, 4)$. Draw the figure and its reflection in the y-axis.

Take the opposite of the x-coordinate.

The y-coordinate does not change.

Vertices of *PQRS*	$(-x, y)$	Vertices of *P′Q′R′S′*
$P(-2, 5)$	$(-(-2), 5)$	$P'(2, 5)$
$Q(-1, -1)$	$(-(-1), -1)$	$Q'(1, -1)$
$R(-4, 2)$	$(-(-4), 2)$	$R'(4, 2)$
$S(-4, 4)$	$(-(-4), 4)$	$S'(4, 4)$

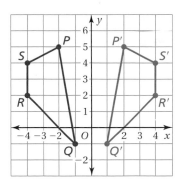

⁝ The figure and its image are shown at the above right.

● **On Your Own**

Now You're Ready
Exercises 10–17

4. The vertices of a rectangle are $A(-4, -3)$, $B(-4, -1)$, $C(-1, -1)$, and $D(-1, -3)$.

 a. Draw the figure and its reflection in the x-axis.

 b. Draw the figure and its reflection in the y-axis.

 c. Are the images in parts (a) and (b) congruent? Explain.

✓ Vocabulary and Concept Check

1. **WHICH ONE DOESN'T BELONG?** Which transformation does *not* belong with the other three? Explain your reasoning.

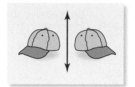   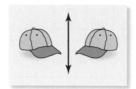   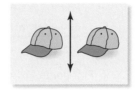   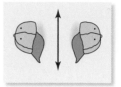

2. **WRITING** How can you tell when one figure is a reflection of another figure?

3. **REASONING** A figure lies entirely in Quadrant I. The figure is reflected in the *x*-axis. In which quadrant is the image?

Practice and Problem Solving

Tell whether the blue figure is a reflection of the red figure.

① **4.**   **5.** 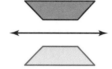   **6.**

7. 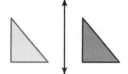   **8.** 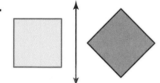   **9.**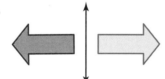

Draw the figure and its reflection in the *x*-axis. Identify the coordinates of the image.

② **10.** $A(3, 2), B(4, 4), C(1, 3)$ **11.** $M(-2, 1), N(0, 3), P(2, 2)$

12. $H(2, -2), J(4, -1), K(6, -3), L(5, -4)$ **13.** $D(-2, -1), E(0, -1), F(0, -5), G(-2, -5)$

Draw the figure and its reflection in the *y*-axis. Identify the coordinates of the image.

③ **14.** $Q(-4, 2), R(-2, 4), S(-1, 1)$ **15.** $T(4, -2), U(4, 2), V(6, -2)$

16. $W(2, -1), X(5, -2), Y(5, -5), Z(2, -4)$ **17.** $J(2, 2), K(7, 4), L(9, -2), M(3, -1)$

18. **ALPHABET** Which letters look the same when reflected in the line?

A B C D E F G H I J K L M N O P Q R S T U V W X Y Z

←————————————————————————————→

The coordinates of a point and its image are given. Is the reflection in the x-axis or y-axis?

19. $(2, -2) \longrightarrow (2, 2)$

20. $(-4, 1) \longrightarrow (4, 1)$

21. $(-2, -5) \longrightarrow (2, -5)$

22. $(-3, -4) \longrightarrow (-3, 4)$

Find the coordinates of the figure after the transformations.

23. Translate the triangle 1 unit right and 5 units down. Then reflect the image in the y-axis.

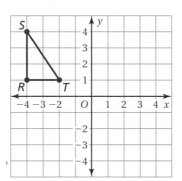

24. Reflect the trapezoid in the x-axis. Then translate the trapezoid 2 units left and 3 units up.

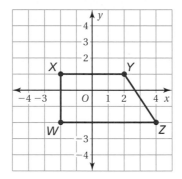

25. REASONING In Exercises 23 and 24, is the original figure congruent to the final image? Explain.

26. NUMBER SENSE You reflect a point (x, y) in the x-axis, and then in the y-axis. What are the coordinates of the final image?

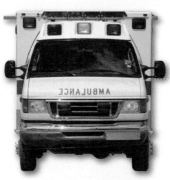

27. EMERGENCY VEHICLE Hold a mirror to the left side of the photo of the vehicle.

    a. What word do you see in the mirror?

    b. Why do you think it is written that way on the front of the vehicle?

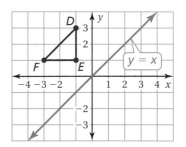

28. Reflect the triangle in the line $y = x$. How are the x- and y-coordinates of the image related to the x- and y-coordinates of the original triangle?

Fair Game Review What you learned in previous grades & lessons

Classify the angle as *acute*, *right*, *obtuse*, or *straight*. *(Skills Review Handbook)*

29.

30.

31.

32.

33. MULTIPLE CHOICE 36 is 75% of what number? *(Section 6.3 and Section 6.4)*

    (A) 27   **(B)** 48   **(C)** 54   **(D)** 63

Essential Question What are the three basic ways to move an object in a plane?

The Meaning of a Word ● Rotate

A bicycle wheel

can **rotate** clockwise

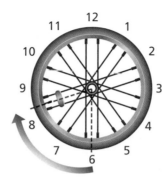

or counterclockwise.

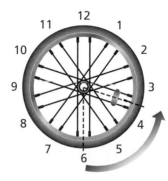

---

1 **ACTIVITY: Three Basic Ways to Move Things**

There are three basic ways to move objects on a flat surface.

_____ the object.   _____ the object.   _____ the object.

COMMON CORE

Geometry

In this lesson, you will
• identify rotations.
• rotate figures in the coordinate plane.
• use more than one transformation to find images of figures.

Learning Standards
8.G.1
8.G.2
8.G.3

Work with a partner.

a. What type of triangle is the blue triangle? Is it congruent to the red triangles? Explain.

b. Decide how you can move the blue triangle to obtain each red triangle.

c. Is each move a *translation*, a *reflection*, or a *rotation*?

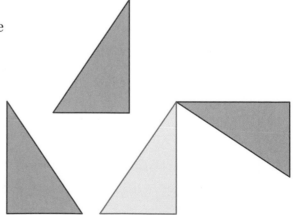

Work with a partner.

a. Draw a rectangle in Quadrant II of a coordinate plane. Find the dimensions of the rectangle.

b. Copy the axes and the rectangle onto a piece of transparent paper.

Align the origin and the vertices of the rectangle on the transparent paper with the coordinate plane. Turn the transparent paper so that the rectangle is in Quadrant I and the axes align.

Draw the new figure in the coordinate plane. List the vertices.

c. Compare the dimensions and the angle measures of the new figure to those of the original rectangle.

d. Are the opposite sides of the new figure still parallel? Explain.

e. Can you conclude that the two figures are congruent? Explain.

f. Turn the transparent paper so that the original rectangle is in Quadrant IV. Draw the new figure in the coordinate plane. List the vertices. Then repeat parts (c)–(e).

g. Compare your results with those of other students in your class. Do you think the results are true for any type of figure?

> **Math Practice** 6
>
> **Calculate Accurately**
> What must you do to rotate the figure correctly?

What Is Your Answer?

3. IN YOUR OWN WORDS What are the three basic ways to move an object in a plane? Draw an example of each.

4. PRECISION Use the results of Activity 2(b).

    a. Draw four angles using the conditions below.

- The origin is the vertex of each angle.
- One side of each angle passes through a vertex of the original rectangle.
- The other side of each angle passes through the corresponding vertex of the rotated rectangle.

    b. Measure each angle in part (a). For each angle, measure the distances between the origin and the vertices of the rectangles. What do you notice?

    c. How can the results of part (b) help you rotate a figure?

5. PRECISION Repeat the procedure in Question 4 using the results of Activity 2(f).

> **Practice**
>
> Use what you learned about transformations to complete Exercises 7–9 on page 489.

Check It Out
Lesson Tutorials
BigIdeasMath **V**com

Key Vocabulary
rotation, *p. 486*
center of rotation,
  p. 486
angle of rotation,
  p. 486

Key Idea

Rotations

A **rotation**, or *turn*, is a transformation in which a figure is rotated about a point called the **center of rotation**. The number of degrees a figure rotates is the **angle of rotation**.

In a rotation, the original figure and its image are congruent.

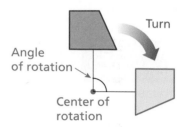

Turn

Angle of rotation

Center of rotation

EXAMPLE ① **Identifying a Rotation**

You must rotate the puzzle piece 270° clockwise about point *P* to fit it into a puzzle. Which piece fits in the puzzle as shown?

•*P*

Ⓐ 　Ⓑ 　Ⓒ 　Ⓓ

Rotate the puzzle piece 270° clockwise about point *P*.

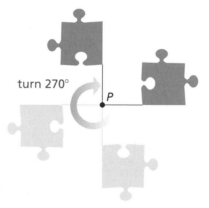

turn 270°

P

Study Tip

When rotating figures, it may help to sketch the rotation in several steps, as shown in Example 1.

∴ So, the correct answer is Ⓒ.

On Your Own

Now You're Ready
Exercises 10–12

1. Which piece is a 90° counterclockwise rotation about point *P*?

2. Is Choice D a rotation of the original puzzle piece? If not, what kind of transformation does the image show?

🔊 Multi-Language Glossary at BigIdeasMath**V**com

EXAMPLE ② Rotating a Figure

The vertices of a trapezoid are $W(-4, 2)$, $X(-3, 4)$, $Y(-1, 4)$, and $Z(-1, 2)$. Rotate the trapezoid 180° about the origin. What are the coordinates of the image?

Study Tip

A 180° clockwise rotation and a 180° counterclockwise rotation have the same image. So, you do not need to specify direction when rotating a figure 180°.

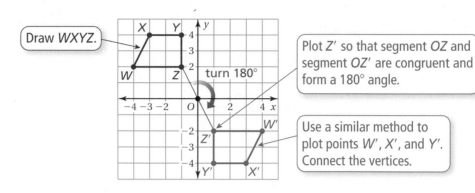

Draw *WXYZ*.

Plot *Z'* so that segment *OZ* and segment *OZ'* are congruent and form a 180° angle.

turn 180°

Use a similar method to plot points *W'*, *X'*, and *Y'*. Connect the vertices.

⫶ The coordinates of the image are $W'(4, -2)$, $X'(3, -4)$, $Y'(1, -4)$, and $Z'(1, -2)$.

EXAMPLE ③ Rotating a Figure

The vertices of a triangle are $J(1, 2)$, $K(4, 2)$, and $L(1, -3)$. Rotate the triangle 90° counterclockwise about vertex L. What are the coordinates of the image?

Common Error

Be sure to pay attention to whether a rotation is clockwise or counterclockwise.

Plot *K'* so that segment *KL* and segment *K'L* are congruent and form a 90° angle.

Use a similar method to plot point *J'*. Connect the vertices.

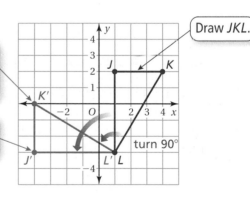

Draw *JKL*.

turn 90°

⫶ The coordinates of the image are $J'(-4, -3)$, $K'(-4, 0)$, and $L'(1, -3)$.

On Your Own

Now You're Ready
Exercises 13–18

3. A triangle has vertices $Q(4, 5)$, $R(4, 0)$, and $S(1, 0)$.

    a. Rotate the triangle 90° counterclockwise about the origin.

    b. Rotate the triangle 180° about vertex S.

    c. Are the images in parts (a) and (b) congruent? Explain.

EXAMPLE 4 **Using More than One Transformation**

The vertices of a rectangle are $A(-3, -3)$, $B(1, -3)$, $C(1, -5)$, and $D(-3, -5)$. Rotate the rectangle 90° clockwise about the origin, and then reflect it in the y-axis. What are the coordinates of the image?

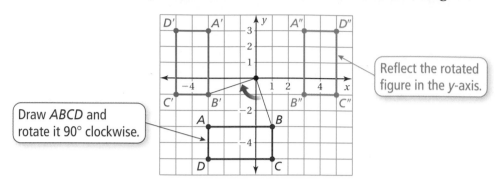

Reflect the rotated figure in the y-axis.

Draw *ABCD* and rotate it 90° clockwise.

⋮ The coordinates of the image are $A''(3, 3)$, $B''(3, -1)$, $C''(5, -1)$ and $D''(5, 3)$.

The image of a translation, reflection, or rotation is congruent to the original figure. So, two figures are congruent when one can be obtained from the other by a sequence of translations, reflections, and rotations.

EXAMPLE 5 **Describing a Sequence of Transformations**

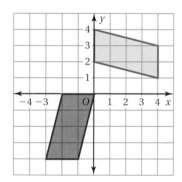

The red figure is congruent to the blue figure. Describe a sequence of transformations in which the blue figure is the image of the red figure.

You can turn the red figure 90° so that it has the same orientation as the blue figure. So, begin with a rotation.

After rotating, you need to slide the figure up.

⋮ So, one possible sequence of transformations is a 90° counterclockwise rotation about the origin followed by a translation 4 units up.

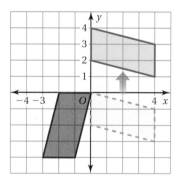

On Your Own

Now You're Ready
Exercises 22–25

4. The vertices of a triangle are $P(-1, 2)$, $Q(-1, 0)$, and $R(2, 0)$. Rotate the triangle 180° about vertex R, and then reflect it in the x-axis. What are the coordinates of the image?

5. In Example 5, describe a different sequence of transformations in which the blue figure is the image of the red figure.

 Vocabulary and Concept Check

1. **VOCABULARY** What are the coordinates of the center of rotation in Example 2? Example 3?

MENTAL MATH A figure lies entirely in Quadrant II. In which quadrant will the figure lie after the given clockwise rotation about the origin?

2. 90°   3. 180°   4. 270°   5. 360°

6. **DIFFERENT WORDS, SAME QUESTION** Which is different? Find "both" answers.

What are the coordinates of the figure after a 90° clockwise rotation about the origin?

What are the coordinates of the figure after a 270° clockwise rotation about the origin?

What are the coordinates of the figure after turning the figure 90° to the right about the origin?

What are the coordinates of the figure after a 270° counterclockwise rotation about the origin?

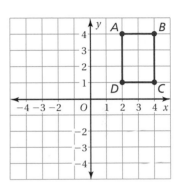

 Practice and Problem Solving

Identify the transformation.

7.

8.

9.

Tell whether the blue figure is a rotation of the red figure about the origin. If so, give the angle and direction of rotation.

10.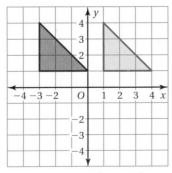

11.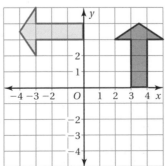

12.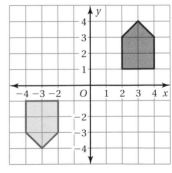

The vertices of a figure are given. Rotate the figure as described. Find the coordinates of the image.

(2) (3) **13.** $A(2, -2), B(4, -1), C(4, -3), D(2, -4)$
90° counterclockwise about the origin

14. $F(1, 2), G(3, 5), H(3, 2)$
180° about the origin

15. $J(-4, 1), K(-2, 1), L(-4, -3)$
90° clockwise about vertex L

16. $P(-3, 4), Q(-1, 4), R(-2, 1), S(-4, 1)$
180° about vertex R

17. $W(-6, -2), X(-2, -2), Y(-2, -6), Z(-5, -6)$
270° counterclockwise about the origin

18. $A(1, -1), B(5, -6), C(1, -6)$
90° counterclockwise about vertex A

A figure has *rotational symmetry* if a rotation of 180° or less produces an image that fits exactly on the original figure. Explain why the figure has rotational symmetry.

19.

20.

21.

The vertices of a figure are given. Find the coordinates of the figure after the transformations given.

(4) **22.** $R(-7, -5), S(-1, -2), T(-1, -5)$

Rotate 90° counterclockwise about the origin. Then translate 3 units left and 8 units up.

23. $J(-4, 4), K(-3, 4), L(-1, 1), M(-4, 1)$

Reflect in the x-axis, and then rotate 180° about the origin.

The red figure is congruent to the blue figure. Describe two different sequences of transformations in which the blue figure is the image of the red figure.

(5) **24.**

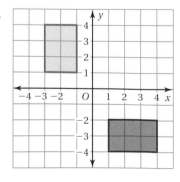

25.

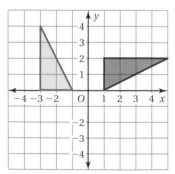

26. **REASONING** A trapezoid has vertices $A(-6, -2)$, $B(-3, -2)$, $C(-1, -4)$, and $D(-6, -4)$.

    a. Rotate the trapezoid 180° about the origin. What are the coordinates of the image?

    b. Describe a way to obtain the same image without using rotations.

27. **TREASURE MAP** You want to find the treasure located on the map at ✕. You are located at ●. The following transformations will lead you to the treasure, but they are not in the correct order. Find the correct order. Use each transformation exactly once.

    ● Rotate 180° about the origin.

    ● Reflect in the y-axis.

    ● Rotate 90° counterclockwise about the origin.

    ● Translate 1 unit right and 1 unit up.

28. **CRITICAL THINKING** Consider $\triangle JKL$.

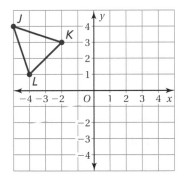

    a. Rotate $\triangle JKL$ 90° clockwise about the origin. How are the x- and y-coordinates of $\triangle J'K'L'$ related to the x- and y-coordinates of $\triangle JKL$?

    b. Rotate $\triangle JKL$ 180° about the origin. How are the x- and y-coordinates of $\triangle J'K'L'$ related to the x- and y-coordinates of $\triangle JKL$?

    c. Do you think your answers to parts (a) and (b) hold true for any figure? Explain.

29. **Reasoning** You rotate a triangle 90° counterclockwise about the origin. Then you translate its image 1 unit left and 2 units down. The vertices of the final image are $(-5, 0)$, $(-2, 2)$, and $(-2, -1)$. What are the vertices of the original triangle?

 Fair Game Review *What you learned in previous grades & lessons*

Tell whether the ratios form a proportion. *(Section 5.2)*

30. $\dfrac{3}{5}, \dfrac{15}{20}$ 31. $\dfrac{2}{3}, \dfrac{12}{18}$ 32. $\dfrac{7}{28}, \dfrac{12}{48}$ 33. $\dfrac{54}{72}, \dfrac{36}{45}$

34. **MULTIPLE CHOICE** What is the solution of the equation $x + 6 \div 2 = 5$? *(Section 3.3)*

    Ⓐ $x = -16$ Ⓑ $x = 2$ Ⓒ $x = 4$ Ⓓ $x = 16$

You can use a **summary triangle** to explain a concept. Here is an example of a summary triangle for translating a figure.

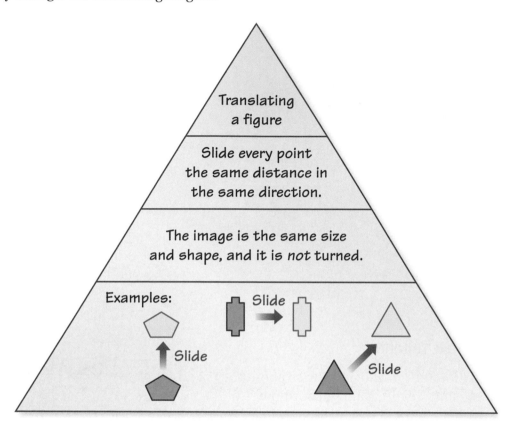

Translating
a figure

Slide every point
the same distance in
the same direction.

The image is the same size
and shape, and it is *not* turned.

Examples:

Slide

Slide

Slide

On Your Own

Make summary triangles to help you study these topics.

1. congruent figures

2. reflecting a figure

3. rotating a figure

After you complete this chapter, make summary triangles for the following topics.

4. similar figures

5. perimeters of similar figures

6. areas of similar figures

7. dilating a figure

8. transforming a figure

Good thing Chuck Berry didn't feel this way.

New Tricks
Can't teach
An Old Newton
Example: Roll over

"I hope my owner sees my summary triangle. I just can't seem to learn 'roll over.'"

Tell whether the two figures are congruent. Explain your reasoning. *(Section 11.1)*

1.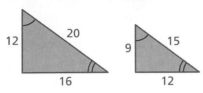

2.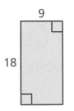

Tell whether the blue figure is a translation of the red figure. *(Section 11.2)*

3.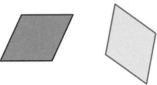

4.

Tell whether the blue figure is a reflection of the red figure. *(Section 11.3)*

5.

6.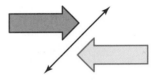

The red figure is congruent to the blue figure. Describe two different sequences of transformations in which the blue figure is the image of the red figure. *(Section 11.4)*

7.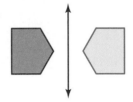

8.

9. AIRPLANE Describe a translation of the airplane from point *A* to point *B*. *(Section 11.2)*

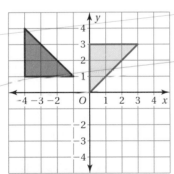

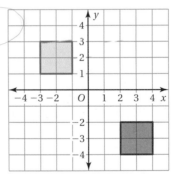

10. MINIGOLF You hit the golf ball along the red path so that its image will be a reflection in the *y*-axis. Does the golf ball land in the hole? Explain. *(Section 11.3)*

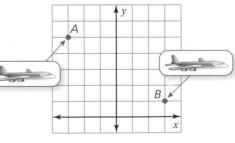

Essential Question How can you use proportions to help make decisions in art, design, and magazine layouts?

In a computer art program, when you click and drag on a side of a photograph, you distort it.

But when you click and drag on a corner of the photograph, the dimensions remain proportional to the original.

Original photograph

Distorted

Distorted

Proportional

1 ACTIVITY: Reducing Photographs

Work with a partner. You are trying to reduce the photograph to the indicated size for a nature magazine. Can you reduce the photograph to the indicated size without distorting or cropping? Explain your reasoning.

a.

5 in.

6 in.

4 in.

5 in.

b.

6 in.

8 in.

3 in.

4 in.

COMMON CORE

Geometry

In this lesson, you will
- name corresponding angles and corresponding sides of similar figures.
- identify similar figures.
- find unknown measures of similar figures.

Preparing for Standard 8.G.4

Work with a partner.

a. Tell whether the dimensions of the new designs are proportional to the
dimensions of the original design. Explain your reasoning.

Original

8   8

7

Design 1

7   7

6

Design 2

$6\frac{6}{7}$ $6\frac{6}{7}$

6

b. Draw two designs whose dimensions are proportional to the given
design. Make one bigger and one smaller. Label the sides of the designs
with their lengths.

5

4

8   10

6

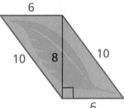

6

10   8   10

6

What Is Your Answer?

3. **IN YOUR OWN WORDS** How can you use proportions to help make
decisions in art, design, and magazine layouts? Give two examples.

4. a. Use a computer art program to draw
two rectangles whose dimensions are
proportional to each other.

"I love this statue. It seems similar to
a big statue I saw in New York."

You've got to hand it to him. He's right.

b. Print the two rectangles on the same
piece of paper.

c. Use a centimeter ruler to measure the
length and the width of each rectangle.

d. Find the following ratios. What can you conclude?

$$\frac{\text{Length of larger}}{\text{Length of smaller}} \qquad \frac{\text{Width of larger}}{\text{Width of smaller}}$$

Practice ▶ Use what you learned about similar figures to complete
Exercises 4 and 5 on page 498.

Check It Out
Lesson Tutorials
BigIdeasMath.com

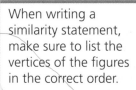

Key Vocabulary
similar figures, *p. 496*

Reading

The symbol ~ means *is similar to.*

Common Error

When writing a similarity statement, make sure to list the vertices of the figures in the correct order.

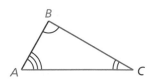 **Key Idea**

Similar Figures

Figures that have the same shape but not necessarily the same size are called **similar figures**.

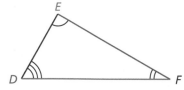

Triangle *ABC* is similar to Triangle *DEF*.

Words   Two figures are similar when

- corresponding side lengths are proportional and
- corresponding angles are congruent.

Symbols   *Side Lengths*   *Angles*   *Figures*

$$\frac{AB}{DE} = \frac{BC}{EF} = \frac{AC}{DF}$$ $\angle A \cong \angle D$ $\triangle ABC \sim \triangle DEF$

$\angle B \cong \angle E$

$\angle C \cong \angle F$

EXAMPLE **1** **Identifying Similar Figures**

Which rectangle is similar to Rectangle A?

Rectangle A   Rectangle B   Rectangle C

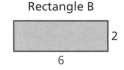

Each figure is a rectangle. So, corresponding angles are congruent. Check to see if corresponding side lengths are proportional.

Rectangle A and Rectangle B

$\dfrac{\text{Length of A}}{\text{Length of B}} = \dfrac{6}{6} = 1$ $\dfrac{\text{Width of A}}{\text{Width of B}} = \dfrac{3}{2}$ Not proportional

Rectangle A and Rectangle C

$\dfrac{\text{Length of A}}{\text{Length of C}} = \dfrac{6}{4} = \dfrac{3}{2}$ $\dfrac{\text{Width of A}}{\text{Width of C}} = \dfrac{3}{2}$ Proportional

∴   So, Rectangle C is similar to Rectangle A.

On Your Own

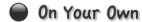
Now You're Ready
Exercises 4–7

1. Rectangle D is 3 units long and 1 unit wide. Which rectangle is similar to Rectangle D?

 Multi-Language Glossary at BigIdeasMath.com

EXAMPLE 2 **Finding an Unknown Measure in Similar Figures**

The triangles are similar. Find x.

Because the triangles are similar, corresponding side lengths are proportional. So, write and solve a proportion to find x.

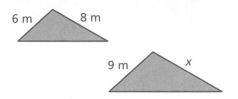
6 m   8 m
9 m   x

$$\frac{6}{9} = \frac{8}{x}$$ Write a proportion.

$6x = 72$ Cross Products Property

$x = 12$ Divide each side by 6.

∴ So, x is 12 meters.

On Your Own

Now You're Ready
Exercises 8–11

The figures are similar. Find x.

2.

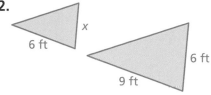
x
6 ft
9 ft
6 ft

3.

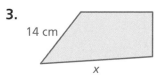
14 cm
x
7 cm
12 cm

EXAMPLE 3 **Real-Life Application**

An artist draws a replica of a painting that is on the Berlin Wall. The painting includes a red trapezoid. The shorter base of the similar trapezoid in the replica is 3.75 inches. What is the height h of the trapezoid in the replica?

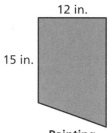
12 in.
15 in.
Painting

Because the trapezoids are similar, corresponding side lengths are proportional. So, write and solve a proportion to find h.

h
3.75 in.
Replica

$$\frac{3.75}{15} = \frac{h}{12}$$ Write a proportion.

$$12 \cdot \frac{3.75}{15} = 12 \cdot \frac{h}{12}$$ Multiplication Property of Equality

$3 = h$ Simplify.

∴ So, the height of the trapezoid in the replica is 3 inches.

On Your Own

4. **WHAT IF?** The longer base in the replica is 4.5 inches. What is the length of the longer base in the painting?

11.5 Exercises

Check It Out
Help with Homework
BigIdeasMath.com

Vocabulary and Concept Check

1. **VOCABULARY** How are corresponding angles of two similar figures related?

2. **VOCABULARY** How are corresponding side lengths of two similar figures related?

3. **CRITICAL THINKING** Are two figures that have the same size and shape similar? Explain.

Practice and Problem Solving

Tell whether the two figures are similar. Explain your reasoning.

4.

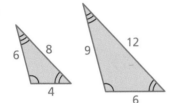

5.

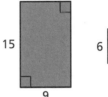

In a coordinate plane, draw the figures with the given vertices. Which figures are similar? Explain your reasoning.

6. Rectangle A: (0, 0), (4, 0), (4, 2), (0, 2)
   Rectangle B: (0, 0), (−6, 0), (−6, 3), (0, 3)
   Rectangle C: (0, 0), (4, 0), (4, 2), (0, 2)

7. Figure A: (−4, 2), (−2, 2), (−2, 0), (−4, 0)
   Figure B: (1, 4), (4, 4), (4, 1), (1, 1)
   Figure C: (2, −1), (5, −1), (5, −3), (2, −3)

The figures are similar. Find x.

8.

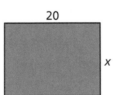

9.

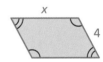

10.

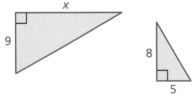

11.

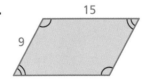

12. **MEXICO** A Mexican flag is 63 inches long and 36 inches wide. Is the drawing at the right similar to the Mexican flag?

13. **DESKS** A student's rectangular desk is 30 inches long and 18 inches wide. The teacher's desk is similar to the student's desk and has a length of 50 inches. What is the width of the teacher's desk?

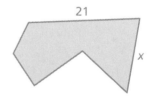
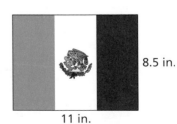

8.5 in.

11 in.

14. LOGIC Are the following figures *always*, *sometimes,* or *never* similar? Explain.

    a. two triangles   **b.** two squares

    c. two rectangles   **d.** a square and a triangle

15. CRITICAL THINKING Can you draw two quadrilaterals each having two 130° angles and two 50° angles that are *not* similar? Justify your answer.

16. SIGN All the angle measures in the sign are 90°.

    a. You increase each side length by 20%. Is the new sign similar to the original?

    b. You increase each side length by 6 inches. Is the new sign similar to the original?

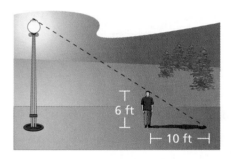

17. STREETLIGHT A person standing 20 feet from a streetlight casts a shadow as shown. How many times taller is the streetlight than the person? Assume the triangles are similar.

18. REASONING Is an object similar to a scale drawing of the object? Explain.

19. GEOMETRY Use a ruler to draw two different isosceles triangles similar to the one shown. Measure the heights of each triangle to the nearest centimeter.

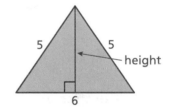

    a. Is the ratio of the corresponding heights proportional to the ratio of the corresponding side lengths?

    b. Do you think this is true for all similar triangles? Explain.

20. **Critical Thinking** Given $\triangle ABC \sim \triangle DEF$ and $\triangle DEF \sim \triangle JKL$, is $\triangle ABC \sim \triangle JKL$? Give an example or a non-example.

Fair Game Review What you learned in previous grades & lessons

Simplify. *(Skills Review Handbook)*

21. $\left(\dfrac{4}{9}\right)^2$ **22.** $\left(\dfrac{3}{8}\right)^2$ **23.** $\left(\dfrac{7}{4}\right)^2$ **24.** $\left(\dfrac{6.5}{2}\right)^2$

25. MULTIPLE CHOICE You solve the equation $S = \ell w + 2wh$ for w. Which equation is correct? *(Topic 3)*

    Ⓐ $w = \dfrac{S - \ell}{2h}$ **Ⓑ** $w = \dfrac{S - 2h}{\ell}$ **Ⓒ** $w = \dfrac{S}{\ell + 2h}$ **Ⓓ** $w = S - \ell - 2h$

Perimeters and Areas of Similar Figures

Essential Question How do changes in dimensions of similar geometric figures affect the perimeters and the areas of the figures?

1 ACTIVITY: Creating Similar Figures

Work with a partner. Use pattern blocks to make a figure whose dimensions are 2, 3, and 4 times greater than those of the original figure.

a. Square

Square with sides labeled 1, 1, 1, 1.

b. Rectangle

Rectangle with sides labeled 2 (top), 1 (left), 1 (right), 2 (bottom).

2 ACTIVITY: Finding Patterns for Perimeters

Work with a partner. Copy and complete the table for the perimeter P of each figure in Activity 1. Describe the pattern.

Figure	Original Side Lengths	Double Side Lengths	Triple Side Lengths	Quadruple Side Lengths
	$P =$			
(dark rectangle)	$P =$			

3 ACTIVITY: Finding Patterns for Areas

Work with a partner. Copy and complete the table for the area A of each figure in Activity 1. Describe the pattern.

Figure	Original Side Lengths	Double Side Lengths	Triple Side Lengths	Quadruple Side Lengths
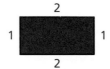	$A =$			
(dark rectangle)	$A =$			

COMMON CORE

Geometry

In this lesson, you will

- understand the relationship between perimeters of similar figures.
- understand the relationship between areas of similar figures.
- find ratios of perimeters and areas for similar figures.

Preparing for Standard 8.G.4

Work with a partner.

a. Find a blue rectangle that is similar to the red rectangle and has one side from $(-1, -6)$ to $(5, -6)$. Label the vertices.

Check that the two rectangles are similar by showing that the ratios of corresponding sides are equal.

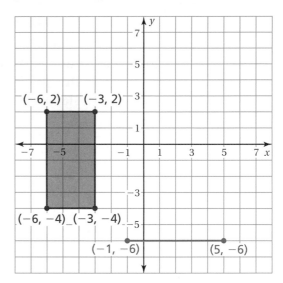

$$\frac{\text{Red Length}}{\text{Blue Length}} \overset{?}{=} \frac{\text{Red Width}}{\text{Blue Width}}$$

$$\frac{\text{change in } y}{\text{change in } y} \overset{?}{=} \frac{\text{change in } x}{\text{change in } x}$$

$$\frac{}{} \overset{?}{=} \frac{}{}$$

$$\frac{}{} \overset{?}{=} \frac{}{}$$

∴ The ratios are equal. So, the rectangles are similar.

b. Compare the perimeters and the areas of the figures. Are the results the same as your results from Activities 2 and 3? Explain.

c. There are three other blue rectangles that are similar to the red rectangle and have the given side.

   ● Draw each one. Label the vertices of each.

   ● Show that each is similar to the original red rectangle.

Math Practice 1

Analyze Givens

What values should you use to fill in the proportion? Does it matter where each value goes? Explain.

What Is Your Answer?

5. **IN YOUR OWN WORDS** How do changes in dimensions of similar geometric figures affect the perimeters and the areas of the figures?

6. What information do you need to know to find the dimensions of a figure that is similar to another figure? Give examples to support your explanation.

Practice

Use what you learned about perimeters and areas of similar figures to complete Exercises 8 and 9 on page 504.

Check It Out
Lesson Tutorials
BigIdeasMath.com

 **Key Idea**

Perimeters of Similar Figures

When two figures are similar, the ratio of their perimeters is equal to the ratio of their corresponding side lengths.

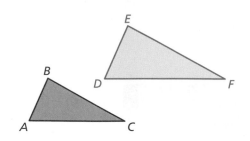

$$\frac{\text{Perimeter of } \triangle ABC}{\text{Perimeter of } \triangle DEF} = \frac{AB}{DE} = \frac{BC}{EF} = \frac{AC}{DF}$$

EXAMPLE 1 **Finding Ratios of Perimeters**

Find the ratio (red to blue) of the perimeters of the similar rectangles.

$$\frac{\text{Perimeter of red rectangle}}{\text{Perimeter of blue rectangle}} = \frac{4}{6} = \frac{2}{3}$$

⠆⠄ The ratio of the perimeters is $\frac{2}{3}$.

● **On Your Own**

1. The height of Figure A is 9 feet. The height of a similar Figure B is 15 feet. What is the ratio of the perimeter of A to the perimeter of B?

 **Key Idea**

Areas of Similar Figures

When two figures are similar, the ratio of their areas is equal to the *square* of the ratio of their corresponding side lengths.

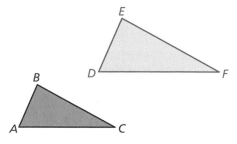

$$\frac{\text{Area of } \triangle ABC}{\text{Area of } \triangle DEF} = \left(\frac{AB}{DE}\right)^2 = \left(\frac{BC}{EF}\right)^2 = \left(\frac{AC}{DF}\right)^2$$

EXAMPLE 2 **Finding Ratios of Areas**

Find the ratio (red to blue) of the areas of the similar triangles.

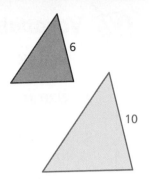

$$\frac{\text{Area of red triangle}}{\text{Area of blue triangle}} = \left(\frac{6}{10}\right)^2$$

$$= \left(\frac{3}{5}\right)^2 = \frac{9}{25}$$

∴ The ratio of the areas is $\frac{9}{25}$.

On Your Own

Now You're Ready
Exercises 4–7

2. The base of Triangle P is 8 meters. The base of a similar Triangle Q is 7 meters. What is the ratio of the area of P to the area of Q?

EXAMPLE 3 **Using Proportions to Find Perimeters and Areas**

A swimming pool is similar in shape to a volleyball court. Find the perimeter P and the area A of the pool.

The rectangular pool and the court are similar. So, use the ratio of corresponding side lengths to write and solve proportions to find the perimeter and the area of the pool.

Perimeter	*Area*

$$\frac{\text{Perimeter of court}}{\text{Perimeter of pool}} = \frac{\text{Width of court}}{\text{Width of pool}}$$

$$\frac{60}{P} = \frac{10}{18}$$

$$1080 = 10P$$

$$108 = P$$

$$\frac{\text{Area of court}}{\text{Area of pool}} = \left(\frac{\text{Width of court}}{\text{Width of pool}}\right)^2$$

$$\frac{200}{A} = \left(\frac{10}{18}\right)^2$$

$$\frac{200}{A} = \frac{100}{324}$$

$$64{,}800 = 100A$$

$$648 = A$$

18 yd

10 yd

Area = 200 yd²
Perimeter = 60 yd

∴ So, the perimeter of the pool is 108 yards, and the area is 648 square yards.

On Your Own

3. WHAT IF? The width of the pool is 16 yards. Find the perimeter P and the area A of the pool.

11.6 Exercises

Check It Out
Help with Homework
BigIdeasMath com

✓ Vocabulary and Concept Check

1. **WRITING** How are the perimeters of two similar figures related?

2. **WRITING** How are the areas of two similar figures related?

3. **NUMBER SENSE** Rectangle *ABCD* is similar to Rectangle *WXYZ*. The area of *ABCD* is 30 square inches. Explain how to find the area of *WXYZ*.

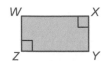

$$\frac{AD}{WZ} = \frac{1}{2} \qquad \frac{AB}{WX} = \frac{1}{2}$$

Practice and Problem Solving

The two figures are similar. Find the ratios (red to blue) of the perimeters and of the areas.

 4.

11   6

5.

5   8

6.

7   4

7.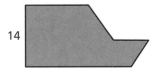

9   14

8. **PERIMETER** How does doubling the side lengths of a right triangle affect its perimeter?

9. **AREA** How does tripling the side lengths of a right triangle affect its area?

The figures are similar. Find *x*.

10. The ratio of the perimeters is $7:10$.

x   12

11. The ratio of the perimeters is $8:5$.

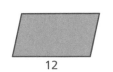 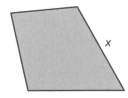

x   16

12. **FOOSBALL** The playing surfaces of two foosball tables are similar. The ratio of the corresponding side lengths is $10:7$. What is the ratio of the areas?

13. **CHEERLEADING** A rectangular school banner has a length of 44 inches, a perimeter of 156 inches, and an area of 1496 square inches. The cheerleaders make signs similar to the banner. The length of a sign is 11 inches. What is its perimeter and its area?

14. **REASONING** The vertices of two rectangles are $A(-5, -1)$, $B(-1, -1)$, $C(-1, -4)$, $D(-5, -4)$ and $W(1, 6)$, $X(7, 6)$, $Y(7, -2)$, $Z(1, -2)$. Compare the perimeters and the areas of the rectangles. Are the rectangles similar? Explain.

21 in.

9 in.

15. **SQUARE** The ratio of the side length of Square A to the side length of Square B is $4:9$. The side length of Square A is 12 yards. What is the perimeter of Square B?

16. **FABRIC** The cost of the fabric is \$1.31. What would you expect to pay for a similar piece of fabric that is 18 inches by 42 inches?

17. **AMUSEMENT PARK** A scale model of a merry-go-round and the actual merry-go-round are similar.

   a. How many times greater is the base area of the actual merry-go-round than the base area of the scale model? Explain.

   b. What is the base area of the actual merry-go-round in square feet?

6 in.

10 ft

Model   450 in.²

18. **STRUCTURE** The circumference of Circle K is π. The circumference of Circle L is 4π.

   a. What is the ratio of their circumferences? of their radii? of their areas?

   b. What do you notice?

Circle K

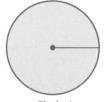

Circle L

19. **GEOMETRY** A triangle with an area of 10 square meters has a base of 4 meters. A similar triangle has an area of 90 square meters. What is the *height* of the larger triangle?

20. **Problem Solving** You need two bottles of fertilizer to treat the flower garden shown. How many bottles do you need to treat a similar garden with a perimeter of 105 feet?

18 ft

4 ft

5 ft

15 ft

 Fair Game Review What you learned in previous grades & lessons

Solve the equation. Check your solution. *(Topic 2)*

21. $4x + 12 = -2x$

22. $2b + 6 = 7b - 2$

23. $8(4n + 13) = 6n$

24. **MULTIPLE CHOICE** Last week, you collected 20 pounds of cans for recycling. This week, you collect 25 pounds of cans for recycling. What is the percent of increase? *(Section 6.5)*

Ⓐ 20%   Ⓑ 25%   Ⓒ 80%   Ⓓ 125%

Essential Question How can you enlarge or reduce a figure in the coordinate plane?

The Meaning of a Word ● Dilate

When you have your eyes checked, the optometrist sometimes **dilates** one or both of the pupils of your eyes.

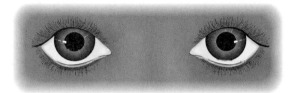

---

1 ACTIVITY: Comparing Triangles in a Coordinate Plane

Work with a partner. Write the coordinates of the vertices of the blue triangle. Then write the coordinates of the vertices of the red triangle.

   a. How are the two sets of coordinates related?

   b. How are the two triangles related? Explain your reasoning.

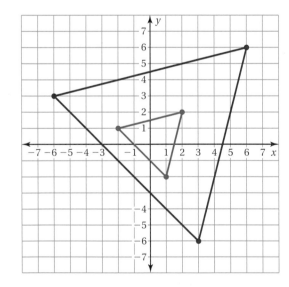

COMMON CORE

Geometry

In this lesson, you will
- identify dilations.
- dilate figures in the coordinate plane.
- use more than one transformation to find images of figures.

Learning Standards
8.G.3
8.G.4

   c. Draw a green triangle whose coordinates are twice the values of the corresponding coordinates of the blue triangle. How are the green and blue triangles related? Explain your reasoning.

   d. How are the coordinates of the red and green triangles related? How are the two triangles related? Explain your reasoning.

2 ACTIVITY: Drawing Triangles in a Coordinate Plane

Work with a partner.

a. Draw the triangle whose vertices are $(0, 2)$, $(-2, 2)$, and $(1, -2)$.

b. Multiply each coordinate of the vertices by 2 to obtain three new vertices. Draw the triangle given by the three new vertices. How are the two triangles related?

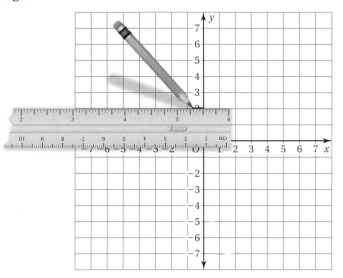

Math Practice 3

Use Prior Results

What are the four types of transformations you studied in this chapter? What information can you use to fill in your table?

c. Repeat part (b) by multiplying by 3 instead of 2.

3 ACTIVITY: Summarizing Transformations

Work with a partner. Make a table that summarizes the relationships between the original figure and its image for the four types of transformations you studied in this chapter.

What Is Your Answer?

4. IN YOUR OWN WORDS How can you enlarge or reduce a figure in the coordinate plane?

5. Describe how knowing how to enlarge or reduce figures in a technical drawing is important in a career such as drafting.

Practice

Use what you learned about dilations to complete Exercises 4–6 on page 511.

Check It Out
Lesson Tutorials
BigIdeasMath.com

A **dilation** is a transformation in which a figure is made larger or smaller with respect to a point called the **center of dilation**.

Center of dilation

EXAMPLE **1** **Identifying a Dilation**

Key Vocabulary
dilation, *p. 508*
center of dilation, *p. 508*
scale factor, *p. 508*

Tell whether the blue figure is a dilation of the red figure.

a.

Lines connecting corresponding vertices meet at a point.

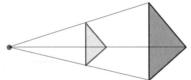

∴ So, the blue figure is a dilation of the red figure.

b.

The figures have the same size and shape. The red figure *slides* to form the blue figure.

∴ So, the blue figure is *not* a dilation of the red figure. It is a translation.

On Your Own

Now You're Ready
Exercises 7–12

Tell whether the blue figure is a dilation of the red figure. Explain.

1.

2.

In a dilation, the original figure and its image are similar. The ratio of the side lengths of the image to the corresponding side lengths of the original figure is the **scale factor** of the dilation.

Key Idea

Dilations in the Coordinate Plane

Words To dilate a figure with respect to the origin, multiply the coordinates of each vertex by the scale factor *k*.

Algebra $(x, y) \rightarrow (kx, ky)$

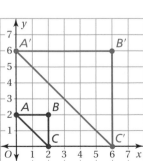

- When $k > 1$, the dilation is an enlargement.
- When $k > 0$ and $k < 1$, the dilation is a reduction.

◀) Multi-Language Glossary at BigIdeasMath.com

EXAMPLE (2) Dilating a Figure

Draw the image of Triangle *ABC* after a dilation with a scale factor of 3. Identify the type of dilation.

Multiply each *x*- and *y*-coordinate by the scale factor 3.

Vertices of *ABC*	(3x, 3y)	Vertices of *A′B′C′*
$A(1, 3)$	$(3 \cdot 1, 3 \cdot 3)$	$A'(3, 9)$
$B(2, 3)$	$(3 \cdot 2, 3 \cdot 3)$	$B'(6, 9)$
$C(2, 1)$	$(3 \cdot 2, 3 \cdot 1)$	$C'(6, 3)$

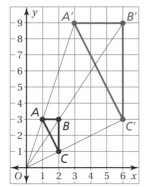

Study Tip

You can check your answer by drawing a line from the origin through each vertex of the original figure. The vertices of the image should lie on these lines.

⋮⋮· The image is shown at the right. The dilation is an *enlargement* because the scale factor is greater than 1.

EXAMPLE (3) Dilating a Figure

Draw the image of Rectangle *WXYZ* after a dilation with a scale factor of 0.5. Identify the type of dilation.

Multiply each *x*- and *y*-coordinate by the scale factor 0.5.

Vertices of *WXYZ*	(0.5x, 0.5y)	Vertices of *W′X′Y′Z′*
$W(-4, -6)$	$(0.5 \cdot (-4), 0.5 \cdot (-6))$	$W'(-2, -3)$
$X(-4, 8)$	$(0.5 \cdot (-4), 0.5 \cdot 8)$	$X'(-2, 4)$
$Y(4, 8)$	$(0.5 \cdot 4, 0.5 \cdot 8)$	$Y'(2, 4)$
$Z(4, -6)$	$(0.5 \cdot 4, 0.5 \cdot (-6))$	$Z'(2, -3)$

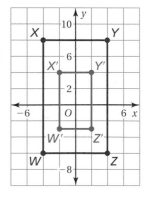

⋮⋮· The image is shown at the right. The dilation is a *reduction* because the scale factor is greater than 0 and less than 1.

On Your Own

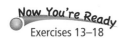
Now You're Ready
Exercises 13–18

3. **WHAT IF?** Triangle *ABC* in Example 2 is dilated by a scale factor of 2. What are the coordinates of the image?

4. **WHAT IF?** Rectangle *WXYZ* in Example 3 is dilated by a scale factor of $\frac{1}{4}$. What are the coordinates of the image?

EXAMPLE 4   Using More than One Transformation

The vertices of a trapezoid are $A(-2, -1)$, $B(-1, 1)$, $C(0, 1)$, and $D(0, -1)$. Dilate the trapezoid with respect to the origin using a scale factor of 2. Then translate it 6 units right and 2 units up. What are the coordinates of the image?

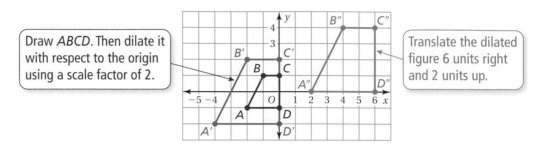

Draw *ABCD*. Then dilate it with respect to the origin using a scale factor of 2.

Translate the dilated figure 6 units right and 2 units up.

⋮• The coordinates of the image are $A''(2, 0)$, $B''(4, 4)$, $C''(6, 4)$, and $D''(6, 0)$.

The image of a translation, reflection, or rotation is congruent to the original figure, and the image of a dilation is similar to the original figure. So, two figures are similar when one can be obtained from the other by a sequence of translations, reflections, rotations, and dilations.

EXAMPLE 5   Describing a Sequence of Transformations

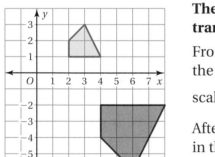

The red figure is similar to the blue figure. Describe a sequence of transformations in which the blue figure is the image of the red figure.

From the graph, you can see that the blue figure is one-half the size of the red figure. So, begin with a dilation with respect to the origin using a scale factor of $\frac{1}{2}$.

After dilating, you need to flip the figure in the *x*-axis.

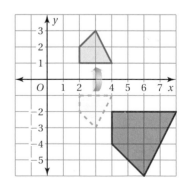

⋮• So, one possible sequence of transformations is a dilation with respect to the origin using a scale factor of $\frac{1}{2}$ followed by a reflection in the *x*-axis.

● On Your Own

Now You're Ready
Exercises 23–28

5. In Example 4, use a scale factor of 3 in the dilation. Then rotate the figure 180° about the image of vertex *C*. What are the coordinates of the image?

6. In Example 5, can you reflect the red figure first, and then perform the dilation to obtain the blue figure? Explain.

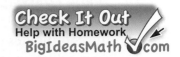

 Vocabulary and Concept Check

1. **VOCABULARY** How is a dilation different from other transformations?

2. **VOCABULARY** For what values of scale factor k is a dilation called an *enlargement*? a *reduction*?

3. **REASONING** Which figure is *not* a dilation of the blue figure? Explain.

   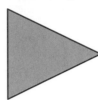

Practice and Problem Solving

Draw the triangle with the given vertices. Multiply each coordinate of the vertices by 3, and then draw the new triangle. How are the two triangles related?

4. $(0, 2), (3, 2), (3, 0)$ 5. $(-1, 1), (-1, -2), (2, -2)$ 6. $(-3, 2), (1, 2), (1, -4)$

Tell whether the blue figure is a dilation of the red figure.

 7.

8.

9.

10.

11.

12.

The vertices of a figure are given. Draw the figure and its image after a dilation with the given scale factor. Identify the type of dilation.

13. $A(1, 1), B(1, 4), C(3, 1); k = 4$ 14. $D(0, 2), E(6, 2), F(6, 4); k = 0.5$

15. $G(-2, -2), H(-2, 6), J(2, 6); k = 0.25$ 16. $M(2, 3), N(5, 3), P(5, 1); k = 3$

17. $Q(-3, 0), R(-3, 6), T(4, 6), U(4, 0); k = \frac{1}{3}$ 18. $V(-2, -2), W(-2, 3), X(5, 3), Y(5, -2); k = 5$

19. **ERROR ANALYSIS** Describe and correct the error in listing the coordinates of the image after a dilation with a scale factor of $\frac{1}{2}$.

Vertices of *ABC*	(2x, 2y)	Vertices of *A'B'C'*
$A(2, 5)$	$(2 \cdot 2, 2 \cdot 5)$	$A'(4, 10)$
$B(2, 0)$	$(2 \cdot 2, 2 \cdot 0)$	$B'(4, 0)$
$C(4, 0)$	$(2 \cdot 4, 2 \cdot 0)$	$C'(8, 0)$

The blue figure is a dilation of the red figure. Identify the type of dilation and find the scale factor.

20.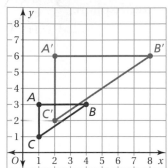

21.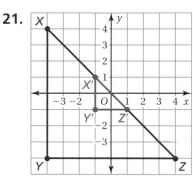

22.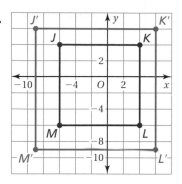

The vertices of a figure are given. Find the coordinates of the figure after the transformations given.

④ 23. $A(-5, 3), B(-2, 3), C(-2, 1), D(-5, 1)$

Reflect in the *y*-axis. Then dilate with respect to the origin using a scale factor of 2.

24. $F(-9, -9), G(-3, -6), H(-3, -9)$

Dilate with respect to the origin using a scale factor of $\frac{2}{3}$. Then translate 6 units up.

25. $J(1, 1), K(3, 4), L(5, 1)$

Rotate 90° clockwise about the origin. Then dilate with respect to the origin using a scale factor of 3.

26. $P(-2, 2), Q(4, 2), R(2, -6), S(-4, -6)$

Dilate with respect to the origin using a scale factor of 5. Then dilate with respect to the origin using a scale factor of 0.5.

The red figure is similar to the blue figure. Describe a sequence of transformations in which the blue figure is the image of the red figure.

⑤ 27.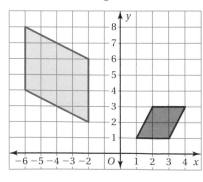

28.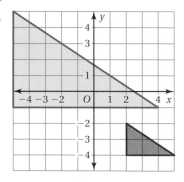

29. **STRUCTURE** In Exercises 27 and 28, is the blue figure still the image of the red figure when you perform the sequence in the opposite order? Explain.

30. OPEN-ENDED Draw a rectangle on a coordinate plane. Choose a scale factor of 2, 3, 4, or 5, and then dilate the rectangle. How many times greater is the area of the image than the area of the original rectangle?

31. SHADOW PUPPET You can use a flashlight and a shadow puppet (your hands) to project shadows on the wall.

   a. Identify the type of dilation.

   b. What does the flashlight represent?

   c. The length of the ears on the shadow puppet is 3 inches. The length of the ears on the shadow is 4 inches. What is the scale factor?

   d. Describe what happens as the shadow puppet moves closer to the flashlight. How does this affect the scale factor?

32. REASONING A triangle is dilated using a scale factor of 3. The image is then dilated using a scale factor of $\frac{1}{2}$. What scale factor could you use to dilate the original triangle to get the final image? Explain.

CRITICAL THINKING The coordinate notation shows how the coordinates of a figure are related to the coordinates of its image after transformations. What are the transformations? Are the figure and its image similar or congruent? Explain.

33. $(x, y) \rightarrow (2x + 4, 2y - 3)$ **34.** $(x, y) \rightarrow (-x - 1, y - 2)$ **35.** $(x, y) \rightarrow \left(\frac{1}{3}x, -\frac{1}{3}y\right)$

36. STRUCTURE How are the transformations $(2x + 3, 2y - 1)$ and $(2(x + 3), 2(y + 1))$ different?

37. **Problem Solving** The vertices of a trapezoid are $A(-2, 3)$, $B(2, 3)$, $C(5, -2)$, and $D(-2, -2)$. Dilate the trapezoid with respect to vertex A using a scale factor of 2. What are the coordinates of the image? Explain the method you used.

Fair Game Review What you learned in previous grades & lessons

Tell whether the angles are *complementary* or *supplementary*. Then find the value of x. *(Section 7.2)*

38.
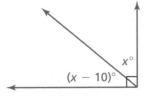
$x°$
$(x - 10)°$

39.
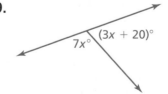
$7x°$ $(3x + 20)°$

40.
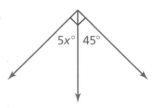
$5x°$ $45°$

41. MULTIPLE CHOICE Which quadrilateral is *not* a parallelogram? *(Section 7.4)*

   A rhombus   **B** trapezoid   **C** square   **D** rectangle

1. Tell whether the two rectangles are similar. Explain your reasoning. *(Section 11.5)*

4 m   8 m   10 m   20 m

The figures are similar. Find x. *(Section 11.5)*

2.

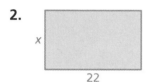

x   22

3   4

3.

8   6   14   x

The two figures are similar. Find the ratios (red to blue) of the perimeters and of the areas. *(Section 11.6)*

4.

12   8

5.

4   15

Tell whether the blue figure is a dilation of the red figure. *(Section 11.7)*

6.

7.

8. **SCREENS** The TV screen is similar to the computer screen. What is the area of the TV screen? *(Section 11.6)*

9. **GEOMETRY** The vertices of a rectangle are $A(2, 4)$, $B(5, 4)$, $C(5, -1)$, and $D(2, -1)$. Dilate the rectangle with respect to the origin using a scale factor of $\frac{1}{2}$. Then translate it 4 units left and 3 units down. What are the coordinates of the image? *(Section 11.7)*

12 in.   20 in.

Area = 108 in.²

10. **TENNIS COURT** The tennis courts for singles and doubles matches are different sizes. Are the courts similar? Explain. *(Section 11.5)*

Singles

27 ft

78 ft

Doubles

36 ft

78 ft

Check It Out
Vocabulary Help
BigIdeasMath ✓com

Review Key Vocabulary

congruent figures, *p. 468*
corresponding angles, *p. 468*
corresponding sides, *p. 468*
transformation, *p. 474*
image, *p. 474*

translation, *p. 474*
reflection, *p. 480*
line of reflection, *p. 480*
rotation, *p. 486*
center of rotation, *p. 486*

angle of rotation, *p. 486*
similar figures, *p. 496*
dilation, *p. 508*
center of dilation, *p. 508*
scale factor, *p. 508*

Review Examples and Exercises

11.1 Congruent Figures *(pp. 466–471)*

Trapezoids *EFGH* and *QRST* are congruent.

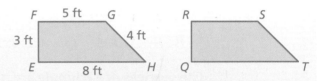

a. What is the length of side *QT*?

Side *QT* corresponds to side *EH*.

∴ So, the length of side *QT* is 8 feet.

b. Which angle of *QRST* corresponds to ∠*H*?

∴ ∠*T* corresponds to ∠*H*.

Exercises

Use the figures above.

1. What is the length of side *QR*?

2. What is the perimeter of *QRST*?

The figures are congruent. Name the corresponding angles and the corresponding sides.

3.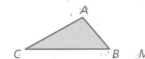

4.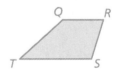

11.2 Translations *(pp. 472–477)*

Translate the red triangle 4 units left and 1 unit down. What are the coordinates of the image?

Move each vertex 4 units left and 1 unit down.

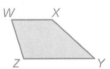

Connect the vertices. Label as *A′*, *B′*, and *C′*.

∴ The coordinates of the image are *A′*(−1, 4), *B′*(2, 2), and *C′*(0, 0).

Exercises

Tell whether the blue figure is a translation of the red figure.

5.

6.

7. The vertices of a quadrilateral are $W(1, 2)$, $X(1, 4)$, $Y(4, 4)$, and $Z(4, 2)$. Draw the figure and its image after a translation 3 units left and 2 units down.

8. The vertices of a triangle are $A(-1, -2)$, $B(-2, 2)$, and $C(-3, 0)$. Draw the figure and its image after a translation 5 units right and 1 unit up.

11.3 Reflections (pp. 478–483)

The vertices of a triangle are $A(-2, 1)$, $B(4, 1)$, and $C(4, 4)$. Draw the figure and its reflection in the x-axis. What are the coordinates of the image?

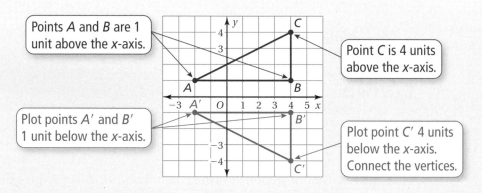

Points A and B are 1 unit above the x-axis.

Point C is 4 units above the x-axis.

Plot points A' and B' 1 unit below the x-axis.

Plot point C' 4 units below the x-axis. Connect the vertices.

The coordinates of the image are $A'(-2, -1)$, $B'(4, -1)$, and $C'(4, -4)$.

Exercises

Tell whether the blue figure is a reflection of the red figure.

9.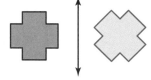

10.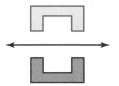

Draw the figure and its reflection in (a) the x-axis and (b) the y-axis.

11. $A(2, 0)$, $B(1, 5)$, $C(4, 3)$

12. $D(-5, -5)$, $E(-5, -1)$, $F(-2, -2)$, $G(-2, -5)$

13. The vertices of a rectangle are $E(-1, 1)$, $F(-1, 3)$, $G(-5, 3)$, and $H(-5, 1)$. Find the coordinates of the figure after reflecting in the x-axis, and then translating 3 units right.

Rotations *(pp. 484–491)*

The vertices of a triangle are *A*(1, 1), *B*(3, 2), and *C*(2, 4). Rotate the triangle 90° counterclockwise about the origin. What are the coordinates of the image?

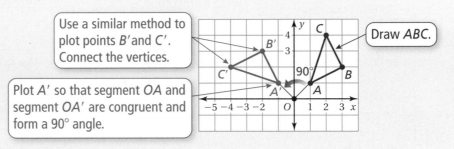

Use a similar method to plot points *B*′ and *C*′. Connect the vertices.

Draw *ABC*.

Plot *A*′ so that segment *OA* and segment *OA*′ are congruent and form a 90° angle.

∴ The coordinates of the image are *A*′(−1, 1), *B*′(−2, 3), and *C*′(−4, 2).

Exercises

Tell whether the blue figure is a rotation of the red figure about the origin. If so, give the angle and the direction of rotation.

14.

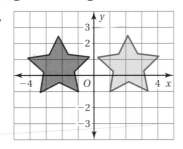

15.

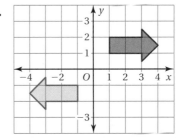

The vertices of a triangle are *A*(−4, 2), *B*(−2, 2), and *C*(−3, 4). Rotate the triangle about the origin as described. Find the coordinates of the image.

16. 180°

17. 270° clockwise

Similar Figures *(pp. 494–499)*

a. Is Rectangle A similar to Rectangle B?

Each figure is a rectangle. So, corresponding angles are congruent. Check to see if corresponding side lengths are proportional.

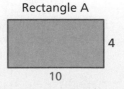

Rectangle A

Rectangle B

$$\frac{\text{Length of A}}{\text{Length of B}} = \frac{10}{5} = 2 \qquad \frac{\text{Width of A}}{\text{Width of B}} = \frac{4}{2} = 2 \qquad \text{Proportional}$$

∴ So, Rectangle A is similar to Rectangle B.

b. The two rectangles are similar. Find _x_.

Because the rectangles are similar, corresponding side lengths are proportional. So, write and solve a proportion to find _x_.

$$\frac{10}{24} = \frac{4}{x}$$ Write a proportion.

$10x = 96$ Cross Products Property

$x = 9.6$ Divide each side by 10.

∴ So, _x_ is 9.6 meters.

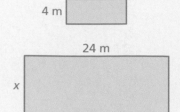

Exercises

Tell whether the two figures are similar. Explain your reasoning.

18.

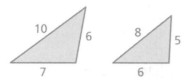

19.

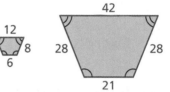

The figures are similar. Find _x_.

20.

21.

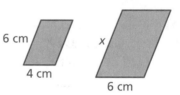

11.6 **Perimeters and Areas of Similar Figures** *(pp. 500–505)*

a. Find the ratio (red to blue) of the perimeters of the similar parallelograms.

$$\frac{\text{Perimeter of red parallelogram}}{\text{Perimeter of blue parallelogram}} = \frac{15}{9}$$

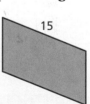

$$= \frac{5}{3}$$

∴ The ratio of the perimeters is $\frac{5}{3}$.

b. Find the ratio (red to blue) of the areas of the similar figures.

$$\frac{\text{Area of red figure}}{\text{Area of blue figure}} = \left(\frac{3}{4}\right)^2$$

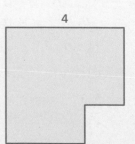

$$= \frac{9}{16}$$

∴ The ratio of the areas is $\frac{9}{16}$.

Exercises

The two figures are similar. Find the ratios (red to blue) of the perimeters and of the areas.

22.

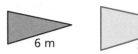

6 m   8 m

23.

16 m   28 m

24. PHOTOS Two photos are similar. The ratio of the corresponding side lengths is 3 : 4. What is the ratio of the areas?

11.7   Dilations   *(pp. 506–513)*

Draw the image of Triangle ABC after a dilation with a scale factor of 2. Identify the type of dilation.

Multiply each *x*- and *y*-coordinate by the scale factor 2.

Vertices of ABC	(2x, 2y)	Vertices of A′B′C′
$A(1, 1)$	$(2 \cdot 1, 2 \cdot 1)$	$A'(2, 2)$
$B(1, 2)$	$(2 \cdot 1, 2 \cdot 2)$	$B'(2, 4)$
$C(3, 2)$	$(2 \cdot 3, 2 \cdot 2)$	$C'(6, 4)$

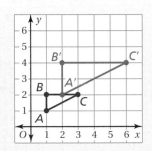

∴ The image is shown at the above right. The dilation is an *enlargement* because the scale factor is greater than 1.

Exercises

Tell whether the blue figure is a dilation of the red figure.

25.

26.

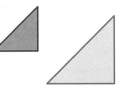

The vertices of a figure are given. Draw the figure and its image after a dilation with the given scale factor. Identify the type of dilation.

27. $P(-3, -2), Q(-3, 0), R(0, 0); k = 4$

28. $B(3, 3), C(3, 6), D(6, 6), E(6, 3); k = \dfrac{1}{3}$

29. The vertices of a rectangle are $Q(-6, 2), R(6, 2), S(6, -4),$ and $T(-6, -4)$. Dilate the rectangle with respect to the origin using a scale factor of $\dfrac{3}{2}$. Then translate it 5 units right and 1 unit down. What are the coordinates of the image?

Check It Out
Test Practice
BigIdeasMath com

Triangles *ABC* and *DEF* are congruent.

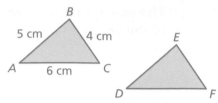

1. Which angle of *DEF* corresponds to $\angle C$?

2. What is the perimeter of *DEF*?

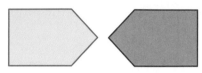

Tell whether the blue figure is a *translation*, *reflection*, *rotation*, or *dilation* of the red figure.

3.

4.

5.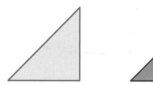

6.

7. The vertices of a triangle are $A(2, 5)$, $B(1, 2)$, and $C(3, 1)$. Reflect the triangle in the *x*-axis, and then rotate the triangle 90° counterclockwise about the origin. What are the coordinates of the image?

8. The vertices of a triangle are $A(2, 4)$, $B(2, 1)$, and $C(5, 1)$. Dilate the triangle with respect to the origin using a scale factor of 2. Then translate the triangle 2 units left and 1 unit up. What are the coordinates of the image?

9. Tell whether the parallelograms are similar. Explain your reasoning.

The two figures are similar. Find the ratios (red to blue) of the perimeters and of the areas.

10.

11.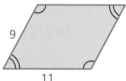

12. **SCREENS** A wide-screen television measures 36 inches by 54 inches. A movie theater screen measures 42 feet by 63 feet. Are the screens similar? Explain.

13. **CURTAINS** You want to use the rectangular piece of fabric shown to make a set of curtains for your window. Name the types of congruent shapes you can make with one straight cut. Draw an example of each type.

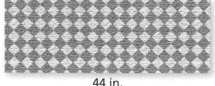

16 in.

44 in.

1. A clockwise rotation of 90° is equivalent to a counterclockwise rotation of how many degrees? *(8.G.2)*

2. The formula $K = C + 273.15$ converts temperatures from Celsius C to Kelvin K. Which of the following formulas is *not* correct? *(8.EE.7a)*

   A. $K - C = 273.15$

   B. $C = K - 273.15$

   C. $C - K = -273.15$

   D. $C = K + 273.15$

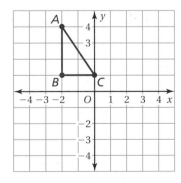

What type of transformation is shown?
Ⓐ rotation Ⓑ translation
Ⓒ dilation Ⓓ reflection

Lookin' good!

"After answering the easy questions, relax and try the harder ones. For this, the image is flipped. So, it's D."

3. Joe wants to solve the equation $-3(x + 2) = 12x$. What should he do first? *(8.EE.7a)*

   F. Subtract 2 from each side.

   G. Add 3 to each side.

   H. Multiply each side by -3.

   I. Divide each side by -3.

4. Which transformation *turns* a figure? *(8.G.1)*

   A. translation

   B. reflection

   C. rotation

   D. dilation

5. A triangle is graphed in the coordinate plane below.

   Translate the triangle 3 units right and 2 units down. What are the coordinates of the image? *(8.G.3)*

   F. $A'(1, 4)$, $B'(1, 1)$, $C'(3, 1)$

   G. $A'(1, 2)$, $B'(1, -1)$, $C'(3, -1)$

   H. $A'(-2, 2)$, $B'(-2, -1)$, $C'(0, -1)$

   I. $A'(0, 1)$, $B'(0, -2)$, $C'(2, -2)$

6. Dale solved the equation in the box shown. What should Dale do to correct the error that he made? *(8.EE.7b)*

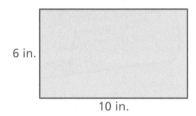

A. Add $\dfrac{2}{5}$ to each side to get $-\dfrac{x}{3} = -\dfrac{1}{15}$.

B. Multiply each side by -3 to get $x + \dfrac{2}{5} = \dfrac{7}{5}$.

C. Multiply each side by -3 to get $x = 2\dfrac{3}{5}$.

D. Subtract $\dfrac{2}{5}$ from each side to get $-\dfrac{x}{3} = -\dfrac{5}{10}$.

7. Jenny dilates the rectangle below using a scale factor of $\dfrac{1}{2}$.

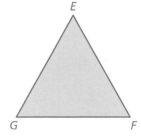

6 in.

10 in.

What is the area of the dilated rectangle in square inches? *(8.G.4)*

8. The vertices of a rectangle are $A(-4, 2)$, $B(3, 2)$, $C(3, -5)$, and $D(-4, -5)$. If the rectangle is dilated by a scale factor of 3, what will be the coordinates of vertex C'? *(8.G.3)*

F. $(9, -15)$ 　　　　　　　　　　H. $(-12, -15)$

G. $(-12, 6)$ 　　　　　　　　　　I. $(9, 6)$

9. In the figures, Triangle *EFG* is a dilation of Triangle *HIJ*.

Which proportion is *not* necessarily correct for Triangle *EFG* and Triangle *HIJ*? *(8.G.4)*

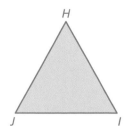

A. $\dfrac{EF}{FG} = \dfrac{HI}{IJ}$ 　　　　　　　　C. $\dfrac{GE}{EF} = \dfrac{JH}{HI}$

B. $\dfrac{EG}{HI} = \dfrac{FG}{IJ}$ 　　　　　　　　D. $\dfrac{EF}{HI} = \dfrac{GE}{JH}$

10. In the figures below, Rectangle *EFGH* is a dilation of Rectangle *IJKL*.

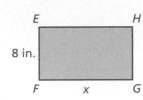

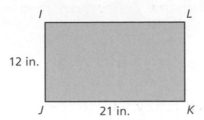

What is *x*?   *(8.G.4)*

F. 14 in.   **H.** 16 in.

G. 15 in.   **I.** 17 in.

11. Several transformations are used to create the pattern.   *(8.G.2, 8.G.4)*

Part A Describe the transformation of Triangle *GLM* to Triangle *DGH*.

Part B Describe the transformation of Triangle *ALQ* to Triangle *GLM*.

Part C Triangle *DFN* is a dilation of Triangle *GHM*. Find the scale factor.

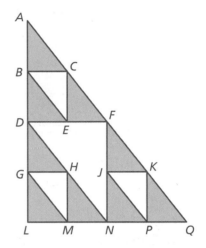

12. A rectangle is graphed in the coordinate plane below.

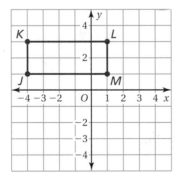

Rotate the triangle 180° about the origin. What are the coordinates of the image?   *(8.G.3)*

A. $J'(4, -1), K'(4, -3), L'(-1, -3), M'(-1, -1)$

B. $J'(-4, -1), K'(-4, -3), L'(1, -3), M'(1, -1)$

C. $J'(1, 4), K'(3, 4), L'(3, -1), M'(1, -1)$

D. $J'(-4, 1), K'(-4, 3), L'(1, 3), M'(1, 1)$

12 Angles and Triangles

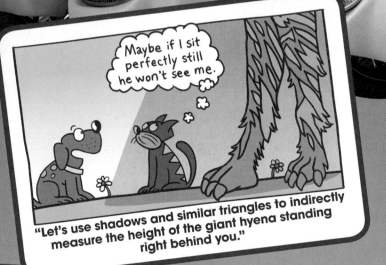

"Let's use shadows and similar triangles to indirectly measure the height of the giant hyena standing right behind you."

"Start with any triangle."

"Tear off the angles. You can always rearrange the angles so that they form a straight line."

"What does that prove?"

What You Learned Before

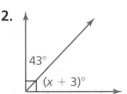

Complementary *Supplementary*

"30 degrees said to 60 degrees, 'It's neat, you make me complete.'"

"I just remember that C comes before S and 90 comes before 180. That makes it easy."

● Adjacent and Vertical Angles
(7.G.5)

Example 1 Tell whether the angles are *adjacent* or *vertical*. Then find the value of *x*.

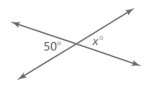

The angles are vertical angles. Because vertical angles are congruent, the angles have the same measure.

∴ So, the value of *x* is 50.

Try It Yourself

Tell whether the angles are *adjacent* or *vertical*. Then find the value of *x*.

1.
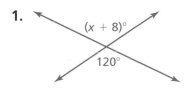

$(x + 8)°$

$120°$

2.

$43°$

$(x + 3)°$

● Complementary and Supplementary Angles (7.G.5)

Example 2 Tell whether the angles are *complementary* or *supplementary*. Then find the value of *x*.

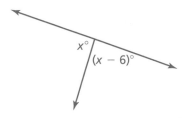

$x°$

$(x - 6)°$

The two angles make up a straight angle. So, the angles are supplementary angles, and the sum of their measures is 180°.

$x + (x - 6) = 180$	Write equation.
$2x - 6 = 180$	Combine like terms.
$2x = 186$	Add 6 to each side.
$x = 93$	Divide each side by 2.

Try It Yourself

Tell whether the angles are *complementary* or *supplementary*. Then find the value of *x*.

3.
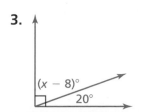

$(x - 8)°$

$20°$

4.
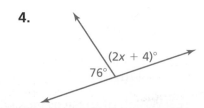

$(2x + 4)°$

$76°$

Essential Question How can you describe angles formed by parallel lines and transversals?

The Meaning of a Word ● Transverse

When an object is **transverse**, it is lying or extending across something.

---

1 **ACTIVITY: A Property of Parallel Lines**

Work with a partner.

- Discuss what it means for two lines to be parallel. Decide on a strategy for drawing two parallel lines. Then draw the two parallel lines.

- Draw a third line that intersects the two parallel lines. This line is called a **transversal**.

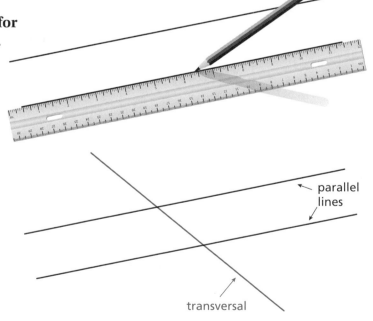

parallel lines

transversal

COMMON CORE

Geometry

In this lesson, you will
- identify the angles formed when parallel lines are cut by a transversal.
- find the measures of angles formed when parallel lines are cut by a transversal.

Learning Standard
8.G.5

a. How many angles are formed by the parallel lines and the transversal? Label the angles.

b. Which of these angles have equal measures? Explain your reasoning.

2 ACTIVITY: Creating Parallel Lines

Work with a partner.

a. If you were building the house in the photograph, how could you make sure that the studs are parallel to each other?

b. Identify sets of parallel lines and transversals in the photograph.

Studs

3 ACTIVITY: Using Technology

Work with a partner. Use geometry software to draw two parallel lines intersected by a transversal.

a. Find all the angle measures.

b. Adjust the figure by moving the parallel lines or the transversal to a different position. Describe how the angle measures and relationships change.

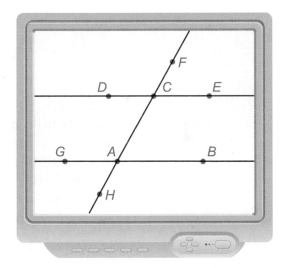

What Is Your Answer?

4. **IN YOUR OWN WORDS** How can you describe angles formed by parallel lines and transversals? Give an example.

5. Use geometry software to draw a transversal that is perpendicular to two parallel lines. What do you notice about the angles formed by the parallel lines and the transversal?

Practice

Use what you learned about parallel lines and transversals to complete Exercises 3–6 on page 531.

Check It Out
Lesson Tutorials
BigIdeasMath.com

Key Vocabulary 🔊
transversal, *p. 528*
interior angles,
  p. 529
exterior angles,
  p. 529

Lines in the same plane that do not intersect are called *parallel lines*. Lines that intersect at right angles are called *perpendicular lines*.

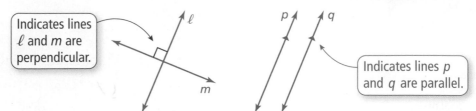

Indicates lines *ℓ* and *m* are perpendicular.

Indicates lines *p* and *q* are parallel.

A line that intersects two or more lines is called a **transversal**. When parallel lines are cut by a transversal, several pairs of congruent angles are formed.

🔑 Key Idea

Study Tip

Corresponding angles lie on the same side of the transversal in corresponding positions.

Corresponding Angles

When a transversal intersects parallel lines, corresponding angles are congruent.

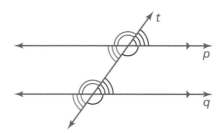

Corresponding angles

EXAMPLE 1 Finding Angle Measures

Use the figure to find the measures of (a) ∠1 and (b) ∠2.

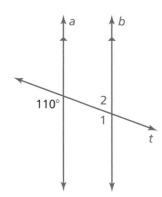

a. ∠1 and the 110° angle are corresponding angles. They are congruent.

  ∴ So, the measure of ∠1 is 110°.

b. ∠1 and ∠2 are supplementary.

$$\angle 1 + \angle 2 = 180°$$ Definition of supplementary angles
$$110° + \angle 2 = 180°$$ Substitute 110° for ∠1.
$$\angle 2 = 70°$$ Subtract 110° from each side.

  ∴ So, the measure of ∠2 is 70°.

On Your Own

Now You're Ready
Exercises 7–9

Use the figure to find the measure of the angle. Explain your reasoning.

  1. ∠1   **2.** ∠2

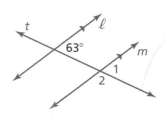

EXAMPLE 2 Using Corresponding Angles

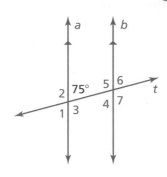

Use the figure to find the measures of the numbered angles.

∠1: ∠1 and the 75° angle are vertical angles. They are congruent.

∴ So, the measure of ∠1 is 75°.

∠2 and ∠3: The 75° angle is supplementary to both ∠2 and ∠3.

$$75° + ∠2 = 180°$$ Definition of supplementary angles

$$∠2 = 105°$$ Subtract 75° from each side.

∴ So, the measures of ∠2 and ∠3 are 105°.

∠4, ∠5, ∠6, and ∠7: Using corresponding angles, the measures of ∠4 and ∠6 are 75°, and the measures of ∠5 and ∠7 are 105°.

On Your Own

Now You're Ready
Exercises 15–17

3. Use the figure to find the measures of the numbered angles.

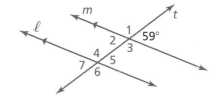

When two parallel lines are cut by a transversal, four **interior angles** are formed on the inside of the parallel lines and four **exterior angles** are formed on the outside of the parallel lines.

∠3, ∠4, ∠5, and ∠6 are interior angles.
∠1, ∠2, ∠7, and ∠8 are exterior angles.

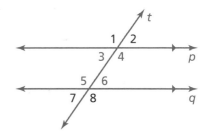

EXAMPLE 3 Using Corresponding Angles

A store owner uses pieces of tape to paint a window advertisement. The letters are slanted at an 80° angle. What is the measure of ∠1?

(A) 80° (B) 100° (C) 110° (D) 120°

Because all the letters are slanted at an 80° angle, the dashed lines are parallel. The piece of tape is the transversal.

Using corresponding angles, the 80° angle is congruent to the angle that is supplementary to ∠1, as shown.

∴ The measure of ∠1 is 180° − 80° = 100°. The correct answer is (B).

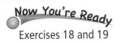 On Your Own

Now You're Ready
Exercises 18 and 19

4. WHAT IF? In Example 3, the letters are slanted at a 65° angle. What is the measure of ∠1?

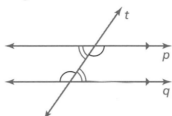 Key Idea

Study Tip

Alternate interior angles and alternate exterior angles lie on opposite sides of the transversal.

Alternate Interior Angles and Alternate Exterior Angles

When a transversal intersects parallel lines, alternate interior angles are congruent and alternate exterior angles are congruent.

Alternate interior angles   **Alternate exterior angles**

EXAMPLE 4   Identifying Alternate Interior and Alternate Exterior Angles

The photo shows a portion of an airport. Describe the relationship between each pair of angles.

a. ∠3 and ∠6

∠3 and ∠6 are alternate exterior angles.

So, ∠3 is congruent to ∠6.

b. ∠2 and ∠7

∠2 and ∠7 are alternate interior angles.

So, ∠2 is congruent to ∠7.

On Your Own

Now You're Ready
Exercises 20 and 21

In Example 4, the measure of ∠4 is 84°. Find the measure of the angle. Explain your reasoning.

5. ∠3   **6.** ∠5   **7.** ∠6

 Vocabulary and Concept Check

1. **VOCABULARY** Draw two parallel lines and a transversal. Label a pair of corresponding angles.

2. **WHICH ONE DOESN'T BELONG?** Which statement does *not* belong with the other three? Explain your reasoning. Refer to the figure for Exercises 3–6.

The measure of ∠2	The measure of ∠5
The measure of ∠6	The measure of ∠8

 Practice and Problem Solving

In Exercises 3–6, use the figure.

3. Identify the parallel lines.

4. Identify the transversal.

5. How many angles are formed by the transversal?

6. Which of the angles are congruent?

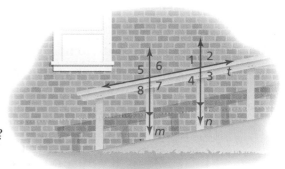

Use the figure to find the measures of the numbered angles.

❶ 7.

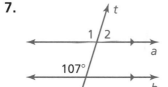

8.

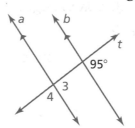

9.

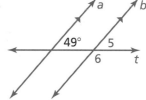

10. **ERROR ANALYSIS** Describe and correct the error in describing the relationship between the angles.

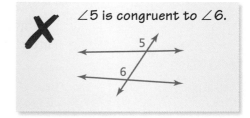

11. **PARKING** The painted lines that separate parking spaces are parallel. The measure of ∠1 is 60°. What is the measure of ∠2? Explain.

12. **OPEN-ENDED** Describe two real-life situations that use parallel lines.

13. **PROJECT** Trace line p and line t on a piece of paper. Label $\angle 1$. Move the paper so that $\angle 1$ aligns with $\angle 8$. Describe the transformations that you used to show that $\angle 1$ is congruent to $\angle 8$.

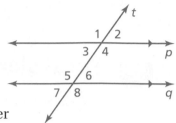

14. **REASONING** Two horizontal lines are cut by a transversal. What is the least number of angle measures you need to know in order to find the measure of every angle? Explain your reasoning.

Use the figure to find the measures of the numbered angles. Explain your reasoning.

② 15.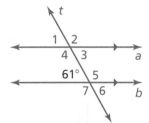

16.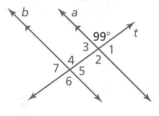

17.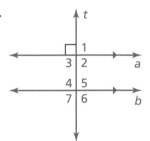

Complete the statement. Explain your reasoning.

③ 18. If the measure of $\angle 1 = 124°$, then the measure of $\angle 4 = $ ▢ .

19. If the measure of $\angle 2 = 48°$, then the measure of $\angle 3 = $ ▢ .

④ 20. If the measure of $\angle 4 = 55°$, then the measure of $\angle 2 = $ ▢ .

21. If the measure of $\angle 6 = 120°$, then the measure of $\angle 8 = $ ▢ .

22. If the measure of $\angle 7 = 50.5°$, then the measure of $\angle 6 = $ ▢ .

23. If the measure of $\angle 3 = 118.7°$, then the measure of $\angle 2 = $ ▢ .

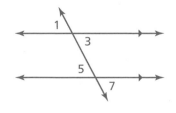

24. **RAINBOW** A rainbow forms when sunlight reflects off raindrops at different angles. For blue light, the measure of $\angle 2$ is 40°. What is the measure of $\angle 1$?

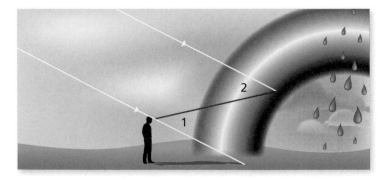

25. **REASONING** When a transversal is perpendicular to two parallel lines, all the angles formed measure 90°. Explain why.

26. **LOGIC** Describe two ways you can show that $\angle 1$ is congruent to $\angle 7$.

CRITICAL THINKING Find the value of *x*.

27.

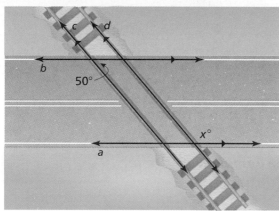

28.

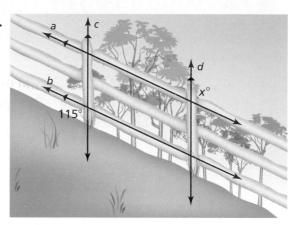

29. OPTICAL ILLUSION Refer to the figure.

   a. Do the horizontal lines appear to be parallel? Explain.

   b. Draw your own optical illusion using parallel lines.

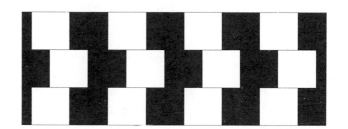

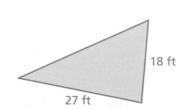

30. ✦**Geometry**✦ The figure shows the angles used to make a double bank shot in an air hockey game.

   a. Find the value of *x*.

   b. Can you still get the red puck in the goal when *x* is increased by a little? by a lot? Explain.

 Fair Game Review What you learned in previous grades & lessons

Evaluate the expression. *(Section 1.4)*

31. $4 + 3^2$ **32.** $5(2)^2 - 6$ **33.** $11 + (-7)^2 - 9$ **34.** $8 \div 2^2 + 1$

35. MULTIPLE CHOICE The triangles are similar. What length does *x* represent? *(Section 11.5)*

   Ⓐ 2 ft   Ⓑ 12 ft

   Ⓒ 15 ft   Ⓓ 27 ft

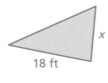

12.2 Angles of Triangles

Essential Question How can you describe the relationships among the angles of a triangle?

1 ACTIVITY: Exploring the Interior Angles of a Triangle

Work with a partner.

a. Draw a triangle. Label the interior angles *A*, *B*, and *C*.

b. Carefully cut out the triangle. Tear off the three corners of the triangle.

c. Arrange angles *A* and *B* so that they share a vertex and are adjacent.

d. How can you place the third angle to determine the sum of the measures of the interior angles? What is the sum?

e. Compare your results with those of others in your class.

f. **STRUCTURE** How does your result in part (d) compare to your conclusion in Lesson 7.3, Activity Question 7?

2 ACTIVITY: Exploring the Interior Angles of a Triangle

Work with a partner.

a. Describe the figure.

b. **LOGIC** Use what you know about parallel lines and transversals to justify your result in part (d) of Activity 1.

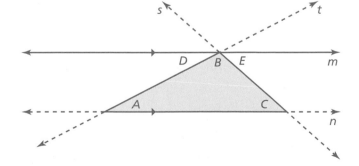

COMMON
CORE

Geometry
In this lesson, you will
- understand that the sum of the interior angle measures of a triangle is 180°.
- find the measures of interior and exterior angles of triangles.

Learning Standard
8.G.5

3 ACTIVITY: Exploring an Exterior Angle of a Triangle

Work with a partner.

a. Draw a triangle. Label the interior angles *A*, *B*, and *C*.

b. Carefully cut out the triangle.

c. Place the triangle on a piece of paper and extend one side to form *exterior angle D*, as shown.

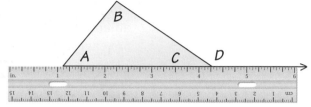

d. Tear off the corners that are not adjacent to the exterior angle. Arrange them to fill the exterior angle, as shown. What does this tell you about the measure of exterior angle *D*?

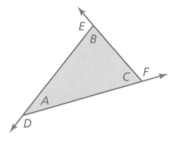

4 ACTIVITY: Measuring the Exterior Angles of a Triangle

Work with a partner.

a. Draw a triangle and label the interior and exterior angles, as shown.

b. Use a protractor to measure all six angles. Copy and complete the table to organize your results. What does the table tell you about the measure of an exterior angle of a triangle?

Exterior Angle	$D =$ °	$E =$ °	$F =$ °
Interior Angle	$B =$ °	$A =$ °	$A =$ °
Interior Angle	$C =$ °	$C =$ °	$B =$ °

What Is Your Answer?

5. **REPEATED REASONING** Draw three triangles that have different shapes. Repeat parts (b)–(d) from Activity 1 for each triangle. Do you get the same results? Explain.

6. **IN YOUR OWN WORDS** How can you describe the relationships among angles of a triangle?

Use what you learned about angles of a triangle to complete Exercises 4–6 on page 538.

12.2 Lesson

Key Vocabulary 🔊
interior angles of a
  polygon, *p. 536*
exterior angles of a
  polygon, *p. 536*

The angles inside a polygon are called **interior angles**. When the sides of a polygon are extended, other angles are formed. The angles outside the polygon that are adjacent to the interior angles are called **exterior angles**.

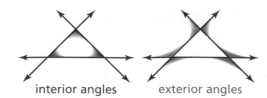

interior angles   exterior angles

🔑 Key Idea

Interior Angle Measures of a Triangle

Words   The sum of the interior angle measures of a triangle is 180°.

Algebra $x + y + z = 180$

EXAMPLE ① Using Interior Angle Measures

Find the value of x.

a.

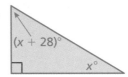

$$x + 32 + 48 = 180$$
$$x + 80 = 180$$
$$x = 100$$

b.

$$x + (x + 28) + 90 = 180$$
$$2x + 118 = 180$$
$$2x = 62$$
$$x = 31$$

⬤ On Your Own

Now You're Ready
Exercises 4–9

Find the value of x.

1.

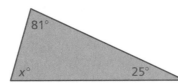

2.

🔑 Key Idea

Exterior Angle Measures of a Triangle

Words   The measure of an exterior angle of a triangle is equal to the sum of the measures of the two nonadjacent interior angles.

Algebra $z = x + y$

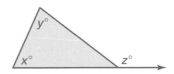

🔊 Multi-Language Glossary at BigIdeasMath ✓com

EXAMPLE 2 Finding Exterior Angle Measures

Study Tip

Each vertex has a pair of congruent exterior angles. However, it is common to show only one exterior angle at each vertex.

Find the measure of the exterior angle.

a.

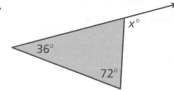

$x = 36 + 72$

$x = 108$

⋮ So, the measure of the exterior angle is 108°.

b.

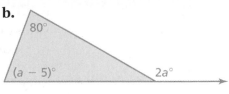

$2a = (a - 5) + 80$

$2a = a + 75$

$a = 75$

⋮ So, the measure of the exterior angle is $2(75)° = 150°$.

EXAMPLE 3 Real-Life Application

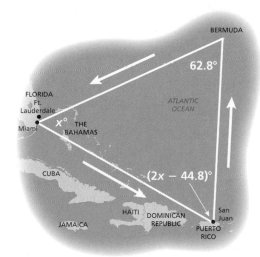

An airplane leaves from Miami and travels around the Bermuda Triangle. What is the value of x?

Ⓐ 26.8 Ⓑ 27.2 Ⓒ 54 Ⓓ 64

Use what you know about the interior angle measures of a triangle to write an equation.

$x + (2x - 44.8) + 62.8 = 180$	Write equation.
$3x + 18 = 180$	Combine like terms.
$3x = 162$	Subtract 18 from each side.
$x = 54$	Divide each side by 3.

⋮ The value of x is 54. The correct answer is Ⓒ.

On Your Own

Now You're Ready
Exercises 12–14

Find the measure of the exterior angle.

3.

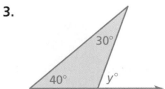

4.

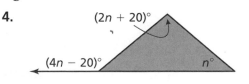

5. In Example 3, the airplane leaves from Fort Lauderdale. The interior angle measure at Bermuda is 63.9°. The interior angle measure at San Juan is $(x + 7.5)°$. Find the value of x.

 Vocabulary and Concept Check

1. **VOCABULARY** You know the measures of two interior angles of a triangle. How can you find the measure of the third interior angle?

2. **VOCABULARY** How many exterior angles does a triangle have at each vertex? Explain.

3. **NUMBER SENSE** List the measures of the exterior angles for the triangle shown at the right.

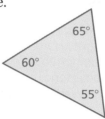

 Practice and Problem Solving

Find the measures of the interior angles.

1. 4.

5.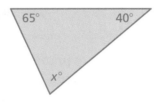

6.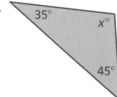

7.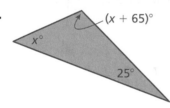

8.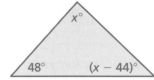

9.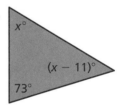

10. **BILLIARD RACK** Find the value of x in the billiard rack.

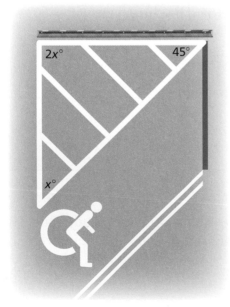

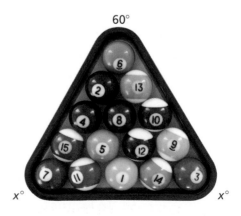

11. **NO PARKING** The triangle with lines through it designates a no parking zone. What is the value of x?

Find the measure of the exterior angle.

② 12.

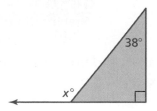

13.

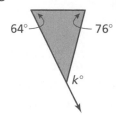

14.

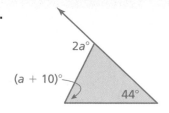

15. ERROR ANALYSIS Describe and correct the error in finding the measure of the exterior angle.

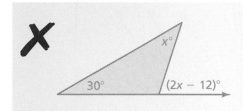

$(2x − 12) + x + 30 = 180$

$3x + 18 = 180$

$x = 54$

The exterior angle is $(2(54) − 12)° = 96°$.

16. RATIO The ratio of the interior angle measures of a triangle is $2 : 3 : 5$. What are the angle measures?

17. CONSTRUCTION The support for a window air-conditioning unit forms a triangle and an exterior angle. What is the measure of the exterior angle?

18. REASONING A triangle has an exterior angle with a measure of $120°$. Can you determine the measures of the interior angles? Explain.

Determine whether the statement is *always*, *sometimes*, or *never* true. Explain your reasoning.

19. Given three angle measures, you can construct a triangle.

20. The acute interior angles of a right triangle are complementary.

21. A triangle has more than one vertex with an acute exterior angle.

22. **Precision** Using the figure at the right, show that $z = x + y$. (*Hint:* Find two equations involving w.)

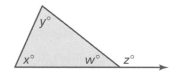

Fair Game Review What you learned in previous grades & lessons

Solve the equation. Check your solution. *(Topic 1)*

23. $-4x + 3 = 19$

24. $2(y − 1) + 6y = −10$

25. $5 + 0.5(6n + 14) = 3$

26. MULTIPLE CHOICE Which transformation moves every point of a figure the same distance and in the same direction? *(Section 11.2)*

    Ⓐ translation   **Ⓑ** reflection   **Ⓒ** rotation   **Ⓓ** dilation

You can use an **example and non-example chart** to list examples and non-examples of a vocabulary word or item. Here is an example and non-example chart for transversals.

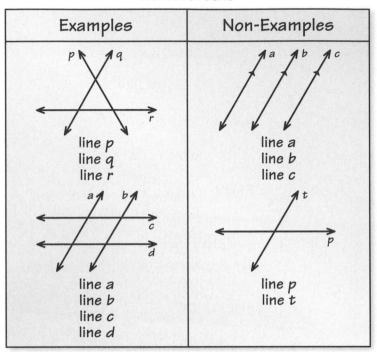

On Your Own

Make example and non-example charts to help you study these topics.

1. interior angles formed by parallel lines and a transversal

2. exterior angles formed by parallel lines and a transversal

After you complete this chapter, make example and non-example charts for the following topics.

3. interior angles of a polygon

4. exterior angles of a polygon

5. regular polygons

6. similar triangles

"What do you think of my example & non-example chart for popular cat toys?"

12.1–12.2 Quiz

Use the figure to find the measure of the angle.
Explain your reasoning. *(Section 12.1)*

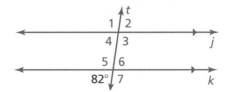

1. $\angle 2$
2. $\angle 6$
3. $\angle 4$
4. $\angle 1$

Complete the statement. Explain your reasoning. *(Section 12.1)*

5. If the measure of $\angle 1 = 123°$, then the measure of $\angle 7 =$ ⬚.

6. If the measure of $\angle 2 = 58°$, then the measure of $\angle 5 =$ ⬚.

7. If the measure of $\angle 5 = 119°$, then the measure of $\angle 3 =$ ⬚.

8. If the measure of $\angle 4 = 60°$, then the measure of $\angle 6 =$ ⬚.

Find the measures of the interior angles. *(Section 12.2)*

9.

10.

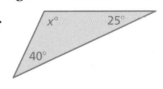

11.

Find the measure of the exterior angle. *(Section 12.2)*

12.

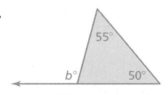

13.

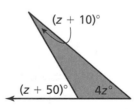

14. **PARK** In a park, a bike path and a horse riding path are parallel. In one part of the park, a hiking trail intersects the two paths. Find the measures of $\angle 1$ and $\angle 2$. Explain your reasoning. *(Section 12.1)*

15. **LADDER** A ladder leaning against a wall forms a triangle and exterior angles with the wall and the ground. What are the measures of the exterior angles? Justify your answer. *(Section 12.2)*

12.3 Angles of Polygons

Essential Question How can you find the sum of the interior angle measures and the sum of the exterior angle measures of a polygon?

1 ACTIVITY: Exploring the Interior Angles of a Polygon

Work with a partner. In parts (a)–(e), identify each polygon and the number of sides *n*. Then find the sum of the interior angle measures of the polygon.

a. Polygon: [ ]   Number of sides: $n =$ [ ]

Draw a line segment on the figure that divides it into two triangles. Is there more than one way to do this? Explain.

What is the sum of the interior angle measures of each triangle?

What is the sum of the interior angle measures of the figure?

b.

c.

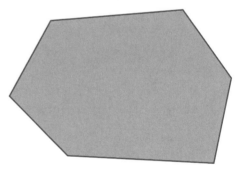

d.

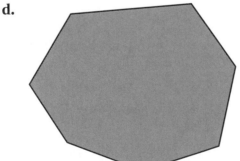

e.

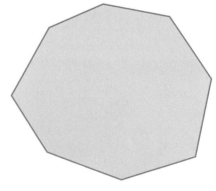

COMMON CORE

Geometry

In this lesson, you will
- find the sum of the interior angle measures of polygons.
- understand that the sum of the exterior angle measures of a polygon is 360°.
- find the measures of interior and exterior angles of polygons.

Applying Standard
8.G.5

f. **REPEATED REASONING** Use your results to complete the table. Then find the sum of the interior angle measures of a polygon with 12 sides.

Number of Sides, *n*	3	4	5	6	7	8
Number of Triangles						
Angle Sum, *S*						

A polygon is **convex** when every line segment connecting any two vertices lies entirely inside the polygon. A polygon is **concave** when at least one line segment connecting any two vertices lies outside the polygon.

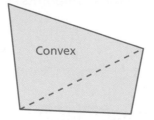

Convex

Concave

2 ACTIVITY: Exploring the Exterior Angles of a Polygon

Work with a partner.

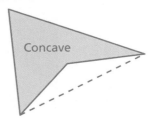

Math Practice 3

Analyze Conjectures

Do your observations about the sum of the exterior angles make sense? Do you think they would hold true for any convex polygon? Explain.

a. Draw a convex pentagon. Extend the sides to form the exterior angles. Label one exterior angle at each vertex *A*, *B*, *C*, *D*, and *E*, as shown.

b. Cut out the exterior angles. How can you join the vertices to determine the sum of the angle measures? What do you notice?

c. **REPEATED REASONING** Repeat the procedure in parts (a) and (b) for each figure below.

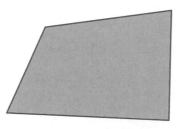 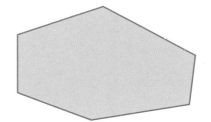

What can you conclude about the sum of the measures of the exterior angles of a convex polygon? Explain.

What Is Your Answer?

3. **STRUCTURE** Use your results from Activity 1 to write an expression that represents the sum of the interior angle measures of a polygon.

4. **IN YOUR OWN WORDS** How can you find the sum of the interior angle measures and the sum of the exterior angle measures of a polygon?

Practice

Use what you learned about angles of polygons to complete Exercises 4–6 on page 547.

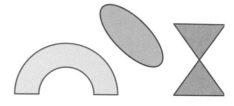

Key Vocabulary
convex polygon,
    p. 543
concave polygon,
    p. 543
regular polygon,
    p. 545

A *polygon* is a closed plane figure made up of three or more line segments that intersect only at their endpoints.

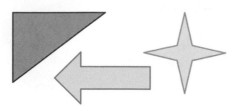

Polygons   **Not polygons**

Key Idea

Interior Angle Measures of a Polygon

The sum S of the interior angle measures of a polygon with n sides is

$$S = (n - 2) \cdot 180°.$$

EXAMPLE 1 **Finding the Sum of Interior Angle Measures**

Find the sum of the interior angle measures of the school crossing sign.

The sign is in the shape of a pentagon. It has 5 sides.

$S = (n - 2) \cdot 180°$	Write the formula.
$= (5 - 2) \cdot 180°$	Substitute 5 for n.
$= 3 \cdot 180°$	Subtract.
$= 540°$	Multiply.

Reading

For polygons whose names you have not learned, you can use the phrase "*n*-gon," where *n* is the number of sides. For example, a 15-gon is a polygon with 15 sides.

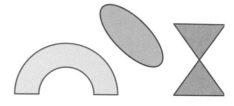

∴ The sum of the interior angle measures is 540°.

On Your Own

Now You're Ready
Exercises 7–9

Find the sum of the interior angle measures of the green polygon.

1.

2.

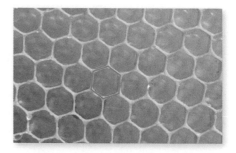

EXAMPLE ② **Finding an Interior Angle Measure of a Polygon**

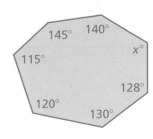

145° 140°

115°

x°

128°

120°

130°

Find the value of x.

Step 1: The polygon has 7 sides. Find the sum of the interior angle measures.

$$S = (n - 2) \cdot 180°$$ Write the formula.

$$= (7 - 2) \cdot 180°$$ Substitute 7 for n.

$$= 900°$$ Simplify. The sum of the interior angle measures is 900°.

Step 2: Write and solve an equation.

$$140 + 145 + 115 + 120 + 130 + 128 + x = 900$$

$$778 + x = 900$$

$$x = 122$$

∴ The value of x is 122.

● **On Your Own**

Now You're Ready
Exercises 12–14

Find the value of x.

3.

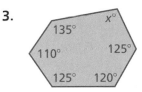

135°

x°

110°

125°

125°   120°

4.

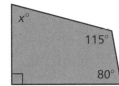

x°

115°

80°

5.

145° 145°

2x°

2x°

110°

In a **regular polygon**, all the sides are congruent, and all the interior angles are congruent.

EXAMPLE ③ **Real-Life Application**

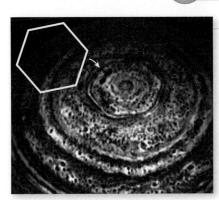

The hexagon is about 15,000 miles across. Approximately four Earths could fit inside it.

A cloud system discovered on Saturn is in the approximate shape of a regular hexagon. Find the measure of each interior angle of the hexagon.

Step 1: A hexagon has 6 sides. Find the sum of the interior angle measures.

$$S = (n - 2) \cdot 180°$$ Write the formula.

$$= (6 - 2) \cdot 180°$$ Substitute 6 for n.

$$= 720°$$ Simplify. The sum of the interior angle measures is 720°.

Step 2: Divide the sum by the number of interior angles, 6.

$$720° \div 6 = 120°$$

∴ The measure of each interior angle is 120°.

Now You're Ready
Exercises 16–18

On Your Own

Find the measure of each interior angle of the regular polygon.

6. octagon   **7.** decagon   **8.** 18-gon

 **Key Idea**

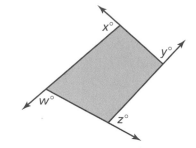

Exterior Angle Measures of a Polygon

Words   The sum of the measures of the exterior angles of a convex polygon is 360°.

Algebra $w + x + y + z = 360$

EXAMPLE 4   **Finding Exterior Angle Measures**

Find the measures of the exterior angles of each polygon.

a.

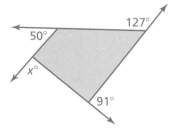

Write and solve an equation for x.

$$x + 50 + 127 + 91 = 360$$
$$x + 268 = 360$$
$$x = 92$$

⋮⋮   So, the measures of the exterior angles are 92°, 50°, 127°, and 91°.

b.

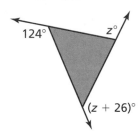

Write and solve an equation for z.

$$124 + z + (z + 26) = 360$$
$$2z + 150 = 360$$
$$z = 105$$

⋮⋮   So, the measures of the exterior angles are 124°, 105°, and $(105 + 26)° = 131°$.

On Your Own

Now You're Ready
Exercises 22–28

9. Find the measures of the exterior angles of the polygon.

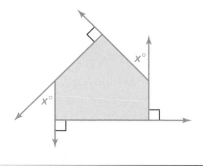

12.3 Exercises

Vocabulary and Concept Check

1. **VOCABULARY** Draw a regular polygon that has three sides.

2. **WHICH ONE DOESN'T BELONG?** Which figure does *not* belong with the other three? Explain your reasoning.

3. **DIFFERENT WORDS, SAME QUESTION** Which is different? Find "both" answers.

What is the measure of an interior angle of a regular pentagon?

What is the sum of the interior angle measures of a convex pentagon?

What is the sum of the interior angle measures of a regular pentagon?

What is the sum of the interior angle measures of a concave pentagon?

Practice and Problem Solving

Use triangles to find the sum of the interior angle measures of the polygon.

4.

5.

6.

Find the sum of the interior angle measures of the polygon.

① 7.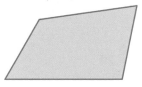

8.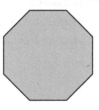

9.

10. **ERROR ANALYSIS** Describe and correct the error in finding the sum of the interior angle measures of a 13-gon.

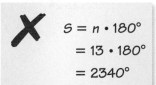
$$S = n \cdot 180°$$
$$= 13 \cdot 180°$$
$$= 2340°$$

11. **NUMBER SENSE** Can a pentagon have interior angles that measure 120°, 105°, 65°, 150°, and 95°? Explain.

Find the measures of the interior angles.

② 12.
137°
x°
25° 155°

13.
x° x°
x°   x°

14.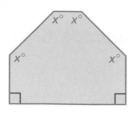
45°   135°
3x°
45°   135°

15. **REASONING** The sum of the interior angle measures in a regular polygon is 1260°. What is the measure of one of the interior angles of the polygon?

Find the measure of each interior angle of the regular polygon.

③ 16.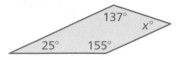
YIELD

17.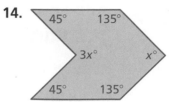

18.

19. **ERROR ANALYSIS** Describe and correct the error in finding the measure of each interior angle of a regular 20-gon.

$S = (n - 2) \cdot 180°$
$= (20 - 2) \cdot 180°$
$= 18 \cdot 180°$
$= 3240°$
$3240° \div 18 = 180$

The measure of each interior angle is 180°.

20. **FIRE HYDRANT** A fire hydrant bolt is in the shape of a regular pentagon.

a. What is the measure of each interior angle?

b. Why are fire hydrants made this way?

21. **PROBLEM SOLVING** The interior angles of a regular polygon each measure 165°. How many sides does the polygon have?

Find the measures of the exterior angles of the polygon.

④ 22.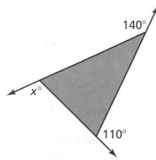
140°
x°
110°

23.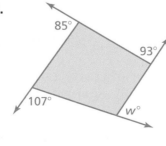
85°
93°
107°
w°

24.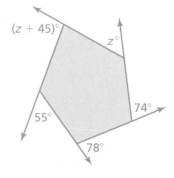
(z + 45)°
z°
55°   74°
78°

25. **REASONING** What is the measure of an exterior angle of a regular hexagon? Explain.

Find the measures of the exterior angles of the polygon.

26.

27.

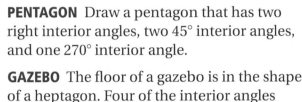

28.

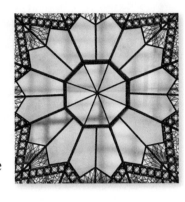

29. **STAINED GLASS** The center of the stained glass window is in the shape of a regular polygon. What is the measure of each interior angle of the polygon? What is the measure of each exterior angle?

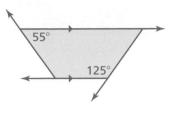

30. **PENTAGON** Draw a pentagon that has two right interior angles, two 45° interior angles, and one 270° interior angle.

31. **GAZEBO** The floor of a gazebo is in the shape of a heptagon. Four of the interior angles measure 135°. The other interior angles have equal measures. Find their measures.

32. **MONEY** The border of a Susan B. Anthony dollar is in the shape of a regular polygon.

   a. How many sides does the polygon have?

   b. What is the measure of each interior angle of the border? Round your answer to the nearest degree.

33. **Geometry** When tiles can be used to cover a floor with no empty spaces, the collection of tiles is called a *tessellation*.

   a. Create a tessellation using equilateral triangles.

   b. Find two more regular polygons that form tessellations.

   c. Create a tessellation that uses two different regular polygons.

   d. Use what you know about interior and exterior angles to explain why the polygons in part (c) form a tessellation.

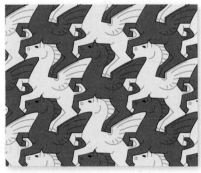

Solve the proportion. *(Section 5.4)*

34. $\dfrac{x}{12} = \dfrac{3}{4}$

35. $\dfrac{14}{21} = \dfrac{x}{3}$

36. $\dfrac{9}{x} = \dfrac{6}{2}$

37. $\dfrac{10}{4} = \dfrac{15}{x}$

38. **MULTIPLE CHOICE** The ratio of tulips to daisies is 3 : 5. Which of the following could be the total number of tulips and daisies? *(Skills Review Handbook)*

   Ⓐ 6   Ⓑ 10   Ⓒ 15   Ⓓ 16

12.4 Using Similar Triangles

Essential Question How can you use angles to tell whether triangles are similar?

1 ACTIVITY: Constructing Similar Triangles

Work with a partner.

- **Use a straightedge to draw a line segment that is 4 centimeters long.**

- **Then use the line segment and a protractor to draw a triangle that has a 60° and a 40° angle, as shown. Label the triangle *ABC*.**

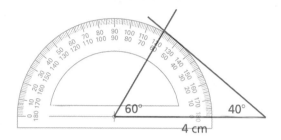

a. Explain how to draw a larger triangle that has the same two angle measures. Label the triangle *JKL*.

b. Explain how to draw a smaller triangle that has the same two angle measures. Label the triangle *PQR*.

c. Are all of the triangles similar? Explain.

2 ACTIVITY: Using Technology to Explore Triangles

Work with a partner. Use geometry software to draw the triangle below.

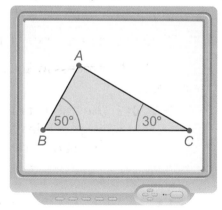

COMMON CORE

Geometry

In this lesson, you will
- understand the concept of similar triangles.
- identify similar triangles.
- use indirect measurement to find missing measures.

Learning Standard
8.G.5

a. Dilate the triangle by the following scale factors.

$$2 \qquad \frac{1}{2} \qquad \frac{1}{4} \qquad 2.5$$

b. Measure the third angle in each triangle. What do you notice?

c. **REASONING** You have two triangles. Two angles in the first triangle are congruent to two angles in the second triangle. Can you conclude that the triangles are similar? Explain.

Math Practice 2

Make Sense of Quantities

What do you know about the sides of the triangles when the triangles are similar?

Work with a partner.

a. Use the fact that two rays from the Sun are parallel to explain why $\triangle ABC$ and $\triangle DEF$ are similar.

b. Explain how to use similar triangles to find the height of the flagpole.

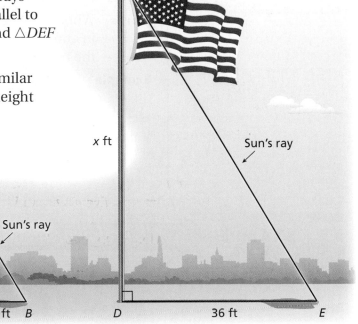

x ft

Sun's ray

Sun's ray

5 ft

A   3 ft   *B*

C

F

D   36 ft   *E*

What Is Your Answer?

4. **IN YOUR OWN WORDS** How can you use angles to tell whether triangles are similar?

5. **PROJECT** Work with a partner or in a small group.

   a. Explain why the process in Activity 3 is called "indirect" measurement.

   b. **CHOOSE TOOLS** Use indirect measurement to measure the height of something outside your school (a tree, a building, a flagpole). Before going outside, decide what materials you need to take with you.

   c. **MODELING** Draw a diagram of the indirect measurement process you used. In the diagram, label the lengths that you actually measured and also the lengths that you calculated.

6. **PRECISION** Look back at Exercise 17 in Section 11.5. Explain how you can show that the two triangles are similar.

Practice ➤ Use what you learned about similar triangles to complete Exercises 4 and 5 on page 554.

Check It Out
Lesson Tutorials
BigIdeasMath .com

Key Vocabulary
indirect measurement,
p. 553

🔑 Key Idea

Angles of Similar Triangles

Words When two angles in one triangle are congruent to two angles in another triangle, the third angles are also congruent and the triangles are similar.

Example

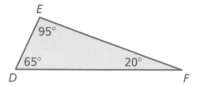

Triangle *ABC* is similar to Triangle *DEF*: $\triangle ABC \sim \triangle DEF$.

EXAMPLE 1 Identifying Similar Triangles

Tell whether the triangles are similar. Explain.

a.
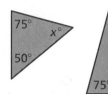

The triangles have two pairs of congruent angles.

∴ So, the third angles are congruent, and the triangles are similar.

b.

Write and solve an equation to find *x*.

$$x + 54 + 63 = 180$$
$$x + 117 = 180$$
$$x = 63$$

The triangles have two pairs of congruent angles.

∴ So, the third angles are congruent, and the triangles are similar.

c.
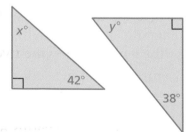

Write and solve an equation to find *x*.

$$x + 90 + 42 = 180$$
$$x + 132 = 180$$
$$x = 48$$

The triangles do not have two pairs of congruent angles.

∴ So, the triangles are not similar.

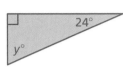

On Your Own

Tell whether the triangles are similar. Explain.

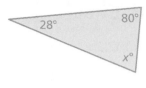

1.

2.

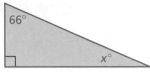

Indirect measurement uses similar figures to find a missing measure when it is difficult to find directly.

EXAMPLE 2 **Using Indirect Measurement**

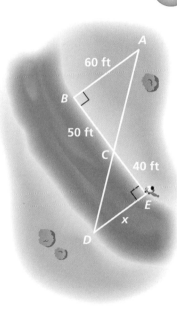

You plan to cross a river and want to know how far it is to the other side. You take measurements on your side of the river and make the drawing shown. **(a)** Explain why $\triangle ABC$ and $\triangle DEC$ are similar. **(b)** What is the distance x across the river?

a. $\angle B$ and $\angle E$ are right angles, so they are congruent. $\angle ACB$ and $\angle DCE$ are vertical angles, so they are congruent.

Because two angles in $\triangle ABC$ are congruent to two angles in $\triangle DEC$, the third angles are also congruent and the triangles are similar.

b. The ratios of the corresponding side lengths in similar triangles are equal. Write and solve a proportion to find x.

$$\frac{x}{60} = \frac{40}{50} \qquad \text{Write a proportion.}$$

$$60 \cdot \frac{x}{60} = 60 \cdot \frac{40}{50} \qquad \text{Multiplication Property of Equality}$$

$$x = 48 \qquad \text{Simplify.}$$

∴ So, the distance across the river is 48 feet.

On Your Own

3. WHAT IF? The distance from vertex A to vertex B is 55 feet. What is the distance across the river?

 Vocabulary and Concept Check

1. **REASONING** How can you use similar triangles to find a missing measurement?

2. **WHICH ONE DOESN'T BELONG?** Which triangle does *not* belong with the other three? Explain your reasoning.

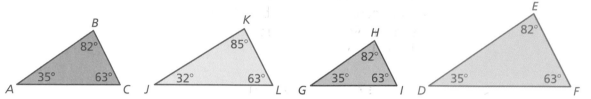

3. **WRITING** Two triangles have two pairs of congruent angles. In your own words, explain why you do not need to find the measures of the third pair of angles to determine that they are congruent.

 Practice and Problem Solving

Make a triangle that is larger or smaller than the one given and has the same angle measures. Find the ratios of the corresponding side lengths.

4.

5.

Tell whether the triangles are similar. Explain.

① 6.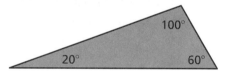

7.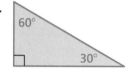

8.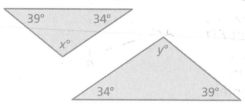

9.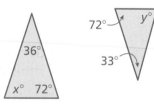

10. **RULERS** Which of the rulers are similar in shape? Explain.

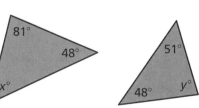

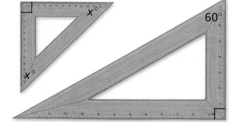

Tell whether the triangles are similar. Explain.

11.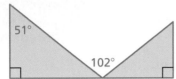

12.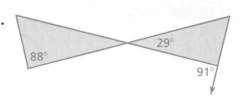

13. TREASURE The map shows the number of steps you must take to get to the treasure. However, the map is old, and the last dimension is unreadable. Explain why the triangles are similar. How many steps do you take from the pyramids to the treasure?

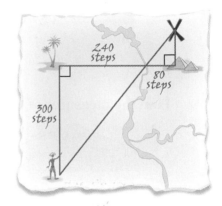

14. CRITICAL THINKING The side lengths of a triangle are increased by 50% to make a similar triangle. Does the area increase by 50% as well? Explain.

15. PINE TREE A person who is 6 feet tall casts a 3-foot-long shadow. A nearby pine tree casts a 15-foot-long shadow. What is the height h of the pine tree?

16. OPEN-ENDED You place a mirror on the ground 6 feet from the lamppost. You move back 3 feet and see the top of the lamppost in the mirror. What is the height of the lamppost?

17. REASONING In each of two right triangles, one angle measure is two times another angle measure. Are the triangles similar? Explain your reasoning.

18. Geometry In the diagram, segments BG, CF, and DE are parallel. The length of segment BD is 6.32 feet, and the length of segment DE is 6 feet. Name all pairs of similar triangles in the diagram. Then find the lengths of segments BG and CF.

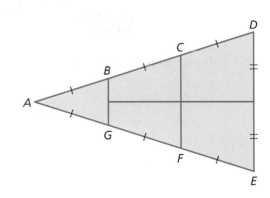

Fair Game Review What you learned in previous grades & lessons

Solve the equation for y. *(Topic 3)*

19. $y - 5x = 3$

20. $4x + 6y = 12$

21. $2x - \frac{1}{4}y = 1$

22. MULTIPLE CHOICE What is the value of x? *(Section 12.2)*

   Ⓐ 17   Ⓑ 62

   Ⓒ 118   Ⓓ 152

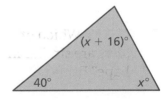

Find the sum of the interior angle measures of the polygon. *(Section 12.3)*

1.

2.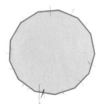

Find the measures of the interior angles of the polygon. *(Section 12.3)*

3.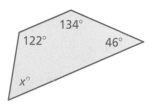

4.

5.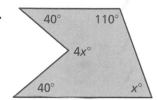

Find the measures of the exterior angles of the polygon. *(Section 12.3)*

6.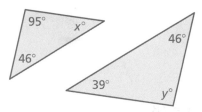

7.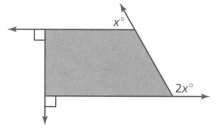

Tell whether the triangles are similar. Explain. *(Section 12.4)*

8.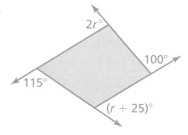

9.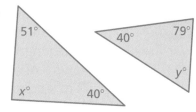

10. **REASONING** The sum of the interior angle measures of a polygon is 4140°. How many sides does the polygon have? *(Section 12.3)*

11. **SWAMP** You are trying to find the distance ℓ across a patch of swamp water. *(Section 12.4)*

    a. Explain why △VWX and △YZX are similar.

    b. What is the distance across the patch of swamp water?

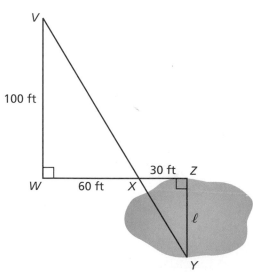

Check It Out
Vocabulary Help
BigIdeasMath ✓com

Review Key Vocabulary

transversal, *p. 528*
interior angles, *p. 529*
exterior angles, *p. 529*

interior angles of a polygon, *p. 536*
exterior angles of a polygon, *p. 536*

convex polygon, *p. 543*
concave polygon, *p. 543*
regular polygon, *p. 545*
indirect measurement, *p. 553*

Review Examples and Exercises

12.1 Parallel Lines and Transversals (pp. 526–533)

Use the figure to find the measure of ∠6.

∠2 and the 55° angle are supplementary.
So, the measure of ∠2 is $180° − 55° = 125°$.

∠2 and ∠6 are corresponding angles.
They are congruent.

⋮• So, the measure of ∠6 is 125°.

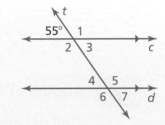

Exercises

**Use the figure to find the measure of the angle.
Explain your reasoning.**

1. ∠8
2. ∠5
3. ∠7
4. ∠2

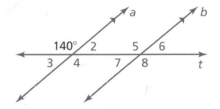

12.2 Angles of Triangles (pp. 534–539)

a. Find the value of x.

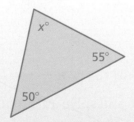

$$x + 50 + 55 = 180$$
$$x + 105 = 180$$
$$x = 75$$

⋮• The value of x is 75.

b. Find the measure of the exterior angle.

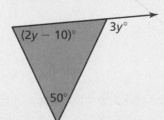

$$3y = (2y − 10) + 50$$
$$3y = 2y + 40$$
$$y = 40$$

⋮• So, the measure of the
exterior angle is $3(40)° = 120°$.

Exercises

Find the measures of the interior angles.

5.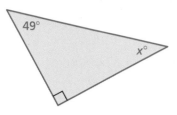

6.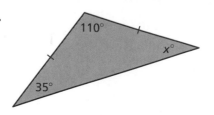

Find the measure of the exterior angle.

7.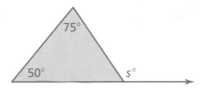

8.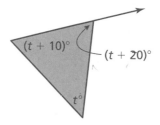

12.3 **Angles of Polygons** *(pp. 542–549)*

a. **Find the value of x.**

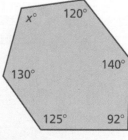

Step 1: The polygon has 6 sides. Find the sum of the interior angle measures.

$$S = (n - 2) \cdot 180°$$ Write the formula.

$$= (6 - 2) \cdot 180°$$ Substitute 6 for n.

$$= 720$$ Simplify. The sum of the interior angle measures is 720°.

Step 2: Write and solve an equation.

$$130 + 125 + 92 + 140 + 120 + x = 720$$

$$607 + x = 720$$

$$x = 113$$

The value of x is 113.

b. **Find the measures of the exterior angles of the polygon.**

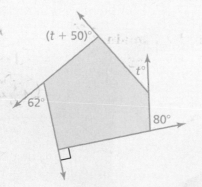

Write and solve an equation for t.

$$t + 80 + 90 + 62 + (t + 50) = 360$$

$$2t + 282 = 360$$

$$2t = 78$$

$$t = 39$$

So, the measures of the exterior angles are 39°, 80°, 90°, 62°, and $(39 + 50)° = 89°$.

Exercises

Find the measures of the interior angles of the polygon.

9.

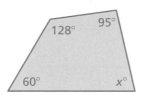

128° 95°

60° x°

10.

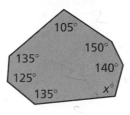

105°
150°
135°
140°
125°
x°
135°

11.

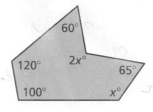

60°
120° 2x°
65°
100° x°

Find the measures of the exterior angles of the polygon.

12.

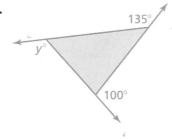

135°
y°
100°

13.

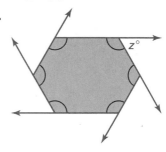

z°

12.4 **Using Similar Triangles** *(pp. 550–555)*

Tell whether the triangles are similar. Explain.

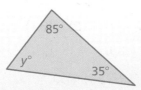

85°
50° x°

85°
y° 35°

Write and solve an equation to find *x*.

$$50 + 85 + x = 180$$
$$135 + x = 180$$
$$x = 45$$

❖ The triangles do not have two pairs of congruent angles. So, the triangles are not similar.

Exercises

Tell whether the triangles are similar. Explain.

14.

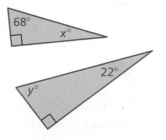

68°
x°

22°
y°

15.

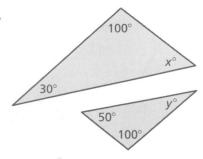

100°
x°
30°
50°
y°
100°

Use the figure to find the measure of the angle. Explain your reasoning.

1. ∠1　　　　　2. ∠8

3. ∠4　　　　　4. ∠5

Find the measures of the interior angles.

5.

6.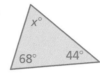

7.

Find the measure of the exterior angle.

8.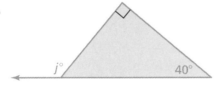

9.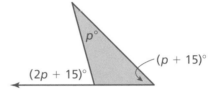

10. Find the measures of the interior angles of the polygon.

11. Find the measures of the exterior angles of the polygon.

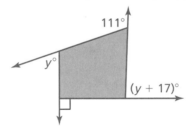

Tell whether the triangles are similar. Explain.

12.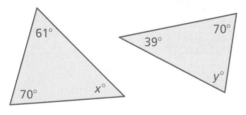

13.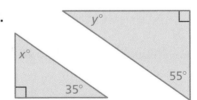

14. **WRITING** Describe two ways you can find the measure of ∠5.

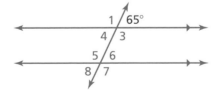

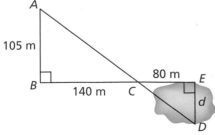

15. **POND** Use the given measurements to find the distance d across the pond.

1. The border of a Canadian one-dollar coin is shaped like an 11-sided regular polygon. The shape was chosen to help visually impaired people identify the coin. How many degrees are in each angle along the border? Round your answer to the nearest degree. *(8.G.5)*

2. A public utility charges its residential customers for natural gas based on the number of therms used each month. The formula below shows how the monthly cost C in dollars is related to the number t of therms used.

$$C = 11 + 1.6t$$

Solve this formula for t. *(8.EE.7b)*

A. $t = \dfrac{C}{12.6}$

B. $t = \dfrac{C - 11}{1.6}$

C. $t = \dfrac{C}{1.6} - 11$

D. $t = C - 12.6$

3. What is the value of x? *(8.EE.7b)*

$$5(x - 4) = 3x$$

F. -10

G. 2

H. $2\dfrac{1}{2}$

I. 10

4. In the figures below, $\triangle PQR$ and $\triangle STU$ are similar.

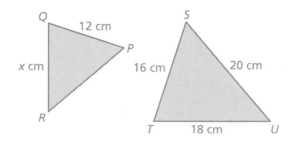

What is the value of x? *(8.G.4)*

A. 9.6

B. $10\dfrac{2}{3}$

C. 13.5

D. 15

5. What is the value of x? *(8.G.5)*

6. Olga was solving an equation in the box shown.

$$-\frac{2}{5}(10x - 15) = -30$$

$$10x - 15 = -30\left(-\frac{2}{5}\right)$$

$$10x - 15 = 12$$

$$10x - 15 + 15 = 12 + 15$$

$$10x = 27$$

$$\frac{10x}{10} = \frac{27}{10}$$

$$x = \frac{27}{10}$$

What should Olga do to correct the error that she made?   *(8.EE.7b)*

F. Multiply both sides by $-\frac{5}{2}$ instead of $-\frac{2}{5}$.

G. Multiply both sides by $\frac{2}{5}$ instead of $-\frac{2}{5}$.

H. Distribute $-\frac{2}{5}$ to get $-4x - 6$.

I. Add 15 to -30.

7. In the coordinate plane below, △*XYZ* is plotted and its vertices are labeled.

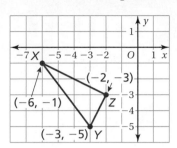

Which of the following shows △*X'Y'Z'*, the image of △*XYZ* after it is reflected in the *y*-axis? *(8.G.3)*

A.

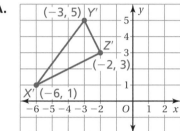

C.

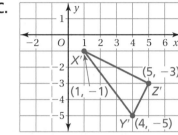

B.

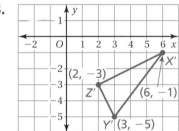

D.

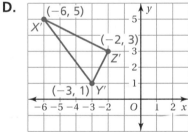

Think
Solve
Explain

8. The sum *S* of the interior angle measures of a polygon with *n* sides can be found by using a formula. *(8.G.5)*

Part A Write the formula.

Part B A quadrilateral has angles measuring 100°, 90°, and 90°. Find the measure of its fourth angle. Show your work and explain your reasoning.

Part C The sum of the measures of the angles of the pentagon shown is 540°. Divide the pentagon into triangles to show why this must be true. Show your work and explain your reasoning.

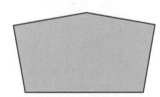

13 Graphing and Writing Linear Equations

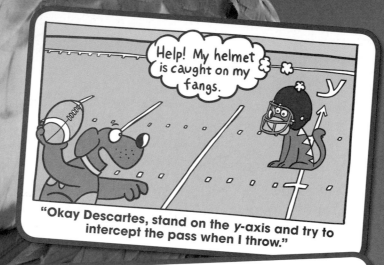

"Okay Descartes, stand on the y-axis and try to intercept the pass when I throw."

"Here's an easy example of a line with a slope of 1."

"You eat one mouse treat the first day. Two treats the second day. And so on. Get it?"

What You Learned Before

"I estimate that we are on a slope of about –0.625. What do you think?"

Evaluating Expressions Using Order of Operations (6.EE.2c)

Example 1 Evaluate $2xy + 3(x + y)$ when $x = 4$ and $y = 7$.

$$2xy + 3(x + y) = 2(4)(7) + 3(4 + 7)$$ Substitute 4 for x and 7 for y.

$$= 8(7) + 3(4 + 7)$$ Use order of operations.

$$= 56 + 3(11)$$ Simplify.

$$= 56 + 33$$ Multiply.

$$= 89$$ Add.

Try It Yourself

Evaluate the expression when $a = \dfrac{1}{4}$ and $b = 6$.

1. $-8ab$

2. $16a^2 - 4b$

3. $\dfrac{5b}{32a^2}$

4. $12a + (b - a - 4)$

Plotting Points (6.NS.6c)

Example 2 Write the ordered pair that corresponds to point U.

Point U is 3 units to the left of the origin and 4 units down. So, the x-coordinate is -3, and the y-coordinate is -4.

∴ The ordered pair $(-3, -4)$ corresponds to point U.

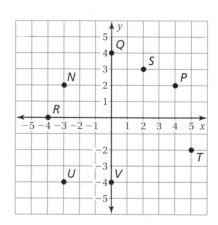

Example 3 Which point is located at $(5, -2)$?

Start at the origin. Move 5 units right and 2 units down.

∴ Point T is located at $(5, -2)$.

Try It Yourself

Use the graph to answer the question.

5. Write the ordered pair that corresponds to point Q.

6. Write the ordered pair that corresponds to point P.

7. Which point is located at $(-4, 0)$?

8. Which point is located in Quadrant II?

Essential Question How can you recognize a linear equation? How can you draw its graph?

1 ACTIVITY: Graphing a Linear Equation

Work with a partner.

a. Use the equation $y = \frac{1}{2}x + 1$ to complete the table. (Choose any two x-values and find the y-values.)

	Solution Points	
x		
$y = \frac{1}{2}x + 1$		

b. Write the two ordered pairs given by the table. These are called *solution points* of the equation.

c. **PRECISION** Plot the two solution points. Draw a line *exactly* through the two points.

d. Find a different point on the line. Check that this point is a solution point of the equation $y = \frac{1}{2}x + 1$.

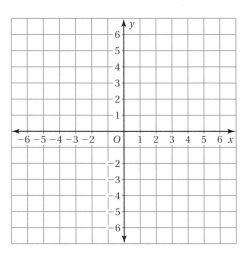

e. **LOGIC** Do you think it is true that *any* point on the line is a solution point of the equation $y = \frac{1}{2}x + 1$? Explain.

f. Choose five additional x-values for the table. (Choose positive and negative x-values.) Plot the five corresponding solution points. Does each point lie on the line?

	Solution Points				
x					
$y = \frac{1}{2}x + 1$					

g. **LOGIC** Do you think it is true that *any* solution point of the equation $y = \frac{1}{2}x + 1$ is a point on the line? Explain.

h. Why do you think $y = ax + b$ is called a *linear equation*?

COMMON CORE

Graphing Equations

In this lesson, you will
- understand that lines represent solutions of linear equations.
- graph linear equations.

Preparing for Standard 8.EE.5

Use a graphing calculator to graph $y = 2x + 5$.

a. Enter the equation $y = 2x + 5$ into your calculator.

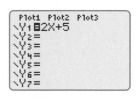

Math Practice 5

Recognize Usefulness of Tools

What are some advantages and disadvantages of using a graphing calculator to graph a linear equation?

b. Check the settings of the *viewing window*. The boundaries of the graph are set by the minimum and the maximum x- and y-values. The numbers of units between the tick marks are set by the x- and y-scales.

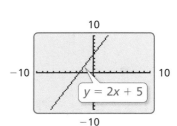

This is the standard viewing window.

c. Graph $y = 2x + 5$ on your calculator.

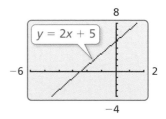

$y = 2x + 5$

d. Change the settings of the viewing window to match those shown.

Compare the two graphs.

$y = 2x + 5$

What Is Your Answer?

3. **IN YOUR OWN WORDS** How can you recognize a linear equation? How can you draw its graph? Write an equation that is linear. Write an equation that is *not* linear.

4. Use a graphing calculator to graph $y = 5x - 12$ in the standard viewing window.

   a. Can you tell where the line crosses the x-axis? Can you tell where the line crosses the y-axis?

   b. How can you adjust the viewing window so that you can determine where the line crosses the x- and y-axes?

5. **CHOOSE TOOLS** You want to graph $y = 2.5x - 3.8$. Would you graph it by hand or by using a graphing calculator? Why?

Practice

Use what you learned about graphing linear equations to complete Exercises 3 and 4 on page 570.

13.1 Lesson

Key Vocabulary
linear equation, p. 568
solution of a linear equation, p. 568

Remember
An ordered pair (x, y) is used to locate a point in a coordinate plane.

Key Idea

Linear Equations

A **linear equation** is an equation whose graph is a line. The points on the line are **solutions** of the equation.

You can use a graph to show the solutions of a linear equation. The graph below represents the equation $y = x + 1$.

x	y	(x, y)
−1	0	(−1, 0)
0	1	(0, 1)
2	3	(2, 3)

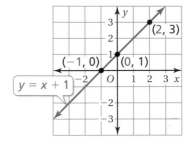

EXAMPLE 1 **Graphing a Linear Equation**

Graph $y = -2x + 1$.

Step 1: Make a table of values.

Check

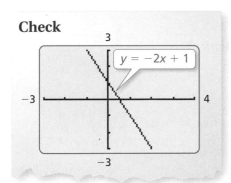

x	y = −2x + 1	y	(x, y)
−1	y = −2(−1) + 1	3	(−1, 3)
0	y = −2(0) + 1	1	(0, 1)
2	y = −2(2) + 1	−3	(2, −3)

Step 2: Plot the ordered pairs.

Step 3: Draw a line through the points.

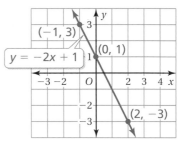

Key Idea

Graphing Horizontal and Vertical Lines

The graph of $y = b$ is a horizontal line passing through $(0, b)$.

The graph of $x = a$ is a vertical line passing through $(a, 0)$.

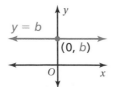

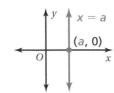

EXAMPLE 2 **Graphing a Horizontal Line and a Vertical Line**

a. Graph $y = -3$.

The graph of $y = -3$ is a horizontal line passing through $(0, -3)$. Draw a horizontal line through this point.

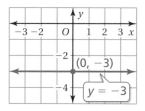

b. Graph $x = 2$.

The graph of $x = 2$ is a vertical line passing through $(2, 0)$. Draw a vertical line through this point.

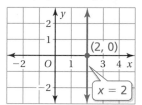

On Your Own

Now You're Ready
Exercises 5–16

Graph the linear equation. Use a graphing calculator to check your graph, if possible.

1. $y = 3x$ **2.** $y = -\dfrac{1}{2}x + 2$ **3.** $x = -4$ **4.** $y = -1.5$

EXAMPLE 3 **Real-Life Application**

The wind speed y (in miles per hour) of a tropical storm is $y = 2x + 66$, where x is the number of hours after the storm enters the Gulf of Mexico.

a. Graph the equation.

b. When does the storm become a hurricane?

A tropical storm becomes a hurricane when wind speeds are at least 74 miles per hour.

a. Make a table of values.

x	$y = 2x + 66$	y	(x, y)
0	$y = 2(0) + 66$	66	$(0, 66)$
1	$y = 2(1) + 66$	68	$(1, 68)$
2	$y = 2(2) + 66$	70	$(2, 70)$
3	$y = 2(3) + 66$	72	$(3, 72)$

Plot the ordered pairs and draw a line through the points.

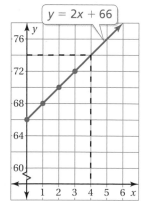

b. From the graph, you can see that $y = 74$ when $x = 4$. So, the storm becomes a hurricane 4 hours after it enters the Gulf of Mexico.

On Your Own

5. WHAT IF? The wind speed of the storm is $y = 1.5x + 62$. When does the storm become a hurricane?

Check It Out
Help with Homework
BigIdeasMath.com

 Vocabulary and Concept Check

1. **VOCABULARY** What type of graph represents the solutions of the equation $y = 2x + 4$?

2. **WHICH ONE DOESN'T BELONG?** Which equation does *not* belong with the other three? Explain your reasoning.

$$y = 0.5x - 0.2 \qquad 4x + 3 = y \qquad y = x^2 + 6 \qquad \frac{3}{4}x + \frac{1}{3} = y$$

 Practice and Problem Solving

PRECISION Copy and complete the table. Plot the two solution points and draw a line *exactly* through the two points. Find a different solution point on the line.

3.

x		
$y = 3x - 1$		

4.

x		
$y = \frac{1}{3}x + 2$		

Graph the linear equation. Use a graphing calculator to check your graph, if possible.

5. $y = -5x$

6. $y = \frac{1}{4}x$

7. $y = 5$

8. $x = -6$

9. $y = x - 3$

10. $y = -7x - 1$

11. $y = -\frac{x}{3} + 4$

12. $y = \frac{3}{4}x - \frac{1}{2}$

13. $y = -\frac{2}{3}$

14. $y = 6.75$

15. $x = -0.5$

16. $x = \frac{1}{4}$

17. **ERROR ANALYSIS** Describe and correct the error in graphing the equation.

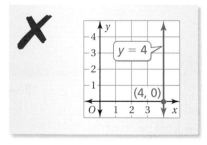

18. **MESSAGING** You sign up for an unlimited text-messaging plan for your cell phone. The equation $y = 20$ represents the cost y (in dollars) for sending x text messages. Graph the equation. What does the graph tell you?

19. **MAIL** The equation $y = 2x + 3$ represents the cost y (in dollars) of mailing a package that weighs x pounds.

   a. Graph the equation.

   b. Use the graph to estimate how much it costs to mail the package.

   c. Use the equation to find exactly how much it costs to mail the package.

Solve for *y*. Then graph the equation. Use a graphing calculator to check your graph.

20. $y - 3x = 1$

21. $5x + 2y = 4$

22. $-\dfrac{1}{3}y + 4x = 3$

23. $x + 0.5y = 1.5$

ACRES OF LAND ON MARS

Acres of land FOR SALE

10 acres for $175

24. SAVINGS You have $100 in your savings account and plan to deposit $12.50 each month.

    a. Graph a linear equation that represents the balance in your account.

    b. How many months will it take you to save enough money to buy 10 acres of land on Mars?

25. GEOMETRY The sum *S* of the interior angle measures of a polygon with *n* sides is $S = (n - 2) \cdot 180°$.

    a. Plot four points (*n*, *S*) that satisfy the equation. Is the equation a linear equation? Explain your reasoning.

    b. Does the value $n = 3.5$ make sense in the context of the problem? Explain your reasoning.

26. SEA LEVEL Along the U.S. Atlantic coast, the sea level is rising about 2 millimeters per year. How many millimeters has sea level risen since you were born? How do you know? Use a linear equation and a graph to justify your answer.

Video time: 1 min. 30 sec.

27. **Problem Solving** One second of video on your digital camera uses the same amount of memory as two pictures. Your camera can store 250 pictures.

    a. Write and graph a linear equation that represents the number *y* of pictures your camera can store when you take *x* seconds of video.

    b. How many pictures can your camera store in addition to the video shown?

 Fair Game Review What you learned in previous grades & lessons

Write the ordered pair corresponding to the point.
(Skills Review Handbook)

28. point *A*

29. point *B*

30. point *C*

31. point *D*

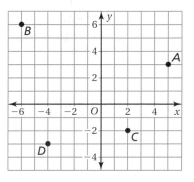

32. MULTIPLE CHOICE A debate team has 15 female members. The ratio of females to males is 3 : 2. How many males are on the debate team? *(Skills Review Handbook)*

    Ⓐ 6   **Ⓑ** 10   **Ⓒ** 22   **Ⓓ** 25

Essential Question How can you use the slope of a line to describe the line?

Slope is the rate of change between any two points on a line. It is the measure of the *steepness* of the line.

To find the slope of a line, find the ratio of the change in y (vertical change) to the change in x (horizontal change).

$$\text{slope} = \frac{\text{change in } y}{\text{change in } x}$$

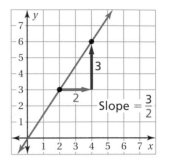

Slope $= \dfrac{3}{2}$

1 ACTIVITY: Finding the Slope of a Line

Work with a partner. Find the slope of each line using two methods.

> **Method 1: Use the two black points.** ●
>
> **Method 2: Use the two pink points.** ●

Do you get the same slope using each method? Why do you think this happens?

a.
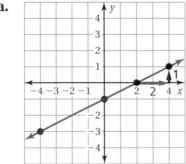

b.
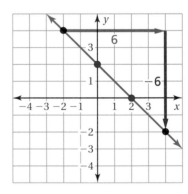

c.
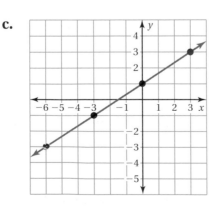

d.
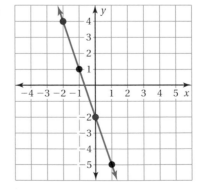

COMMON CORE

Graphing Equations

In this lesson, you will
● find slopes of lines by using two points.
● find slopes of lines from tables.

Learning Standard
8.EE.6

Work with a partner. Use the figure shown.

a. △*ABC* is a right triangle formed by drawing a horizontal line segment from point *A* and a vertical line segment from point *B*. Use this method to draw another right triangle, △*DEF*.

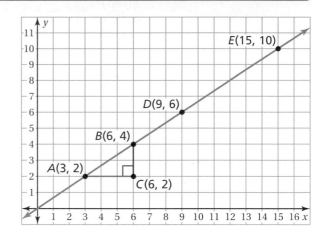

b. What can you conclude about △*ABC* and △*DEF*? Justify your conclusion.

c. For each triangle, find the ratio of the length of the vertical side to the length of the horizontal side. What do these ratios represent?

d. What can you conclude about the slope between any two points on the line?

Work with a partner.

a. Draw two lines with slope $\frac{3}{4}$. One line passes through $(-4, 1)$, and the other line passes through $(4, 0)$. What do you notice about the two lines?

b. Draw two lines with slope $-\frac{4}{3}$. One line passes through $(2, 1)$, and the other line passes through $(-1, -1)$. What do you notice about the two lines?

c. **CONJECTURE** Make a conjecture about two different nonvertical lines in the same plane that have the same slope.

d. Graph one line from part (a) and one line from part (b) in the same coordinate plane. Describe the angle formed by the two lines. What do you notice about the product of the slopes of the two lines?

e. **REPEATED REASONING** Repeat part (d) for the two lines you did *not* choose. Based on your results, make a conjecture about two lines in the same plane whose slopes have a product of -1.

Math Practice 1

Interpret a Solution

What does the slope tell you about the graph of the line? Explain.

What Is Your Answer?

4. **IN YOUR OWN WORDS** How can you use the slope of a line to describe the line?

Practice

Use what you learned about the slope of a line to complete Exercises 4–6 on page 577.

Check It Out
Lesson Tutorials
BigIdeasMath.com

Key Vocabulary
slope, p. 574
rise, p. 574
run, p. 574

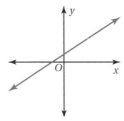 **Key Idea**

Slope

The **slope** m of a line is a ratio of the change in y (the **rise**) to the change in x (the **run**) between any two points, (x_1, y_1) and (x_2, y_2), on the line.

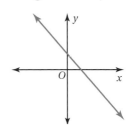

$$m = \frac{\text{rise}}{\text{run}} = \frac{\text{change in } y}{\text{change in } x} = \frac{y_2 - y_1}{x_2 - x_1}$$

Reading

In the slope formula, x_1 is read as "x sub one," and y_2 is read as "y sub two." The numbers 1 and 2 in x_1 and y_2 are called *subscripts*.

Positive Slope

The line rises from left to right.

Negative Slope

The line falls from left to right.

EXAMPLE 1 **Finding the Slope of a Line**

Describe the slope of the line. Then find the slope.

a.
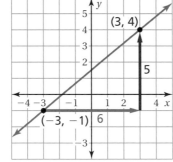

b.
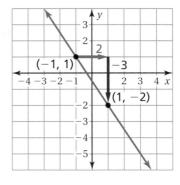

The line rises from left to right. So, the slope is positive.
Let $(x_1, y_1) = (-3, -1)$ and $(x_2, y_2) = (3, 4)$.

$$m = \frac{y_2 - y_1}{x_2 - x_1}$$

$$= \frac{4 - (-1)}{3 - (-3)}$$

$$= \frac{5}{6}$$

The line falls from left to right. So, the slope is negative.
Let $(x_1, y_1) = (-1, 1)$ and $(x_2, y_2) = (1, -2)$.

$$m = \frac{y_2 - y_1}{x_2 - x_1}$$

$$= \frac{-2 - 1}{1 - (-1)}$$

$$= \frac{-3}{2}, \text{ or } -\frac{3}{2}$$

Study Tip

When finding slope, you can label either point as (x_1, y_1) and the other point as (x_2, y_2).

On Your Own

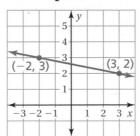

Now You're Ready
Exercises 7–9

Find the slope of the line.

1.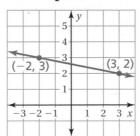

2.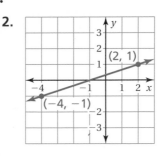

3.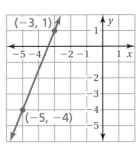

EXAMPLE 2 | **Finding the Slope of a Horizontal Line**

Find the slope of the line.

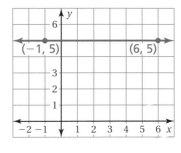

$$m = \frac{y_2 - y_1}{x_2 - x_1}$$

$$= \frac{5 - 5}{6 - (-1)}$$

$$= \frac{0}{7}, \text{ or } 0$$

∴ The slope is 0.

EXAMPLE 3 | **Finding the Slope of a Vertical Line**

Find the slope of the line.

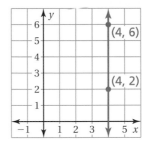

$$m = \frac{y_2 - y_1}{x_2 - x_1}$$

$$= \frac{6 - 2}{4 - 4}$$

$$= \frac{4}{0} \quad ✗$$

Study Tip

The slope of every horizontal line is 0. The slope of every vertical line is undefined.

∴ Because division by zero is undefined, the slope of the line is undefined.

On Your Own

Now You're Ready
Exercises 13–15

Find the slope of the line through the given points.

4. $(1, -2), (7, -2)$

5. $(-2, 4), (3, 4)$

6. $(-3, -3), (-3, -5)$

7. $(0, 8), (0, 0)$

8. How do you know that the slope of every horizontal line is 0? How do you know that the slope of every vertical line is undefined?

EXAMPLE **4** **Finding Slope from a Table**

The points in the table lie on a line. How can you find the slope of the line from the table? What is the slope?

x	1	4	7	10
y	8	6	4	2

Choose any two points from the table and use the slope formula.

Use the points $(x_1, y_1) = (1, 8)$ and $(x_2, y_2) = (4, 6)$.

$$m = \frac{y_2 - y_1}{x_2 - x_1}$$

$$= \frac{6 - 8}{4 - 1}$$

$$= \frac{-2}{3}, \text{ or } -\frac{2}{3}$$

∴ The slope is $-\frac{2}{3}$.

Check

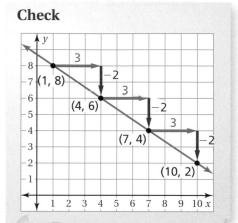

On Your Own

Now You're Ready
Exercises 21–24

The points in the table lie on a line. Find the slope of the line.

9.

x	1	3	5	7
y	2	5	8	11

10.

x	−3	−2	−1	0
y	6	4	2	0

Summary

Slope

Positive Slope	*Negative Slope*	*Slope of 0*	*Undefined Slope*
The line rises from left to right.	The line falls from left to right.	The line is horizontal.	The line is vertical.

13.2 Exercises

Check It Out
Help with Homework
BigIdeasMath ✓com

✓ Vocabulary and Concept Check

1. **CRITICAL THINKING** Refer to the graph.

   a. Which lines have positive slopes?

   b. Which line has the steepest slope?

   c. Do any lines have an undefined slope? Explain.

2. **OPEN-ENDED** Describe a real-life situation in which you need to know the slope.

3. **REASONING** The slope of a line is 0. What do you know about the line?

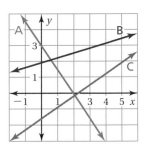

Practice and Problem Solving

Draw a line through each point using the given slope. What do you notice about the two lines?

4. slope = 1

5. slope = −3

6. slope = $\frac{1}{4}$

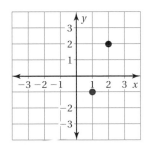

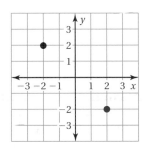

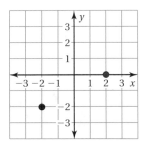

Find the slope of the line.

 7.

8.

9.

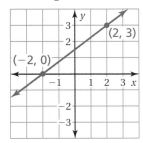

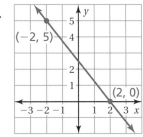

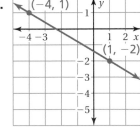

10.

11.

12.

 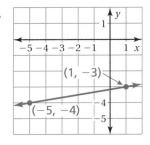

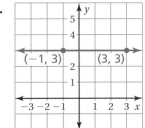

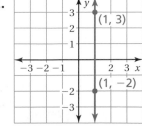

Find the slope of the line through the given points.

② ③ **13.** $(4, -1), (-2, -1)$ **14.** $(5, -3), (5, 8)$ **15.** $(-7, 0), (-7, -6)$

16. $(-3, 1), (-1, 5)$ **17.** $(10, 4), (4, 15)$ **18.** $(-3, 6), (2, 6)$

19. ERROR ANALYSIS Describe and correct the error in finding the slope of the line.

20. CRITICAL THINKING Is it more difficult to walk up the ramp or the hill? Explain.

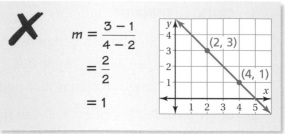

$$m = \frac{3-1}{4-2}$$
$$= \frac{2}{2}$$
$$= 1$$

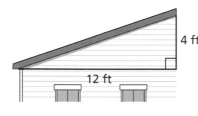

The points in the table lie on a line. Find the slope of the line.

④ **21.**

x	1	3	5	7
y	2	10	18	26

22.

x	-3	2	7	12
y	0	2	4	6

23.

x	-6	-2	2	6
y	8	5	2	-1

24.

x	-8	-2	4	10
y	8	1	-6	-13

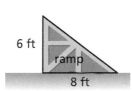

25. PITCH Carpenters refer to the slope of a roof as the *pitch* of the roof. Find the pitch of the roof.

26. PROJECT The guidelines for a wheelchair ramp suggest that the ratio of the rise to the run be no greater than $1 : 12$.

     a. CHOOSE TOOLS Find a wheelchair ramp in your school or neighborhood. Measure its slope. Does the ramp follow the guidelines?

     b. Design a wheelchair ramp that provides access to a building with a front door that is 2.5 feet above the sidewalk. Illustrate your design.

Use an equation to find the value of k so that the line that passes through the given points has the given slope.

27. $(1, 3), (5, k); m = 2$ **28.** $(-2, k), (2, 0); m = -1$

29. $(-4, k), (6, -7); m = -\dfrac{1}{5}$ **30.** $(4, -4), (k, -1); m = \dfrac{3}{4}$

31. TURNPIKE TRAVEL The graph shows the cost of traveling by car on a turnpike.

   a. Find the slope of the line.

   b. Explain the meaning of the slope as a rate of change.

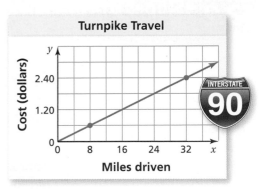

Turnpike Travel

32. BOAT RAMP Which is steeper: the boat ramp or a road with a 12% grade? Explain. (*Note:* Road grade is the vertical increase divided by the horizontal distance.)

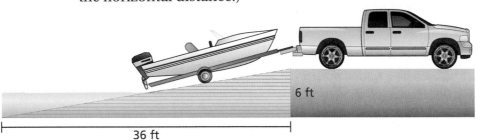

6 ft

36 ft

33. REASONING Do the points $A(-2, -1)$, $B(1, 5)$, and $C(4, 11)$ lie on the same line? Without using a graph, how do you know?

34. BUSINESS A small business earns a profit of $6500 in January and $17,500 in May. What is the rate of change in profit for this time period?

35. STRUCTURE Choose two points in the coordinate plane. Use the slope formula to find the slope of the line that passes through the two points. Then find the slope using the formula $\dfrac{y_1 - y_2}{x_1 - x_2}$. Explain why your results are the same.

36. **Critical Thinking** The top and the bottom of the slide are level with the ground, which has a slope of 0.

   a. What is the slope of the main portion of the slide?

   b. How does the slope change when the bottom of the slide is only 12 inches above the ground? Is the slide steeper? Explain.

1 ft

8 ft

1 ft

18 in.

12 ft

 Fair Game Review What you learned in previous grades & lessons

Solve the proportion. (*Section 5.4*)

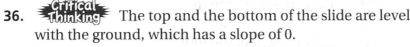

37. $\dfrac{b}{30} = \dfrac{5}{6}$

38. $\dfrac{7}{4} = \dfrac{n}{32}$

39. $\dfrac{3}{8} = \dfrac{x}{20}$

40. MULTIPLE CHOICE What is the prime factorization of 84? (*Skills Review Handbook*)

   Ⓐ $2 \times 3 \times 7$ **Ⓑ** $2^2 \times 3 \times 7$ **Ⓒ** $2 \times 3^2 \times 7$ **Ⓓ** $2^2 \times 21$

Check It Out
Lesson Tutorials
BigIdeasMath ✓.com

Key Idea

COMMON CORE

Graphing Equations

In this extension, you will

- identify parallel and perpendicular lines.

Applying Standard
8.EE.6

Parallel Lines and Slopes

Lines in the same plane that do not intersect are parallel lines. Nonvertical parallel lines have the same slope.

All vertical lines are parallel.

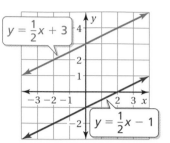

EXAMPLE **1** **Identifying Parallel Lines**

Which two lines are parallel? How do you know?

Find the slope of each line.

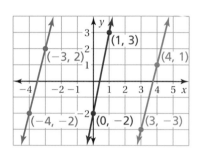

Blue Line

$$m = \frac{y_2 - y_1}{x_2 - x_1}$$

$$= \frac{-2 - 2}{-4 - (-3)}$$

$$= \frac{-4}{-1}, \text{ or } 4$$

Red Line

$$m = \frac{y_2 - y_1}{x_2 - x_1}$$

$$= \frac{-2 - 3}{0 - 1}$$

$$= \frac{-5}{-1}, \text{ or } 5$$

Green Line

$$m = \frac{y_2 - y_1}{x_2 - x_1}$$

$$= \frac{-3 - 1}{3 - 4}$$

$$= \frac{-4}{-1}, \text{ or } 4$$

The slopes of the blue and green lines are 4. The slope of the red line is 5.

∴ The blue and green lines have the same slope, so they are parallel.

Practice

Which lines are parallel? How do you know?

1.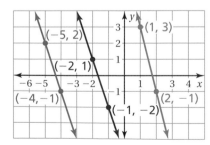

2.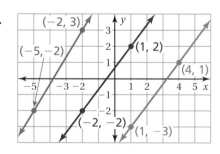

Are the given lines parallel? Explain your reasoning.

3. $y = -5, y = 3$ 　　　**4.** $y = 0, x = 0$ 　　　**5.** $x = -4, x = 1$

6. GEOMETRY The vertices of a quadrilateral are $A(-5, 3)$, $B(2, 2)$, $C(4, -3)$, and $D(-2, -2)$. How can you use slope to determine whether the quadrilateral is a parallelogram? Is it a parallelogram? Justify your answer.

 **Key Idea**

Perpendicular Lines and Slope

Lines in the same plane that intersect at right angles are perpendicular lines. Two nonvertical lines are perpendicular when the product of their slopes is -1.

Vertical lines are perpendicular to horizontal lines.

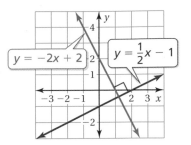

EXAMPLE 2 **Identifying Perpendicular Lines**

Which two lines are perpendicular? How do you know?

Find the slope of each line.

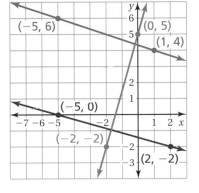

Blue Line

$$m = \frac{y_2 - y_1}{x_2 - x_1}$$

$$= \frac{4 - 6}{1 - (-5)}$$

$$= \frac{-2}{6}, \text{ or } -\frac{1}{3}$$

Red Line

$$m = \frac{y_2 - y_1}{x_2 - x_1}$$

$$= \frac{-2 - 0}{2 - (-5)}$$

$$= -\frac{2}{7}$$

Green Line

$$m = \frac{y_2 - y_1}{x_2 - x_1}$$

$$= \frac{5 - (-2)}{0 - (-2)}$$

$$= \frac{7}{2}$$

The slope of the red line is $-\frac{2}{7}$. The slope of the green line is $\frac{7}{2}$.

∴ Because $-\frac{2}{7} \cdot \frac{7}{2} = -1$, the red and green lines are perpendicular.

Practice

Which lines are perpendicular? How do you know?

7.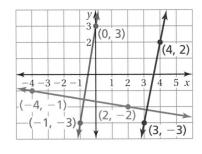

8.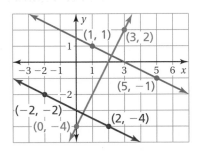

Are the given lines perpendicular? Explain your reasoning.

9. $x = -2, y = 8$

10. $x = -8, x = 7$

11. $y = 0, x = 0$

12. **GEOMETRY** The vertices of a parallelogram are $J(-5, 0)$, $K(1, 4)$, $L(3, 1)$, and $M(-3, -3)$. How can you use slope to determine whether the parallelogram is a rectangle? Is it a rectangle? Justify your answer.

Essential Question How can you describe the graph of the equation $y = mx$?

1 ACTIVITY: Identifying Proportional Relationships

Work with a partner. Tell whether x and y are in a proportional relationship. Explain your reasoning.

a. **Money**

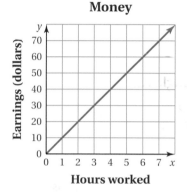

Hours worked

b. **Helicopter**

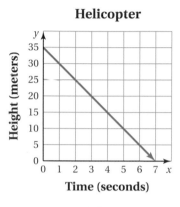

Time (seconds)

c. **Tickets**

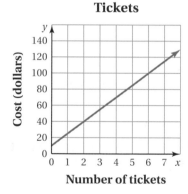

Number of tickets

d. **Pizzas**

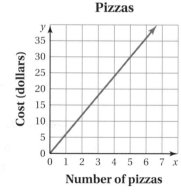

Number of pizzas

e.

Laps, x	1	2	3	4
Time (seconds), y	90	200	325	480

f.

Cups of Sugar, x	$\frac{1}{2}$	1	$1\frac{1}{2}$	2
Cups of Flour, y	1	2	3	4

2 ACTIVITY: Analyzing Proportional Relationships

Work with a partner. Use only the proportional relationships in Activity 1 to do the following.

- Find the slope of the line.
- Find the value of y for the ordered pair $(1, y)$.

What do you notice? What does the value of y represent?

COMMON CORE

Graphing Equations

In this lesson, you will

- write and graph proportional relationships.

Learning Standards
8.EE.5
8.EE.6

Work with a partner. Let (x, y) represent any point on the graph of a proportional relationship.

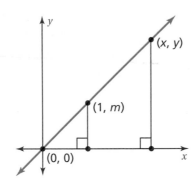

a. Explain why the two triangles are similar.

b. Because the triangles are similar, the corresponding side lengths are proportional. Use the vertical and horizontal side lengths to complete the steps below.

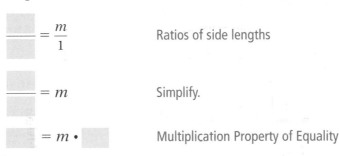

What does the final equation represent?

c. Use your result in part (b) to write an equation that represents each proportional relationship in Activity 1.

Math Practice 7

View as Components

What part of the graph can you use to find the side lengths?

What Is Your Answer?

4. **IN YOUR OWN WORDS** How can you describe the graph of the equation $y = mx$? How does the value of m affect the graph of the equation?

5. Give a real-life example of two quantities that are in a proportional relationship. Write an equation that represents the relationship and sketch its graph.

Practice

Use what you learned about proportional relationships to complete Exercises 3–6 on page 586.

Key Idea

Direct Variation

Study Tip

In the direct variation equation $y = mx$, m represents the constant of proportionality, the slope, and the unit rate.

Words When two quantities x and y are proportional, the relationship can be represented by the direct variation equation $y = mx$, where m is the constant of proportionality.

Graph The graph of $y = mx$ is a line with a slope of m that passes through the origin.

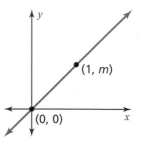

EXAMPLE 1 **Graphing a Proportional Relationship**

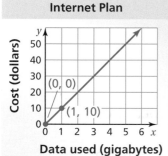

Internet Plan

Cost (dollars)

(0, 0)
(1, 10)

Data used (gigabytes)

The cost y (in dollars) for x gigabytes of data on an Internet plan is represented by $y = 10x$. Graph the equation and interpret the slope.

The equation shows that the slope m is 10. So, the graph passes through (0, 0) and (1, 10).

Plot the points and draw a line through the points. Because negative values of x do not make sense in this context, graph in the first quadrant only.

∴ The slope indicates that the unit cost is $10 per gigabyte.

EXAMPLE 2 **Writing and Using a Direct Variation Equation**

The weight y of an object on Titan, one of Saturn's moons, is proportional to the weight x of the object on Earth. An object that weighs 105 pounds on Earth would weigh 15 pounds on Titan.

a. Write an equation that represents the situation.

Study Tip

In Example 2, the slope indicates that the weight of an object on Titan is one-seventh its weight on Earth.

Use the point (105, 15) to find the slope of the line.

$y = mx$ Direct variation equation

$15 = m(105)$ Substitute 15 for y and 105 for x.

$\dfrac{1}{7} = m$ Simplify.

∴ So, an equation that represents the situation is $y = \dfrac{1}{7}x$.

b. How much would a chunk of ice that weighs 3.5 pounds on Titan weigh on Earth?

$3.5 = \dfrac{1}{7}x$ Substitute 3.5 for y.

$24.5 = x$ Multiply each side by 7.

∴ So, the chunk of ice would weigh 24.5 pounds on Earth.

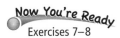

On Your Own

Now You're Ready
Exercises 7–8

1. **WHAT IF?** In Example 1, the cost is represented by $y = 12x$. Graph the equation and interpret the slope.

2. In Example 2, how much would a spacecraft that weighs 3500 kilograms on Earth weigh on Titan?

EXAMPLE ③ **Comparing Proportional Relationships**

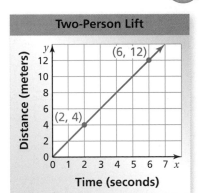

Two-Person Lift

The distance y (in meters) that a four-person ski lift travels in x seconds is represented by the equation $y = 2.5x$. The graph shows the distance that a two-person ski lift travels.

a. Which ski lift is faster?

Interpret each slope as a unit rate.

Four-Person Lift

$$y = 2.5x$$

The slope is 2.5.

The four-person lift travels 2.5 meters per second.

Two-Person Lift

$$\text{slope} = \frac{\text{change in } y}{\text{change in } x}$$

$$= \frac{8}{4} = 2$$

The two-person lift travels 2 meters per second.

So, the four-person lift is faster than the two-person lift.

b. Graph the equation that represents the four-person lift in the same coordinate plane as the two-person lift. Compare the steepness of the graphs. What does this mean in the context of the problem?

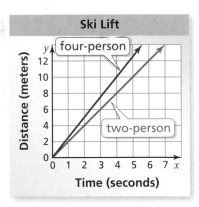

Ski Lift

The graph that represents the four-person lift is steeper than the graph that represents the two-person lift. So, the four-person lift is faster.

On Your Own

Now You're Ready
Exercise 9

3. The table shows the distance y (in meters) that a T-bar ski lift travels in x seconds. Compare its speed to the ski lifts in Example 3.

x (seconds)	1	2	3	4
y (meters)	$2\frac{1}{4}$	$4\frac{1}{2}$	$6\frac{3}{4}$	9

✓ Vocabulary and Concept Check

1. **VOCABULARY** What point is on the graph of every direct variation equation?

2. **REASONING** Does the equation $y = 2x + 3$ represent a proportional relationship? Explain.

Practice and Problem Solving

Tell whether *x* and *y* are in a proportional relationship. Explain your reasoning. If so, write an equation that represents the relationship.

3.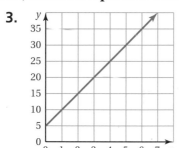

4.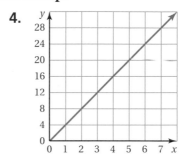

5.

x	3	6	9	12
y	1	2	3	4

6.

x	2	5	8	10
y	4	8	13	23

① 7. **TICKETS** The amount y (in dollars) that you raise by selling x fundraiser tickets is represented by the equation $y = 5x$. Graph the equation and interpret the slope.

② 8. **KAYAK** The cost y (in dollars) to rent a kayak is proportional to the number x of hours that you rent the kayak. It costs $27 to rent the kayak for 3 hours.

   a. Write an equation that represents the situation.

   b. Interpret the slope.

   c. How much does it cost to rent the kayak for 5 hours?

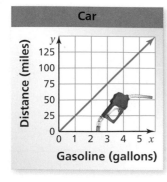

③ 9. **MILEAGE** The distance y (in miles) that a truck travels on x gallons of gasoline is represented by the equation $y = 18x$. The graph shows the distance that a car travels.

   a. Which vehicle gets better gas mileage? Explain how you found your answer.

   b. How much farther can the vehicle you chose in part (a) travel than the other vehicle on 8 gallons of gasoline?

10. **BIOLOGY** Toenails grow about 13 millimeters per year. The table shows fingernail growth.

Weeks	1	2	3	4
Fingernail Growth (millimeters)	0.7	1.4	2.1	2.8

   a. Do fingernails or toenails grow faster? Explain.

   b. In the same coordinate plane, graph equations that represent the growth rates of toenails and fingernails. Compare the steepness of the graphs. What does this mean in the context of the problem?

11. **REASONING** The quantities x and y are in a proportional relationship. What do you know about the ratio of y to x for any point (x, y) on the line?

12. **PROBLEM SOLVING** The graph relates the temperature change y (in degrees Fahrenheit) to the altitude change x (in thousands of feet).

   a. Is the relationship proportional? Explain.

   b. Write an equation of the line. Interpret the slope.

   c. You are at the bottom of a mountain where the temperature is 74°F. The top of the mountain is 5500 feet above you. What is the temperature at the top of the mountain?

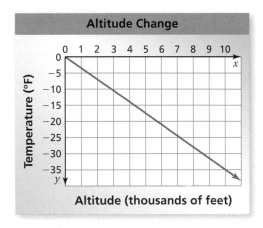

13. **Critical Thinking** Consider the distance equation $d = rt$, where d is the distance (in feet), r is the rate (in feet per second), and t is the time (in seconds).

   a. You run 6 feet per second. Are distance and time proportional? Explain. Graph the equation.

   b. You run for 50 seconds. Are distance and rate proportional? Explain. Graph the equation.

   c. You run 300 feet. Are rate and time proportional? Explain. Graph the equation.

   d. One of these situations represents *inverse variation*. Which one is it? Why do you think it is called inverse variation?

Fair Game Review What you learned in previous grades & lessons

Graph the linear equation. *(Section 13.1)*

14. $y = -\dfrac{1}{2}x$

15. $y = 3x - \dfrac{3}{4}$

16. $y = -\dfrac{x}{3} - \dfrac{3}{2}$

17. **MULTIPLE CHOICE** What is the value of x? *(Section 12.3)*

   Ⓐ 110

   Ⓑ 135

   Ⓒ 315

   Ⓓ 522

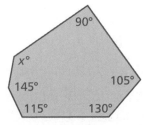

You can use a **process diagram** to show the steps involved in a procedure. Here is an example of a process diagram for graphing a linear equation.

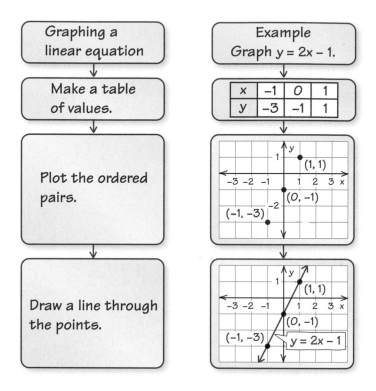

On Your Own

Make process diagrams with examples to help you study these topics.

1. finding the slope of a line

2. graphing a proportional relationship

After you complete this chapter, make process diagrams for the following topics.

3. graphing a linear equation using
   a. slope and *y*-intercept
   b. *x*- and *y*-intercepts

4. writing equations in slope-intercept form

5. writing equations in point-slope form

"Here is a process diagram with suggestions for what to do if a hyena knocks on your door."

Graph the linear equation. *(Section 13.1)*

1. $y = -x + 8$ **2.** $y = \dfrac{x}{3} - 4$ **3.** $x = -1$ **4.** $y = 3.5$

Find the slope of the line. *(Section 13.2)*

5.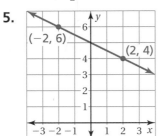

6.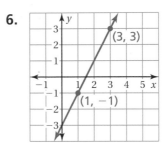

7.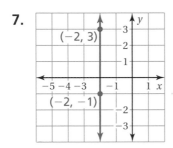

8.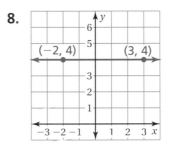

9. What is the slope of a line that is parallel to the line in Exercise 5? What is the slope of a line that is perpendicular to the line in Exercise 5? *(Section 13.2)*

10. Are the lines $y = -1$ and $x = 1$ parallel? Are they perpendicular? Justify your answer. *(Section 13.2)*

11. **BANKING** A bank charges $3 each time you use an out-of-network ATM. At the beginning of the month, you have $1500 in your bank account. You withdraw $60 from your bank account each time you use an out-of-network ATM. Graph a linear equation that represents the balance in your account after you use an out-of-network ATM x times. *(Section 13.1)*

12. **MUSIC** The number y of hours of cello lessons that you take after x weeks is represented by the equation $y = 3x$. Graph the equation and interpret the slope. *(Section 13.3)*

13. **DINNER PARTY** The cost y (in dollars) to provide food for guests at a dinner party is proportional to the number x of guests attending the party. It costs $30 to provide food for 4 guests. *(Section 13.3)*

    a. Write an equation that represents the situation.

    b. Interpret the slope.

    c. How much does it cost to provide food for 10 guests?

Essential Question How can you describe the graph of the equation $y = mx + b$?

1 ACTIVITY: Analyzing Graphs of Lines

Work with a partner.

- **Graph each equation.**

- **Find the slope of each line.**

- **Find the point where each line crosses the y-axis.**

- **Complete the table.**

Equation	Slope of Graph	Point of Intersection with y-axis
a. $y = -\dfrac{1}{2}x + 1$		
b. $y = -x + 2$		
c. $y = -x - 2$		
d. $y = \dfrac{1}{2}x + 1$		
e. $y = x + 2$		
f. $y = x - 2$		
g. $y = \dfrac{1}{2}x - 1$		
h. $y = -\dfrac{1}{2}x - 1$		
i. $y = 3x + 2$		
j. $y = 3x - 2$		

k. Do you notice any relationship between the slope of the graph and its equation? between the point of intersection with the y-axis and its equation? Compare the results with those of other students in your class.

COMMON CORE

Graphing Equations

In this lesson, you will

- find slopes and y-intercepts of graphs of linear equations.
- graph linear equations written in slope-intercept form.

Learning Standard
8.EE.6

ACTIVITY: Deriving an Equation

Work with a partner.

a. Look at the graph of each equation in Activity 1. Do any of the graphs represent a proportional relationship? Explain.

b. For a nonproportional linear relationship, the graph crosses the y-axis at some point $(0, b)$, where b does not equal 0. Let (x, y) represent any other point on the graph. You can use the formula for slope to write the equation for a nonproportional linear relationship.

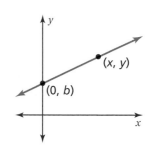

Use the graph to complete the steps.

$$\frac{y_2 - y_1}{x_2 - x_1} = m \qquad \text{Slope formula}$$

$$\frac{y - \boxed{}}{x - \boxed{}} = m \qquad \text{Substitute values.}$$

$$\frac{\boxed{}}{\boxed{}} = m \qquad \text{Simplify.}$$

$$\frac{\boxed{}}{\boxed{}} \cdot \boxed{} = m \cdot \boxed{} \qquad \text{Multiplication Property of Equality}$$

$$y - \boxed{} = m \cdot \boxed{} \qquad \text{Simplify.}$$

$$y = m \boxed{} + \boxed{} \qquad \text{Addition Property of Equality}$$

Math Practice 3

Use Prior Results

How can you use the results of Activity 1 to help support your answer?

c. What do m and b represent in the equation?

What Is Your Answer?

3. IN YOUR OWN WORDS How can you describe the graph of the equation $y = mx + b$?

a. How does the value of m affect the graph of the equation?

b. How does the value of b affect the graph of the equation?

c. Check your answers to parts (a) and (b) with three equations that are not in Activity 1.

4. LOGIC Why do you think $y = mx + b$ is called the *slope-intercept form* of the equation of a line? Use drawings or diagrams to support your answer.

Practice

Use what you learned about graphing linear equations in slope-intercept form to complete Exercises 4–6 on page 594.

Check It Out
Lesson Tutorials
BigIdeasMath com

Key Vocabulary
x-intercept, *p. 592*
y-intercept, *p. 592*
slope-intercept form,
   p. 592

 **Key Ideas**

Intercepts

The **x-intercept** of a line is the
x-coordinate of the point where
the line crosses the x-axis. It occurs
when $y = 0$.

The **y-intercept** of a line is the
y-coordinate of the point where
the line crosses the y-axis. It occurs
when $x = 0$.

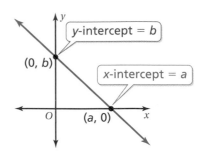

Slope-Intercept Form

Words A linear equation written in the form $y = mx + b$ is
         in **slope-intercept form**. The slope of the line is m,
         and the y-intercept of the line is b.

Algebra $y = mx + b$

                          slope   y-intercept

Study Tip

Linear equations can,
but do not always,
pass through the origin.
So, proportional
relationships are a
special type of linear
equation in which
$b = 0$.

EXAMPLE 1   Identifying Slopes and y-Intercepts

Find the slope and the y-intercept of the graph of each linear equation.

a. $y = -4x - 2$

  $y = -4x + (-2)$ Write in slope-intercept form.

  ∴ The slope is -4, and the y-intercept is -2.

b. $y - 5 = \dfrac{3}{2}x$

  $y = \dfrac{3}{2}x + 5$ Add 5 to each side.

  ∴ The slope is $\dfrac{3}{2}$, and the y-intercept is 5.

On Your Own

Exercises 7–15

Find the slope and the y-intercept of the graph of the linear equation.

  1. $y = 3x - 7$ **2.** $y - 1 = -\dfrac{2}{3}x$

EXAMPLE 2 **Graphing a Linear Equation in Slope-Intercept Form**

Graph $y = -3x + 3$. Identify the x-intercept.

Step 1: Find the slope and the y-intercept.

$$y = -3x + 3$$

slope — y-intercept

Check

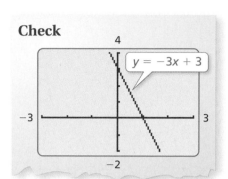

Step 2: The y-intercept is 3. So, plot $(0, 3)$.

Step 3: Use the slope to find another point and draw the line.

$$m = \frac{\text{rise}}{\text{run}} = \frac{-3}{1}$$

Plot the point that is 1 unit right and 3 units down from $(0, 3)$. Draw a line through the two points.

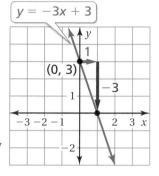

∴ The line crosses the x-axis at $(1, 0)$. So, the x-intercept is 1.

EXAMPLE 3 **Real-Life Application**

The cost y (in dollars) of taking a taxi x miles is $y = 2.5x + 2$.
(a) Graph the equation. (b) Interpret the y-intercept and the slope.

a. The slope of the line is $2.5 = \frac{5}{2}$. Use the slope and the y-intercept to graph the equation.

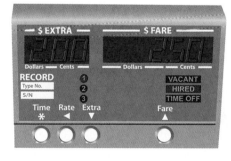

The y-intercept is 2. So, plot $(0, 2)$.

Use the slope to plot another point, $(2, 7)$. Draw a line through the points.

b. The slope is 2.5. So, the cost per mile is \$2.50. The y-intercept is 2. So, there is an initial fee of \$2 to take the taxi.

● **On Your Own**

Now You're Ready
Exercises 18–23

Graph the linear equation. Identify the x-intercept. Use a graphing calculator to check your answer.

3. $y = x - 4$

4. $y = -\frac{1}{2}x + 1$

5. In Example 3, the cost y (in dollars) of taking a different taxi x miles is $y = 2x + 1.5$. Interpret the y-intercept and the slope.

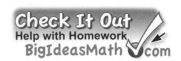
Vocabulary and Concept Check

1. **VOCABULARY** How can you find the x-intercept of the graph of $2x + 3y = 6$?

2. **CRITICAL THINKING** Is the equation $y = 3x$ in slope-intercept form? Explain.

3. **OPEN-ENDED** Describe a real-life situation that you can model with a linear equation. Write the equation. Interpret the y-intercept and the slope.

Practice and Problem Solving

Match the equation with its graph. Identify the slope and the y-intercept.

4. $y = 2x + 1$

5. $y = \dfrac{1}{3}x - 2$

6. $y = -\dfrac{2}{3}x + 1$

A.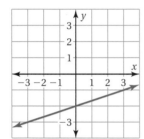

B.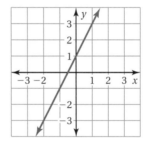

C.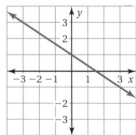

Find the slope and the y-intercept of the graph of the linear equation.

7. $y = 4x - 5$

8. $y = -7x + 12$

9. $y = -\dfrac{4}{5}x - 2$

10. $y = 2.25x + 3$

11. $y + 1 = \dfrac{4}{3}x$

12. $y - 6 = \dfrac{3}{8}x$

13. $y - 3.5 = -2x$

14. $y = -5 - \dfrac{1}{2}x$

15. $y = 11 + 1.5x$

16. **ERROR ANALYSIS** Describe and correct the error in finding the slope and the y-intercept of the graph of the linear equation.

$$y = 4x - 3$$

The slope is 4, and the y-intercept is 3.

17. **SKYDIVING** A skydiver parachutes to the ground. The height y (in feet) of the skydiver after x seconds is $y = -10x + 3000$.

   a. Graph the equation.

   b. Interpret the x-intercept and the slope.

Graph the linear equation. Identify the x-intercept. Use a graphing calculator to check your answer.

② 18. $y = \frac{1}{5}x + 3$

19. $y = 6x - 7$

20. $y = -\frac{8}{3}x + 9$

21. $y = -1.4x - 1$

22. $y + 9 = -3x$

23. $y = 4 - \frac{3}{5}x$

24. APPLES You go to a harvest festival and pick apples.

   a. Which equation represents the cost (in dollars) of going to the festival and picking x pounds of apples? Explain.

      $y = 5x + 0.75$ $y = 0.75x + 5$

   b. Graph the equation you chose in part (a).

Admission: \$5.00
Apples: \$0.75 per lb

25. REASONING Without graphing, identify the equations of the lines that are (a) parallel and (b) perpendicular. Explain your reasoning.

$y = 2x + 4$ $y = -\frac{1}{3}x - 1$ $y = -3x - 2$ $y = \frac{1}{2}x + 1$

$y = 3x + 3$ $y = -\frac{1}{2}x + 2$ $y = -3x + 5$ $y = 2x - 3$

26. **Critical Thinking** Six friends create a website. The website earns money by selling banner ads. The site has 5 banner ads. It costs \$120 a month to operate the website.

   a. A banner ad earns \$0.005 per click. Write a linear equation that represents the monthly income y (in dollars) for x clicks.

   b. Graph the equation in part (a). On the graph, label the number of clicks needed for the friends to start making a profit.

Ⓐ **Fair Game Review** What you learned in previous grades & lessons

Solve the equation for y. *(Topic 3)*

27. $y - 2x = 3$ **28.** $4x + 5y = 13$ **29.** $2x - 3y = 6$ **30.** $7x + 4y = 8$

31. MULTIPLE CHOICE Which point is a solution of the equation $3x - 8y = 11$? *(Section 13.1)*

   Ⓐ $(1, 1)$ **Ⓑ** $(1, -1)$ **Ⓒ** $(-1, 1)$ **Ⓓ** $(-1, -1)$

Essential Question How can you describe the graph of the equation $ax + by = c$?

1 ACTIVITY: Using a Table to Plot Points

Work with a partner. You sold a total of $16 worth of tickets to a school concert. You lost track of how many of each type of ticket you sold.

$$\boxed{} \cdot \frac{\text{Number of}}{\text{adult tickets}} + \boxed{} \cdot \frac{\text{Number of}}{\text{student tickets}} = \boxed{}$$

a. Let x represent the number of adult tickets.

Let y represent the number of student tickets.

Write an equation that relates x and y.

b. Copy and complete the table showing the different combinations of tickets you might have sold.

Number of Adult Tickets, x					
Number of Student Tickets, y					

c. Plot the points from the table. Describe the pattern formed by the points.

d. If you remember how many adult tickets you sold, can you determine how many student tickets you sold? Explain your reasoning.

COMMON CORE

Graphing Equations

In this lesson, you will

• graph linear equations written in standard form.

Applying Standard 8.EE.6

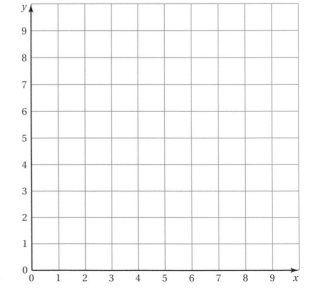

ACTIVITY: Rewriting an Equation

Work with a partner. You sold a total of $16 worth of cheese. You forgot how many pounds of each type of cheese you sold.

CHEESE FOR SALE
Swiss: $4/lb   Cheddar: $2/lb

$$\boxed{} \over \text{pound}} \cdot \begin{array}{c}\text{Pounds}\\\text{of swiss}\end{array} + {\boxed{} \over \text{pound}} \cdot \begin{array}{c}\text{Pounds of}\\\text{cheddar}\end{array} = \boxed{}$$

Math Practice 2

Understand Quantities

What do the equation and the graph represent? How can you use this information to solve the problem?

a. Let x represent the number of pounds of swiss cheese.

Let y represent the number of pounds of cheddar cheese.

Write an equation that relates x and y.

b. Rewrite the equation in slope-intercept form. Then graph the equation.

c. You sold 2 pounds of cheddar cheese. How many pounds of swiss cheese did you sell?

d. Does the value $x = 2.5$ make sense in the context of the problem? Explain.

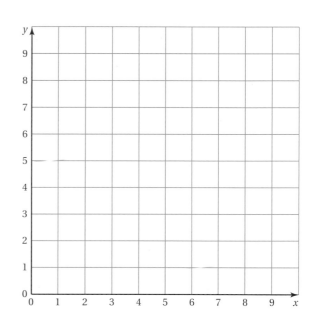

What Is Your Answer?

3. **IN YOUR OWN WORDS** How can you describe the graph of the equation $ax + by = c$?

4. Activities 1 and 2 show two different methods for graphing $ax + by = c$. Describe the two methods. Which method do you prefer? Explain.

5. Write a real-life problem that is similar to those shown in Activities 1 and 2.

6. Why do you think it might be easier to graph $x + y = 10$ without rewriting it in slope-intercept form and then graphing?

Practice

Use what you learned about graphing linear equations in standard form to complete Exercises 3 and 4 on page 600.

Check It Out
Lesson Tutorials
BigIdeasMath com

Key Vocabulary
standard form, *p. 598*

Study Tip

Any linear equation can be written in standard form.

Key Idea

Standard Form of a Linear Equation
The **standard form** of a linear equation is

$$ax + by = c$$

where *a* and *b* are not both zero.

EXAMPLE 1 **Graphing a Linear Equation in Standard Form**

Graph $-2x + 3y = -6$.

Step 1: Write the equation in slope-intercept form.

$-2x + 3y = -6$	Write the equation.
$3y = 2x - 6$	Add 2x to each side.
$y = \frac{2}{3}x - 2$	Divide each side by 3.

Step 2: Use the slope and the *y*-intercept to graph the equation.

$$y = \frac{2}{3}x + (-2)$$

slope   *y*-intercept

Check

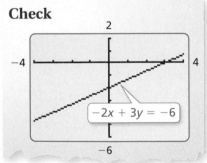

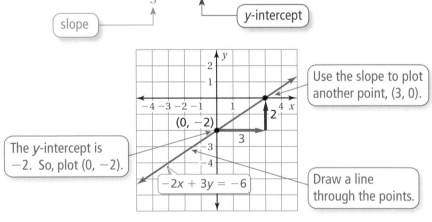

Use the slope to plot another point, (3, 0).

The *y*-intercept is −2. So, plot (0, −2).

$-2x + 3y = -6$

Draw a line through the points.

On Your Own

Now You're Ready
Exercises 5–10

Graph the linear equation. Use a graphing calculator to check your graph.

1. $x + y = -2$

2. $-\frac{1}{2}x + 2y = 6$

3. $-\frac{2}{3}x + y = 0$

4. $2x + y = 5$

EXAMPLE **2** **Graphing a Linear Equation in Standard Form**

Graph $x + 3y = -3$ using intercepts.

Step 1: To find the x-intercept, substitute 0 for y.

$$x + 3y = -3$$
$$x + 3(0) = -3$$
$$x = -3$$

To find the y-intercept, substitute 0 for x.

$$x + 3y = -3$$
$$0 + 3y = -3$$
$$y = -1$$

Step 2: Graph the equation.

Check

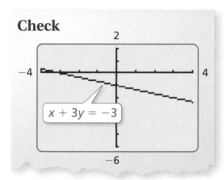

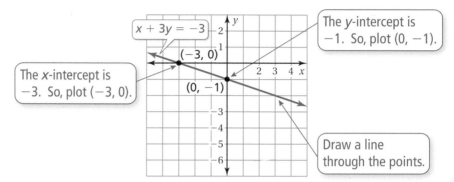

The y-intercept is -1. So, plot $(0, -1)$.

The x-intercept is -3. So, plot $(-3, 0)$.

Draw a line through the points.

EXAMPLE **3** **Real-Life Application**

Bananas
$0.60/pound

Apples
$1.50/pound

You have $6 to spend on apples and bananas. **(a) Graph the equation $1.5x + 0.6y = 6$, where x is the number of pounds of apples and y is the number of pounds of bananas. (b) Interpret the intercepts.**

a. Find the intercepts and graph the equation.

x-intercept	**y-intercept**
$1.5x + 0.6y = 6$	$1.5x + 0.6y = 6$
$1.5x + 0.6(0) = 6$	$1.5(0) + 0.6y = 6$
$x = 4$	$y = 10$

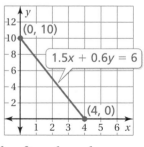

b. The x-intercept shows that you can buy 4 pounds of apples when you do not buy any bananas. The y-intercept shows that you can buy 10 pounds of bananas when you do not buy any apples.

On Your Own

Now You're Ready
Exercises 16–18

Graph the linear equation using intercepts. Use a graphing calculator to check your graph.

5. $2x - y = 8$

6. $x + 3y = 6$

7. **WHAT IF?** In Example 3, you buy y pounds of oranges instead of bananas. Oranges cost $1.20 per pound. Graph the equation $1.5x + 1.2y = 6$. Interpret the intercepts.

Vocabulary and Concept Check

1. **VOCABULARY** Is the equation $y = -2x + 5$ in standard form? Explain.

2. **WRITING** Describe two ways to graph the equation $4x + 2y = 6$.

Practice and Problem Solving

Define two variables for the verbal model. Write an equation in slope-intercept form that relates the variables. Graph the equation.

3. $\dfrac{\$2.00}{\text{pound}}$ · Pounds of peaches $+$ $\dfrac{\$1.50}{\text{pound}}$ · Pounds of apples $=$ $\$15$

4. $\dfrac{16 \text{ miles}}{\text{hour}}$ · Hours biked $+$ $\dfrac{2 \text{ miles}}{\text{hour}}$ · Hours walked $=$ $\dfrac{32}{\text{miles}}$

Write the linear equation in slope-intercept form.

  5. $2x + y = 17$

6. $5x - y = \dfrac{1}{4}$

7. $-\dfrac{1}{2}x + y = 10$

Graph the linear equation. Use a graphing calculator to check your graph.

8. $-18x + 9y = 72$

9. $16x - 4y = 2$

10. $\dfrac{1}{4}x + \dfrac{3}{4}y = 1$

Match the equation with its graph.

11. $15x - 12y = 60$

12. $5x + 4y = 20$

13. $10x + 8y = -40$

A.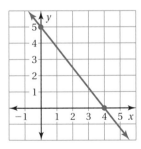

B.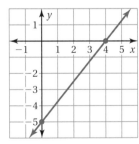

C.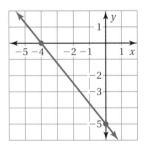

14. **ERROR ANALYSIS** Describe and correct the error in finding the x-intercept.

15. **BRACELET** A charm bracelet costs $65, plus $25 for each charm. The equation $-25x + y = 65$ represents the cost y of the bracelet, where x is the number of charms.

    a. Graph the equation.

    b. How much does the bracelet shown cost?

$-2x + 3y = 12$
$-2(0) + 3y = 12$
$3y = 12$
$y = 4$

Graph the linear equation using intercepts. Use a graphing calculator to check your graph.

② 16. $3x - 4y = -12$

17. $2x + y = 8$

18. $\frac{1}{3}x - \frac{1}{6}y = -\frac{2}{3}$

19. SHOPPING The amount of money you spend on x CDs and y DVDs is given by the equation $14x + 18y = 126$. Find the intercepts and graph the equation.

20. SCUBA Five friends go scuba diving. They rent a boat for x days and scuba gear for y days. The total spent is $1000.

Boat: $250/day
Gear: $50/day

a. Write an equation in standard form that represents the situation.

b. Graph the equation and interpret the intercepts.

21. MODELING You work at a restaurant as a host and a server. You earn $9.45 for each hour you work as a host and $7.65 for each hour you work as a server.

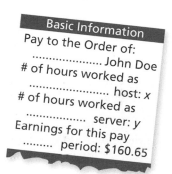

Basic Information
Pay to the Order of:
.... John Doe
# of hours worked as
.... host: x
# of hours worked as
.... server: y
Earnings for this pay
.... period: $160.65

a. Write an equation in standard form that models your earnings.

b. Graph the equation.

22. LOGIC Does the graph of every linear equation have an x-intercept? Explain your reasoning. Include an example.

23. **Critical Thinking** For a house call, a veterinarian charges $70, plus $40 an hour.

a. Write an equation that represents the total fee y (in dollars) the veterinarian charges for a visit lasting x hours.

b. Find the x-intercept. Does this value make sense in this context? Explain your reasoning.

c. Graph the equation.

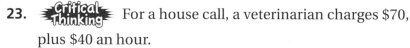

Fair Game Review *What you learned in previous grades & lessons*

The points in the table lie on a line. Find the slope of the line. *(Section 13.2)*

24.

x	-2	-1	0	1
y	-10	-6	-2	2

25.

x	2	4	6	8
y	2	3	4	5

26. MULTIPLE CHOICE Which value of x makes the equation $4x - 12 = 3x - 9$ true? *(Topic 2)*

Ⓐ -1　　　Ⓑ 0　　　Ⓒ 1　　　Ⓓ 3

Essential Question

How can you write an equation of a line when you are given the slope and the *y*-intercept of the line?

---

1 ACTIVITY: Writing Equations of Lines

Work with a partner.

- **Find the slope of each line.**

- **Find the *y*-intercept of each line.**

- **Write an equation for each line.**

- **What do the three lines have in common?**

a.
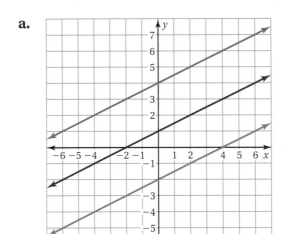

b.
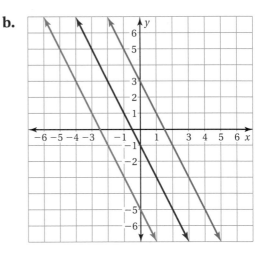

c.
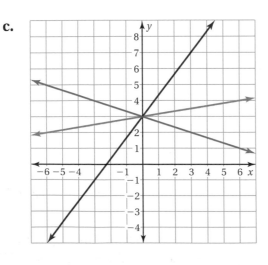

d.
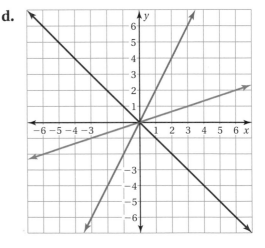

COMMON CORE

Writing Equations
In this lesson, you will
- write equations of lines in slope-intercept form.

Applying Standard
8.EE.6

ACTIVITY: Describing a Parallelogram

Math Practice **1**

Analyze Givens
What do you need to know to write an equation?

Work with a partner.

- Find the area of each parallelogram.
- Write an equation that represents each side of each parallelogram.

a.
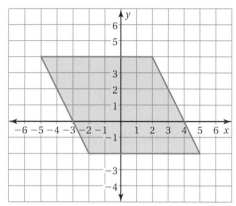

b.
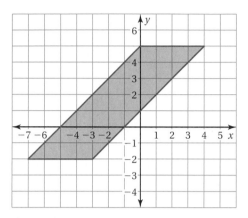

3 **ACTIVITY: Interpreting the Slope and the y-Intercept**

Work with a partner. The graph shows a trip taken by a car, where t is the time (in hours) and y is the distance (in miles) from Phoenix.

a. Find the y-intercept of the graph. What does it represent?

b. Find the slope of the graph. What does it represent?

c. How long did the trip last?

d. How far from Phoenix was the car at the end of the trip?

e. Write an equation that represents the graph.

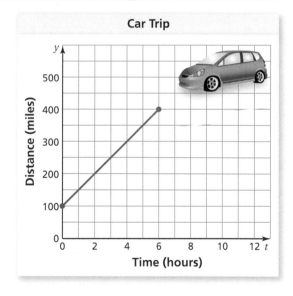

Car Trip

What Is Your Answer?

4. **IN YOUR OWN WORDS** How can you write an equation of a line when you are given the slope and the y-intercept of the line? Give an example that is different from those in Activities 1, 2, and 3.

5. Two sides of a parallelogram are represented by the equations $y = 2x + 1$ and $y = -x + 3$. Give two equations that can represent the other two sides.

Practice

Use what you learned about writing equations in slope-intercept form to complete Exercises 3 and 4 on page 606.

13.6 Lesson

EXAMPLE **1** **Writing Equations in Slope-Intercept Form**

Write an equation of the line in slope-intercept form.

a.

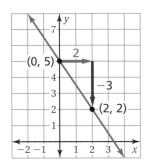

Find the slope and the y-intercept.

$$m = \frac{y_2 - y_1}{x_2 - x_1}$$

$$= \frac{2 - 5}{2 - 0}$$

$$= \frac{-3}{2}, \text{ or } -\frac{3}{2}$$

Because the line crosses the y-axis at $(0, 5)$, the y-intercept is 5.

Study Tip

After writing an equation, check that the given points are solutions of the equation.

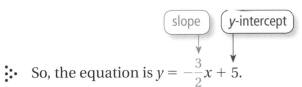

slope y-intercept

∴ So, the equation is $y = -\frac{3}{2}x + 5$.

b.

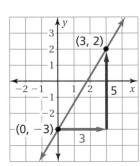

Find the slope and the y-intercept.

$$m = \frac{y_2 - y_1}{x_2 - x_1}$$

$$= \frac{-3 - 2}{0 - 3}$$

$$= \frac{-5}{-3}, \text{ or } \frac{5}{3}$$

Because the line crosses the y-axis at $(0, -3)$, the y-intercept is -3.

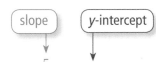

slope y-intercept

∴ So, the equation is $y = \frac{5}{3}x + (-3)$, or $y = \frac{5}{3}x - 3$.

On Your Own

Now You're Ready
Exercises 5–10

Write an equation of the line in slope-intercept form.

1.

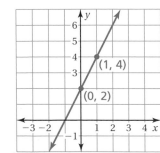

2.

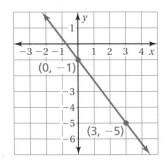

EXAMPLE 2 Writing an Equation

Which equation is shown in the graph?

(A) $y = -4$ (B) $y = -3$

(C) $y = 0$ (D) $y = -3x$

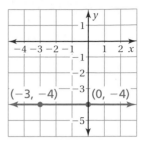

Remember

The graph of $y = a$ is a horizontal line that passes through $(0, a)$.

Find the slope and the y-intercept.

The line is horizontal, so the change in y is 0.

$$m = \frac{\text{change in } y}{\text{change in } x} = \frac{0}{3} = 0$$

Because the line crosses the y-axis at $(0, -4)$, the y-intercept is -4.

So, the equation is $y = 0x + (-4)$, or $y = -4$. The correct answer is (A).

EXAMPLE 3 Real-Life Application

The graph shows the distance remaining to complete a tunnel.
(a) Write an equation that represents the distance y (in feet) remaining after x months. (b) How much time does it take to complete the tunnel?

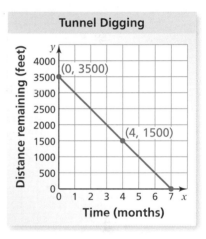

Tunnel Digging

a. Find the slope and the y-intercept.

$$m = \frac{\text{change in } y}{\text{change in } x} = \frac{-2000}{4} = -500$$

Because the line crosses the y-axis at $(0, 3500)$, the y-intercept is 3500.

So, the equation is $y = -500x + 3500$.

Engineers used tunnel boring machines like the ones shown above to dig an extension of the Metro Gold Line in Los Angeles. The new tunnels are 1.7 miles long and 21 feet wide.

b. The tunnel is complete when the distance remaining is 0 feet. So, find the value of x when $y = 0$.

$y = -500x + 3500$	Write the equation.
$0 = -500x + 3500$	Substitute 0 for y.
$-3500 = -500x$	Subtract 3500 from each side.
$7 = x$	Divide each side by -500.

It takes 7 months to complete the tunnel.

On Your Own

Exercises 13–15

3. Write an equation of the line that passes through $(0, 5)$ and $(4, 5)$.

4. **WHAT IF?** In Example 3, the points are $(0, 3500)$ and $(5, 1500)$. How long does it take to complete the tunnel?

 Vocabulary and Concept Check

1. **PRECISION** Explain how to find the slope of a line given the intercepts of the line.

2. **WRITING** Explain how to write an equation of a line using its graph.

 Practice and Problem Solving

Write an equation that represents each side of the figure.

3.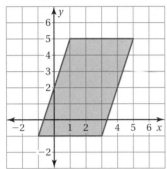

4.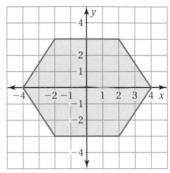

Write an equation of the line in slope-intercept form.

5.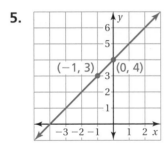

6.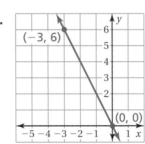

7.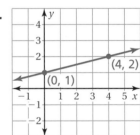

8.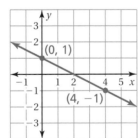

9.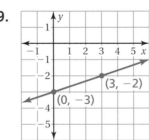

10.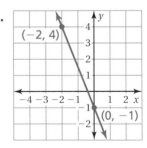

11. **ERROR ANALYSIS** Describe and correct the error in writing an equation of the line.

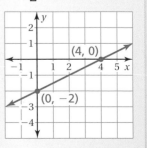
$$y = \frac{1}{2}x + 4$$

12. **BOA** A boa constrictor is 18 inches long at birth and grows 8 inches per year. Write an equation that represents the length y (in feet) of a boa constrictor that is x years old.

Write an equation of the line that passes through the points.

② **13.** $(2, 5), (0, 5)$ **14.** $(-3, 0), (0, 0)$ **15.** $(0, -2), (4, -2)$

16. WALKATHON One of your friends gives you $10 for a charity walkathon. Another friend gives you an amount per mile. After 5 miles, you have raised $13.50 total. Write an equation that represents the amount y of money you have raised after x miles.

17. BRAKING TIME During each second of braking, an automobile slows by about 10 miles per hour.

     a. Plot the points $(0, 60)$ and $(6, 0)$. What do the points represent?

     b. Draw a line through the points. What does the line represent?

     c. Write an equation of the line.

18. PAPER You have 500 sheets of notebook paper. After 1 week, you have 72% of the sheets left. You use the same number of sheets each week. Write an equation that represents the number y of pages remaining after x weeks.

19. **Critical Thinking** The palm tree on the left is 10 years old. The palm tree on the right is 8 years old. The trees grow at the same rate.

     a. Estimate the height y (in feet) of each tree.

     b. Plot the two points (x, y), where x is the age of each tree and y is the height of each tree.

     c. What is the rate of growth of the trees?

     d. Write an equation that represents the height of a palm tree in terms of its age.

6 ft

 Fair Game Review *What you learned in previous grades & lessons*

Plot the ordered pair in a coordinate plane. *(Skills Review Handbook)*

20. $(1, 4)$ **21.** $(-1, -2)$ **22.** $(0, 1)$ **23.** $(2, 7)$

24. MULTIPLE CHOICE Which of the following statements is true? *(Section 13.4)*

     Ⓐ The x-intercept is 5.

     Ⓑ The x-intercept is -2.

     Ⓒ The y-intercept is 5.

     Ⓓ The y-intercept is -2.

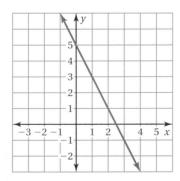

Essential Question
How can you write an equation of a line when you are given the slope and a point on the line?

> **1** **ACTIVITY: Writing Equations of Lines**

Work with a partner.

- Sketch the line that has the given slope and passes through the given point.
- Find the y-intercept of the line.
- Write an equation of the line.

a. $m = -2$

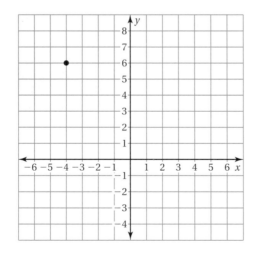

b. $m = \dfrac{1}{3}$

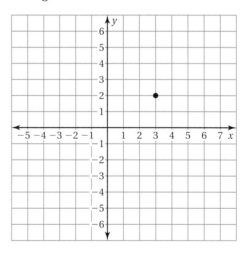

c. $m = -\dfrac{2}{3}$

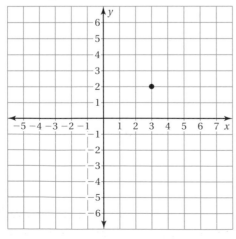

d. $m = \dfrac{5}{2}$

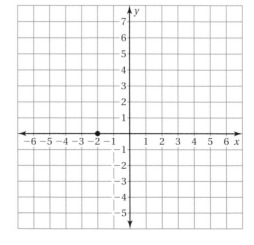

COMMON CORE

Writing Equations

In this lesson, you will
- write equations of lines using a slope and a point.
- write equations of lines using two points.

Applying Standard
8.EE.6

ACTIVITY: Deriving an Equation

Work with a partner.

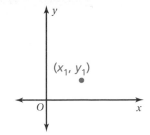

a. Draw a nonvertical line that passes through the point (x_1, y_1).

b. Plot another point on your line. Label this point as (x, y). This point represents any other point on the line.

Math Practice 3

Construct Arguments

How does a graph help you derive an equation?

c. Label the rise and the run of the line through the points (x_1, y_1) and (x, y).

d. The rise can be written as $y - y_1$. The run can be written as $x - x_1$. Explain why this is true.

e. Write an equation for the slope m of the line using the expressions from part (d).

f. Multiply each side of the equation by the expression in the denominator. Write your result. What does this result represent?

---

3 **ACTIVITY: Writing an Equation**

Work with a partner.

For 4 months, you saved $25 a month. You now have $175 in your savings account.

- Draw a graph that shows the balance in your account after t months.

- Use your result from Activity 2 to write an equation that represents the balance A after t months.

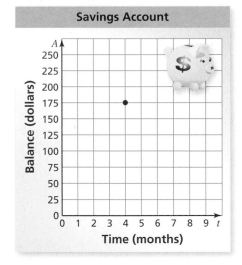

Savings Account

What Is Your Answer?

4. Redo Activity 1 using the equation you found in Activity 2. Compare the results. What do you notice?

5. Why do you think $y - y_1 = m(x - x_1)$ is called the *point-slope form* of the equation of a line? Why do you think it is important?

6. **IN YOUR OWN WORDS** How can you write an equation of a line when you are given the slope and a point on the line? Give an example that is different from those in Activity 1.

Practice

Use what you learned about writing equations using a slope and a point to complete Exercises 3–5 on page 612.

Key Vocabulary 🔊
point-slope form,
   p. 610

 Key Idea

Point-Slope Form

Words A linear equation written in the form $y - y_1 = m(x - x_1)$ is in **point-slope form**. The line passes through the point (x_1, y_1), and the slope of the line is m.

Algebra

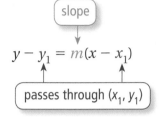

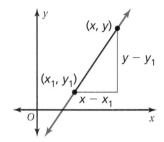

slope

$$y - y_1 = m(x - x_1)$$

passes through (x_1, y_1)

EXAMPLE ① **Writing an Equation Using a Slope and a Point**

Write in point-slope form an equation of the line that passes through the point $(-6, 1)$ with slope $\frac{2}{3}$.

$y - y_1 = m(x - x_1)$ Write the point-slope form.

$y - 1 = \frac{2}{3}[x - (-6)]$ Substitute $\frac{2}{3}$ for m, -6 for x_1, and 1 for y_1.

$y - 1 = \frac{2}{3}(x + 6)$ Simplify.

∴ So, the equation is $y - 1 = \frac{2}{3}(x + 6)$.

Check Check that $(-6, 1)$ is a solution of the equation.

$y - 1 = \frac{2}{3}(x + 6)$ Write the equation.

$1 - 1 \overset{?}{=} \frac{2}{3}(-6 + 6)$ Substitute.

$0 = 0$ ✔   Simplify.

On Your Own

Now You're Ready
Exercises 6–11

Write in point-slope form an equation of the line that passes through the given point and has the given slope.

1. $(1, 2)$; $m = -4$ **2.** $(7, 0)$; $m = 1$ **3.** $(-8, -5)$; $m = -\frac{3}{4}$

EXAMPLE 2

Writing an Equation Using Two Points

Study Tip

You can use either of the given points to write the equation of the line.

Use $m = -2$ and $(5, -2)$.

$y - (-2) = -2(x - 5)$
$y + 2 = -2x + 10$
$y = -2x + 8$ ✔

Write in slope-intercept form an equation of the line that passes through the points $(2, 4)$ and $(5, -2)$.

Find the slope: $m = \dfrac{y_2 - y_1}{x_2 - x_1} = \dfrac{-2 - 4}{5 - 2} = \dfrac{-6}{3} = -2$

Then use the slope $m = -2$ and the point $(2, 4)$ to write an equation of the line.

$y - y_1 = m(x - x_1)$	Write the point-slope form.
$y - 4 = -2(x - 2)$	Substitute -2 for m, 2 for x_1, and 4 for y_1.
$y - 4 = -2x + 4$	Distributive Property
$y = -2x + 8$	Write in slope-intercept form.

EXAMPLE 3

Real-Life Application

You finish parasailing and are being pulled back to the boat. After 2 seconds, you are 25 feet above the boat. (a) Write and graph an equation that represents your height y (in feet) above the boat after x seconds. (b) At what height were you parasailing?

a. You are being pulled down at the rate of 10 feet per second. So, the slope is -10. You are 25 feet above the boat after 2 seconds. So, the line passes through $(2, 25)$. Use the point-slope form.

$y - 25 = -10(x - 2)$	Substitute for m, x_1, and y_1.
$y - 25 = -10x + 20$	Distributive Property
$y = -10x + 45$	Write in slope-intercept form.

⋮ So, the equation is $y = -10x + 45$.

b. You start descending when $x = 0$. The y-intercept is 45. So, you were parasailing at a height of 45 feet.

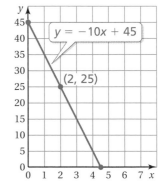

On Your Own

Now You're Ready
Exercises 12–17

Write in slope-intercept form an equation of the line that passes through the given points.

4. $(-2, 1), (3, -4)$ **5.** $(-5, -5), (-3, 3)$ **6.** $(-8, 6), (-2, 9)$

7. WHAT IF? In Example 3, you are 35 feet above the boat after 2 seconds. Write and graph an equation that represents your height y (in feet) above the boat after x seconds.

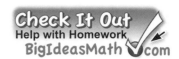

Vocabulary and Concept Check

1. **VOCABULARY** From the equation $y - 3 = -2(x + 1)$, identify the slope and a point on the line.

2. **WRITING** Describe how to write an equation of a line using (a) its slope and a point on the line and (b) two points on the line.

Practice and Problem Solving

Use the point-slope form to write an equation of the line with the given slope that passes through the given point.

3. $m = \dfrac{1}{2}$

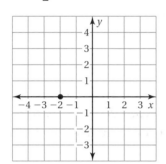

4. $m = -\dfrac{3}{4}$

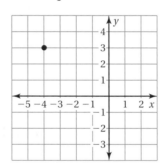

5. $m - {-}3$

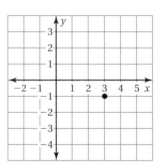

Write in point-slope form an equation of the line that passes through the given point and has the given slope.

1   6. $(3, 0)$; $m = -\dfrac{2}{3}$

7. $(4, 8)$; $m = \dfrac{3}{4}$

8. $(1, -3)$; $m = 4$

9. $(7, -5)$; $m = -\dfrac{1}{7}$

10. $(3, 3)$; $m = \dfrac{5}{3}$

11. $(-1, -4)$; $m = -2$

Write in slope-intercept form an equation of the line that passes through the given points.

2   12. $(-1, -1), (1, 5)$

13. $(2, 4), (3, 6)$

14. $(-2, 3), (2, 7)$

15. $(4, 1), (8, 2)$

16. $(-9, 5), (-3, 3)$

17. $(1, 2), (-2, -1)$

18. **CHEMISTRY** At $0\,°C$, the volume of a gas is 22 liters. For each degree the temperature T (in degrees Celsius) increases, the volume V (in liters) of the gas increases by $\dfrac{2}{25}$. Write an equation that represents the volume of the gas in terms of the temperature.

19. CARS After it is purchased, the value of a new car decreases $4000 each year. After 3 years, the car is worth $18,000.

    a. Write an equation that represents the value V (in dollars) of the car x years after it is purchased.

    b. What was the original value of the car?

20. REASONING Write an equation of a line that passes through the point (8, 2) that is (a) parallel and (b) perpendicular to the graph of the equation $y = 4x - 3$.

21. CRICKETS According to Dolbear's law, you can predict the temperature T (in degrees Fahrenheit) by counting the number x of chirps made by a snowy tree cricket in 1 minute. For each rise in temperature of 0.25°F, the cricket makes an additional chirp each minute.

    a. A cricket chirps 40 times in 1 minute when the temperature is 50°F. Write an equation that represents the temperature in terms of the number of chirps in 1 minute.

    b. You count 100 chirps in 1 minute. What is the temperature?

    c. The temperature is 96°F. How many chirps would you expect the cricket to make?

Leaning Tower of Pisa

(10.75, 42)

7.75 m

22. WATERING CAN You water the plants in your classroom at a constant rate. After 5 seconds, your watering can contains 58 ounces of water. Fifteen seconds later, the can contains 28 ounces of water.

    a. Write an equation that represents the amount y (in ounces) of water in the can after x seconds.

    b. How much water was in the can when you started watering the plants?

    c. When is the watering can empty?

23. **Problem Solving** The Leaning Tower of Pisa in Italy was built between 1173 and 1350.

    a. Write an equation for the yellow line.

    b. The tower is 56 meters tall. How far off center is the top of the tower?

 Fair Game Review *What you learned in previous grades & lessons*

Graph the linear equation. *(Section 13.4)*

24. $y = 4x$ **25.** $y = -2x + 1$ **26.** $y = 3x - 5$

27. MULTIPLE CHOICE What is the x-intercept of the equation $3x + 5y = 30$? *(Section 13.5)*

    A -10 **B** -6 **C** 6 **D** 10

Find the slope and the *y*-intercept of the graph of the linear equation. *(Section 13.4)*

1. $y = \frac{1}{4}x - 8$

2. $y = -x + 3$

Find the *x*- and *y*-intercepts of the graph of the equation. *(Section 13.5)*

3. $3x - 2y = 12$

4. $x + 5y = 15$

Write an equation of the line in slope-intercept form. *(Section 13.6)*

5.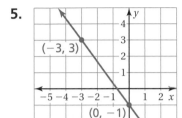

6.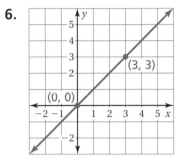

7.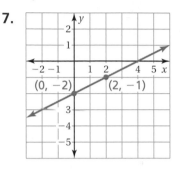

Write in point-slope form an equation of the line that passes through the given point and has the given slope. *(Section 13.7)*

8. $(1, 3)$; $m = 2$

9. $(-3, -2)$; $m = \frac{1}{3}$

10. $(-1, 4)$; $m = -1$

11. $(8, -5)$; $m = -\frac{1}{8}$

Write in slope-intercept form an equation of the line that passes through the given points. *(Section 13.7)*

12. $\left(0, -\frac{2}{3}\right), \left(-3, -\frac{2}{3}\right)$

13. $(4, 0), (0, 4)$

14. STATE FAIR The cost *y* (in dollars) of one person buying admission to a fair and going on *x* rides is $y = x + 12$. *(Section 13.4)*

   a. Graph the equation.

   b. Interpret the *y*-intercept and the slope.

15. PAINTING You used $90 worth of paint for a school float. *(Section 13.5)*

   a. Graph the equation $18x + 15y = 90$, where *x* is the number of gallons of blue paint and *y* is the number of gallons of white paint.

   b. Interpret the intercepts.

16. CONSTRUCTION A construction crew is extending a highway sound barrier that is 13 miles long. The crew builds $\frac{1}{2}$ of a mile per week. Write an equation that represents the length *y* (in miles) of the barrier after *x* weeks. *(Section 13.6)*

Review Key Vocabulary

linear equation *p. 568*
solution of a linear equation, *p. 568*
slope, *p. 574*
rise, *p. 574*
run, *p. 574*

x-intercept, *p. 592*
y-intercept, *p. 592*
slope-intercept form, *p. 592*
standard form, *p. 598*
point-slope form, *p. 610*

Review Examples and Exercises

13.1 **Graphing Linear Equations** *(pp. 566–571)*

Graph $y = 3x - 1$.

Step 1: Make a table of values.

x	y = 3x − 1	y	(x, y)
−2	$y = 3(-2) - 1$	−7	(−2, −7)
−1	$y = 3(-1) - 1$	−4	(−1, −4)
0	$y = 3(0) - 1$	−1	(0, −1)
1	$y = 3(1) - 1$	2	(1, 2)

Step 2: Plot the ordered pairs.   **Step 3:** Draw a line through the points.

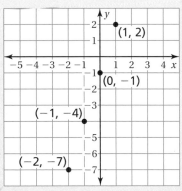
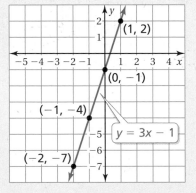

Exercises

Graph the linear equation.

1. $y = \dfrac{3}{5}x$

2. $y = -2$

3. $y = 9 - x$

4. $y = 1$

5. $y = \dfrac{2}{3}x + 2$

6. $x = -5$

Slope of a Line   (pp. 572–581)

Find the slope of each line in the graph.

Red Line: $m = \dfrac{y_2 - y_1}{x_2 - x_1} = \dfrac{5 - (-3)}{2 - 2} = \dfrac{8}{0}$

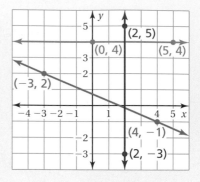

∴ The slope of the red line is undefined.

Blue Line: $m = \dfrac{y_2 - y_1}{x_2 - x_1} = \dfrac{-1 - 2}{4 - (-3)} = \dfrac{-3}{7}$, or $-\dfrac{3}{7}$

Green Line: $m = \dfrac{y_2 - y_1}{x_2 - x_1} = \dfrac{4 - 4}{5 - 0} = \dfrac{0}{5}$, or 0

Exercises

The points in the table lie on a line. Find the slope of the line.

7.

x	0	1	2	3
y	−1	0	1	2

8.

x	−2	0	2	4
y	3	4	5	6

9. Are the lines $x = 2$ and $y = 4$ parallel? Are they perpendicular? Explain.

Graphing Proportional Relationships   (pp. 582–587)

The cost y (in dollars) for x tickets to a movie is represented by the equation $y = 7x$. Graph the equation and interpret the slope.

The equation shows that the slope m is 7. So, the graph passes through (0, 0) and (1, 7).

Plot the points and draw a line through the points. Because negative values of x do not make sense in this context, graph in the first quadrant only.

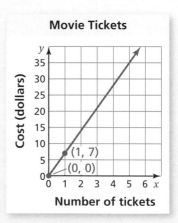

Movie Tickets

∴ The slope indicates that the unit cost is $7 per ticket.

Exercises

10. **RUNNING** The number y of miles you run after x weeks is represented by the equation $y = 8x$. Graph the equation and interpret the slope.

11. **STUDYING** The number y of hours that you study after x days is represented by the equation $y = 1.5x$. Graph the equation and interpret the slope.

13.4 Graphing Linear Equations in Slope-Intercept Form (pp. 590–595)

Graph $y = 0.5x - 3$. Identify the x-intercept.

Step 1: Find the slope and the y-intercept.

$$y = 0.5x + (-3)$$

slope ↑　　　　↑ y-intercept

Step 2: The y-intercept is -3. So, plot $(0, -3)$.

Step 3: Use the slope to find another point and draw the line.

$$m = \frac{\text{rise}}{\text{run}} = \frac{1}{2}$$

Plot the point that is 2 units right and 1 unit up from $(0, -3)$. Draw a line through the two points.

∴ The line crosses the x-axis at $(6, 0)$. So, the x-intercept is 6.

Exercises

Graph the linear equation. Identify the x-intercept. Use a graphing calculator to check your answer.

12. $y = 2x - 6$　　　**13.** $y = -4x + 8$　　　**14.** $y = -x - 8$

13.5 Graphing Linear Equations in Standard Form (pp. 596–601)

Graph $8x + 4y = 16$.

Step 1: Write the equation in slope-intercept form.

$8x + 4y = 16$	Write the equation.
$4y = -8x + 16$	Subtract $8x$ from each side.
$y = -2x + 4$	Divide each side by 4.

Step 2: Use the slope and the y-intercept to graph the equation.

$$y = -2x + 4$$

slope ↑　　　　↑ y-intercept

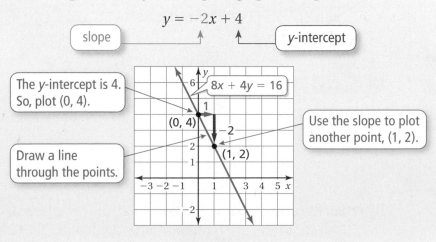

The y-intercept is 4. So, plot $(0, 4)$.

Use the slope to plot another point, $(1, 2)$.

Draw a line through the points.

Exercises

Graph the linear equation.

15. $\frac{1}{4}x + y = 3$

16. $-4x + 2y = 8$

17. $x + 5y = 10$

18. $-\frac{1}{2}x + \frac{1}{8}y = \frac{3}{4}$

19. A dog kennel charges \$30 per night to board your dog and \$6 for each hour of playtime. The amount of money you spend is given by $30x + 6y = 180$, where x is the number of nights and y is the number of hours of playtime. Graph the equation and interpret the intercepts.

13.6 **Writing Equations in Slope-Intercept Form** *(pp. 602–607)*

Write an equation of the line in slope-intercept form.

a.

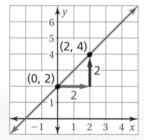

Find the slope and the y-intercept.

$$m = \frac{y_2 - y_1}{x_2 - x_1} = \frac{4 - 2}{2 - 0} = \frac{2}{2}, \text{ or } 1$$

Because the line crosses the y-axis at $(0, 2)$, the y-intercept is 2.

∴ So, the equation is $y = 1x + 2$, or $y = x + 2$.

b.

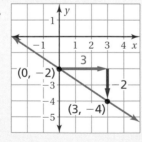

Find the slope and the y-intercept.

$$m = \frac{y_2 - y_1}{x_2 - x_1} = \frac{-4 - (-2)}{3 - 0} = \frac{-2}{3}, \text{ or } -\frac{2}{3}$$

Because the line crosses the y-axis at $(0, -2)$, the y-intercept is -2.

slope y-intercept

∴ So, the equation is $y = -\frac{2}{3}x + (-2)$, or $y = -\frac{2}{3}x - 2$.

Exercises

Write an equation of the line in slope-intercept form.

20.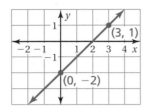

21.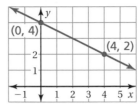

22.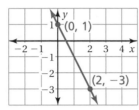

23.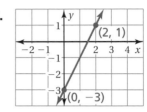

24. Write an equation of the line that passes through (0, 8) and (6, 8).

25. Write an equation of the line that passes through (0, −5) and (−5, −5).

13.7 **Writing Equations in Point-Slope Form** *(pp. 608–613)*

Write in slope-intercept form an equation of the line that passes through the points (2, 1) and (3, 5).

Find the slope.

$$m = \frac{y_2 - y_1}{x_2 - x_1} = \frac{5 - 1}{3 - 2} = \frac{4}{1}, \text{ or } 4$$

Then use the slope and one of the given points to write an equation of the line.

Use $m = 4$ and (2, 1).

$y - y_1 = m(x - x_1)$	Write the point-slope form.
$y - 1 = 4(x - 2)$	Substitute 4 for m, 2 for x_1, and 1 for y_1.
$y - 1 = 4x - 8$	Distributive Property
$y = 4x - 7$	Write in slope-intercept form.

So, the equation is $y = 4x - 7$.

Exercises

26. Write in point-slope form an equation of the line that passes through the point (4, 4) with slope 3.

27. Write in slope-intercept form an equation of the line that passes through the points (−4, 2) and (6, −3).

Find the slope and the y-intercept of the graph of the linear equation.

1. $y = 6x - 5$

2. $y = 20x + 15$

3. $y = -5x - 16$

4. $y - 1 = 3x + 8.4$

5. $y + 4.3 = 0.1x$

6. $-\dfrac{1}{2}x + 2y = 7$

Graph the linear equation.

7. $y = 2x + 4$

8. $y = -\dfrac{1}{2}x - 5$

9. $-3x + 6y = 12$

10. Which lines are parallel? Which lines are perpendicular? Explain.

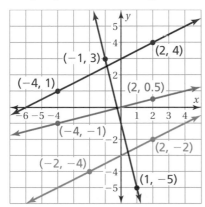

11. The points in the table lie on a line. Find the slope of the line.

x	y
-1	-4
0	-1
1	2
2	5

Write an equation of the line in slope-intercept form.

12.
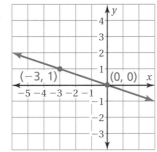

13.
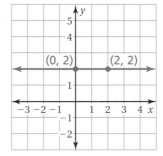

Write in slope-intercept form an equation of the line that passes through the given points.

14. $(-1, 5), (3, -3)$

15. $(-4, 1), (4, 3)$

16. $(-2, 5), (-1, 1)$

17. VOCABULARY The number y of new vocabulary words that you learn after x weeks is represented by the equation $y = 15x$.

    a. Graph the equation and interpret the slope.

    b. How many new vocabulary words do you learn after 5 weeks?

    c. How many more vocabulary words do you learn after 6 weeks than after 4 weeks?

1. Which equation matches the line shown in the graph? *(8.EE.6)*

   A. $y = 2x - 2$

   B. $y = 2x + 1$

   C. $y = x - 2$

   D. $y = x + 1$

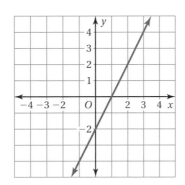

2. The equation $6x - 5y = 14$ is written in standard form. Which point lies on the graph of this equation? *(8.EE.6)*

   F. $(-4, -1)$

   G. $(-2, 4)$

   H. $(-1, -4)$

   I. $(4, -2)$

3. Which line has a slope of 0? *(8.EE.6)*

   A.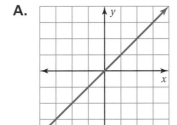

   B.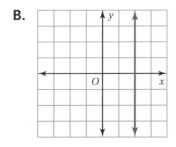

   C.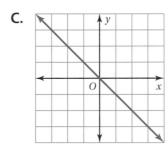

   D.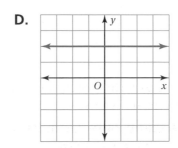

4. Which of the following is the equation of a line perpendicular to the line shown in the graph? *(8.EE.6)*

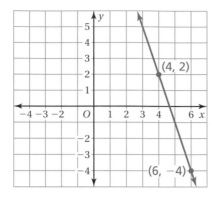

F. $y - 3x - 10$

G. $y = \dfrac{1}{3}x + 12$

H. $y = -3x + 5$

I. $y = -\dfrac{1}{3}x - 18$

5. What is the slope of the line that passes through the points $(2, -2)$ and $(8, 1)$? *(8.EE.6)*

6. A cell phone plan costs $10 per month plus $0.10 for each minute used. Last month, you spent $18.50 using this plan. This can be modeled by the equation below, where m represents the number of minutes used.

$$0.1m + 10 = 18.5$$

How many minutes did you use last month? *(8.EE.7b)*

A. 8.4 min

B. 85 min

C. 185 min

D. 285 min

7. It costs $40 to rent a car for one day. In addition, the rental agency charges you for each mile driven, as shown in the graph. *(8.EE.6)*

Part A Determine the slope of the line joining the points on the graph.

Part B Explain what the slope represents.

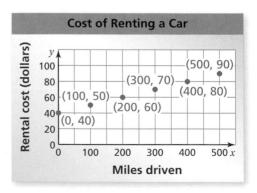

8. What value of x makes the equation below true?   *(8.EE.7a)*

$$7 + 2x = 4x - 5$$

9. Trapezoid *KLMN* is graphed in the coordinate plane shown.

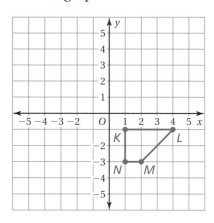

Rotate Trapezoid *KLMN* 90° clockwise about the origin. What are the coordinates of point M', the image of point M after the rotation?   *(8.G.3)*

F. $(-3, -2)$ **H.** $(-2, 3)$

G. $(-2, -3)$ **I.** $(3, 2)$

10. Solve the formula $K = 3M - 7$ for M.   *(8.EE.7b)*

A. $M = K + 7$ **C.** $M = \dfrac{K}{3} + 7$

B. $M = \dfrac{K + 7}{3}$ **D.** $M = \dfrac{K - 7}{3}$

11. What is the distance d across the canyon?   *(8.G.5)*

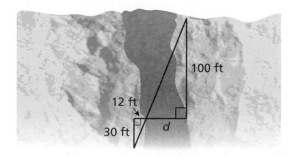

F. 3.6 ft   **H.** 40 ft

G. 12 ft   **I.** 250 ft

14 Real Numbers and the Pythagorean Theorem

"I'm pretty sure that Pythagoras was a Greek."

"I said 'Greek,' not 'Geek.'"

"Here's how I remember the square root of 2."

"February is the 2nd month. It has 28 days. Split 28 into 14 and 14. Move the decimal to get 1.414."

What You Learned Before

Pythagorean Theorem

$(shorter)^2 + (shorter)^2 = (longest)^2$

Speaking of being a square...

"I just remember that the sum of the squares of the shorter sides is equal to the square of the longest side."

● Comparing Decimals (5.NBT.3b)

Complete the number sentence with <, >, or =.

Example 1 1.1 ☐ 1.01

Because $\frac{110}{100}$ is greater than $\frac{101}{100}$, 1.1 is greater than 1.01.

∴ So, 1.1 > 1.01.

Example 2 −0.3 ☐ −0.003

Because $-\frac{300}{1000}$ is less than $-\frac{3}{1000}$, −0.3 is less than −0.003.

∴ So, −0.3 < −0.003.

Example 3 Find three decimals that make the number sentence −5.12 > ☐ true.

Any decimal less than −5.12 will make the sentence true.

∴ *Sample answer:* −10.1, −9.05, −8.25

Try It Yourself

Complete the number sentence with <, >, or =.

1. 2.10 ☐ 2.1

2. −4.5 ☐ −4.25

3. π ☐ 3.2

Find three decimals that make the number sentence true.

4. −0.01 ≤ ☐

5. 1.75 > ☐

6. 0.75 ≥ ☐

● Using Order of Operations (7.NS.1, 7.NS.2)

Example 4 Evaluate $8^2 \div (32 \div 2) - 2(3 - 5)$.

First:	Parentheses	$8^2 \div (32 \div 2) - 2(3 - 5) = 8^2 \div 16 - 2(-2)$
Second:	Exponents	$= 64 \div 16 - 2(-2)$
Third:	Multiplication and Division (from left to right)	$= 4 + 4$
Fourth:	Addition and Subtraction (from left to right)	$= 8$

Try It Yourself

Evaluate the expression.

7. $15\left(\dfrac{12}{3}\right) - 7^2 - 2 \cdot 7$

8. $3^2 \cdot 4 \div 18 + 30 \cdot 6 - 1$

9. $-1 + \left(\dfrac{4}{2}(6 - 1)\right)^2$

14.1 Finding Square Roots

Essential Question How can you find the dimensions of a square or a circle when you are given its area?

When you multiply a number by itself, you square the number.

> Symbol for squaring is the exponent 2. ⟶ $4^2 = 4 \cdot 4$
>
> $= 16$ 4 squared is 16.

To "undo" this, take the *square root* of the number.

> Symbol for square root is a *radical sign*, $\sqrt{}$. ⟶ $\sqrt{16} = \sqrt{4^2} = 4$ The square root of 16 is 4.

1 ACTIVITY: Finding Square Roots

Work with a partner. Use a square root symbol to write the side length of the square. Then find the square root. Check your answer by multiplying.

a. Sample: $s = \sqrt{121} = 11$ ft

Area = 121 ft²

s

s

Check
```
    11
  × 11
    11
   110
   121  ✓
```

:: The side length of the square is 11 feet.

b. Area = 81 yd²

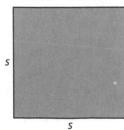

s

s

c. Area = 324 cm²

s

s

d. Area = 361 mi²

s

s

e. Area = 225 mi²

s

s

f. Area = 2.89 in.²

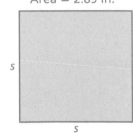

s

s

g. Area = $\frac{4}{9}$ ft²

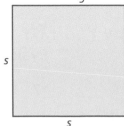

s

s

COMMON CORE

Square Roots
In this lesson, you will
• find square roots of perfect squares.
• evaluate expressions involving square roots.
• use square roots to solve equations.
Learning Standard
8.EE.2

2 ACTIVITY: Using Square Roots

Work with a partner. Find the radius of each circle.

a.

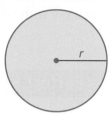

Area = 36π in.^{2}

b.

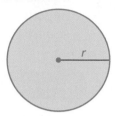

Area = π yd^2

c.

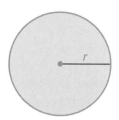

Area = 0.25π ft^2

d.

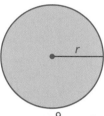

Area = $\dfrac{9}{16}\pi$ m^2

3 ACTIVITY: The Period of a Pendulum

Math Practice 6

Calculate Accurately

How can you use the graph to help you determine whether you calculated the values of *T* correctly?

Work with a partner.

The **period of a pendulum** is the time (in seconds) it takes the pendulum to swing back *and* forth.

The period *T* is represented by $T = 1.1\sqrt{L}$, where *L* is the length of the pendulum (in feet).

Copy and complete the table. Then graph the ordered pairs. Is the equation linear?

L	1.00	1.96	3.24	4.00	4.84	6.25	7.29	7.84	9.00
T									

What Is Your Answer?

4. IN YOUR OWN WORDS How can you find the dimensions of a square or a circle when you are given its area? Give an example of each. How can you check your answers?

Use what you learned about finding square roots to complete Exercises 4–6 on page 630.

14.1 Lesson

Check It Out
Lesson Tutorials
BigIdeasMath com

Key Vocabulary
square root, *p. 628*
perfect square, *p. 628*
radical sign, *p. 628*
radicand, *p. 628*

A **square root** of a number is a number that, when multiplied by itself, equals the given number. Every positive number has a positive *and* a negative square root. A **perfect square** is a number with integers as its square roots.

EXAMPLE 1 — Finding Square Roots of a Perfect Square

Find the two square roots of 49.

$$7 \cdot 7 = 49 \text{ and } (-7) \cdot (-7) = 49$$

Study Tip

Zero has one square root, which is 0.

∴ So, the square roots of 49 are 7 and -7.

The symbol $\sqrt{}$ is called a **radical sign**. It is used to represent a square root. The number under the radical sign is called the **radicand**.

Positive Square Root, $\sqrt{}$	Negative Square Root, $-\sqrt{}$	Both Square Roots, $\pm\sqrt{}$
$\sqrt{16} = 4$	$-\sqrt{16} = -4$	$\pm\sqrt{16} = \pm 4$

EXAMPLE 2 — Finding Square Roots

Find the square root(s).

a. $\sqrt{25}$

> $\sqrt{25}$ represents the *positive* square root.

∴ Because $5^2 = 25$, $\sqrt{25} = \sqrt{5^2} = 5$.

b. $-\sqrt{\dfrac{9}{16}}$

> $-\sqrt{\dfrac{9}{16}}$ represents the *negative* square root.

∴ Because $\left(\dfrac{3}{4}\right)^2 = \dfrac{9}{16}$, $-\sqrt{\dfrac{9}{16}} = -\sqrt{\left(\dfrac{3}{4}\right)^2} = -\dfrac{3}{4}$.

c. $\pm\sqrt{2.25}$

> $\pm\sqrt{2.25}$ represents both the *positive* and the *negative* square roots.

∴ Because $1.5^2 = 2.25$, $\pm\sqrt{2.25} = \pm\sqrt{1.5^2} = 1.5$ and -1.5.

On Your Own

Now You're Ready
Exercises 7–18

Find the two square roots of the number.

1. 36   **2.** 100   **3.** 121

Find the square root(s).

4. $-\sqrt{1}$ **5.** $\pm\sqrt{\dfrac{4}{25}}$ **6.** $\sqrt{12.25}$

Squaring a positive number and finding a square root are inverse operations. You can use this relationship to evaluate expressions and solve equations involving squares.

EXAMPLE 3 **Evaluating Expressions Involving Square Roots**

Evaluate each expression.

a. $5\sqrt{36} + 7 = 5(6) + 7$ Evaluate the square root.

$\qquad\qquad\qquad = 30 + 7$ Multiply.

$\qquad\qquad\qquad = 37$ Add.

b. $\dfrac{1}{4} + \sqrt{\dfrac{18}{2}} = \dfrac{1}{4} + \sqrt{9}$ Simplify.

$\qquad\qquad\quad = \dfrac{1}{4} + 3$ Evaluate the square root.

$\qquad\qquad\quad = 3\dfrac{1}{4}$ Add.

c. $\left(\sqrt{81}\right)^2 - 5 = 81 - 5$ Evaluate the power using inverse operations.

$\qquad\qquad\qquad = 76$ Subtract.

EXAMPLE 4 **Real-Life Application**

The area of a crop circle is 45,216 square feet. What is the radius of the crop circle? Use 3.14 for π.

$A = \pi r^2$ Write the formula for the area of a circle.

$45{,}216 \approx 3.14 r^2$ Substitute 45,216 for A and 3.14 for π.

$14{,}400 = r^2$ Divide each side by 3.14.

$\sqrt{14{,}400} = \sqrt{r^2}$ Take positive square root of each side.

$120 = r$ Simplify.

∴ The radius of the crop circle is about 120 feet.

On Your Own

Now You're Ready
Exercises 20–27

Evaluate the expression.

7. $12 - 3\sqrt{25}$ **8.** $\sqrt{\dfrac{28}{7}} + 2.4$ **9.** $15 - \left(\sqrt{4}\right)^2$

10. The area of a circle is 2826 square feet. Write and solve an equation to find the radius of the circle. Use 3.14 for π.

Vocabulary and Concept Check

1. **VOCABULARY** Is 26 a perfect square? Explain.

2. **REASONING** Can the square of an integer be a negative number? Explain.

3. **NUMBER SENSE** Does $\sqrt{256}$ represent the positive square root of 256, the negative square root of 256, or both? Explain.

Practice and Problem Solving

Find the dimensions of the square or circle. Check your answer.

4.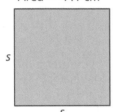
Area = 441 cm²
s
s
s

5.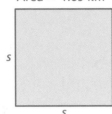
Area = 1.69 km²
s
s

6.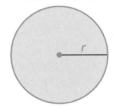
Area = 64 π in.²
r

Find the two square roots of the number.

① 7. 9

8. 64

9. 4

10. 144

Find the square root(s).

② 11. $\sqrt{625}$

12. $\pm\sqrt{196}$

13. $\pm\sqrt{\dfrac{1}{961}}$

14. $-\sqrt{\dfrac{9}{100}}$

15. $\pm\sqrt{4.84}$

16. $\sqrt{7.29}$

17. $-\sqrt{361}$

18. $-\sqrt{2.25}$

19. **ERROR ANALYSIS** Describe and correct the error in finding the square roots.

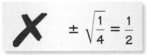

$$\times \quad \pm\sqrt{\dfrac{1}{4}} = \dfrac{1}{2}$$

Evaluate the expression.

③ 20. $\left(\sqrt{9}\right)^2 + 5$

21. $28 - \left(\sqrt{144}\right)^2$

22. $3\sqrt{16} - 5$

23. $10 - 4\sqrt{\dfrac{1}{16}}$

24. $\sqrt{6.76} + 5.4$

25. $8\sqrt{8.41} + 1.8$

26. $2\left(\sqrt{\dfrac{80}{5}} - 5\right)$

27. $4\left(\sqrt{\dfrac{147}{3}} + 3\right)$

28. **NOTEPAD** The area of the base of a square notepad is 2.25 square inches. What is the length of one side of the base of the notepad?

29. **CRITICAL THINKING** There are two square roots of 25. Why is there only one answer for the radius of the button?

$A = 25\pi$ mm²

Copy and complete the statement with <, >, or =.

30. $\sqrt{81}$ ▭ 8

31. 0.5 ▭ $\sqrt{0.25}$

32. $\dfrac{3}{2}$ ▭ $\sqrt{\dfrac{25}{4}}$

33. SAILBOAT The area of a sail is $40\frac{1}{2}$ square feet. The base and the height of the sail are equal. What is the height of the sail (in feet)?

34. REASONING Is the product of two perfect squares always a perfect square? Explain your reasoning.

35. ENERGY The kinetic energy K (in joules) of a falling apple is represented by $K = \dfrac{v^2}{2}$, where v is the speed of the apple (in meters per second). How fast is the apple traveling when the kinetic energy is 32 joules?

Area = 4π cm^2

36. PRECISION The areas of the two watch faces have a ratio of $16:25$.

    a. What is the ratio of the radius of the smaller watch face to the radius of the larger watch face?

    b. What is the radius of the larger watch face?

37. WINDOW The cost C (in dollars) of making a square window with a side length of n inches is represented by $C = \dfrac{n^2}{5} + 175$. A window costs \$355. What is the length (in feet) of the window?

38. Geometry The area of the triangle is represented by the formula $A = \sqrt{s(s-21)(s-17)(s-10)}$, where s is equal to half the perimeter. What is the height of the triangle?

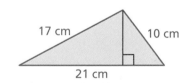

17 cm   10 cm

21 cm

 Fair Game Review *What you learned in previous grades & lessons*

Write in slope-intercept form an equation of the line that passes through the given points. *(Section 13.7)*

39. $(2, 4)$, $(5, 13)$

40. $(-1, 7)$, $(3, -1)$

41. $(-5, -2)$, $(5, 4)$

42. MULTIPLE CHOICE What is the value of x? *(Section 12.2)*

    A 41   **B** 44

    C 88   **D** 134

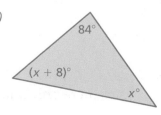

84°

$(x + 8)°$

$x°$

14.2 Finding Cube Roots

Essential Question How is the cube root of a number different from the square root of a number?

When you multiply a number by itself twice, you cube the number.

> Symbol for cubing is the exponent 3.

$4^3 = 4 \cdot 4 \cdot 4$

$= 64$ 4 cubed is 64.

To "undo" this, take the *cube root* of the number.

> Symbol for cube root is $\sqrt[3]{}$.

$\sqrt[3]{64} = \sqrt[3]{4^3} = 4$ The cube root of 64 is 4.

1 ACTIVITY: Finding Cube Roots

Work with a partner. Use a cube root symbol to write the edge length of the cube. Then find the cube root. Check your answer by multiplying.

a. **Sample:**   Volume = 343 in.³

$s = \sqrt[3]{343} = \sqrt[3]{7^3} = 7$ inches

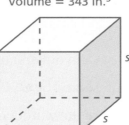

Check

$7 \cdot 7 \cdot 7 = 49 \cdot 7$

$= 343$ ✓

∴ The edge length of the cube is 7 inches.

b.   Volume = 27 ft³

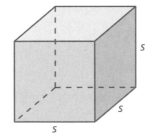

c.   Volume = 125 m³

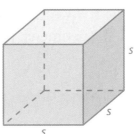

d.   Volume = 0.001 cm³

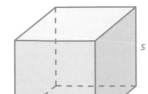

e.   Volume = $\frac{1}{8}$ yd³

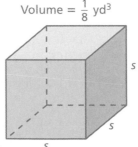

COMMON CORE

Cube Roots

In this lesson, you will
- find cube roots of perfect cubes.
- evaluate expressions involving cube roots.
- use cube roots to solve equations.

Learning Standard 8.EE.2

ACTIVITY: Using Prime Factorizations to Find Cube Roots

Math Practice 7

View as Components

When writing the prime factorizations in Activity 2, how many times do you expect to see each factor? Why?

Work with a partner. Write the prime factorization of each number. Then use the prime factorization to find the cube root of the number.

a. 216

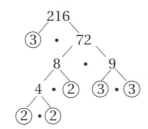

$216 = 3 \cdot 2 \cdot 3 \cdot 3 \cdot 2 \cdot 2$ Prime factorization

$= \left(3 \cdot \boxed{}\right) \cdot \left(3 \cdot \boxed{}\right) \cdot \left(3 \cdot \boxed{}\right)$ Commutative Property of Multiplication

$= \boxed{} \cdot \boxed{} \cdot \boxed{}$ Simplify.

∴ The cube root of 216 is $\boxed{}$.

b. 1000   **c.** 3375

d. **STRUCTURE** Does this procedure work for every number? Explain why or why not.

What Is Your Answer?

3. Complete each statement using *positive* or *negative*.

   a. A positive number times a positive number is a _____ number.

   b. A negative number times a negative number is a _____ number.

   c. A positive number multiplied by itself twice is a _____ number.

   d. A negative number multiplied by itself twice is a _____ number.

4. **REASONING** Can a negative number have a cube root? Give an example to support your explanation.

5. **IN YOUR OWN WORDS** How is the cube root of a number different from the square root of a number?

6. Give an example of a number whose square root and cube root are equal.

7. A cube has a volume of 13,824 cubic meters. Use a calculator to find the edge length.

Practice

Use what you learned about cube roots to complete Exercises 3–5 on page 636.

Key Vocabulary 🔊
cube root, *p. 634*
perfect cube, *p. 634*

A **cube root** of a number is a number that, when multiplied by itself, and then multiplied by itself again, equals the given number. A **perfect cube** is a number that can be written as the cube of an integer. The symbol $\sqrt[3]{}$ is used to represent a cube root.

EXAMPLE 1   **Finding Cube Roots**

Find each cube root.

a. $\sqrt[3]{8}$

    Because $2^3 = 8$, $\sqrt[3]{8} = \sqrt[3]{2^3} = 2$.

b. $\sqrt[3]{-27}$

    Because $(-3)^3 = -27$, $\sqrt[3]{-27} = \sqrt[3]{(-3)^3} = -3$.

c. $\sqrt[3]{\dfrac{1}{64}}$

    Because $\left(\dfrac{1}{4}\right)^3 = \dfrac{1}{64}$, $\sqrt[3]{\dfrac{1}{64}} = \sqrt[3]{\left(\dfrac{1}{4}\right)^3} = \dfrac{1}{4}$.

Cubing a number and finding a cube root are inverse operations. You can use this relationship to evaluate expressions and solve equations involving cubes.

EXAMPLE 2   **Evaluating Expressions Involving Cube Roots**

Evaluate each expression.

a. $2\sqrt[3]{-216} - 3 = 2(-6) - 3$ Evaluate the cube root.

              $= -12 - 3$ Multiply.

              $= -15$ Subtract.

b. $\left(\sqrt[3]{125}\right)^3 + 21 = 125 + 21$ Evaluate the power using inverse operations.

                    $= 146$ Add.

On Your Own

Now You're Ready
Exercises 6–17

Find the cube root.

1. $\sqrt[3]{1}$ **2.** $\sqrt[3]{-343}$ **3.** $\sqrt[3]{-\dfrac{27}{1000}}$

Evaluate the expression.

4. $18 - 4\sqrt[3]{8}$ **5.** $\left(\sqrt[3]{-64}\right)^3 + 43$ **6.** $5\sqrt[3]{512} - 19$

EXAMPLE 3 **Evaluating an Algebraic Expression**

Evaluate $\dfrac{x}{4} + \sqrt[3]{\dfrac{x}{3}}$ when $x = 192$.

$$\dfrac{x}{4} + \sqrt[3]{\dfrac{x}{3}} = \dfrac{192}{4} + \sqrt[3]{\dfrac{192}{3}}$$ Substitute 192 for x.

$$= 48 + \sqrt[3]{64}$$ Simplify.

$$= 48 + 4$$ Evaluate the cube root.

$$= 52$$ Add.

On Your Own

Now You're Ready
Exercises 18–20

Evaluate the expression for the given value of the variable.

7. $\sqrt[3]{8y} + y, \ y = 64$

8. $2b - \sqrt[3]{9b}, \ b = -3$

EXAMPLE 4 **Real-Life Application**

Find the surface area of the baseball display case.

The baseball display case is in the shape of a cube. Use the formula for the volume of a cube to find the edge length s.

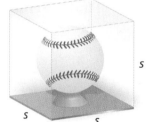

Volume = 125 in.³

$$V = s^3$$ Write formula for volume.

$$125 = s^3$$ Substitute 125 for V.

$$\sqrt[3]{125} = \sqrt[3]{s^3}$$ Take the cube root of each side.

$$5 = s$$ Simplify.

> **Remember**
>
> The volume V of a cube with edge length s is given by $V = s^3$. The surface area S is given by $S = 6s^2$.

The edge length is 5 inches. Use a formula to find the surface area of the cube.

$$S = 6s^2$$ Write formula for surface area.

$$= 6(5)^2$$ Substitute 5 for s.

$$= 150$$ Simplify.

⋮ So, the surface area of the baseball display case is 150 square inches.

On Your Own

9. The volume of a music box that is shaped like a cube is 512 cubic centimeters. Find the surface area of the music box.

Check It Out
Help with Homework
BigIdeasMath ✓com

Vocabulary and Concept Check

1. **VOCABULARY** Is 25 a perfect cube? Explain.

2. **REASONING** Can the cube of an integer be a negative number? Explain.

Practice and Problem Solving

Find the edge length of the cube.

3. Volume = 125,000 in.³

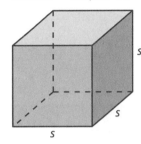

4. Volume = $\frac{1}{27}$ ft³

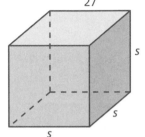

5. Volume = 0.064 m³

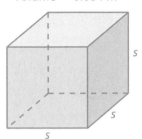

Find the cube root.

① 6. $\sqrt[3]{729}$

7. $\sqrt[3]{-125}$

8. $\sqrt[3]{-1000}$

9. $\sqrt[3]{1728}$

10. $\sqrt[3]{-\frac{1}{512}}$

11. $\sqrt[3]{\frac{343}{64}}$

Evaluate the expression.

② 12. $18 - \left(\sqrt[3]{27}\right)^3$

13. $\left(\sqrt[3]{-\frac{1}{8}}\right)^3 + 3\frac{3}{4}$

14. $5\sqrt[3]{729} - 24$

15. $\frac{1}{4} - 2\sqrt[3]{-\frac{1}{216}}$

16. $54 + \sqrt[3]{-4096}$

17. $4\sqrt[3]{8000} - 6$

Evaluate the expression for the given value of the variable.

③ 18. $\sqrt[3]{\frac{n}{4}} + \frac{n}{10}$, $n = 500$

19. $\sqrt[3]{6w} - w$, $w = 288$

20. $2d + \sqrt[3]{-45d}$, $d = 75$

21. **STORAGE CUBE** The volume of a plastic storage cube is 27,000 cubic centimeters. What is the edge length of the storage cube?

22. **ICE SCULPTURE** The volume of a cube of ice for an ice sculpture is 64,000 cubic inches.

   a. What is the edge length of the cube of ice?

   b. What is the surface area of the cube of ice?

Copy and complete the statement with <, >, or =.

23. $-\dfrac{1}{4}$ ⬜ $\sqrt[3]{-\dfrac{8}{125}}$

24. $\sqrt[3]{0.001}$ ⬜ 0.01

25. $\sqrt[3]{64}$ ⬜ $\sqrt{64}$

26. DRAG RACE The estimated velocity v (in miles per hour) of a car at the end of a drag race is $v = 234\sqrt[3]{\dfrac{p}{w}}$, where p is the horsepower of the car and w is the weight (in pounds) of the car. A car has a horsepower of 1311 and weighs 2744 pounds. Find the velocity of the car at the end of a drag race. Round your answer to the nearest whole number.

27. NUMBER SENSE There are three numbers that are their own cube roots. What are the numbers?

28. LOGIC Each statement below is true for square roots. Determine whether the statement is also true for cube roots. Explain your reasoning and give an example to support your explanation.

   a. You cannot find the square root of a negative number.

   b. Every positive number has a positive square root and a negative square root.

29. GEOMETRY The pyramid has a volume of 972 cubic inches. What are the dimensions of the pyramid?

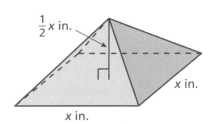

$\frac{1}{2}x$ in.

x in.

x in.

30. RATIOS The ratio $125 : x$ is equivalent to the ratio $x^2 : 125$. What is the value of x?

 Solve the equation.

31. $(3x + 4)^3 = 2197$

32. $(8x^3 - 9)^3 = 5832$

33. $\big((5x - 16)^3 - 4\big)^3 = 216{,}000$

 Fair Game Review *What you learned in previous grades & lessons*

Evaluate the expression. *(Skills Review Handbook)*

34. $3^2 + 4^2$

35. $8^2 + 15^2$

36. $13^2 - 5^2$

37. $25^2 - 24^2$

38. MULTIPLE CHOICE Which linear equation is shown by the table? *(Section 13.6)*

x	0	1	2	3
y	1	4	7	10

  A $y = \dfrac{1}{3}x + 1$
  B $y = 4x$
  C $y = 3x + 1$
  D $y = \dfrac{1}{4}x$

Essential Question How are the lengths of the sides of a right triangle related?

Pythagoras was a Greek mathematician and philosopher who discovered one of the most famous rules in mathematics. In mathematics, a rule is called a **theorem**. So, the rule that Pythagoras discovered is called the Pythagorean Theorem.

Pythagoras
(c. 570–c. 490 B.C.)

1 ACTIVITY: Discovering the Pythagorean Theorem

Work with a partner.

a. On grid paper, draw any right triangle. Label the lengths of the two shorter sides a and b.

b. Label the length of the longest side c.

c. Draw squares along each of the three sides. Label the areas of the three squares a^2, b^2, and c^2.

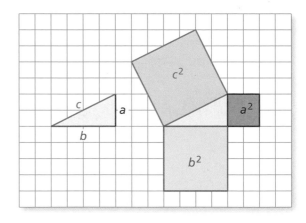

d. Cut out the three squares. Make eight copies of the right triangle and cut them out. Arrange the figures to form two identical larger squares.

e. **MODELING** The Pythagorean Theorem describes the relationship among a^2, b^2, and c^2. Use your result from part (d) to write an equation that describes this relationship.

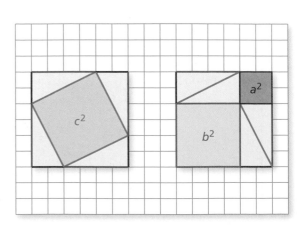

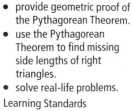

COMMON CORE

Pythagorean Theorem

In this lesson, you will
- provide geometric proof of the Pythagorean Theorem.
- use the Pythagorean Theorem to find missing side lengths of right triangles.
- solve real-life problems.

Learning Standards
8.EE.2
8.G.6
8.G.7
8.G.8

2 ACTIVITY: Using the Pythagorean Theorem in Two Dimensions

Work with a partner. Use a ruler to measure the longest side of each right triangle. Verify the result of Activity 1 for each right triangle.

a.

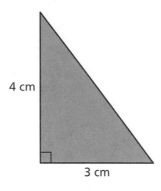

4 cm

3 cm

b.

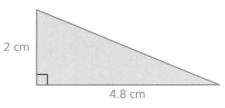

2 cm

4.8 cm

c.

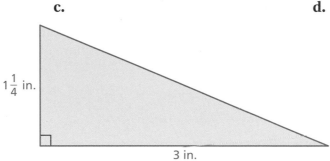

$1\frac{1}{4}$ in.

3 in.

d.

1$\frac{1}{2}$ in.

2 in.

3 ACTIVITY: Using the Pythagorean Theorem in Three Dimensions

Math Practice 3

Use Definitions

How can you use what you know about the Pythagorean Theorem to describe the procedure for finding the length of the guy wire?

Work with a partner. A guy wire attached 24 feet above ground level on a telephone pole provides support for the pole.

a. **PROBLEM SOLVING** Describe a procedure that you could use to find the length of the guy wire without directly measuring the wire.

b. Find the length of the wire when it meets the ground 10 feet from the base of the pole.

guy wire

What Is Your Answer?

4. **IN YOUR OWN WORDS** How are the lengths of the sides of a right triangle related? Give an example using whole numbers.

Practice

Use what you learned about the Pythagorean Theorem to complete Exercises 3 and 4 on page 642.

Key Vocabulary ◀))
theorem, *p. 638*
legs, *p. 640*
hypotenuse, *p. 640*
Pythagorean
  Theorem, *p. 640*

 **Key Ideas**

Sides of a Right Triangle

The sides of a right triangle have special names.

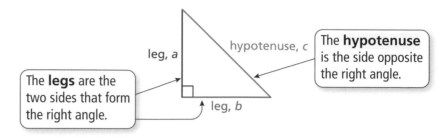

The **legs** are the two sides that form the right angle.

leg, *a*

hypotenuse, *c*

The **hypotenuse** is the side opposite the right angle.

leg, *b*

Study Tip

In a right triangle, the legs are the shorter sides and the hypotenuse is always the longest side.

The Pythagorean Theorem

Words In any right triangle, the sum of the squares of the lengths of the legs is equal to the square of the length of the hypotenuse.

Algebra $a^2 + b^2 = c^2$

EXAMPLE **1** **Finding the Length of a Hypotenuse**

Find the length of the hypotenuse of the triangle.

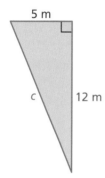

5 m

c 12 m

$a^2 + b^2 = c^2$	Write the Pythagorean Theorem.
$5^2 + 12^2 = c^2$	Substitute 5 for *a* and 12 for *b*.
$25 + 144 = c^2$	Evaluate powers.
$169 = c^2$	Add.
$\sqrt{169} = \sqrt{c^2}$	Take positive square root of each side.
$13 = c$	Simplify.

∴ The length of the hypotenuse is 13 meters.

On Your Own

Now You're Ready
Exercises 3 and 4

Find the length of the hypotenuse of the triangle.

1.

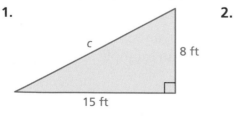

c

8 ft

15 ft

2.

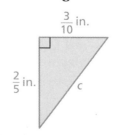

$\frac{3}{10}$ in.

$\frac{2}{5}$ in.

c

EXAMPLE 2 Finding the Length of a Leg

Find the missing length of the triangle.

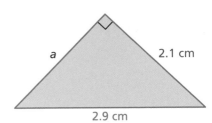

$$a^2 + b^2 = c^2$$ Write the Pythagorean Theorem.

$$a^2 + 2.1^2 = 2.9^2$$ Substitute 2.1 for *b* and 2.9 for *c*.

$$a^2 + 4.41 = 8.41$$ Evaluate powers.

$$a^2 = 4$$ Subtract 4.41 from each side.

$$a = 2$$ Take positive square root of each side.

∴ The missing length is 2 centimeters.

EXAMPLE 3 Real-Life Application

You are playing capture the flag. You are 50 yards north and 20 yards east of your team's base. The other team's base is 80 yards north and 60 yards east of your base. How far are you from the other team's base?

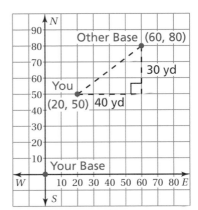

Step 1: Draw the situation in a coordinate plane. Let the origin represent your team's base. From the descriptions, you are at (20, 50) and the other team's base is at (60, 80).

Step 2: Draw a right triangle with a hypotenuse that represents the distance between you and the other team's base. The lengths of the legs are 30 yards and 40 yards.

Step 3: Use the Pythagorean Theorem to find the length of the hypotenuse.

$$a^2 + b^2 = c^2$$ Write the Pythagorean Theorem.

$$30^2 + 40^2 = c^2$$ Substitute 30 for *a* and 40 for *b*.

$$900 + 1600 = c^2$$ Evaluate powers.

$$2500 = c^2$$ Add.

$$50 = c$$ Take positive square root of each side.

∴ So, you are 50 yards from the other team's base.

● On Your Own

Now You're Ready
Exercises 5–8

Find the missing length of the triangle.

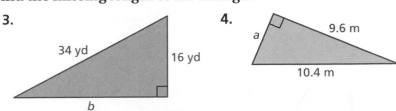

3.

34 yd   16 yd

b

4.

a   9.6 m

10.4 m

5. In Example 3, what is the distance between the bases?

Check It Out
Help with Homework
BigIdeasMath.com

 Vocabulary and Concept Check

1. **VOCABULARY** In a right triangle, how can you tell which sides are the legs and which side is the hypotenuse?

2. **DIFFERENT WORDS, SAME QUESTION** Which is different? Find "both" answers.

Which side is the hypotenuse?

Which side is the longest?

Which side is a leg?

Which side is opposite the right angle?

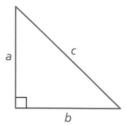

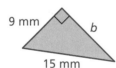 **Practice and Problem Solving**

Find the missing length of the triangle.

 3.
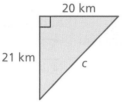
20 km
21 km
c

4.
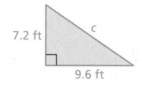
7.2 ft
c
9.6 ft

5.

5.6 in.
a
10.6 in.

6.
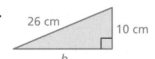
9 mm
b
15 mm

7.
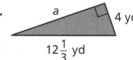
26 cm
10 cm
b

8.
a
4 yd
$12\frac{1}{3}$ yd

9. **ERROR ANALYSIS** Describe and correct the error in finding the missing length of the triangle.

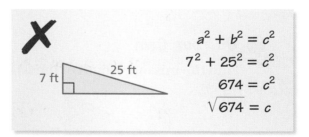

✗
7 ft
25 ft

$$a^2 + b^2 = c^2$$
$$7^2 + 25^2 = c^2$$
$$674 = c^2$$
$$\sqrt{674} = c$$

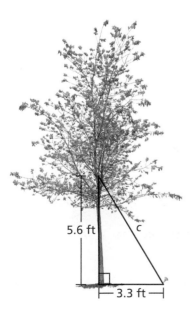
5.6 ft
c
3.3 ft

10. **TREE SUPPORT** How long is the wire that supports the tree?

Find the missing length of the figure.

11.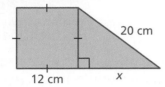

20 cm

12 cm   x

12.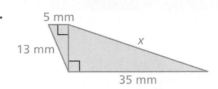

5 mm

13 mm   x

35 mm

13. GOLF The figure shows the location of a golf ball after a tee shot. How many feet from the hole is the ball?

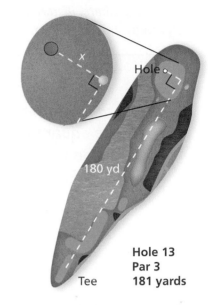

Hole

180 yd

Hole 13
Par 3
Tee   181 yards

14. TENNIS A tennis player asks the referee a question. The sound of the player's voice travels only 30 feet. Can the referee hear the question? Explain.

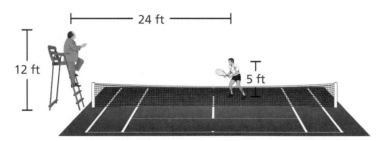

24 ft

12 ft

5 ft

15. PROJECT Measure the length, width, and height of a rectangular room. Use the Pythagorean Theorem to find length BC and length AB.

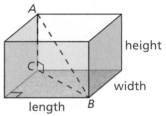

A

height

C

width

length   B

16. ALGEBRA The legs of a right triangle have lengths of 28 meters and 21 meters. The hypotenuse has a length of $5x$ meters. What is the value of x?

17. SNOWBALLS You and a friend stand back-to-back. You run 20 feet forward, then 15 feet to your right. At the same time, your friend runs 16 feet forward, then 12 feet to her right. She stops and hits you with a snowball.

  a. Draw the situation in a coordinate plane.

  b. How far does your friend throw the snowball?

18. Precision A box has a length of 6 inches, a width of 8 inches, and a height of 24 inches. Can a cylindrical rod with a length of 63.5 centimeters fit in the box? Explain your reasoning.

 Fair Game Review *What you learned in previous grades & lessons*

Find the square root(s). *(Section 14.1)*

19. $\pm\sqrt{36}$ **20.** $-\sqrt{121}$ **21.** $\sqrt{169}$ **22.** $-\sqrt{225}$

23. MULTIPLE CHOICE Which equation represents a proportional relationship? *(Section 13.3)*

  A $y = x + 1$ **B** $y = x - 1$ **C** $y = 0.5x$ **D** $y = 5 - 0.5x$

14 Study Help

You can use a **four square** to organize information about a topic. Each of the four squares can be a category, such as *definition, vocabulary, example, non-example, words, algebra, table, numbers, visual, graph,* or *equation.* Here is an example of a four square for the Pythagorean Theorem.

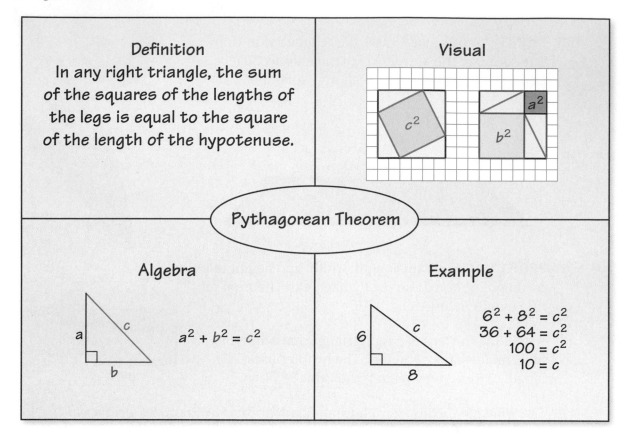

Definition
In any right triangle, the sum of the squares of the lengths of the legs is equal to the square of the length of the hypotenuse.

Visual

Pythagorean Theorem

Algebra
$a^2 + b^2 = c^2$

Example
$6^2 + 8^2 = c^2$
$36 + 64 = c^2$
$100 = c^2$
$10 = c$

On Your Own

Make four squares to help you study these topics.

1. square roots

2. cube roots

After you complete this chapter, make four squares for the following topics.

3. irrational numbers

4. real numbers

5. converse of the Pythagorean Theorem

6. distance formula

"I'm taking a survey for my four square. How many fleas do you have?"

Find the square root(s). *(Section 14.1)*

1. $-\sqrt{4}$

2. $\sqrt{\dfrac{16}{25}}$

3. $\pm\sqrt{6.25}$

Find the cube root. *(Section 14.2)*

4. $\sqrt[3]{64}$

5. $\sqrt[3]{-216}$

6. $\sqrt[3]{-\dfrac{343}{1000}}$

Evaluate the expression. *(Section 14.1 and Section 14.2)*

7. $3\sqrt{49} + 5$

8. $10 - 4\sqrt{16}$

9. $\dfrac{1}{4} + \sqrt{\dfrac{100}{4}}$

10. $\left(\sqrt[3]{-27}\right)^3 + 61$

11. $15 + 3\sqrt[3]{125}$

12. $2\sqrt[3]{-729} - 5$

Find the missing length of the triangle. *(Section 14.3)*

13.
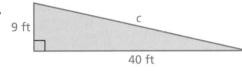
9 ft   c
40 ft

14.
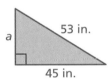
a   53 in.
45 in.

15.
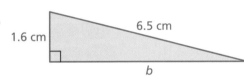
6.5 cm
1.6 cm
b

16.
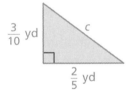
$\dfrac{3}{10}$ yd   c
$\dfrac{2}{5}$ yd

17. **POOL** The area of a circular pool cover is 314 square feet. Write and solve an equation to find the diameter of the pool cover. Use 3.14 for π. *(Section 14.1)*

18. **PACKAGE** A cube-shaped package has a volume of 5832 cubic inches. What is the edge length of the package? *(Section 14.2)*

19. **FABRIC** You are cutting a rectangular piece of fabric in half along the diagonal. The fabric measures 28 inches wide and $1\dfrac{1}{4}$ yards long. What is the length (in inches) of the diagonal? *(Section 14.3)*

14.4 Approximating Square Roots

Essential Question
How can you find decimal approximations of square roots that are not rational?

1 ACTIVITY: Approximating Square Roots

Work with a partner. Archimedes was a Greek mathematician, physicist, engineer, inventor, and astronomer. He tried to find a rational number whose square is 3. Two that he tried were $\frac{265}{153}$ and $\frac{1351}{780}$.

square root key

a. Are either of these numbers equal to $\sqrt{3}$? Explain.

b. Use a calculator to approximate $\sqrt{3}$. Write the number on a piece of paper. Enter it into the calculator and square it. Then subtract 3. Do you get 0? What does this mean?

c. The value of $\sqrt{3}$ is between which two integers?

d. Tell whether the value of $\sqrt{3}$ is between the given numbers. Explain your reasoning.

1.7 and 1.8	1.72 and 1.73	1.731 and 1.732

2 ACTIVITY: Approximating Square Roots Geometrically

Work with a partner. Refer to the square on the number line below.

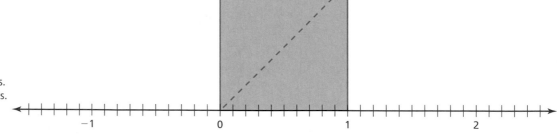

COMMON CORE

Square Roots

In this lesson, you will
- define irrational numbers.
- approximate square roots.
- approximate values of expressions involving irrational numbers.

Learning Standards
8.NS.1
8.NS.2
8.EE.2

a. What is the length of the diagonal of the square?

b. Copy the square and its diagonal onto a piece of transparent paper. Rotate it about zero on the number line so that the diagonal aligns with the number line. Use the number line to estimate the length of the diagonal.

c. **STRUCTURE** How do you think your answers in parts (a) and (b) are related?

Math Practice 5

Recognize Usefulness of Tools

Why is the Pythagorean Theorem a useful tool when approximating a square root?

Work with a partner.

a. Use grid paper and the given scale to draw a horizontal line segment 1 unit in length. Label this segment AC.

b. Draw a vertical line segment 2 units in length. Label this segment DC.

c. Set the point of a compass on A. Set the compass to 2 units. Swing the compass to intersect segment DC. Label this intersection as B.

d. Use the Pythagorean Theorem to find the length of segment BC.

e. Use the grid paper to approximate $\sqrt{3}$ to the nearest tenth.

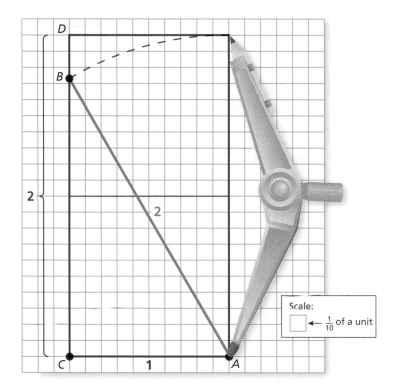

Scale:
☐ ← $\frac{1}{10}$ of a unit

What Is Your Answer?

4. Compare your approximation in Activity 3 with your results from Activity 1.

5. Repeat Activity 3 for a triangle in which segment AC is 2 units and segment BA is 3 units. Use the Pythagorean Theorem to find the length of segment BC. Use the grid paper to approximate $\sqrt{5}$ to the nearest tenth.

6. **IN YOUR OWN WORDS** How can you find decimal approximations of square roots that are not rational?

Practice ➤ Use what you learned about approximating square roots to complete Exercises 5–8 on page 651.

Check It Out
Lesson Tutorials
BigIdeasMath.com

Key Vocabulary
irrational number,
  p. 648
real numbers, p. 648

A rational number is a number that can be written as the ratio of two integers. An **irrational number** cannot be written as the ratio of two integers.

- The square root of any whole number that is not a perfect square is irrational. The cube root of any integer that is not a perfect cube is irrational.

- The decimal form of an irrational number neither terminates nor repeats.

 **Key Idea**

Real Numbers

Rational numbers and irrational numbers together form the set of **real numbers**.

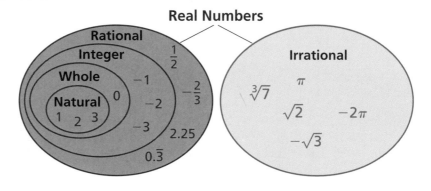

Remember

The decimal form of a rational number either terminates or repeats.

EXAMPLE 1 **Classifying Real Numbers**

Classify each real number.

	Number	Subset(s)	Reasoning
a.	$\sqrt{12}$	Irrational	12 is not a perfect square.
b.	$-0.\overline{25}$	Rational	$-0.\overline{25}$ is a repeating decimal.
c.	$-\sqrt{9}$	Integer, Rational	$-\sqrt{9}$ is equal to -3.
d.	$\dfrac{72}{4}$	Natural, Whole, Integer, Rational	$\dfrac{72}{4}$ is equal to 18.
e.	π	Irrational	The decimal form of π neither terminates nor repeats.

Study Tip

When classifying a real number, list all the subsets in which the number belongs.

On Your Own

Classify the real number.

Now You're Ready
Exercises 9–16

1. $0.121221222\ldots$ **2.** $-\sqrt{196}$ **3.** $\sqrt[3]{2}$

EXAMPLE 2

Approximating a Square Root

Estimate $\sqrt{71}$ to the nearest (a) integer and (b) tenth.

a. Make a table of numbers whose squares are close to 71.

Number	7	8	9	10
Square of Number	49	64	81	100

The table shows that 71 is between the perfect squares 64 and 81. Because 71 is closer to 64 than to 81, $\sqrt{71}$ is closer to 8 than to 9.

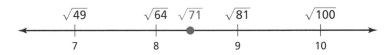

So, $\sqrt{71} \approx 8$.

b. Make a table of numbers between 8 and 9 whose squares are close to 71.

Number	8.3	8.4	8.5	8.6
Square of Number	68.89	70.56	72.25	73.96

Because 71 is closer to 70.56 than to 72.25, $\sqrt{71}$ is closer to 8.4 than to 8.5.

So, $\sqrt{71} \approx 8.4$.

Study Tip

You can continue the process shown in Example 2 to approximate square roots using more decimal places.

On Your Own

Now You're Ready
Exercises 20–25

Estimate the square root to the nearest (a) integer and (b) tenth.

4. $\sqrt{8}$ **5.** $-\sqrt{13}$ **6.** $-\sqrt{24}$ **7.** $\sqrt{110}$

EXAMPLE 3   Comparing Real Numbers

Which is greater, $\sqrt{5}$ or $2\frac{2}{3}$?

Estimate $\sqrt{5}$ to the nearest integer. Then graph the numbers on a number line.

$\sqrt{5} \approx 2$ $2\frac{2}{3} = 2.\overline{6}$

$\sqrt{4} = 2$ $\sqrt{9} = 3$

$2\frac{2}{3}$ is to the right of $\sqrt{5}$. So, $2\frac{2}{3}$ is greater.

EXAMPLE 4 **Approximating the Value of an Expression**

The radius of a circle with area A is approximately $\sqrt{\dfrac{A}{3}}$. The area of a circular mouse pad is 51 square inches. Estimate its radius to the nearest integer.

$$\sqrt{\dfrac{A}{3}} = \sqrt{\dfrac{51}{3}} \qquad \text{Substitute 51 for } A.$$

$$= \sqrt{17} \qquad \text{Divide.}$$

The nearest perfect square less than 17 is 16. The nearest perfect square greater than 17 is 25.

$$\sqrt{17}$$

$$\sqrt{16} = 4 \qquad\qquad \sqrt{25} = 5$$

Because 17 is closer to 16 than to 25, $\sqrt{17}$ is closer to 4 than to 5.

∴ So, the radius is about 4 inches.

EXAMPLE 5 **Real-Life Application**

The distance (in nautical miles) you can see with a periscope is $1.17\sqrt{h}$, where h is the height of the periscope above the water. Can you see twice as far with a periscope that is 6 feet above the water than with a periscope that is 3 feet above the water? Explain.

Use a calculator to find the distances.

3 Feet Above Water		*6 Feet Above Water*
$1.17\sqrt{h} = 1.17\sqrt{3}$	Substitute for h.	$1.17\sqrt{h} = 1.17\sqrt{6}$
≈ 2.03	Use a calculator.	≈ 2.87

```
1.17√(3)
       2.026499445
1.17√(6)
       2.865902999
```

You can see $\dfrac{2.87}{2.03} \approx 1.41$ times farther with the periscope that is 6 feet above the water than with the periscope that is 3 feet above the water.

∴ No, you cannot see twice as far with the periscope that is 6 feet above the water.

● **On Your Own**

Now You're Ready
Exercises 26–31

Which number is greater? Explain.

8. $4\dfrac{1}{5}, \sqrt{23}$

9. $\sqrt{10}, -\sqrt{5}$

10. $-\sqrt{2}, -2$

11. The area of a circular mouse pad is 64 square inches. Estimate its radius to the nearest integer.

12. In Example 5, you use a periscope that is 10 feet above the water. Can you see farther than 4 nautical miles? Explain.

14.4 Exercises

✓ Vocabulary and Concept Check

1. **VOCABULARY** How are rational numbers and irrational numbers different?

2. **WRITING** Describe a method of approximating $\sqrt{32}$.

3. **VOCABULARY** What are real numbers? Give three examples.

4. **WHICH ONE DOESN'T BELONG?** Which number does *not* belong with the other three? Explain your reasoning.

$$-\frac{11}{12} \qquad 25.075 \qquad \sqrt{8} \qquad -3.\overline{3}$$

Practice and Problem Solving

Tell whether the rational number is a reasonable approximation of the square root.

5. $\dfrac{559}{250}, \sqrt{5}$

6. $\dfrac{3021}{250}, \sqrt{11}$

7. $\dfrac{678}{250}, \sqrt{28}$

8. $\dfrac{1677}{250}, \sqrt{45}$

Classify the real number.

 9. 0

10. $\sqrt[3]{343}$

11. $\dfrac{\pi}{6}$

12. $-\sqrt{81}$

13. -1.125

14. $\dfrac{52}{13}$

15. $\sqrt[3]{-49}$

16. $\sqrt{15}$

17. **ERROR ANALYSIS** Describe and correct the error in classifying the number.

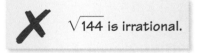
✗ $\sqrt{144}$ is irrational.

18. **SCRAPBOOKING** You cut a picture into a right triangle for your scrapbook. The lengths of the legs of the triangle are 4 inches and 6 inches. Is the length of the hypotenuse a rational number? Explain.

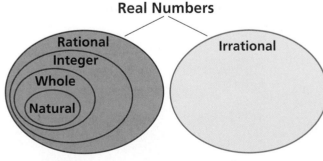
Real Numbers

19. **VENN DIAGRAM** Place each number in the correct area of the Venn Diagram.

   a. the last digit of your phone number

   b. the square root of any prime number

   c. the ratio of the circumference of a circle to its diameter

Estimate the square root to the nearest (a) integer and (b) tenth.

20. $\sqrt{46}$

21. $\sqrt{685}$

22. $-\sqrt{61}$

23. $-\sqrt{105}$

24. $\sqrt{\dfrac{27}{4}}$

25. $-\sqrt{\dfrac{335}{2}}$

Which number is greater? Explain.

③ **26.** $\sqrt{20}$, 10

27. $\sqrt{15}$, −3.5

28. $\sqrt{133}$, $10\frac{3}{4}$

29. $\frac{2}{3}$, $\sqrt{\frac{16}{81}}$

30. $-\sqrt{0.25}$, −0.25

31. $-\sqrt{182}$, $-\sqrt{192}$

Use the graphing calculator screen to determine whether the statement is *true* or *false*.

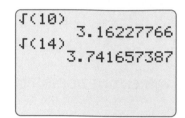

32. To the nearest tenth, $\sqrt{10} = 3.1$.

33. The value of $\sqrt{14}$ is between 3.74 and 3.75.

34. $\sqrt{10}$ lies between 3.1 and 3.16 on a number line.

35. FOUR SQUARE The area of a four square court is 66 square feet. Estimate the side length *s* to the nearest tenth of a foot.

36. CHECKERS A checkers board is 8 squares long and 8 squares wide. The area of each square is 14 square centimeters. Estimate the perimeter of the checkers board to the nearest tenth of a centimeter.

Approximate the length of the diagonal of the square or rectangle to the nearest tenth.

37.

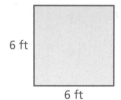

6 ft

6 ft

38.

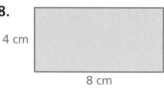

4 cm

8 cm

39.

10 in.

18 in.

40. WRITING Explain how to continue the method in Example 2 to estimate $\sqrt{71}$ to the nearest hundredth.

41. REPEATED REASONING Describe a method that you can use to estimate a cube root to the nearest tenth. Use your method to estimate $\sqrt[3]{14}$ to the nearest tenth.

42. RADIO SIGNAL The maximum distance (in nautical miles) that a radio transmitter signal can be sent is represented by the expression $1.23\sqrt{h}$, where *h* is the height (in feet) above the transmitter.

Estimate the maximum distance *x* (in nautical miles) between the plane that is receiving the signal and the transmitter. Round your answer to the nearest tenth.

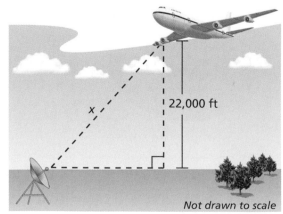

x

22,000 ft

Not drawn to scale

43. OPEN-ENDED Find two numbers a and b that satisfy the diagram.

Estimate the square root to the nearest tenth.

44. $\sqrt{0.39}$ **45.** $\sqrt{1.19}$ **46.** $\sqrt{1.52}$

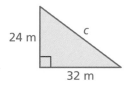

$r = 16.764$ m

47. ROLLER COASTER The speed s (in meters per second) of a roller-coaster car is approximated by the equation $s = 3\sqrt{6r}$, where r is the radius of the loop. Estimate the speed of a car going around the loop. Round your answer to the nearest tenth.

48. STRUCTURE Is $\sqrt{\dfrac{1}{4}}$ a rational number? Is $\sqrt{\dfrac{3}{16}}$ a rational number? Explain.

49. WATER BALLOON The time t (in seconds) it takes a water balloon to fall d meters is represented by the equation $t = \sqrt{\dfrac{d}{4.9}}$. Estimate the time it takes the balloon to fall to the ground from a window that is 14 meters above the ground. Round your answer to the nearest tenth.

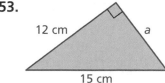

50. **Number Sense** Determine if the statement is *sometimes*, *always*, or *never* true. Explain your reasoning and give an example of each.

a. A rational number multiplied by a rational number is rational.

b. A rational number multiplied by an irrational number is rational.

c. An irrational number multiplied by an irrational number is rational.

Fair Game Review What you learned in previous grades & lessons

Find the missing length of the triangle. *(Section 14.3)*

51.

24 m, c, 32 m

52.

10 in., 26 in., b

53.

12 cm, a, 15 cm

54. MULTIPLE CHOICE What is the ratio (red to blue) of the corresponding side lengths of the similar triangles? *(Section 11.5)*

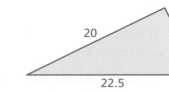

8, 4, 9

20, 10, 22.5

Ⓐ 1 : 3 **Ⓑ** 5 : 2

Ⓒ 3 : 4 **Ⓓ** 2 : 5

You have written terminating decimals as fractions. Because repeating decimals are rational numbers, you can also write repeating decimals as fractions.

 **Key Idea**

COMMON CORE

Rational Numbers

In this extension, you will

- write a repeating decimal as a fraction.

Learning Standard
8.NS.1

Writing a Repeating Decimal as a Fraction

Let a variable x equal the repeating decimal d.

Step 1: Write the equation $x = d$.

Step 2: Multiply each side of the equation by 10^n to form a new equation, where n is the number of repeating digits.

Step 3: Subtract the original equation from the new equation.

Step 4: Solve for x.

EXAMPLE 1 **Writing a Repeating Decimal as a Fraction (1 Digit Repeats)**

Write $0.\overline{4}$ as a fraction in simplest form.

Let $x = 0.\overline{4}$.

$$x = 0.\overline{4}$$ 　　Step 1: Write the equation.

$$10 \cdot x = 10 \cdot 0.\overline{4}$$ 　　Step 2: There is 1 repeating digit, so multiply each side by $10^1 = 10$.

$$10x = 4.\overline{4}$$ 　　Simplify.

$$- (x = 0.\overline{4})$$ 　　Step 3: Subtract the original equation.

$$9x = 4$$ 　　Simplify.

$$x = \frac{4}{9}$$ 　　Step 4: Solve for x.

∴ So, $0.\overline{4} = \frac{4}{9}$.

Check

```
      0.44...
  9 ) 4.00      ✓
      36
      ──
      40
      36
      ──
      40
```

● Practice

Write the decimal as a fraction or a mixed number.

1. $0.\overline{1}$ 　　　　**2.** $-0.\overline{5}$ 　　　　**3.** $-1.\overline{2}$ 　　　　**4.** $5.\overline{8}$

5. **STRUCTURE** In Example 1, why can you subtract the original equation from the new equation after multiplying by 10? Explain why these two steps are performed.

6. **REPEATED REASONING** Compare the repeating decimals and their equivalent fractions in Exercises 1–4. Describe the pattern. Use the pattern to explain how to write a repeating decimal as a fraction when only the tenths digit repeats.

EXAMPLE 2 — Writing a Repeating Decimal as a Fraction (1 Digit Repeats)

Write $-0.2\overline{3}$ as a fraction in simplest form.

Let $x = -0.2\overline{3}$.

$x = -0.2\overline{3}$	**Step 1:** Write the equation.
$10 \cdot x = 10 \cdot (-0.2\overline{3})$	**Step 2:** There is 1 repeating digit, so multiply each side by $10^1 = 10$.
$10x = -2.\overline{3}$	Simplify.
$\underline{-(x = -0.2\overline{3})}$	**Step 3:** Subtract the original equation.
$9x = -2.1$	Simplify.
$x = \dfrac{-2.1}{9}$	**Step 4:** Solve for x.

Check

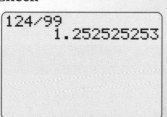

-7/30
      -.2333333333

So, $-0.2\overline{3} = \dfrac{-2.1}{9} = -\dfrac{21}{90} = -\dfrac{7}{30}$.

EXAMPLE 3 — Writing a Repeating Decimal as a Fraction (2 Digits Repeat)

Write $1.\overline{25}$ as a mixed number.

Let $x = 1.\overline{25}$.

$x = 1.\overline{25}$	**Step 1:** Write the equation.
$100 \cdot x = 100 \cdot 1.\overline{25}$	**Step 2:** There are 2 repeating digits, so multiply each side by $10^2 = 100$.
$100x = 125.\overline{25}$	Simplify.
$\underline{-(x = 1.\overline{25})}$	**Step 3:** Subtract the original equation.
$99x = 124$	Simplify.
$x = \dfrac{124}{99}$	**Step 4:** Solve for x.

Check

124/99
      1.252525253

So, $1.\overline{25} = \dfrac{124}{99} = 1\dfrac{25}{99}$.

Practice

Write the decimal as a fraction or a mixed number.

7. $-0.4\overline{3}$ **8.** $2.0\overline{6}$ **9.** $0.\overline{27}$ **10.** $-4.\overline{50}$

11. REPEATED REASONING Find a pattern in the fractional representations of repeating decimals in which only the tenths and hundredths digits repeat. Use the pattern to explain how to write $9.\overline{04}$ as a mixed number.

Essential Question In what other ways can you use the Pythagorean Theorem?

The *converse* of a statement switches the hypothesis and the conclusion.

Statement:	Converse of the statement:
If p, then q.	If q, then p.

1 ACTIVITY: Analyzing Converses of Statements

Work with a partner. Write the converse of the true statement. Determine whether the converse is *true* or *false*. If it is true, justify your reasoning. If it is false, give a counterexample.

a. If $a = b$, then $a^2 = b^2$.

b. If $a = b$, then $a^3 = b^3$.

c. If one figure is a translation of another figure, then the figures are congruent.

d. If two triangles are similar, then the triangles have the same angle measures.

Is the converse of a true statement always true? always false? Explain.

2 ACTIVITY: The Converse of the Pythagorean Theorem

Work with a partner. The converse of the Pythagorean Theorem states: "If the equation $a^2 + b^2 = c^2$ is true for the side lengths of a triangle, then the triangle is a right triangle."

a. Do you think the converse of the Pythagorean Theorem is *true* or *false*? How could you use deductive reasoning to support your answer?

b. Consider $\triangle DEF$ with side lengths a, b, and c, such that $a^2 + b^2 = c^2$. Also consider $\triangle JKL$ with leg lengths a and b, where $\angle K = 90°$.

- What does the Pythagorean Theorem tell you about $\triangle JKL$?

- What does this tell you about c and x?

- What does this tell you about $\triangle DEF$ and $\triangle JKL$?

- What does this tell you about $\angle E$?

- What can you conclude?

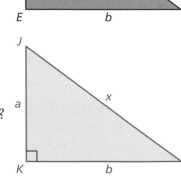

COMMON CORE

Pythagorean Theorem
In this lesson, you will
- use the converse of the Pythagorean Theorem to identify right triangles.
- use the Pythagorean Theorem to find distances in a coordinate plane.
- solve real-life problems.

Learning Standards
8.EE.2
8.G.6
8.G.7
8.G.8

ACTIVITY: Developing the Distance Formula

Work with a partner. Follow the steps below to write a formula that you can use to find the distance between any two points in a coordinate plane.

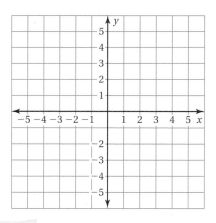

Step 1: Choose two points in the coordinate plane that do not lie on the same horizontal or vertical line. Label the points (x_1, y_1) and (x_2, y_2).

Step 2: Draw a line segment connecting the points. This will be the hypotenuse of a right triangle.

Step 3: Draw horizontal and vertical line segments from the points to form the legs of the right triangle.

Step 4: Use the x-coordinates to write an expression for the length of the horizontal leg.

Step 5: Use the y-coordinates to write an expression for the length of the vertical leg.

Step 6: Substitute the expressions for the lengths of the legs into the Pythagorean Theorem.

Step 7: Solve the equation in Step 6 for the hypotenuse c.

What does the length of the hypotenuse tell you about the two points?

Math Practice 6

Communicate Precisely

What steps can you take to make sure that you have written the distance formula accurately?

What Is Your Answer?

4. **IN YOUR OWN WORDS** In what other ways can you use the Pythagorean Theorem?

5. What kind of real-life problems do you think the converse of the Pythagorean Theorem can help you solve?

Practice

Use what you learned about the converse of a true statement to complete Exercises 3 and 4 on page 660.

Check It Out
Lesson Tutorials
BigIdeasMath ✓com

 **Key Ideas**

Converse of the Pythagorean Theorem
If the equation $a^2 + b^2 = c^2$ is true for the side lengths of a triangle, then the triangle is a right triangle.

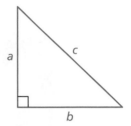

EXAMPLE ① **Identifying a Right Triangle**

Study Tip

A *Pythagorean triple* is a set of three positive integers a, b, and c, where $a^2 + b^2 = c^2$.

Common Error ⚠

When using the converse of the Pythagorean Theorem, always substitute the length of the longest side for c.

Tell whether each triangle is a right triangle.

a. *(triangle: 41 cm, 9 cm, 40 cm)*

$$a^2 + b^2 = c^2$$
$$9^2 + 40^2 \stackrel{?}{=} 41^2$$
$$81 + 1600 \stackrel{?}{=} 1681$$
$$1681 = 1681 \quad ✓$$

⋮ It *is* a right triangle.

b. *(triangle: 18 ft, 12 ft, 24 ft)*

$$a^2 + b^2 = c^2$$
$$12^2 + 18^2 \stackrel{?}{=} 24^2$$
$$144 + 324 \stackrel{?}{=} 576$$
$$468 \neq 576 \quad ✗$$

⋮ It is *not* a right triangle.

 **On Your Own**

Now You're Ready
Exercises 5–10

Tell whether the triangle with the given side lengths is a right triangle.

1. 28 in., 21 in., 20 in.   2. 1.25 mm, 1 mm, 0.75 mm

On page 657, you used the Pythagorean Theorem to develop the *distance formula*. You can use the **distance formula** to find the distance between any two points in a coordinate plane.

 Key Idea

Distance Formula
The distance d between any two points (x_1, y_1) and (x_2, y_2) is given by the formula
$$d = \sqrt{(x_2 - x_1)^2 + (y_2 - y_1)^2}.$$

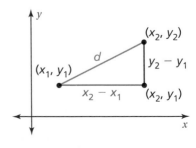

EXAMPLE 2 **Finding the Distance Between Two Points**

Find the distance between $(1, 5)$ and $(-4, -2)$.

Let $(x_1, y_1) = (1, 5)$ and $(x_2, y_2) = (-4, -2)$.

$$d = \sqrt{(x_2 - x_1)^2 + (y_2 - y_1)^2}$$ Write the distance formula.

$$= \sqrt{(-4 - 1)^2 + (-2 - 5)^2}$$ Substitute.

$$= \sqrt{(-5)^2 + (-7)^2}$$ Simplify.

$$= \sqrt{25 + 49}$$ Evaluate powers.

$$= \sqrt{74}$$ Add.

EXAMPLE 3 **Real-Life Application**

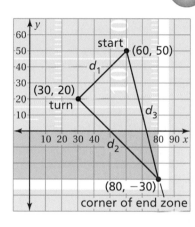

You design a football play in which a player runs down the field, makes a 90° turn, and runs to the corner of the end zone. Your friend runs the play as shown. Did your friend make a 90° turn? Each unit of the grid represents 10 feet.

Use the distance formula to find the lengths of the three sides.

$$d_1 = \sqrt{(60 - 30)^2 + (50 - 20)^2} = \sqrt{30^2 + 30^2} = \sqrt{1800} \text{ feet}$$

$$d_2 = \sqrt{(80 - 30)^2 + (-30 - 20)^2} = \sqrt{50^2 + (-50)^2} = \sqrt{5000} \text{ feet}$$

$$d_3 = \sqrt{(80 - 60)^2 + (-30 - 50)^2} = \sqrt{20^2 + (-80)^2} = \sqrt{6800} \text{ feet}$$

Use the converse of the Pythagorean Theorem to determine if the side lengths form a right triangle.

$$\left(\sqrt{1800}\right)^2 + \left(\sqrt{5000}\right)^2 \overset{?}{=} \left(\sqrt{6800}\right)^2$$

$$1800 + 5000 \overset{?}{=} 6800$$

$$6800 = 6800 \checkmark$$

The sides form a right triangle.

⋮ So, your friend made a 90° turn.

On Your Own

Now You're Ready
Exercises 11–16

Find the distance between the two points.

3. $(0, 0), (4, 5)$ **4.** $(7, -3), (9, 6)$ **5.** $(-2, -3), (-5, 1)$

6. **WHAT IF?** In Example 3, your friend made the turn at $(20, 10)$. Did your friend make a 90° turn?

 Vocabulary and Concept Check

1. **WRITING** Describe two ways to find the distance between two points in a coordinate plane.

2. **WHICH ONE DOESN'T BELONG?** Which set of numbers does *not* belong with the other three? Explain your reasoning.

$$3, 6, 8 \qquad 6, 8, 10 \qquad 5, 12, 13 \qquad 7, 24, 25$$

 Practice and Problem Solving

Write the converse of the true statement. Determine whether the converse is *true* or *false*. If it is true, justify your reasoning. If it is false, give a counterexample.

3. If a is an odd number, then a^2 is odd.

4. If $ABCD$ is a square, then $ABCD$ is a parallelogram.

Tell whether the triangle with the given side lengths is a right triangle.

① 5.
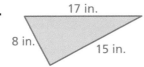
17 in.
8 in.
15 in.

6.
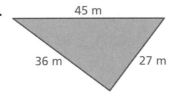
45 m
36 m 27 m

7.
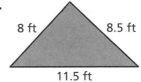
8 ft 8.5 ft
11.5 ft

8. 14 mm, 19 mm, 23 mm

9. $\frac{9}{10}$ mi, $1\frac{1}{5}$ mi, $1\frac{1}{2}$ mi

10. 1.4 m, 4.8 m, 5 m

Find the distance between the two points.

② 11. $(1, 2), (7, 6)$

12. $(4, -5), (-1, 7)$

13. $(2, 4), (7, 2)$

14. $(-1, -3), (1, 3)$

15. $(-6, -7), (0, 0)$

16. $(12, 5), (-12, -2)$

17. **ERROR ANALYSIS** Describe and correct the error in finding the distance between the points $(-3, -2)$ and $(7, 4)$.

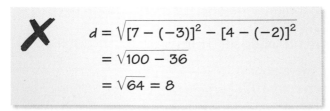

$$d = \sqrt{[7 - (-3)]^2 - [4 - (-2)]^2}$$
$$= \sqrt{100 - 36}$$
$$= \sqrt{64} = 8$$

18. **CONSTRUCTION** A post and beam frame for a shed is shown in the diagram. Does the brace form a right triangle with the post and beam? Explain.

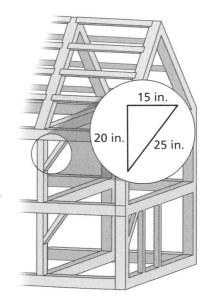

15 in.
20 in.
25 in.

Tell whether a triangle with the given side lengths is a right triangle.

19. $\sqrt{63}, 9, 12$

20. $4, \sqrt{15}, 6$

21. $\sqrt{18}, \sqrt{24}, \sqrt{42}$

22. REASONING Plot the points $(-1, 3)$, $(4, -2)$, and $(1, -5)$ in a coordinate plane. Are the points the vertices of a right triangle? Explain.

23. GEOCACHING You spend the day looking for hidden containers in a wooded area using a Global Positioning System (GPS). You park your car on the side of the road, and then locate Container 1 and Container 2 before going back to the car. Does your path form a right triangle? Explain. Each unit of the grid represents 10 yards.

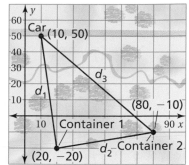

24. REASONING Your teacher wants the class to find the distance between the two points $(2, 4)$ and $(9, 7)$. You use $(2, 4)$ for (x_1, y_1), and your friend uses $(9, 7)$ for (x_1, y_1). Do you and your friend obtain the same result? Justify your answer.

25. AIRPORT Which plane is closer to the base of the airport tower? Explain.

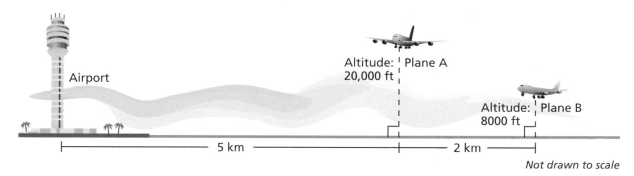

Not drawn to scale

26. **Structure** Consider the two points (x_1, y_1) and (x_2, y_2) in the coordinate plane. How can you find the point (x_m, y_m) located in the middle of the two given points? Justify your answer using the distance formula.

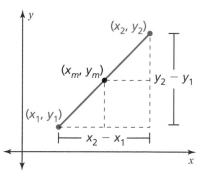

 Fair Game Review What you learned in previous grades & lessons

Find the mean, median, and mode of the data. *(Skills Review Handbook)*

27. 12, 9, 17, 15, 12, 13

28. 21, 32, 16, 27, 22, 19, 10

29. 67, 59, 34, 71, 59

30. MULTIPLE CHOICE What is the sum of the interior angle measures of an octagon? *(Section 12.3)*

Ⓐ 720°   Ⓑ 1080°   Ⓒ 1440°   Ⓓ 1800°

14.4–14.5 Quiz

Classify the real number. *(Section 14.4)*

1. $-\sqrt{225}$

2. $-1\dfrac{1}{9}$

3. $\sqrt{41}$

4. $\sqrt{17}$

Estimate the square root to the nearest (a) integer and (b) tenth. *(Section 14.4)*

5. $\sqrt{38}$

6. $-\sqrt{99}$

7. $\sqrt{172}$

8. $\sqrt{115}$

Which number is greater? Explain. *(Section 14.4)*

9. $\sqrt{11}, 3\dfrac{3}{5}$

10. $\sqrt{1.44}, 1.1\overline{8}$

Write the decimal as a fraction or a mixed number. *(Section 14.4)*

11. $0.\overline{7}$

12. $-1.\overline{63}$

Tell whether the triangle with the given side lengths is a right triangle. *(Section 14.5)*

13.

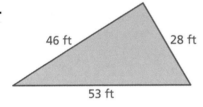

46 ft   28 ft
53 ft

14.

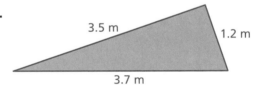

3.5 m   1.2 m
3.7 m

Find the distance between the two points. *(Section 14.5)*

15. $(-3, -1), (-1, -5)$

16. $(-4, 2), (5, 1)$

17. $(1, -2), (4, -5)$

18. $(-1, 1), (7, 4)$

19. $(-6, 5), (-4, -6)$

20. $(-1, 4), (1, 3)$

Use the figure to answer Exercises 21–24. Round your answer to the nearest tenth. *(Section 14.5)*

21. How far is the cabin from the peak?

22. How far is the fire tower from the lake?

23. How far is the lake from the peak?

24. You are standing at $(-5, -6)$. How far are you from the lake?

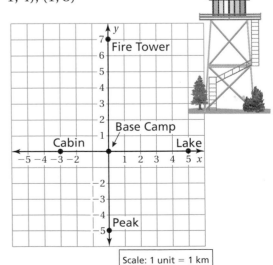

Scale: 1 unit = 1 km

14 Chapter Review

Check It Out
Vocabulary Help
BigIdeasMath ✓.com

Review Key Vocabulary

square root, *p. 628*

perfect square, *p. 628*

radical sign, *p. 628*

radicand, *p. 628*

cube root, *p. 634*

perfect cube, *p. 634*

theorem, *p. 638*

legs, *p. 640*

hypotenuse, *p. 640*

Pythagorean Theorem, *p. 640*

irrational number, *p. 648*

real numbers, *p. 648*

distance formula, *p. 658*

Review Examples and Exercises

14.1 Finding Square Roots *(pp. 626–631)*

Find $-\sqrt{36}$.

$-\sqrt{36}$ represents the *negative* square root.

∴ Because $6^2 = 36$, $-\sqrt{36} = -\sqrt{6^2} = -6$.

Exercises

Find the square root(s).

1. $\sqrt{1}$

2. $-\sqrt{\dfrac{9}{25}}$

3. $\pm\sqrt{1.69}$

Evaluate the expression.

4. $15 - 4\sqrt{36}$

5. $\sqrt{\dfrac{54}{6}} + \dfrac{2}{3}$

6. $10\left(\sqrt{81} - 12\right)$

14.2 Finding Cube Roots *(pp. 632–637)*

Find $\sqrt[3]{\dfrac{125}{216}}$.

∴ Because $\left(\dfrac{5}{6}\right)^3 = \dfrac{125}{216}$, $\sqrt[3]{\dfrac{125}{216}} = \sqrt[3]{\left(\dfrac{5}{6}\right)^3} = \dfrac{5}{6}$.

Exercises

Find the cube root.

7. $\sqrt[3]{729}$

8. $\sqrt[3]{\dfrac{64}{343}}$

9. $\sqrt[3]{-\dfrac{8}{27}}$

Evaluate the expression.

10. $\sqrt[3]{27} - 16$

11. $25 + 2\sqrt[3]{-64}$

12. $3\sqrt[3]{-125} - 27$

14.3 The Pythagorean Theorem (pp. 638–643)

Find the length of the hypotenuse of the triangle.

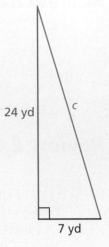

$$a^2 + b^2 = c^2 \qquad \text{Write the Pythagorean Theorem.}$$
$$7^2 + 24^2 = c^2 \qquad \text{Substitute.}$$
$$49 + 576 = c^2 \qquad \text{Evaluate powers.}$$
$$625 = c^2 \qquad \text{Add.}$$
$$\sqrt{625} = \sqrt{c^2} \qquad \text{Take positive square root of each side.}$$
$$25 = c \qquad \text{Simplify.}$$

∴ The length of the hypotenuse is 25 yards.

Exercises

Find the missing length of the triangle.

13.

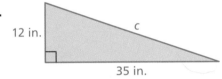

12 in. c 35 in.

14.

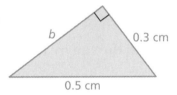

b 0.3 cm   0.5 cm

14.4 Approximating Square Roots (pp. 646–655)

a. Classify $\sqrt{19}$.

∴ The number $\sqrt{19}$ is irrational because 19 is not a perfect square.

b. Estimate $\sqrt{34}$ to the nearest integer.

Make a table of numbers whose squares are close to the radicand, 34.

Number	4	5	6	7
Square of Number	16	25	36	49

The table shows that 34 is between the perfect squares 25 and 36. Because 34 is closer to 36 than to 25, $\sqrt{34}$ is closer to 6 than to 5.

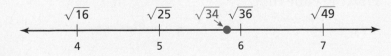

∴ So, $\sqrt{34} \approx 6$.

Exercises

Classify the real number.

15. $0.81\overline{5}$

16. $\sqrt{101}$

17. $\sqrt{4}$

Estimate the square root to the nearest (a) integer and (b) tenth.

18. $\sqrt{14}$

19. $\sqrt{90}$

20. $\sqrt{175}$

Write the decimal as a fraction.

21. $0.\overline{8}$

22. $0.\overline{36}$

23. $-1.\overline{6}$

14.5 Using the Pythagorean Theorem *(pp. 656–661)*

a. Is the triangle formed by the rope and the tent a right triangle?

$$a^2 + b^2 = c^2$$

$$64^2 + 48^2 \stackrel{?}{=} 80^2$$

$$4096 + 2304 \stackrel{?}{=} 6400$$

$$6400 = 6400 \checkmark$$

∴ It *is* a right triangle.

b. Find the distance between $(-3, 1)$ and $(4, 7)$.

Let $(x_1, y_1) = (-3, 1)$ and $(x_2, y_2) = (4, 7)$.

$$d = \sqrt{(x_2 - x_1)^2 + (y_2 - y_1)^2} \qquad \text{Write the distance formula.}$$

$$= \sqrt{[4 - (-3)]^2 + (7 - 1)^2} \qquad \text{Substitute.}$$

$$= \sqrt{7^2 + 6^2} \qquad \text{Simplify.}$$

$$= \sqrt{49 + 36} \qquad \text{Evaluate powers.}$$

$$= \sqrt{85} \qquad \text{Add.}$$

Exercises

Tell whether the triangle is a right triangle.

24.

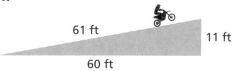

25.

Find the distance between the two points.

26. $(-2, -5), (3, 5)$

27. $(-4, 7), (4, 0)$

Check It Out
Test Practice
BigIdeasMath.com

Find the square root(s).

1. $-\sqrt{1600}$

2. $\sqrt{\dfrac{25}{49}}$

3. $\pm\sqrt{\dfrac{100}{9}}$

Find the cube root.

4. $\sqrt[3]{-27}$

5. $\sqrt[3]{\dfrac{8}{125}}$

6. $\sqrt[3]{-\dfrac{729}{64}}$

Evaluate the expression.

7. $12 + 8\sqrt{16}$

8. $\dfrac{1}{2} + \sqrt{\dfrac{72}{2}}$

9. $\left(\sqrt[3]{-125}\right)^3 + 75$

10. $50\sqrt[3]{\dfrac{512}{1000}} + 14$

11. Find the missing length of the triangle.

26 in.

a

24 in.

Classify the real number.

12. 16π

13. $-\sqrt{49}$

Estimate the square root to the nearest (a) integer and (b) tenth.

14. $\sqrt{58}$

15. $\sqrt{83}$

Write the decimal as a fraction or a mixed number.

16. $-0.\overline{3}$

17. $1.\overline{24}$

18. Tell whether the triangle is a right triangle.

80 mm   39 mm

89 mm

Find the distance between the two points.

19. $(-2, 3), (6, 9)$

20. $(0, -5), (4, 1)$

21. SUPERHERO Find the altitude of the superhero balloon.

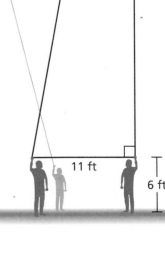

61 ft x

11 ft

6 ft

1. The period *T* of a pendulum is the time, in seconds, it takes the pendulum to swing back and forth. The period can be found using the formula $T = 1.1\sqrt{L}$, where *L* is the length, in feet, of the pendulum. A pendulum has a length of 4 feet. Find its period. *(8.EE.2)*

   A. 5.1 sec   C. 3.1 sec

   B. 4.4 sec   D. 2.2 sec

2. Which parallelogram is a dilation of parallelogram *JKLM*? (Figures not drawn to scale.) *(8.G.4)*

   F.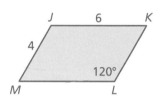

   H.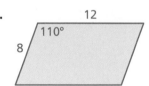

   G.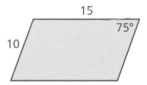

   I.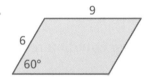

3. The point (1, 3) is on the graph of a proportional relationship. Which point is *not* on the graph? *(8.EE.5)*

   A. (0, 0)   C. (2, 6)

   B. (2, 4)   D. (6, 18)

4. Which linear equation matches the line shown in the graph? *(8.EE.6)*

   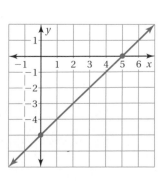

   F. $y = x - 5$ H. $y = -x - 5$

   G. $y = x + 5$ I. $y = -x + 5$

5. A football field is 40 yards wide and 120 yards long. Find the distance between opposite corners of the football field. Show your work and explain your reasoning. *(8.G.7)*

6. A computer consultant charges $50 plus $40 for each hour she works. The consultant charged $650 for one job. This can be represented by the equation below, where h represents the number of hours worked.

$$40h + 50 = 650$$

How many hours did the consultant work? *(8.EE.7b)*

7. You can use the formula below to find the sum S of the interior angle measures of a polygon with n sides. Solve the formula for n. *(8.EE.7b)*

$$S = 180(n - 2)$$

A. $n = 180(S - 2)$

C. $n = \dfrac{S}{180} - 2$

B. $n = \dfrac{S}{180} + 2$

D. $n = \dfrac{S}{180} + \dfrac{1}{90}$

8. Which linear equation relates y to x? *(8.EE.6)*

x	1	2	3	4	5
y	4	2	0	−2	−4

F. $y = 2x + 2$

H. $y = -2x + 2$

G. $y = 4x$

I. $y = -2x + 6$

9. An airplane flies from City 1 at $(0, 0)$ to City 2 at $(33, 56)$ and then to City 3 at $(23, 32)$. What is the total number of miles it flies? Each unit of the coordinate grid represents 1 mile. *(8.G.8)*

10. What is the missing length of the right triangle shown? *(8.G.7)*

A. 16 cm

C. 24 cm

B. 18 cm

D. $\sqrt{674}$ cm

11. Which sequence of transformations shows that the triangles in the coordinate plane are similar? *(8.G.4)*

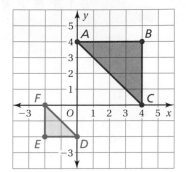

F. Dilate △*ABC* using a scale factor of 2 and then rotate 180° about the origin.

G. Dilate △*DEF* using a scale factor of $\frac{1}{2}$ and then rotate 180° about the origin.

H. Reflect △*DEF* in both axes and then dilate using a scale factor of 2.

I. Rotate △*ABC* 180° about the origin and then dilate using a scale factor of 2.

12. In the diagram, lines ℓ and m are parallel. Which angle has the same measure as ∠1? *(8.G.5)*

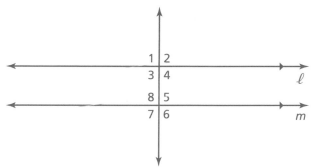

A. ∠2

B. ∠5

C. ∠7

D. ∠8

13. Which graph represents the linear equation $y = -2x - 2$? *(8.EE.6)*

F.

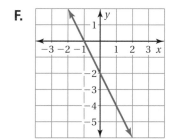

H.

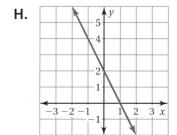

G.

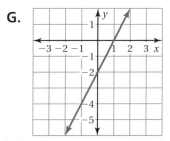

I.

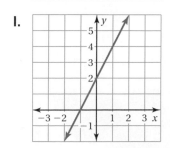

15 Volume and Similar Solids

"Dear Sir: Why do you sell dog food in tall cans and sell cat food in short cans?"

"Neither of these shapes is the optimal use of surface area when compared to volume."

"Do you know why the volume of a cone is one-third the volume of a cylinder with the same height and base?"

What You Learned Before

"I just figured out how to find your volume. We'll immerse you in a barrel of water and measure the water that overflows."

Number one on America's list of 10 worst ideas.

● **Finding the Area of a Composite Figure** (7.G.6)

Example 1 Find the area of the figure.

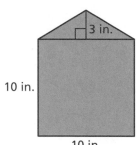

3 in.

10 in.

10 in.

Area = Area of square + Area of triangle

$$A = s^2 + \frac{1}{2}bh$$

$$= 10^2 + \left(\frac{1}{2} \cdot 10 \cdot 3\right)$$

$$= 100 + 15$$

$$= 115 \text{ in.}^2$$

Try It Yourself

Find the area of the figure.

1.

8 m

15 m

2.

9 cm

4 cm

14 cm

5 cm

● **Finding the Areas of Circles** (7.G.4)

Example 2 Find the area of the circle.

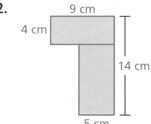

7 mm

$$A = \pi r^2$$

$$\approx \frac{22}{7} \cdot 7^2$$

$$= \frac{22}{7} \cdot 49$$

$$= 154 \text{ mm}^2$$

Example 3 Find the area of the circle.

24 yd

$$A = \pi r^2$$

$$\approx 3.14 \cdot 12^2$$

$$= 3.14 \cdot 144$$

$$= 452.16 \text{ yd}^2$$

Try It Yourself

Find the area of the circle.

3.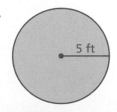

5 ft

4.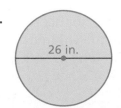

26 in.

5.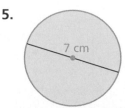

7 cm

Essential Question How can you find the volume of a cylinder?

1 ACTIVITY: Finding a Formula Experimentally

Work with a partner.

a. Find the area of the face of a coin.

b. Find the volume of a stack of a dozen coins.

c. Write a formula for the volume of a cylinder.

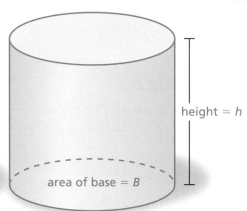

height = h

area of base = B

2 ACTIVITY: Making a Business Plan

COMMON CORE

Geometry

In this lesson, you will
• find the volumes of cylinders.
• find the heights of cylinders given the volumes.
• solve real-life problems.
Learning Standard
8.G.9

Work with a partner. You are planning to make and sell three different sizes of cylindrical candles. You buy 1 cubic foot of candle wax for $20 to make 8 candles of each size.

a. Design the candles. What are the dimensions of each size of candle?

b. You want to make a profit of $100. Decide on a price for each size of candle.

c. Did you set the prices so that they are proportional to the volume of each size of candle? Why or why not?

3 ACTIVITY: Science Experiment

Work with a partner. Use the diagram to describe how you can find the volume of a small object.

4 ACTIVITY: Comparing Cylinders

Math Practice 1

Consider Similar Problems

How can you use the results of Activity 1 to find the volumes of the cylinders?

Work with a partner.

a. Just by looking at the two cylinders, which one do you think has the greater volume? Explain your reasoning.

b. Find the volume of each cylinder. Was your prediction in part (a) correct? Explain your reasoning.

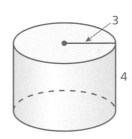

What Is Your Answer?

5. **IN YOUR OWN WORDS** How can you find the volume of a cylinder?

6. Compare your formula for the volume of a cylinder with the formula for the volume of a prism. How are they the same?

"Here's how I remember how to find the volume of <u>any</u> prism or cylinder."

"Base times tall, will fill 'em all."

Practice

Use what you learned about the volumes of cylinders to complete Exercises 3–5 on page 676.

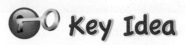 **Key Idea**

Volume of a Cylinder

Words The volume V of a cylinder is the product of the area of the base and the height of the cylinder.

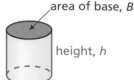

area of base, B

height, h

Algebra $V = Bh$

Area of base — Height of cylinder

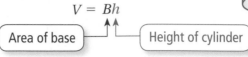 **Finding the Volume of a Cylinder**

Find the volume of the cylinder. Round your answer to the nearest tenth.

$$V = Bh$$ Write formula for volume.

$$= \pi(3)^2(6)$$ Substitute.

$$= 54\pi \approx 169.6$$ Use a calculator.

∴ The volume is about 169.6 cubic meters.

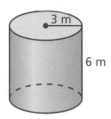

3 m

6 m

Study Tip

Because $B = \pi r^2$, you can use $V = \pi r^2 h$ to find the volume of a cylinder.

EXAMPLE 2 **Finding the Height of a Cylinder**

Find the height of the cylinder. Round your answer to the nearest whole number.

The diameter is 10 inches. So, the radius is 5 inches.

$$V = Bh$$ Write formula for volume.

$$314 = \pi(5)^2(h)$$ Substitute.

$$314 = 25\pi h$$ Simplify.

$$4 \approx h$$ Divide each side by 25π.

∴ The height is about 4 inches.

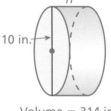

h

10 in.

Volume = 314 in.^{3}

On Your Own

Now You're Ready
Exercises 3–11
and 13–15

Find the volume V or height h of the cylinder. Round your answer to the nearest tenth.

1.

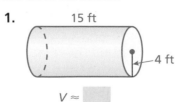

15 ft

4 ft

$V \approx$ ▭

2.

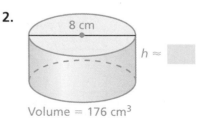

8 cm

$h \approx$ ▭

Volume = 176 cm³

EXAMPLE **3** **Real-Life Application**

How much salsa is missing from the jar?

The empty space in the jar is a cylinder with a height of $10 - 4 = 6$ centimeters and a radius of 5 centimeters.

5 cm

10 cm

4 cm

$$V = Bh \qquad \text{Write formula for volume.}$$
$$= \pi(5)^2(6) \qquad \text{Substitute.}$$
$$= 150\pi \approx 471 \qquad \text{Use a calculator.}$$

∴ So, about 471 cubic centimeters of salsa are missing from the jar.

EXAMPLE **4** **Real-Life Application**

1.7 ft

1 ft

About how many gallons of water does the watercooler bottle contain? (1 ft³ ≈ 7.5 gal)

(A) 5.3 gallons   **(B)** 10 gallons   **(C)** 17 gallons   **(D)** 40 gallons

Find the volume of the cylinder. The diameter is 1 foot. So, the radius is 0.5 foot.

$$V = Bh \qquad \text{Write formula for volume.}$$
$$= \pi(0.5)^2(1.7) \qquad \text{Substitute.}$$
$$= 0.425\pi \approx 1.3352 \qquad \text{Use a calculator.}$$

So, the bottle contains about 1.3352 cubic feet of water. To find the number of gallons it contains, multiply by the conversion factor $\dfrac{7.5 \text{ gal}}{1 \text{ ft}^3}$.

$$1.3352 \ \cancel{\text{ft}^3} \times \frac{7.5 \text{ gal}}{1 \ \cancel{\text{ft}^3}} \approx 10 \text{ gal}$$

∴ The watercooler bottle contains about 10 gallons of water. So, the correct answer is **(B)**.

● **On Your Own**

Now You're Ready
Exercise 12

3. WHAT IF? In Example 3, the height of the salsa in the jar is 5 centimeters. How much salsa is missing from the jar?

4. A cylindrical water tower has a diameter of 15 meters and a height of 5 meters. About how many gallons of water can the tower contain? (1 m³ ≈ 264 gal)

15.1 Exercises

✓ Vocabulary and Concept Check

1. **DIFFERENT WORDS, SAME QUESTION** Which is different? Find "both" answers.

How much does it take to fill the cylinder?

What is the capacity of the cylinder?

How much does it take to cover the cylinder?

How much does the cylinder contain?

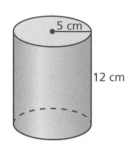
5 cm
12 cm

2. **REASONING** Without calculating, which of the solids has the greater volume? Explain.

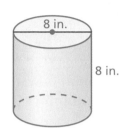
8 in.
8 in.

8 in.
8 in.
8 in.
8 in.

✏️ Practice and Problem Solving

Find the volume of the cylinder. Round your answer to the nearest tenth.

 3.
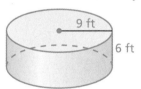
9 ft
6 ft

4.
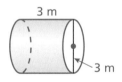
3 m
3 m

5.
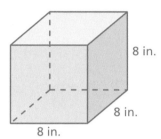
7 ft
5 ft

6.
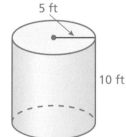
5 ft
10 ft

7.
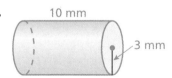
10 mm
3 mm

8.
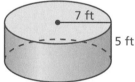
2 ft
1 ft

9.
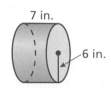
7 in.
6 in.

10.
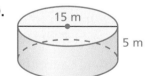
15 m
5 m

11.
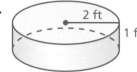
16 cm
8 cm

④ 12. **SWIMMING POOL** A cylindrical swimming pool has a diameter of 16 feet and a height of 4 feet. About how many gallons of water can the pool contain? Round your answer to the nearest whole number. (1 ft^3 ≈ 7.5 gal)

Find the missing dimension of the cylinder. Round your answer to the nearest whole number.

② 13. Volume = 250 ft³

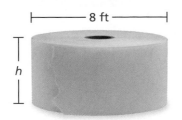

8 ft

h

14. Volume = 10,000π in.³

32 in.

h

15. Volume = 600,000 cm³

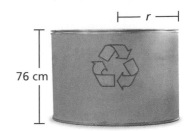

r

76 cm

16. CRITICAL THINKING How does the volume of a cylinder change when its diameter is halved? Explain.

5 ft

4 ft

Round hay bale

17. MODELING A traditional "square" bale of hay is actually in the shape of a rectangular prism. Its dimensions are 2 feet by 2 feet by 4 feet. How many square bales contain the same amount of hay as one large "round" bale?

18. ROAD ROLLER A tank on a road roller is filled with water to make the roller heavy. The tank is a cylinder that has a height of 6 feet and a radius of 2 feet. One cubic foot of water weighs 62.5 pounds. Find the weight of the water in the tank.

19. VOLUME A cylinder has a surface area of 1850 square meters and a radius of 9 meters. Estimate the volume of the cylinder to the nearest whole number.

20. **Problem Solving** Water flows at 2 feet per second through a pipe with a diameter of 8 inches. A cylindrical tank with a diameter of 15 feet and a height of 6 feet collects the water.

a. What is the volume, in cubic inches, of water flowing out of the pipe every second?

b. What is the height, in inches, of the water in the tank after 5 minutes?

c. How many minutes will it take to fill 75% of the tank?

 Fair Game Review *What you learned in previous grades & lessons*

Tell whether the triangle with the given side lengths is a right triangle. *(Section 14.5)*

21. 20 m, 21 m, 29 m

22. 1 in., 2.4 in., 2.6 in.

23. 5.6 ft, 8 ft, 10.6 ft

24. MULTIPLE CHOICE What is the area of a circle with a diameter of 10 meters? *(Section 8.3)*

Ⓐ $10\pi\,\text{m}^2$ Ⓑ $20\pi\,\text{m}^2$ Ⓒ $25\pi\,\text{m}^2$ Ⓓ $100\pi\,\text{m}^2$

Essential Question How can you find the volume of a cone?

You already know how the volume of a pyramid relates to the volume of a prism. In this activity, you will discover how the volume of a cone relates to the volume of a cylinder.

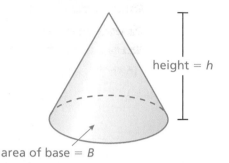

height = h

area of base = B

1 ACTIVITY: Finding a Formula Experimentally

Work with a partner. Use a paper cup that is shaped like a cone.

- Estimate the height of the cup.
- Trace the top of the cup on a piece of paper. Find the diameter of the circle.
- Use these measurements to draw a net for a cylinder with the same base and height as the paper cup.
- Cut out the net. Then fold and tape it to form an open cylinder.
- Fill the paper cup with rice. Then pour the rice into the cylinder. Repeat this until the cylinder is full. How many cones does it take to fill the cylinder?
- Use your result to write a formula for the volume of a cone.

2 ACTIVITY: Summarizing Volume Formulas

COMMON CORE

Geometry

In this lesson, you will
- find the volumes of cones.
- find the heights of cones given the volumes.
- solve real-life problems.

Learning Standard
8.G.9

Work with a partner. You can remember the volume formulas for prisms, cylinders, pyramids, and cones with just two concepts.

Volumes of Prisms and Cylinders

Volume = Area of base ×

Volumes of Pyramids and Cones

Volume = Volume of prism or cylinder with same base and height

Make a list of all the formulas you need to remember to find the area of a base. Talk about strategies for remembering these formulas.

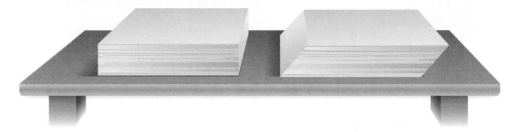

Work with a partner. Think of a stack of paper. When you adjust the stack so that the sides are oblique (slanted), do you change the volume of the stack? If the volume of the stack does not change, then the formulas for volumes of right solids also apply to oblique solids.

Math Practice 2

Use Equations
What equation would you use to find the volume of the oblique solid? Explain.

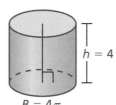

$h = 4$

$B = 4\pi$

Right cylinder

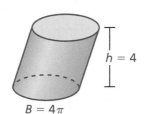

$h = 4$

$B = 4\pi$

Oblique cylinder

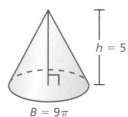

$h = 5$

$B = 9\pi$

Right cone

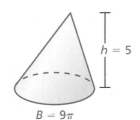

$h = 5$

$B = 9\pi$

Oblique cone

What Is Your Answer?

4. **IN YOUR OWN WORDS** How can you find the volume of a cone?

5. Describe the intersection of the plane and the cone. Then explain how to find the volume of each section of the solid.

a.

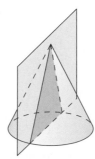

b.

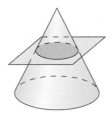

Practice ➤ Use what you learned about the volumes of cones to complete Exercises 4–6 on page 682.

 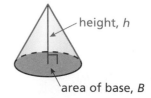
 **Key Idea**

Volume of a Cone

Words The volume *V* of a cone is one-third the product of the area of the base and the height of the cone.

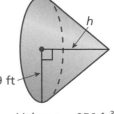
height, *h*

area of base, *B*

Study Tip

The *height* of a cone is the perpendicular distance from the base to the vertex.

Algebra $V = \dfrac{1}{3}Bh$

Area of base ⟶

Height of cone ⟶

EXAMPLE 1 Finding the Volume of a Cone

Find the volume of the cone. Round your answer to the nearest tenth.

The diameter is 4 meters. So, the radius is 2 meters.

Study Tip

Because $B = \pi r^2$, you can use $V = \dfrac{1}{3}\pi r^2 h$ to find the volume of a cone.

$V = \dfrac{1}{3}Bh$ Write formula for volume.

$\quad = \dfrac{1}{3}\pi(2)^2(6)$ Substitute.

$\quad = 8\pi \approx 25.1$ Use a calculator.

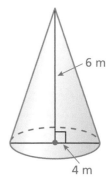
6 m

4 m

∴ The volume is about 25.1 cubic meters.

EXAMPLE 2 Finding the Height of a Cone

Find the height of the cone. Round your answer to the nearest tenth.

$V = \dfrac{1}{3}Bh$ Write formula for volume.

$956 = \dfrac{1}{3}\pi(9)^2(h)$ Substitute.

$956 = 27\pi h$ Simplify.

$11.3 \approx h$ Divide each side by 27π.

h

9 ft

Volume = 956 ft³

∴ The height is about 11.3 feet.

Now You're Ready
Exercises 4–12
and 15–17

On Your Own

Find the volume V or height h of the cone. Round your answer to the nearest tenth.

1.

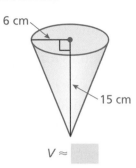

6 cm

15 cm

$V \approx$

2.

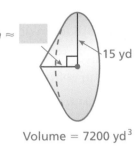

$h \approx$

15 yd

Volume = 7200 yd³

EXAMPLE 3 — **Real-Life Application**

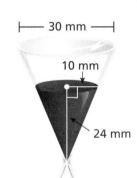

├── 30 mm ──┤

10 mm

24 mm

You must answer a trivia question before the sand in the timer falls to the bottom. The sand falls at a rate of 50 cubic millimeters per second. How much time do you have to answer the question?

Use the formula for the volume of a cone to find the volume of the sand in the timer.

$$V = \frac{1}{3}Bh \qquad \text{Write formula for volume.}$$

$$= \frac{1}{3}\pi(10)^2(24) \qquad \text{Substitute.}$$

$$= 800\pi \approx 2513 \qquad \text{Use a calculator.}$$

The volume of the sand is about 2513 cubic millimeters. To find the amount of time you have to answer the question, multiply the volume by the rate at which the sand falls.

$$2513 \text{ mm}^3 \times \frac{1 \text{ sec}}{50 \text{ mm}^3} = 50.26 \text{ sec}$$

So, you have about 50 seconds to answer the question.

On Your Own

3. WHAT IF? The sand falls at a rate of 60 cubic millimeters per second. How much time do you have to answer the question?

4. WHAT IF? The height of the sand in the timer is 12 millimeters, and the radius is 5 millimeters. How much time do you have to answer the question?

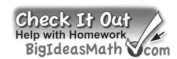

Vocabulary and Concept Check

1. **VOCABULARY** Describe the height of a cone.

2. **WRITING** Compare and contrast the formulas for the volume of a pyramid and the volume of a cone.

3. **REASONING** You know the volume of a cylinder. How can you find the volume of a cone with the same base and height?

Practice and Problem Solving

Find the volume of the cone. Round your answer to the nearest tenth.

4.
4 in.
2 in.

5.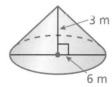
3 m
6 m

6.
10 mm
5 mm

7.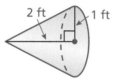
2 ft   1 ft

8.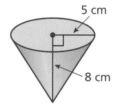
5 cm
8 cm

9.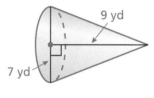
9 yd
7 yd

10.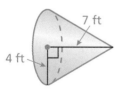
7 ft
4 ft

11.
10 in.
5 in.

12. 4 cm
8 cm

13. **ERROR ANALYSIS** Describe and correct the error in finding the volume of the cone.

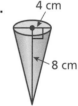
3 m
2 m

$$✗ \quad V = \frac{1}{3}Bh$$
$$= \frac{1}{3}(\pi)(2)^2(3)$$
$$= 4\pi \text{ m}^3$$

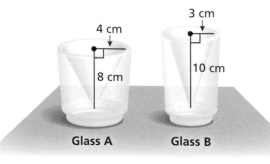
4 cm
8 cm
3 cm
10 cm
Glass A   Glass B

14. **GLASS** The inside of each glass is shaped like a cone. Which glass can hold more liquid? How much more?

Find the missing dimension of the cone. Round your answer to the nearest tenth.

② 15. Volume = $\frac{1}{18}\pi$ ft^3

16. Volume = 225 cm^3

17. Volume = 3.6 in.^{3}

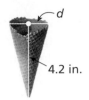

18. REASONING The volume of a cone is 20π cubic meters. What is the volume of a cylinder with the same base and height?

19. VASE Water leaks from a crack in a vase at a rate of 0.5 cubic inch per minute. How long does it take for 20% of the water to leak from a full vase?

20. LEMONADE STAND You have 10 gallons of lemonade to sell. (1 gal ≈ 3785 cm^3)

a. Each customer uses one paper cup. How many paper cups will you need?

b. The cups are sold in packages of 50. How many packages should you buy?

c. How many cups will be left over if you sell 80% of the lemonade?

21. STRUCTURE The cylinder and the cone have the same volume. What is the height of the cone?

22. 🌟 **Critical Thinking** In Example 3, you use a different timer with the same dimensions. The sand in this timer has a height of 30 millimeters. How much time do you have to answer the question?

Fair Game Review *What you learned in previous grades & lessons*

The vertices of a figure are given. Rotate the figure as described. Find the coordinates of the image. *(Section 11.4)*

23. $A(-1, 1)$, $B(2, 3)$, $C(2, 1)$
90° counterclockwise about vertex A

24. $E(-4, 1)$, $F(-3, 3)$, $G(-2, 3)$, $H(-1, 1)$
180° about the origin

25. MULTIPLE CHOICE $\triangle ABC$ is similar to $\triangle XYZ$. How many times greater is the area of $\triangle XYZ$ than the area of $\triangle ABC$? *(Section 11.6)*

Ⓐ $\frac{1}{9}$

Ⓑ $\frac{1}{3}$

Ⓒ 3

Ⓓ 9

 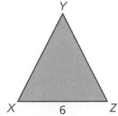

Check It Out
Graphic Organizer
BigIdeasMath ✓com

You can use a **formula triangle** to arrange variables and operations of a formula. Here is an example of a formula triangle for the volume of a cylinder.

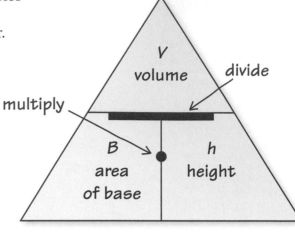

To find an unknown variable, use the other variables and the operation between them. For example, to find the area B of the base, cover up the B. Then you can see that you divide the volume V by the height h.

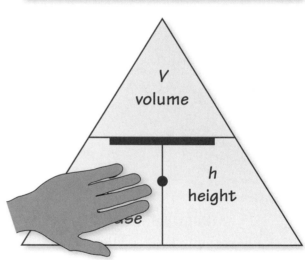

On Your Own

Make a formula triangle to help you study this topic. (*Hint:* Your formula triangle may have a different form than what is shown in the example.)

1. volume of a cone

After you complete this chapter, make formula triangles for the following topics.

2. volume of a sphere

3. volume of a composite solid

4. surface areas of similar solids

5. volumes of similar solids

"See how a formula triangle works? Cover any variable and you get its formula."

Find the volume of the solid. Round your answer to the nearest tenth. *(Section 15.1 and Section 15.2)*

1. 4 yd

3.5 yd

2. 3 ft

4 ft

3. 5 cm

6 cm

4. 11 in.

12 in.

Find the missing dimension of the solid. Round your answer to the nearest tenth.
(Section 15.1 and Section 15.2)

5. h

3 ft

Volume = 340 ft³

6. 4.7 cm

r

Volume = 938 cm³

7. PAPER CONE The paper cone can hold 84.78 cubic centimeters of water. What is the height of the cone? *(Section 15.2)*

6 cm

h

8. GEOMETRY Triple both dimensions of the cylinder. How many times greater is the volume of the new cylinder than the volume of the original cylinder? *(Section 15.1)*

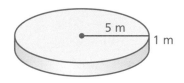

5 m

1 m

9. SAND ART There are 42.39 cubic inches of blue sand and 28.26 cubic inches of red sand in the cylindrical container. How many cubic inches of white sand are in the container? *(Section 15.1)*

1.5 in.

16 in.

10. JUICE CAN You are buying two cylindrical cans of juice. Each can holds the same amount of juice. What is the height of Can B? *(Section 15.1)*

4 in.

6 in.

6 in.

h

Can A   Can B

Essential Question How can you find the volume of a sphere?

A **sphere** is the set of all points in space that are the same distance from a point called the *center*. The *radius r* is the distance from the center to any point on the sphere.

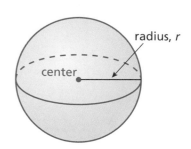

radius, *r*

center

A sphere is different from the other solids you have studied so far because it does not have a base. To discover the volume of a sphere, you can use an activity similar to the one in the previous section.

---

1 ACTIVITY: Exploring the Volume of a Sphere

Work with a partner. Use a plastic ball similar to the one shown.

- Estimate the diameter and the radius of the ball.

- Use these measurements to draw a net for a cylinder with a diameter and a height equal to the diameter of the ball. How is the height *h* of the cylinder related to the radius *r* of the ball? Explain.

- Cut out the net. Then fold and tape it to form an open cylinder. Make two marks on the cylinder that divide it into thirds, as shown.

COMMON CORE

Geometry

In this lesson, you will
- find the volumes of spheres.
- find the radii of spheres given the volumes.
- solve real-life problems.

Learning Standard
8.G.9

- Cover the ball with aluminum foil or tape. Leave one hole open. Fill the ball with rice. Then pour the rice into the cylinder. What fraction of the cylinder is filled with rice?

Work with a partner. Use the results from Activity 1 and the formula for the volume of a cylinder to complete the steps.

Math Practice 4

Analyze Relationships

What is the relationship between the volume of a sphere and the volume of a cylinder? How does this help you derive a formula for the volume of a sphere?

$$V = \pi r^2 h$$ Write formula for volume of a cylinder.

$$= \frac{\boxed{}}{\boxed{}}\, \pi r^2 h$$ Multiply by $\dfrac{\boxed{}}{\boxed{}}$ because the volume of a sphere

is $\dfrac{\boxed{}}{\boxed{}}$ of the volume of the cylinder.

$$= \frac{\boxed{}}{\boxed{}}\, \pi r^2 \boxed{}$$ Substitute $\boxed{}$ for h.

$$= \frac{\boxed{}}{\boxed{}}\, \pi \boxed{}$$ Simplify.

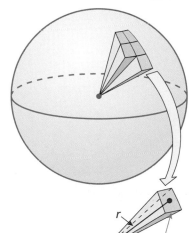

r

area of base, B

Work with a partner. Imagine filling the inside of a sphere with n small pyramids. The vertex of each pyramid is at the center of the sphere. The height of each pyramid is approximately equal to r, as shown. Complete the steps. (The surface area of a sphere is equal to $4\pi r^2$.)

$$V = \frac{1}{3}Bh$$ Write formula for volume of a pyramid.

$$= n\frac{1}{3}B\,\boxed{}$$ Multiply by the number of small pyramids n and substitute $\boxed{}$ for h.

$$= \frac{1}{3}\left(4\pi r^2\right)\boxed{}$$ $4\pi r^2 \approx n \cdot \boxed{}$

Show how this result is equal to the result in Activity 2.

What Is Your Answer?

4. **IN YOUR OWN WORDS** How can you find the volume of a sphere?

5. Describe the intersection of the plane and the sphere. Then explain how to find the volume of each section of the solid.

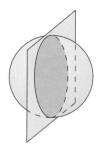

Practice

Use what you learned about the volumes of spheres to complete Exercises 3–5 on page 690.

Check It Out
Lesson Tutorials
BigIdeasMath com

Key Vocabulary 🔊
sphere, *p. 686*
hemisphere, *p. 689*

 **Key Idea**

Volume of a Sphere

Words The volume V of a sphere is the product of $\frac{4}{3}\pi$ and the cube of the radius of the sphere.

radius, r

Algebra $V = \frac{4}{3}\pi r^3$

$\boxed{\text{Cube of radius of sphere}}$

EXAMPLE ① **Finding the Volume of a Sphere**

Find the volume of the sphere. Round your answer to the nearest tenth.

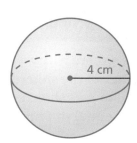

4 cm

$V = \frac{4}{3}\pi r^3$ Write formula for volume.

$= \frac{4}{3}\pi (4)^3$ Substitute 4 for r.

$= \frac{256}{3}\pi$ Simplify.

≈ 268.1 Use a calculator.

⋮ The volume is about 268.1 cubic centimeters.

EXAMPLE ② **Finding the Radius of a Sphere**

Find the radius of the sphere.

Volume = 288π in.^{3}

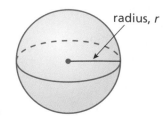

r

$V = \frac{4}{3}\pi r^3$ Write formula.

$288\pi = \frac{4}{3}\pi r^3$ Substitute.

$288\pi = \frac{4\pi}{3}r^3$ Multiply.

$\frac{3}{4\pi} \cdot 288\pi = \frac{3}{4\pi} \cdot \frac{4\pi}{3}r^3$ Multiplication Property of Equality

$216 = r^3$ Simplify.

$6 = r$ Take the cube root of each side.

⋮ The radius is 6 inches.

On Your Own

Now You're Ready
Exercises 3–11

Find the volume *V* or radius *r* of the sphere. Round your answer to the nearest tenth, if necessary.

1.

16 ft

$V \approx$

2.

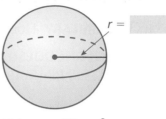

$r =$

Volume = 36π m³

EXAMPLE 3 **Finding the Volume of a Composite Solid**

52 ft

A hemisphere is one-half of a sphere. The top of the silo is a hemisphere with a radius of 12 feet. What is the volume of the silo? Round your answer to the nearest thousand.

The silo is made up of a cylinder and a hemisphere. Find the volume of each solid.

Cylinder

12 ft

40 ft

Hemisphere

12 ft

Study Tip

In Example 3, the height of the cylindrical part of the silo is the difference of the silo height and the radius of the hemisphere.
52 − 12 = 40 ft

$V = Bh$

$= \pi(12)^2(40)$

$= 5760\pi$

$V = \dfrac{1}{2} \cdot \dfrac{4}{3}\pi r^3$

$= \dfrac{1}{2} \cdot \dfrac{4}{3}\pi(12)^3$

$= 1152\pi$

⁑ So, the volume is $5760\pi + 1152\pi = 6912\pi \approx 22{,}000$ cubic feet.

On Your Own

Now You're Ready
Exercises 14–16

Find the volume of the composite solid. Round your answer to the nearest tenth.

3.

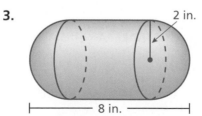

2 in.

8 in.

4.

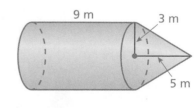

9 m

3 m

5 m

 ## Vocabulary and Concept Check

1. **VOCABULARY** How is a sphere different from a hemisphere?

2. **WHICH ONE DOESN'T BELONG?** Which figure does *not* belong with the other three? Explain your reasoning.

 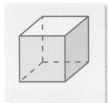  

 ## Practice and Problem Solving

Find the volume of the sphere. Round your answer to the nearest tenth.

3.
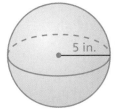
5 in.

4.
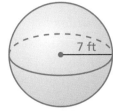
7 ft

5.

18 mm

6.

12 yd

7.
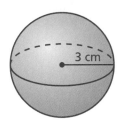
3 cm

8.

28 m

Find the radius of the sphere with the given volume.

9. Volume $= 972\pi \text{ mm}^3$

10. Volume $= 4.5\pi \text{ cm}^3$

11. Volume $= 121.5\pi \text{ ft}^3$

12. **GLOBE** The globe of the Moon has a radius of 10 inches. Find the volume of the globe. Round your answer to the nearest whole number.

13. **SOFTBALL** A softball has a volume of $\dfrac{125}{6}\pi$ cubic inches. Find the radius of the softball.

Find the volume of the composite solid. Round your answer to the nearest tenth.

❸ 14.

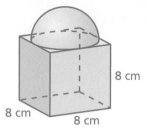

8 cm

8 cm

8 cm

15.

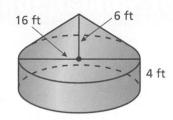

16 ft

6 ft

4 ft

16.

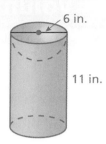

6 in.

11 in.

17. **REASONING** A sphere and a right cylinder have the same radius and volume. Find the radius *r* in terms of the height *h* of the cylinder.

18. **PACKAGING** A cylindrical container of three rubber balls has a height of 18 centimeters and a diameter of 6 centimeters. Each ball in the container has a radius of 3 centimeters. Find the amount of space in the container that is not occupied by rubber balls. Round your answer to the nearest whole number.

Volume = 4500π in.³

19. **BASKETBALL** The basketball shown is packaged in a box that is in the shape of a cube. The edge length of the box is equal to the diameter of the basketball. What is the surface area and the volume of the box?

20. **Logic** Your friend says that the volume of a sphere with radius *r* is four times the volume of a cone with radius *r*. When is this true? Justify your answer.

Fair Game Review What you learned in previous grades & lessons

The blue figure is a dilation of the red figure. Identify the type of dilation and find the scale factor. *(Section 11.7)*

21.

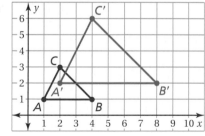

22.

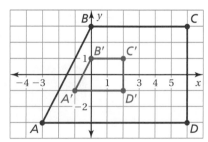

23. **MULTIPLE CHOICE** A person who is 5 feet tall casts a 6-foot-long shadow. A nearby flagpole casts a 30-foot-long shadow. What is the height of the flagpole? *(Section 12.4)*

(A) 25 ft
(B) 29 ft
(C) 36 ft
(D) 40 ft

Essential Question

When the dimensions of a solid increase by a factor of k, how does the surface area change? How does the volume change?

1 ACTIVITY: Comparing Surface Areas and Volumes

Work with a partner. Copy and complete the table. Describe the pattern. Are the dimensions proportional? Explain your reasoning.

a.

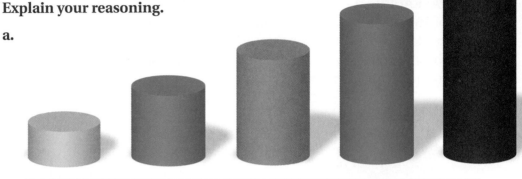

Radius	1	1	1	1	1
Height	1	2	3	4	5
Surface Area					
Volume					

b.

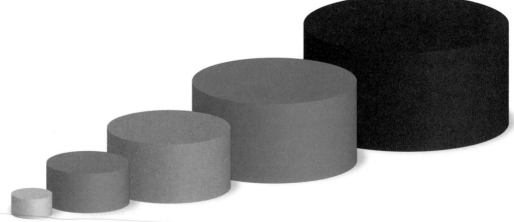

Radius	1	2	3	4	5
Height	1	2	3	4	5
Surface Area					
Volume					

COMMON CORE

Geometry

In this lesson, you will

- identify similar solids.
- use properties of similar solids to find missing measures.
- understand the relationship between surface areas of similar solids.
- understand the relationship between volumes of similar solids.
- solve real-life problems.

Applying Standard
8.G.9

Work with a partner. Copy and complete the table. Describe the pattern. Are the dimensions proportional? Explain.

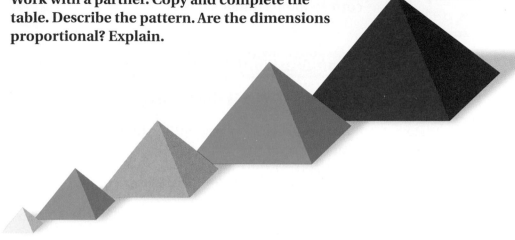

Math Practice 8

Repeat Calculations

Which calculations are repeated? How does this help you describe the pattern?

Base Side	6	12	18	24	30
Height	4	8	12	16	20
Slant Height	5	10	15	20	25
Surface Area					
Volume					

What Is Your Answer?

3. **IN YOUR OWN WORDS** When the dimensions of a solid increase by a factor of k, how does the surface area change?

4. **IN YOUR OWN WORDS** When the dimensions of a solid increase by a factor of k, how does the volume change?

5. **REPEATED REASONING** All the dimensions of a prism increase by a factor of 5.

   a. How many times greater is the surface area? Explain.

   | 5 | 10 | 25 | 125 |

   b. How many times greater is the volume? Explain.

   | 5 | 10 | 25 | 125 |

Practice

Use what you learned about surface areas and volumes of similar solids to complete Exercise 3 on page 697.

Check It Out
Lesson Tutorials
BigIdeasMath ✓com

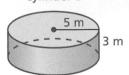

Key Vocabulary
similar solids, *p. 694*

Similar solids are solids that have the same shape and proportional corresponding dimensions.

EXAMPLE 1 Identifying Similar Solids

Cylinder B

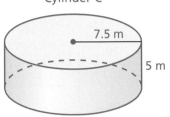
5 m
3 m

Cylinder A

6 m
4 m

Which cylinder is similar to Cylinder A?

Check to see if corresponding dimensions are proportional.

Cylinder A and Cylinder B

$$\frac{\text{Height of A}}{\text{Height of B}} = \frac{4}{3} \qquad \frac{\text{Radius of A}}{\text{Radius of B}} = \frac{6}{5}$$

Not proportional

Cylinder C

7.5 m
5 m

Cylinder A and Cylinder C

$$\frac{\text{Height of A}}{\text{Height of C}} = \frac{4}{5} \qquad \frac{\text{Radius of A}}{\text{Radius of C}} = \frac{6}{7.5} = \frac{4}{5}$$

Proportional

So, Cylinder C is similar to Cylinder A.

EXAMPLE 2 Finding Missing Measures in Similar Solids

Cone X

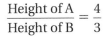
13 yd
5 yd

Cone Y

ℓ
7 yd

The cones are similar. Find the missing slant height ℓ.

$$\frac{\text{Radius of X}}{\text{Radius of Y}} = \frac{\text{Slant height of X}}{\text{Slant height of Y}}$$

$$\frac{5}{7} = \frac{13}{\ell} \qquad \text{Substitute.}$$

$$5\ell = 91 \qquad \text{Cross Products Property}$$

$$\ell = 18.2 \qquad \text{Divide each side by 5.}$$

The slant height is 18.2 yards.

On Your Own

Now You're Ready
Exercises 4–9

1. Cylinder D has a radius of 7.5 meters and a height of 4.5 meters. Which cylinder in Example 1 is similar to Cylinder D?

2. The prisms at the right are similar. Find the missing width and length.

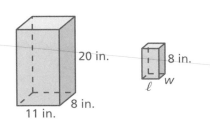
20 in.
8 in.
11 in.
8 in.
ℓ
w

🔊 Multi-Language Glossary at BigIdeasMath✓com

 **Key Ideas**

Linear Measures

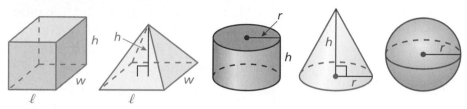

Surface Areas of Similar Solids

When two solids are similar, the ratio of their surface areas is equal to the square of the ratio of their corresponding linear measures.

$$\frac{\text{Surface Area of A}}{\text{Surface Area of B}} = \left(\frac{a}{b}\right)^2$$

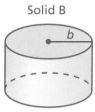

Solid A

Solid B

EXAMPLE 3 **Finding Surface Area**

Pyramid A

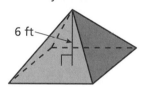

6 ft

Pyramid B

10 ft

Surface Area = 600 ft²

The pyramids are similar. What is the surface area of Pyramid A?

$$\frac{\text{Surface Area of A}}{\text{Surface Area of B}} = \left(\frac{\text{Height of A}}{\text{Height of B}}\right)^2$$

$$\frac{S}{600} = \left(\frac{6}{10}\right)^2 \qquad \text{Substitute.}$$

$$\frac{S}{600} = \frac{36}{100} \qquad \text{Evaluate.}$$

$$\frac{S}{600} \cdot 600 = \frac{36}{100} \cdot 600 \qquad \text{Multiplication Property of Equality}$$

$$S = 216 \qquad \text{Simplify.}$$

⋮ The surface area of Pyramid A is 216 square feet.

● **On Your Own**

The solids are similar. Find the surface area of the red solid. Round your answer to the nearest tenth.

3.

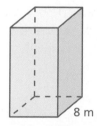

8 m

Surface Area = 608 m²

4.

 5 cm

 4 cm

5 m

Surface Area = 110 cm²

 **Key Idea**

Volumes of Similar Solids

When two solids are similar, the ratio of their volumes is equal to the cube of the ratio of their corresponding linear measures.

$$\frac{\text{Volume of A}}{\text{Volume of B}} = \left(\frac{a}{b}\right)^3$$

 Solid A

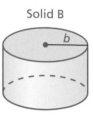 Solid B

EXAMPLE **4** **Finding Volume**

Original Tank

Volume = 2000 ft³

The dimensions of the touch tank at an aquarium are doubled. What is the volume of the new touch tank?

(A) 150 ft³   **(B)** 4000 ft³

(C) 8000 ft³   **(D)** 16,000 ft³

The dimensions are doubled, so the ratio of the dimensions of the original tank to the dimensions of the new tank is 1 : 2.

$$\frac{\text{Original volume}}{\text{New volume}} = \left(\frac{\text{Original dimension}}{\text{New dimension}}\right)^3$$

$$\frac{2000}{V} = \left(\frac{1}{2}\right)^3 \qquad \text{Substitute.}$$

$$\frac{2000}{V} = \frac{1}{8} \qquad \text{Evaluate.}$$

$$16{,}000 = V \qquad \text{Cross Products Property}$$

Study Tip

When the dimensions of a solid are multiplied by k, the surface area is multiplied by k^2 and the volume is multiplied by k^3.

⋮ The volume of the new tank is 16,000 cubic feet. So, the correct answer is **(D)**.

On Your Own

Now You're Ready
Exercises 10–13

The solids are similar. Find the volume of the red solid. Round your answer to the nearest tenth.

5.

5 cm

12 cm

Volume = 288 cm³

6.

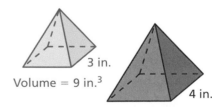

3 in.

Volume = 9 in.³

4 in.

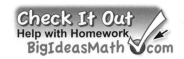

Vocabulary and Concept Check

1. **VOCABULARY** What are similar solids?

2. **OPEN-ENDED** Draw two similar solids and label their corresponding linear measures.

Practice and Problem Solving

3. **NUMBER SENSE** All the dimensions of a cube increase by a factor of $\frac{3}{2}$.

   a. How many times greater is the surface area? Explain.

   b. How many times greater is the volume? Explain.

Determine whether the solids are similar.

4.

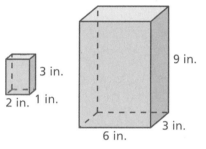

3 in.
9 in.
2 in. 1 in.
6 in. 3 in.
3 in.

5.

4 in.
4 in.
4 in. 2 in.
2 in. 1 in.

6.

6 ft 6.5 ft 12 ft 13 ft
5 ft
5 ft
10 ft
10 ft

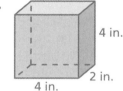

7.

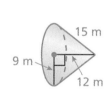

15 m
9 m
12 m
21 m 20 m
29 m

The solids are similar. Find the missing dimension(s).

8.

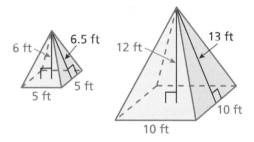

10 ft
4 ft
d
10 in.

9.

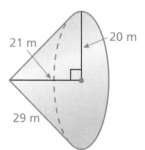

13 m
5 m
6 m
12 m
c

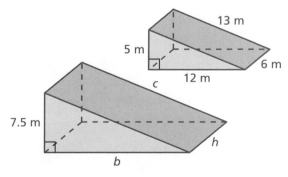

7.5 m
h
b

The solids are similar. Find the surface area S or volume V of the red solid. Round your answer to the nearest tenth.

③ ④ 10.

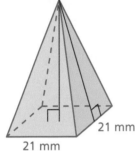

4 m
Surface Area = 336 m²

6 m

11.

20 in.

15 in.

Surface Area = 1800 in.²

12.

21 mm

7 mm
7 mm

21 mm
Volume = 5292 mm³

13.

12 ft

10 ft

Volume = 7850 ft³

14. ERROR ANALYSIS The ratio of the corresponding linear measures of two similar solids is $3 : 5$. The volume of the smaller solid is 108 cubic inches. Describe and correct the error in finding the volume of the larger solid.

$$✗ \quad \frac{108}{V} = \left(\frac{3}{5}\right)^2$$

$$\frac{108}{V} = \frac{9}{25}$$

$$300 = V$$

The volume of the larger solid is 300 cubic inches.

15. MIXED FRUIT The ratio of the corresponding linear measures of two similar cans of fruit is 4 to 7. The smaller can has a surface area of 220 square centimeters. Find the surface area of the larger can.

16. CLASSIC MUSTANG The volume of a 1968 Ford Mustang GT engine is 390 cubic inches. Which scale model of the Mustang has the greater engine volume, a $1 : 18$ scale model or a $1 : 24$ scale model? How much greater is it?

17. **MARBLE STATUE** You have a small marble statue of Wolfgang Mozart. It is 10 inches tall and weighs 16 pounds. The original statue is 7 feet tall.

a. Estimate the weight of the original statue. Explain your reasoning.

b. If the original statue were 20 feet tall, how much would it weigh?

Wolfgang Mozart

18. **REPEATED REASONING** The largest doll is 7 inches tall. Each of the other dolls is 1 inch shorter than the next larger doll. Make a table that compares the surface areas and the volumes of the seven dolls.

19. **⟨Precision⟩** You and a friend make paper cones to collect beach glass. You cut out the largest possible three-fourths circle from each piece of paper.

a. Are the cones similar? Explain your reasoning.

b. Your friend says that because your sheet of paper is twice as large, your cone will hold exactly twice the volume of beach glass. Is this true? Explain your reasoning.

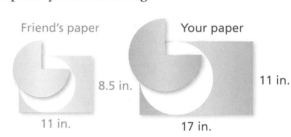

Friend's paper — 8.5 in., 11 in.

Your paper — 11 in., 17 in.

Fair Game Review What you learned in previous grades & lessons

Draw the figure and its reflection in the *x*-axis. Identify the coordinates of the image. *(Section 11.3)*

20. $A(1, 1)$, $B(3, 4)$, $C(4, 2)$

21. $J(-3, 0)$, $K(-4, 3)$, $L(-1, 4)$

22. **MULTIPLE CHOICE** Which pair of lines have the same slope, but different *y*-intercepts? *(Section 13.4 and Section 13.5)*

Ⓐ $y = 4x + 1$
  $y = -4x + 1$

Ⓑ $y = 2x - 7$
  $y = 2x + 7$

Ⓒ $3x + y = 1$
  $6x + 2y = 2$

Ⓓ $5x + y = 3$
  $x + 5y = 15$

Check It Out
Progress Check
BigIdeasMath ✓com

Find the volume of the sphere. Round your answer to the nearest tenth. *(Section 15.3)*

1.

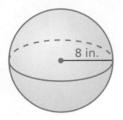

8 in.

2.

32 cm

Find the radius of the sphere with the given volume. *(Section 15.3)*

3. Volume = 4500π yd^3

4. Volume = $\dfrac{32}{3}\pi$ ft^3

5. Find the volume of the composite solid. Round your answer to the nearest tenth. *(Section 15.3)*

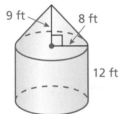

9 ft   8 ft

12 ft

6. Determine whether the solids are similar. *(Section 15.4)*

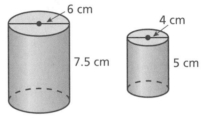

6 cm   4 cm

7.5 cm   5 cm

7. The prisms are similar. Find the missing width and height. *(Section 15.4)*

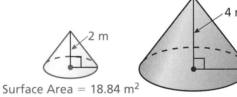

h
10 in.
w

2 in.
4 in.   1 in.

8. The solids are similar. Find the surface area of the red solid. *(Section 15.4)*

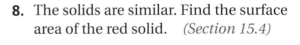

4 m

2 m

Surface Area = 18.84 m^2

2 cm

9. HAMSTER A hamster toy is in the shape of a sphere. What is the volume of the toy? Round your answer to the nearest whole number. *(Section 15.3)*

10. JEWELRY BOXES The ratio of the corresponding linear measures of two similar jewelry boxes is 2 to 3. The larger box has a volume of 162 cubic inches. Find the volume of the smaller jewelry box. *(Section 15.4)*

11. ARCADE You win a token after playing an arcade game. What is the volume of the gold ring? Round your answer to the nearest tenth. *(Section 15.3)*

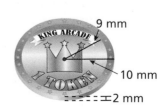

9 mm

10 mm

2 mm

Check It Out
Vocabulary Help
BigIdeasMath ✔com

Review Key Vocabulary

sphere, *p. 686*   hemisphere, *p. 689*   similar solids, *p. 694*

Review Examples and Exercises

 Volumes of Cylinders   *(pp. 672–677)*

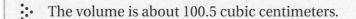

Find the volume of the cylinder. Round your answer to the nearest tenth.

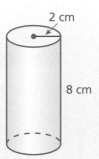

$$V = Bh \qquad \text{Write formula for volume.}$$
$$= \pi(2)^2(8) \qquad \text{Substitute.}$$
$$= 32\pi \approx 100.5 \qquad \text{Use a calculator.}$$

2 cm

8 cm

∴ The volume is about 100.5 cubic centimeters.

Exercises

Find the volume of the cylinder. Round your answer to the nearest tenth.

1.

15 ft

7 ft

2.

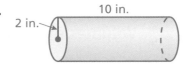

10 in.

2 in.

3.

3 yd

12 yd

4.

9 in.

18 in.

Find the missing dimension of the cylinder. Round your answer to the nearest whole number.

5. Volume = 25 in.^{3}

3 in.

h

6. Volume = 7599 m^3

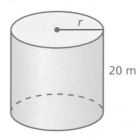

r

20 m

15.2 Volumes of Cones (pp. 678–683)

Find the height of the cone. Round your answer to the nearest tenth.

$$V = \frac{1}{3}Bh \qquad \text{Write formula for volume.}$$

$$900 = \frac{1}{3}\pi(6)^2(h) \qquad \text{Substitute.}$$

$$900 = 12\pi h \qquad \text{Simplify.}$$

$$23.9 \approx h \qquad \text{Divide each side by } 12\pi.$$

∴ The height is about 23.9 millimeters.

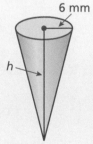

6 mm

h

Volume = 900 mm³

Exercises

Find the volume V or height h of the cone. Round your answer to the nearest tenth.

7.

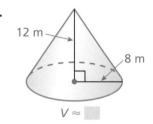

12 m

8 m

$V \approx$ ▭

8.

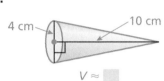

4 cm

10 cm

$V \approx$ ▭

9.

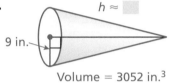

$h \approx$ ▭

9 in.

Volume = 3052 in.³

15.3 Volumes of Spheres (pp. 686–691)

a. **Find the volume of the sphere. Round your answer to the nearest tenth.**

$$V = \frac{4}{3}\pi r^3 \qquad \text{Write formula for volume.}$$

$$= \frac{4}{3}\pi(11)^3 \qquad \text{Substitute 11 for } r.$$

$$= \frac{5324}{3}\pi \qquad \text{Simplify.}$$

$$\approx 5575.3 \qquad \text{Use a calculator.}$$

∴ The volume is about 5575.3 cubic meters.

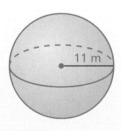

11 m

b. **Find the volume of the composite solid. Round your answer to the nearest tenth.**

Square Prism	Cylinder
$V = Bh$	$V = Bh$
$= (12)(12)(9)$	$= \pi(5)^2(9)$
$= 1296$	$= 225\pi \approx 706.9$

∴ So, the volume is about $1296 + 706.9 = 2002.9$ cubic feet.

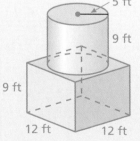

5 ft

9 ft

9 ft

12 ft

12 ft

Exercises

Find the volume *V* or radius *r* of the sphere. Round your answer to the nearest tenth, if necessary.

10.

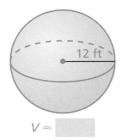

$V \approx$ ▭

11.

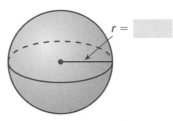

$r =$ ▭

Volume $= 12,348\pi$ in.^{3}

Find the volume of the composite solid. Round your answer to the nearest tenth.

12.

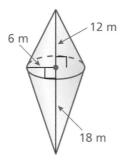

6 m
12 m
18 m

13.

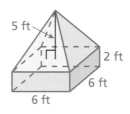

5 ft
2 ft
6 ft
6 ft

14.

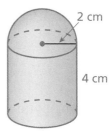

2 cm
4 cm

15.4 **Surface Areas and Volumes of Similar Solids** *(pp. 692–699)*

The cones are similar. What is the volume of the red cone? Round your answer to the nearest tenth.

$$\frac{\text{Volume of A}}{\text{Volume of B}} = \left(\frac{\text{Height of A}}{\text{Height of B}}\right)^3$$

$\dfrac{V}{157} = \left(\dfrac{4}{6}\right)^3$ Substitute.

$\dfrac{V}{157} = \dfrac{64}{216}$ Evaluate.

$V \approx 46.5$ Solve for *V*.

❖ The volume is about 46.5 cubic inches.

Cone A
4 in.

Cone B
6 in.

Volume $= 157$ in.^{3}

Exercises

The solids are similar. Find the surface area *S* or volume *V* of the red solid. Round your answer to the nearest tenth.

15.

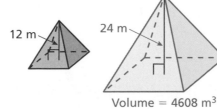

12 m
24 m
Volume $= 4608$ m^3

16.

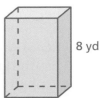

6 yd
8 yd
Surface Area $= 154$ yd^2

Check It Out
Test Practice
BigIdeasMath  .com

Find the volume of the solid. Round your answer to the nearest tenth.

1.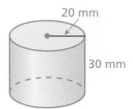
20 mm
30 mm

2.
6 cm
3 cm

3.
26 ft

4.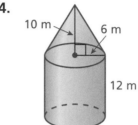
10 m   6 m
12 m

5. The pyramids are similar.

   a. Find the missing dimension.

   b. Find the surface area of the red pyramid.

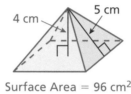
4 cm   5 cm   6 cm   ℓ
Surface Area = 96 cm²

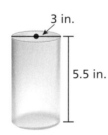
5 in.   3 in.
5 in.   5.5 in.

6. SMOOTHIES You are making smoothies. You will use either the cone-shaped glass or the cylindrical glass. Which glass holds more? About how much more?

7. WAFFLE CONES The ratio of the corresponding linear measures of two similar waffle cones is 3 to 4. The smaller cone has a volume of about 18 cubic inches. Find the volume of the larger cone. Round your answer to the nearest tenth.

8. OPEN-ENDED Draw two different composite solids that have the same volume but different surface areas. Explain your reasoning.

9. MILK Glass A has a diameter of 3.5 inches and a height of 4 inches. Glass B has a radius of 1.5 inches and a height of 5 inches. Which glass can hold more milk?

10. REASONING Which solid has the greater volume? Explain your reasoning.

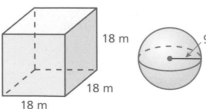
18 m   9 m   18 m   18 m

1. What value of w makes the equation below true? *(8.EE.7b)*

$$\frac{w}{3} = 3(w - 1) - 1$$

   A. $\frac{1}{2}$ **C.** $\frac{5}{4}$

   B. $\frac{3}{4}$ **D.** $\frac{3}{2}$

2. A right circular cone and its dimensions are shown below.

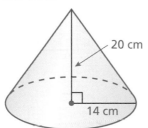

   20 cm

   14 cm

   What is the volume of the right circular cone? $\left(\text{Use } \frac{22}{7} \text{ for } \pi.\right)$ *(8.G.9)*

   F. $1{,}026\frac{2}{3}$ cm³   **H.** $4{,}106\frac{2}{3}$ cm³

   G. $3{,}080$ cm³   **I.** $12{,}320$ cm³

3. Patricia solved the equation in the box shown.

   What should Patricia do to correct the error that she made?   *(8.EE.7b)*

   A. Add 10 to -20.

   B. Distribute $-\frac{3}{2}$ to get $-12x - 15$.

   C. Multiply both sides by $-\frac{2}{3}$ instead of $-\frac{3}{2}$.

   D. Multiply both sides by $\frac{3}{2}$ instead of $-\frac{3}{2}$.

$$-\frac{3}{2}(8x - 10) = -20$$
$$8x - 10 = -20\left(-\frac{3}{2}\right)$$
$$8x - 10 = 30$$
$$8x - 10 + 10 = 30 + 10$$
$$8x = 40$$
$$\frac{8x}{8} = \frac{40}{8}$$
$$x = 5$$

4. On the grid below, Rectangle *EFGH* is plotted and its vertices are labeled.

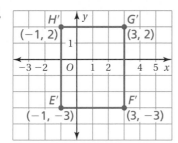

Which of the following shows Rectangle *E'F'G'H'*, the image of Rectangle *EFGH* after it is reflected in the *x*-axis? *(8.G.3)*

F.

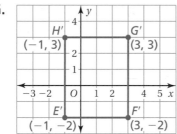

H.

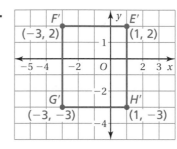

G.

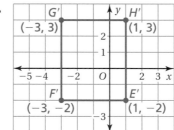

I.

5. What is the measure of the exterior angle in the triangle below? *(8.G.5)*

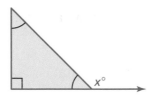

   A. 45°   **C.** 135°

   B. 90°   **D.** 145°

6. The temperature fell from 54 degrees Fahrenheit to 36 degrees Fahrenheit over a 6-hour period. The temperature fell by the same number of degrees each hour. How many degrees Fahrenheit did the temperature fall each hour? *(8.EE.7b)*

7. Solve the formula below for I.   *(8.EE.7b)*

$$A = P + PI$$

F. $I = A - 2P$

H. $I = A - \dfrac{P}{P}$

G. $I = \dfrac{A}{P} - P$

I. $I = \dfrac{A - P}{P}$

8. A right circular cylinder has a volume of 1296 cubic inches. If you divide the radius of the cylinder by 12, what would be the volume, in cubic inches, of the smaller cylinder?   *(8.G.9)*

9. Which line has a slope of -2?   *(8.EE.6)*

A.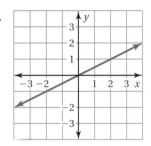

C.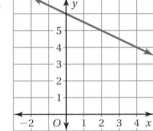

B.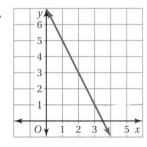

D.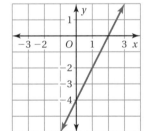

10. The figure below is a diagram for making a tin lantern.

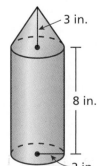

3 in.

8 in.

2 in.

The figure consists of a right circular cylinder without its top base and a right circular cone without its base. What is the volume, in cubic inches, of the entire lantern? Show your work and explain your reasoning. (Use 3.14 for π.)   *(8.G.9)*

16 Exponents and Scientific Notation

"Here's how it goes, Descartes."

"The friends of my friends are my friends. The friends of my enemies are my enemies."

"The enemies of my friends are my enemies. The enemies of my enemies are my friends."

"If one flea had 100 babies, and each baby grew up and had 100 babies, ..."

"... and each of those babies grew up and had 100 babies, you would have 1,010,101 fleas."

What You Learned Before

"It's called the Power of Negative One, Descartes!"

Using Order of Operations (6.EE.1)

Example 1 Evaluate $6^2 \div 4 - 2(9 - 5)$.

First:	Parentheses	$6^2 \div 4 - 2(9 - 5) = 6^2 \div 4 - 2 \cdot 4$
Second:	Exponents	$= 36 \div 4 - 2 \cdot 4$
Third:	Multiplication and Division (from left to right)	$= 9 - 8$
Fourth:	Addition and Subtraction (from left to right)	$= 1$

Try It Yourself
Evaluate the expression.

1. $15\left(\dfrac{8}{4}\right) + 2^2 - 3 \cdot 7$

2. $5^2 \cdot 2 \div 10 + 3 \cdot 2 - 1$

3. $3^2 - 1 + 2(4(3 + 2))$

Multiplying and Dividing Decimals (6.NS.3)

Example 2 Find $2.1 \cdot 0.35$.

$$
\begin{array}{r}
2.1 \leftarrow \text{1 decimal place} \\
\times\ 0.3\,5 \leftarrow +\text{ 2 decimal places} \\
\hline
1\,0\,5 \\
6\,3 \\
\hline
0.7\,3\,5 \leftarrow \text{3 decimal places}
\end{array}
$$

Example 3 Find $1.08 \div 0.9$.

$0.9)\overline{1.08}$ Multiply each number by 10.

$$
\begin{array}{r}
1.2 \\
9)\overline{10.8} \\
-\ 9 \\
\hline
1\,8 \\
-\,1\,8 \\
\hline
0
\end{array}
$$

Place the decimal point above the decimal point in the dividend 10.8.

Try It Yourself
Find the product or quotient.

4. $1.75 \cdot 0.2$

5. $1.4 \cdot 0.6$

6. $\begin{array}{r} 7.03 \\ \times\ 4.3 \\ \hline \end{array}$

7. $\begin{array}{r} 0.894 \\ \times\ \ 0.2 \\ \hline \end{array}$

8. $5.40 \div 0.09$

9. $4.17 \div 0.3$

10. $0.15)\overline{3.6}$

11. $0.004)\overline{7.2}$

16.1 Exponents

Essential Question How can you use exponents to write numbers?

The expression 3^5 is called a *power*. The *base* is 3. The *exponent* is 5.

$$\text{base} \longrightarrow 3^5 \longleftarrow \text{exponent}$$

1 ACTIVITY: Using Exponent Notation

Work with a partner.

a. Copy and complete the table.

Power	Repeated Multiplication Form	Value
$(-3)^1$	-3	-3
$(-3)^2$	$(-3) \cdot (-3)$	9
$(-3)^3$		
$(-3)^4$		
$(-3)^5$		
$(-3)^6$		
$(-3)^7$		

b. **REPEATED REASONING** Describe what is meant by the expression $(-3)^n$. How can you find the value of $(-3)^n$?

2 ACTIVITY: Using Exponent Notation

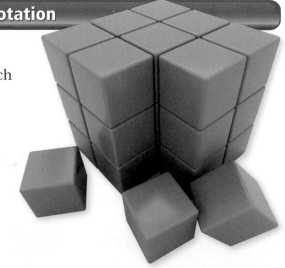

Work with a partner.

a. The cube at the right has \$3 in each of its small cubes. Write a power that represents the total amount of money in the large cube.

b. Evaluate the power to find the total amount of money in the large cube.

COMMON CORE

Exponents

In this lesson, you will
• write expressions using integer exponents.
• evaluate expressions involving integer exponents.

Learning Standard
8.EE.1

3 ACTIVITY: Writing Powers as Whole Numbers

Work with a partner. Write each distance as a whole number. Which numbers do you know how to write in words? For instance, in words, 10^3 is equal to *one thousand.*

a. 10^{26} meters: diameter of observable universe

b. 10^{21} meters: diameter of Milky Way galaxy

c. 10^{16} meters: diameter of solar system

d. 10^7 meters: diameter of Earth

e. 10^6 meters: length of Lake Erie shoreline

f. 10^5 meters: width of Lake Erie

4 ACTIVITY: Writing a Power

Math Practice 1

Analyze Givens

What information is given in the poem? What are you trying to find?

Work with a partner. Write the numbers of kits, cats, sacks, and wives as powers.

As I was going to St. Ives
I met a man with seven wives
Each wife had seven sacks
Each sack had seven cats
Each cat had seven kits
Kits, cats, sacks, wives
How many were going to St. Ives?

Nursery Rhyme, 1730

What Is Your Answer?

5. IN YOUR OWN WORDS How can you use exponents to write numbers? Give some examples of how exponents are used in real life.

Practice ▸ Use what you learned about exponents to complete Exercises 3–5 on page 714.

Check It Out
Lesson Tutorials
BigIdeasMath com

Key Vocabulary
power, *p. 712*
base, *p. 712*
exponent, *p. 712*

A **power** is a product of repeated factors. The **base** of a power is the common factor. The **exponent** of a power indicates the number of times the base is used as a factor.

base → ← exponent

$$\left(\frac{1}{2}\right)^5 = \frac{1}{2} \cdot \frac{1}{2} \cdot \frac{1}{2} \cdot \frac{1}{2} \cdot \frac{1}{2}$$

power $\frac{1}{2}$ is used as a factor 5 times.

EXAMPLE 1 Writing Expressions Using Exponents

Study Tip

Use parentheses to write powers with negative bases.

Write each product using exponents.

a. $(-7) \cdot (-7) \cdot (-7)$

Because -7 is used as a factor 3 times, its exponent is 3.

So, $(-7) \cdot (-7) \cdot (-7) = (-7)^3$.

b. $\pi \cdot \pi \cdot r \cdot r \cdot r$

Because π is used as a factor 2 times, its exponent is 2. Because r is used as a factor 3 times, its exponent is 3.

So, $\pi \cdot \pi \cdot r \cdot r \cdot r = \pi^2 r^3$.

On Your Own

Now You're Ready
Exercises 3–10

Write the product using exponents.

1. $\frac{1}{4} \cdot \frac{1}{4} \cdot \frac{1}{4} \cdot \frac{1}{4} \cdot \frac{1}{4}$

2. $0.3 \cdot 0.3 \cdot 0.3 \cdot 0.3 \cdot x \cdot x$

EXAMPLE 2 Evaluating Expressions

Evaluate each expression.

a. $(-2)^4$

The base is -2.

$(-2)^4 = (-2) \cdot (-2) \cdot (-2) \cdot (-2)$ Write as repeated multiplication.

$= 16$ Simplify.

b. -2^4

The base is 2.

$-2^4 = -(2 \cdot 2 \cdot 2 \cdot 2)$ Write as repeated multiplication.

$= -16$ Simplify.

Multi-Language Glossary at BigIdeasMath.com

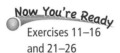

EXAMPLE **3** **Using Order of Operations**

Evaluate each expression.

a. $3 + 2 \cdot 3^4 = 3 + 2 \cdot 81$ Evaluate the power.

$= 3 + 162$ Multiply.

$= 165$ Add.

b. $3^3 - 8^2 \div 2 = 27 - 64 \div 2$ Evaluate the powers.

$= 27 - 32$ Divide.

$= -5$ Subtract.

On Your Own

Now You're Ready
Exercises 11–16
and 21–26

Evaluate the expression.

3. -5^4 **4.** $\left(-\dfrac{1}{6}\right)^3$ **5.** $\left| -3^3 \div 27 \right|$ **6.** $9 - 2^5 \cdot 0.5$

EXAMPLE **4** **Real-Life Application**

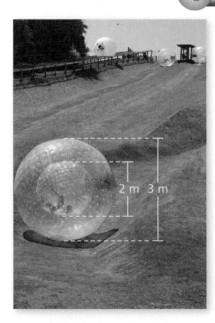

2 m   3 m

In sphering, a person is secured inside a small, hollow sphere that is surrounded by a larger sphere. The space between the spheres is inflated with air. What is the volume of the inflated space?

You can find the radius of each sphere by dividing each diameter given in the diagram by 2.

Outer Sphere		*Inner Sphere*
$V = \dfrac{4}{3}\pi r^3$	Write formula.	$V = \dfrac{4}{3}\pi r^3$
$= \dfrac{4}{3}\pi\left(\dfrac{3}{2}\right)^3$	Substitute.	$= \dfrac{4}{3}\pi(1)^3$
$= \dfrac{4}{3}\pi\left(\dfrac{27}{8}\right)$	Evaluate the power.	$= \dfrac{4}{3}\pi(1)$
$= \dfrac{9}{2}\pi$	Multiply.	$= \dfrac{4}{3}\pi$

So, the volume of the inflated space is $\dfrac{9}{2}\pi - \dfrac{4}{3}\pi = \dfrac{19}{6}\pi$, or about 10 cubic meters.

On Your Own

7. WHAT IF? The diameter of the inner sphere is 1.8 meters. What is the volume of the inflated space?

✓ ## Vocabulary and Concept Check

1. **NUMBER SENSE** Describe the difference between -3^4 and $(-3)^4$.

2. **WHICH ONE DOESN'T BELONG?** Which one does *not* belong with the other three? Explain your reasoning.

| 5^3 The exponent is 3. | 5^3 The power is 5. | 5^3 The base is 5. | 5^3 Five is used as a factor 3 times. |

 ## Practice and Problem Solving

Write the product using exponents.

① 3. $3 \cdot 3 \cdot 3 \cdot 3$

4. $(-6) \cdot (-6)$

5. $\left(-\dfrac{1}{2}\right) \cdot \left(-\dfrac{1}{2}\right) \cdot \left(-\dfrac{1}{2}\right)$

6. $\dfrac{1}{3} \cdot \dfrac{1}{3} \cdot \dfrac{1}{3}$

7. $\pi \cdot \pi \cdot \pi \cdot x \cdot x \cdot x \cdot x$

8. $(-4) \cdot (-4) \cdot (-4) \cdot y \cdot y$

9. $6.4 \cdot 6.4 \cdot 6.4 \cdot 6.4 \cdot b \cdot b \cdot b$

10. $(-t) \cdot (-t) \cdot (-t) \cdot (-t) \cdot (-t)$

Evaluate the expression.

② 11. 5^2

12. -11^3

13. $(-1)^6$

14. $\left(\dfrac{1}{2}\right)^6$

15. $\left(-\dfrac{1}{12}\right)^2$

16. $-\left(\dfrac{1}{9}\right)^3$

17. **ERROR ANALYSIS** Describe and correct the error in evaluating the expression.

✗ $-6^2 = (-6) \cdot (-6) = 36$

18. **PRIME FACTORIZATION** Write the prime factorization of 675 using exponents.

19. **STRUCTURE** Write $-\left(\dfrac{1}{4} \cdot \dfrac{1}{4} \cdot \dfrac{1}{4} \cdot \dfrac{1}{4}\right)$ using exponents.

20. **RUSSIAN DOLLS** The largest doll is 12 inches tall. The height of each of the other dolls is $\dfrac{7}{10}$ the height of the next larger doll. Write an expression involving a power for the height of the smallest doll. What is the height of the smallest doll?

Evaluate the expression.

③ 21. $5 + 3 \cdot 2^3$

22. $2 + 7 \cdot (-3)^2$

23. $(13^2 - 12^2) \div 5$

24. $\frac{1}{2}(4^3 - 6 \cdot 3^2)$

25. $\left| \frac{1}{2}(7 + 5^3) \right|$

26. $\left| \left(-\frac{1}{2} \right)^3 \div \left(\frac{1}{4} \right)^2 \right|$

27. MONEY You have a part-time job. One day your boss offers to pay you either $2^h - 1$ or 2^{h-1} dollars for each hour h you work that day. Copy and complete the table. Which option should you choose? Explain.

h	1	2	3	4	5
$2^h - 1$					
2^{h-1}					

28. CARBON-14 DATING Scientists use carbon-14 dating to determine the age of a sample of organic material.

    a. The amount C (in grams) of a 100-gram sample of carbon-14 remaining after t years is represented by the equation $C = 100(0.99988)^t$. Use a calculator to find the amount of carbon-14 remaining after 4 years.

    b. What percent of the carbon-14 remains after 4 years?

29. *Critical Thinking* The frequency (in vibrations per second) of a note on a piano is represented by the equation $F = 440(1.0595)^n$, where n is the number of notes above A-440. Each black or white key represents one note.

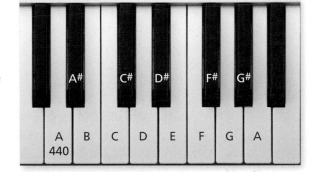

    a. How many notes do you take to travel from A-440 to A?

    b. What is the frequency of A?

    c. Describe the relationship between the number of notes between A-440 and A and the increase in frequency.

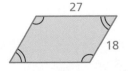

Fair Game Review   What you learned in previous grades & lessons

Tell which property is illustrated by the statement. *(Skills Review Handbook)*

30. $8 \cdot x = x \cdot 8$

31. $(2 \cdot 10)x = 2(10 \cdot x)$

32. $3(x \cdot 1) = 3x$

33. MULTIPLE CHOICE The polygons are similar. What is the value of x? *(Section 11.5)*

    Ⓐ 15   **Ⓑ** 16

    Ⓒ 17   **Ⓓ** 36

16.2 Product of Powers Property

Essential Question
How can you use inductive reasoning to observe patterns and write general rules involving properties of exponents?

1 ACTIVITY: Finding Products of Powers

Work with a partner.

a. Copy and complete the table.

Product	Repeated Multiplication Form	Power
$2^2 \cdot 2^4$		
$(-3)^2 \cdot (-3)^4$		
$7^3 \cdot 7^2$		
$5.1^1 \cdot 5.1^6$		
$(-4)^2 \cdot (-4)^2$		
$10^3 \cdot 10^5$		
$\left(\dfrac{1}{2}\right)^5 \cdot \left(\dfrac{1}{2}\right)^5$		

b. **INDUCTIVE REASONING** Describe the pattern in the table. Then write a *general rule* for multiplying two powers that have the same base.

$$a^m \cdot a^n = a^{\boxed{}}$$

c. Use your rule to simplify the products in the first column of the table above. Does your rule give the results in the third column?

d. Most calculators have *exponent* keys that you can use to evaluate powers. Use a calculator with an exponent key to evaluate the products in part (a).

2 ACTIVITY: Writing a Rule for Powers of Powers

Work with a partner. Write the expression as a single power. Then write a *general rule* for finding a power of a power.

a. $(3^2)^3 = (3 \cdot 3)(3 \cdot 3)(3 \cdot 3) = \boxed{}^{\boxed{}}$

b. $(2^2)^4 = \boxed{}$ c. $(7^3)^2 = \boxed{}$

d. $(y^3)^3 = \boxed{}$ e. $(x^4)^2 = \boxed{}$

COMMON CORE

Exponents

In this lesson, you will
- multiply powers with the same base.
- find a power of a power.
- find a power of a product.

Learning Standard
8.EE.1

Work with a partner. Write the expression as the product of two powers. Then write a *general rule* for finding a power of a product.

a. $(2 \cdot 3)^3 = (2 \cdot 3)(2 \cdot 3)(2 \cdot 3) = \boxed{}^{\boxed{}} \cdot \boxed{}^{\boxed{}}$

b. $(2 \cdot 5)^2 = \boxed{}$ **c.** $(5 \cdot 4)^3 = \boxed{}$

d. $(6a)^4 = \boxed{}$ **e.** $(3x)^2 = \boxed{}$

④ **ACTIVITY: The Penny Puzzle**

Math Practice 7

Look for Patterns

What patterns do you notice? How does this help you determine which stack is the tallest?

Work with a partner.

- The rows y and columns x of a chessboard are numbered as shown.
- Each position on the chessboard has a stack of pennies. (Only the first row is shown.)
- The number of pennies in each stack is $2^x \cdot 2^y$.

a. How many pennies are in the stack in location (3, 5)?

b. Which locations have 32 pennies in their stacks?

c. How much money (in dollars) is in the location with the tallest stack?

d. A penny is about 0.06 inch thick. About how tall (in inches) is the tallest stack?

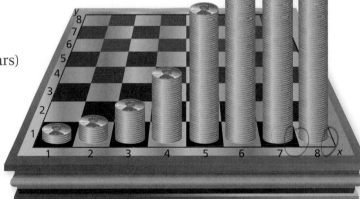

What Is Your Answer?

5. IN YOUR OWN WORDS How can you use inductive reasoning to observe patterns and write general rules involving properties of exponents?

Practice ➤ Use what you learned about properties of exponents to complete Exercises 3–5 on page 720.

Key Ideas

Product of Powers Property

Words To multiply powers with the same base, add their exponents.

Numbers $4^2 \cdot 4^3 = 4^{2+3} = 4^5$ **Algebra** $a^m \cdot a^n = a^{m+n}$

Power of a Power Property

Words To find a power of a power, multiply the exponents.

Numbers $(4^6)^3 = 4^{6 \cdot 3} = 4^{18}$ **Algebra** $(a^m)^n = a^{mn}$

Power of a Product Property

Words To find a power of a product, find the power of each factor and multiply.

Numbers $(3 \cdot 2)^5 = 3^5 \cdot 2^5$ **Algebra** $(ab)^m = a^m b^m$

EXAMPLE **1** **Multiplying Powers with the Same Base**

a. $2^4 \cdot 2^5 = 2^{4+5}$ Product of Powers Property

$ = 2^9$ Simplify.

Study Tip

When a number is written without an exponent, its exponent is 1.

b. $-5 \cdot (-5)^6 = (-5)^1 \cdot (-5)^6$ Rewrite -5 as $(-5)^1$.

$ = (-5)^{1+6}$ Product of Powers Property

$ = (-5)^7$ Simplify.

c. $x^3 \cdot x^7 = x^{3+7}$ Product of Powers Property

$ = x^{10}$ Simplify.

EXAMPLE **2** **Finding a Power of a Power**

a. $(3^4)^3 = 3^{4 \cdot 3}$ Power of a Power Property

$ = 3^{12}$ Simplify.

b. $(w^5)^4 = w^{5 \cdot 4}$ Power of a Power Property

$ = w^{20}$ Simplify.

EXAMPLE **3** **Finding a Power of a Product**

a. $(2x)^3 = 2^3 \cdot x^3$ Power of a Product Property

       $= 8x^3$ Simplify.

b. $(3xy)^2 = 3^2 \cdot x^2 \cdot y^2$ Power of a Product Property

         $= 9x^2y^2$ Simplify.

On Your Own

Now You're Ready
Exercises 3–14
and 17–22

Simplify the expression.

1. $6^2 \cdot 6^4$ **2.** $\left(-\dfrac{1}{2}\right)^3 \cdot \left(-\dfrac{1}{2}\right)^6$ **3.** $z \cdot z^{12}$

4. $(4^4)^3$ **5.** $(y^2)^4$ **6.** $((-4)^3)^2$

7. $(5y)^4$ **8.** $(ab)^5$ **9.** $(0.5mn)^2$

EXAMPLE **4** **Simplifying an Expression**

Details	⊗
Local Disk (C:)	
Local Disk	
Free Space: 16GB	
Total Space: 64GB	

A gigabyte (GB) of computer storage space is 2^{30} bytes. The details of a computer are shown. How many bytes of total storage space does the computer have?

     Ⓐ 2^{34} Ⓑ 2^{36} Ⓒ 2^{180} Ⓓ 128^{30}

The computer has 64 gigabytes of total storage space. Notice that you can write 64 as a power, 2^6. Use a model to solve the problem.

$$\dfrac{\text{Total number}}{\text{of bytes}} = \dfrac{\text{Number of bytes}}{\text{in a gigabyte}} \cdot \dfrac{\text{Number of}}{\text{gigabytes}}$$

       $= 2^{30} \cdot 2^6$ Substitute.

       $= 2^{30+6}$ Product of Powers Property

       $= 2^{36}$ Simplify.

⋰ The computer has 2^{36} bytes of total storage space. The correct answer is Ⓑ.

On Your Own

10. How many bytes of free storage space does the computer have?

Vocabulary and Concept Check

1. **REASONING** When should you use the Product of Powers Property?

2. **CRITICAL THINKING** Can you use the Product of Powers Property to multiply $5^2 \cdot 6^4$? Explain.

Practice and Problem Solving

Simplify the expression. Write your answer as a power.

1 2 **3.** $3^2 \cdot 3^2$

4. $8^{10} \cdot 8^4$

5. $(-4)^5 \cdot (-4)^7$

6. $a^3 \cdot a^3$

7. $h^6 \cdot h$

8. $\left(\dfrac{2}{3}\right)^2 \cdot \left(\dfrac{2}{3}\right)^6$

9. $\left(-\dfrac{5}{7}\right)^8 \cdot \left(-\dfrac{5}{7}\right)^9$

10. $(-2.9) \cdot (-2.9)^7$

11. $\left(5^4\right)^3$

12. $\left(b^{12}\right)^3$

13. $\left(3.8^3\right)^4$

14. $\left(\left(-\dfrac{3}{4}\right)^5\right)^2$

ERROR ANALYSIS Describe and correct the error in simplifying the expression.

15.
$$5^2 \cdot 5^9 = (5 \cdot 5)^{2+9}$$
$$= 25^{11}$$

16.
$$\left(r^6\right)^4 = r^{6+4}$$
$$= r^{10}$$

Simplify the expression.

3 **17.** $(6g)^3$

18. $(-3v)^5$

19. $\left(\dfrac{1}{5}k\right)^2$

20. $(1.2m)^4$

21. $(rt)^{12}$

22. $\left(-\dfrac{3}{4}p\right)^3$

23. **PRECISION** Is $3^2 + 3^3$ equal to 3^5? Explain.

24. **ARTIFACT** A display case for the artifact is in the shape of a cube. Each side of the display case is three times longer than the width of the artifact.

   a. Write an expression for the volume of the case. Write your answer as a power.

   b. Simplify the expression.

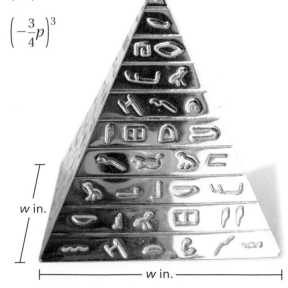

w in.

w in.

Simplify the expression.

25. $2^4 \cdot 2^5 - (2^2)^2$

26. $16\left(\dfrac{1}{2}x\right)^4$

27. $5^2(5^3 \cdot 5^2)$

28. CLOUDS The lowest altitude of an altocumulus cloud is about 3^8 feet. The highest altitude of an altocumulus cloud is about 3 times the lowest altitude. What is the highest altitude of an altocumulus cloud? Write your answer as a power.

29. PYTHON EGG The volume V of a python egg is given by the formula $V = \dfrac{4}{3}\pi abc$. For the python eggs shown, $a = 2$ inches, $b = 2$ inches, and $c = 3$ inches.

   a. Find the volume of a python egg.

   b. Square the dimensions of the python egg. Then evaluate the formula. How does this volume compare to your answer in part (a)?

30. PYRAMID A square pyramid has a height h and a base with side length b. The side lengths of the base increase by 50%. Write a formula for the volume of the new pyramid in terms of b and h.

31. MAIL The United States Postal Service delivers about $2^8 \cdot 5^2$ pieces of mail each second. There are $2^8 \cdot 3^4 \cdot 5^2$ seconds in 6 days. How many pieces of mail does the United States Postal Service deliver in 6 days? Write your answer as an expression involving powers.

32. ⭐**Critical Thinking** Find the value of x in the equation without evaluating the power.

   a. $2^5 \cdot 2^x = 256$

   b. $\left(\dfrac{1}{3}\right)^2 \cdot \left(\dfrac{1}{3}\right)^x = \dfrac{1}{729}$

 Fair Game Review *What you learned in previous grades & lessons*

Simplify. *(Skills Review Handbook)*

33. $\dfrac{4 \cdot 4}{4}$

34. $\dfrac{5 \cdot 5 \cdot 5}{5}$

35. $\dfrac{2 \cdot 3}{2}$

36. $\dfrac{8 \cdot 6 \cdot 6}{6 \cdot 8}$

37. MULTIPLE CHOICE What is the measure of each interior angle of the regular polygon? *(Section 12.3)*

   Ⓐ   45°

   Ⓑ   135°

   Ⓒ   1080°

   Ⓓ   1440°

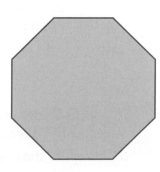

Essential Question How can you divide two powers that have the same base?

1 ACTIVITY: Finding Quotients of Powers

Work with a partner.

a. Copy and complete the table.

Quotient	Repeated Multiplication Form	Power
$\dfrac{2^4}{2^2}$		
$\dfrac{(-4)^5}{(-4)^2}$		
$\dfrac{7^7}{7^3}$		
$\dfrac{8.5^9}{8.5^6}$		
$\dfrac{10^8}{10^5}$		
$\dfrac{3^{12}}{3^4}$		
$\dfrac{(-5)^7}{(-5)^5}$		
$\dfrac{11^4}{11^1}$		

b. **INDUCTIVE REASONING** Describe the pattern in the table. Then write a rule for dividing two powers that have the same base.

$$\frac{a^m}{a^n} = a^{\boxed{}}$$

c. Use your rule to simplify the quotients in the first column of the table above. Does your rule give the results in the third column?

COMMON CORE

Exponents

In this lesson, you will
- divide powers with the same base.
- simplify expressions involving the quotient of powers.

Learning Standard
8.EE.1

ACTIVITY: Comparing Volumes

Math Practice 8

Repeat Calculations

What calculations are repeated in the table?

Work with a partner.

How many of the smaller cubes will fit inside the larger cube? Record your results in the table. Describe the pattern in the table.

a.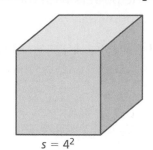

$s = 4$ $s = 4^2$

b.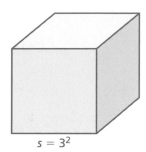

$s = 3$ $s = 3^2$

c.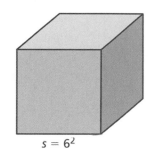

$s = 6$ $s = 6^2$

d.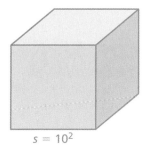

$s = 10$ $s = 10^2$

	Volume of Smaller Cube	Volume of Larger Cube	$\dfrac{\text{Larger Volume}}{\text{Smaller Volume}}$	Answer
a.				
b.				
c.				
d.				

What Is Your Answer?

3. **IN YOUR OWN WORDS** How can you divide two powers that have the same base? Give two examples of your rule.

Practice

Use what you learned about dividing powers with the same base to complete Exercises 3–6 on page 726.

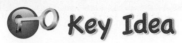 **Key Idea**

Quotient of Powers Property

Words To divide powers with the same base, subtract their exponents.

Numbers $\dfrac{4^5}{4^2} = 4^{5-2} = 4^3$ **Algebra** $\dfrac{a^m}{a^n} = a^{m-n}$, where $a \neq 0$

EXAMPLE 1 **Dividing Powers with the Same Base**

a. $\dfrac{2^6}{2^4} = 2^{6-4}$ Quotient of Powers Property

    $= 2^2$ Simplify.

Common Error ⚠

When dividing powers, do not divide the bases.
$\dfrac{2^6}{2^4} = 2^2$, not 1^2.

b. $\dfrac{(-7)^9}{(-7)^3} = (-7)^{9-3}$ Quotient of Powers Property

    $= (-7)^6$ Simplify.

c. $\dfrac{h^7}{h^6} = h^{7-6}$ Quotient of Powers Property

    $= h^1 = h$ Simplify.

On Your Own

Now You're Ready
Exercises 7–14

Simplify the expression. Write your answer as a power.

1. $\dfrac{9^7}{9^4}$ 2. $\dfrac{4.2^6}{4.2^5}$ 3. $\dfrac{(-8)^8}{(-8)^4}$ 4. $\dfrac{x^8}{x^3}$

EXAMPLE 2 **Simplifying an Expression**

Simplify $\dfrac{3^4 \cdot 3^2}{3^3}$. Write your answer as a power.

The numerator is a product of powers. Add the exponents in the numerator.

$\dfrac{3^4 \cdot 3^2}{3^3} = \dfrac{3^{4+2}}{3^3}$ Product of Powers Property

    $= \dfrac{3^6}{3^3}$ Simplify.

    $= 3^{6-3}$ Quotient of Powers Property

    $= 3^3$ Simplify.

EXAMPLE 3

Simplifying an Expression

Study Tip

You can also simplify the expression in Example 3 as follows.

$$\frac{a^{10}}{a^6} \cdot \frac{a^7}{a^4} = \frac{a^{10} \cdot a^7}{a^6 \cdot a^4}$$

$$= \frac{a^{17}}{a^{10}}$$

$$= a^{17-10}$$

$$= a^7$$

Simplify $\dfrac{a^{10}}{a^6} \cdot \dfrac{a^7}{a^4}$. Write your answer as a power.

$$\frac{a^{10}}{a^6} \cdot \frac{a^7}{a^4} = a^{10-6} \cdot a^{7-4} \qquad \text{Quotient of Powers Property}$$

$$= a^4 \cdot a^3 \qquad\qquad \text{Simplify.}$$

$$= a^{4+3} \qquad\qquad \text{Product of Powers Property}$$

$$= a^7 \qquad\qquad \text{Simplify.}$$

On Your Own

Now You're Ready
Exercises 16–21

Simplify the expression. Write your answer as a power.

5. $\dfrac{2^{15}}{2^3 \cdot 2^5}$

6. $\dfrac{d^5}{d} \cdot \dfrac{d^9}{d^8}$

7. $\dfrac{5^9}{5^4} \cdot \dfrac{5^5}{5^2}$

EXAMPLE 4

Real-Life Application

The projected population of Tennessee in 2030 is about $5 \cdot 5.9^8$. Predict the average number of people per square mile in 2030.

Use a model to solve the problem.

$$\frac{\text{People per}}{\text{square mile}} = \frac{\text{Population in 2030}}{\text{Land area}}$$

Land area: about 5.9^6 mi^2

$$= \frac{5 \cdot 5.9^8}{5.9^6} \qquad \text{Substitute.}$$

$$= 5 \cdot \frac{5.9^8}{5.9^6} \qquad \text{Rewrite.}$$

$$= 5 \cdot 5.9^2 \qquad \text{Quotient of Powers Property}$$

$$= 174.05 \qquad \text{Evaluate.}$$

So, there will be about 174 people per square mile in Tennessee in 2030.

On Your Own

Now You're Ready
Exercises 23–28

8. The projected population of Alabama in 2030 is about $2.25 \cdot 2^{21}$. The land area of Alabama is about 2^{17} square kilometers. Predict the average number of people per square kilometer in 2030.

Check It Out
Help with Homework
BigIdeasMath ✓.com

✓ **Vocabulary and Concept Check**

1. **WRITING** Describe in your own words how to divide powers.

2. **WHICH ONE DOESN'T BELONG?** Which quotient does *not* belong with the other three? Explain your reasoning.

$$\frac{(-10)^7}{(-10)^2} \qquad \frac{6^3}{6^2} \qquad \frac{(-4)^8}{(-3)^4} \qquad \frac{5^6}{5^3}$$

Practice and Problem Solving

Simplify the expression. Write your answer as a power.

3. $\dfrac{6^{10}}{6^4}$

4. $\dfrac{8^9}{8^7}$

5. $\dfrac{(-3)^4}{(-3)^1}$

6. $\dfrac{4.5^5}{4.5^3}$

7. $\dfrac{5^9}{5^3}$

8. $\dfrac{64^4}{64^3}$

9. $\dfrac{(-17)^5}{(-17)^2}$

10. $\dfrac{(-7.9)^{10}}{(-7.9)^4}$

11. $\dfrac{(-6.4)^8}{(-6.4)^6}$

12. $\dfrac{\pi^{11}}{\pi^7}$

13. $\dfrac{b^{24}}{b^{11}}$

14. $\dfrac{n^{18}}{n^7}$

15. **ERROR ANALYSIS** Describe and correct the error in simplifying the quotient.

$$\mathbf{X} \quad \frac{6^{15}}{6^5} = 6^{\frac{15}{5}}$$
$$= 6^3$$

Simplify the expression. Write your answer as a power.

16. $\dfrac{7^5 \cdot 7^3}{7^2}$

17. $\dfrac{2^{19} \cdot 2^5}{2^{12} \cdot 2^3}$

18. $\dfrac{(-8.3)^8}{(-8.3)^7} \cdot \dfrac{(-8.3)^4}{(-8.3)^3}$

19. $\dfrac{\pi^{30}}{\pi^{18} \cdot \pi^4}$

20. $\dfrac{c^{22}}{c^8 \cdot c^9}$

21. $\dfrac{k^{13}}{k^5} \cdot \dfrac{k^{17}}{k^{11}}$

22. **SOUND INTENSITY** The sound intensity of a normal conversation is 10^6 times greater than the quietest noise a person can hear. The sound intensity of a jet at takeoff is 10^{14} times greater than the quietest noise a person can hear. How many times more intense is the sound of a jet at takeoff than the sound of a normal conversation?

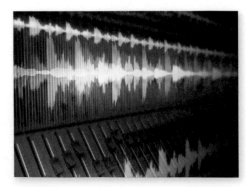

Simplify the expression.

④ **23.** $\dfrac{x \cdot 4^8}{4^5}$

24. $\dfrac{6^3 \cdot w}{6^2}$

25. $\dfrac{a^3 \cdot b^4 \cdot 5^4}{b^2 \cdot 5}$

26. $\dfrac{5^{12} \cdot c^{10} \cdot d^2}{5^9 \cdot c^9}$

27. $\dfrac{x^{15}y^9}{x^8y^3}$

28. $\dfrac{m^{10}n^7}{m^1n^6}$

29. MEMORY The memory capacities and prices of five MP3 players are shown in the table.

MP3 Player	Memory (GB)	Price
A	2^1	$70
B	2^2	$120
C	2^3	$170
D	2^4	$220
E	2^5	$270

a. How many times more memory does MP3 Player D have than MP3 Player B?

b. Do memory and price show a linear relationship? Explain.

30. CRITICAL THINKING Consider the equation $\dfrac{9^m}{9^n} = 9^2$.

a. Find two numbers m and n that satisfy the equation.

b. Describe the number of solutions that satisfy the equation. Explain your reasoning.

Milky Way galaxy
$10 \cdot 10^{10}$ stars

31. STARS There are about 10^{24} stars in the universe. Each galaxy has approximately the same number of stars as the Milky Way galaxy. About how many galaxies are in the universe?

32. *Number Sense* Find the value of x that makes $\dfrac{8^{3x}}{8^{2x + 1}} = 8^9$ true. Explain how you found your answer.

 Fair Game Review *What you learned in previous grades & lessons*

Subtract. *(Section 1.3)*

33. $-4 - 5$

34. $-23 - (-15)$

35. $33 - (-28)$

36. $18 - 22$

37. MULTIPLE CHOICE What is the value of x? *(Section 7.1 and Section 7.2)*

Ⓐ 20

Ⓑ 30

Ⓒ 45

Ⓓ 60

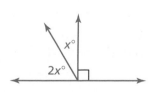

16.4 Zero and Negative Exponents

Essential Question How can you evaluate a nonzero number with an exponent of zero? How can you evaluate a nonzero number with a negative integer exponent?

1 ACTIVITY: Using the Quotient of Powers Property

Work with a partner.

a. Copy and complete the table.

Quotient	Quotient of Powers Property	Power
$\dfrac{5^3}{5^3}$		
$\dfrac{6^2}{6^2}$		
$\dfrac{(-3)^4}{(-3)^4}$		
$\dfrac{(-4)^5}{(-4)^5}$		

b. **REPEATED REASONING** Evaluate each expression in the first column of the table. What do you notice?

c. How can you use these results to define a^0 where $a \neq 0$?

2 ACTIVITY: Using the Product of Powers Property

Work with a partner.

COMMON CORE

Exponents
In this lesson, you will
- evaluate expressions involving numbers with zero as an exponent.
- evaluate expressions involving negative integer exponents.

Learning Standard
8.EE.1

a. Copy and complete the table.

Product	Product of Powers Property	Power
$3^0 \cdot 3^4$		
$8^2 \cdot 8^0$		
$(-2)^3 \cdot (-2)^0$		
$\left(-\dfrac{1}{3}\right)^0 \cdot \left(-\dfrac{1}{3}\right)^5$		

b. Do these results support your definition in Activity 1(c)?

ACTIVITY: Using the Product of Powers Property

Work with a partner.

a. Copy and complete the table.

Product	Product of Powers Property	Power
$5^{-3} \cdot 5^3$		
$6^2 \cdot 6^{-2}$		
$(-3)^4 \cdot (-3)^{-4}$		
$(-4)^{-5} \cdot (-4)^5$		

b. According to your results from Activities 1 and 2, the products in the first column are equal to what value?

c. **REASONING** How does the Multiplicative Inverse Property help you rewrite the numbers with negative exponents?

d. **STRUCTURE** Use these results to define a^{-n} where $a \neq 0$ and n is an integer.

4 **ACTIVITY: Using a Place Value Chart**

Math Practice 2

Use Operations

What operations are used when writing the expanded form?

Work with a partner. Use the place value chart that shows the number 3452.867.

Place Value Chart

thousands	hundreds	tens	ones	and	tenths	hundredths	thousandths
10^3	10^2	10^1	$10^{}$		$10^{}$	$10^{}$	$10^{}$
3	4	5	2	.	8	6	7

a. **REPEATED REASONING** What pattern do you see in the exponents? Continue the pattern to find the other exponents.

b. **STRUCTURE** Show how to write the expanded form of 3452.867.

What Is Your Answer?

5. **IN YOUR OWN WORDS** How can you evaluate a nonzero number with an exponent of zero? How can you evaluate a nonzero number with a negative integer exponent?

Practice

Use what you learned about zero and negative exponents to complete Exercises 5–8 on page 732.

 **Key Ideas**

Zero Exponents

Words For any nonzero number a, $a^0 = 1$. The power 0^0 is *undefined*.

Numbers $4^0 = 1$ **Algebra** $a^0 = 1$, where $a \neq 0$

Negative Exponents

Words For any integer n and any nonzero number a, a^{-n} is the reciprocal of a^n.

Numbers $4^{-2} = \dfrac{1}{4^2}$ **Algebra** $a^{-n} = \dfrac{1}{a^n}$, where $a \neq 0$

EXAMPLE 1   **Evaluating Expressions**

a. $3^{-4} = \dfrac{1}{3^4}$ Definition of negative exponent

$= \dfrac{1}{81}$ Evaluate power.

b. $(-8.5)^{-4} \cdot (-8.5)^4 = (-8.5)^{-4+4}$ Product of Powers Property

$= (-8.5)^0$ Simplify.

$= 1$ Definition of zero exponent

c. $\dfrac{2^6}{2^8} = 2^{6-8}$ Quotient of Powers Property

$= 2^{-2}$ Simplify.

$= \dfrac{1}{2^2}$ Definition of negative exponent

$= \dfrac{1}{4}$ Evaluate power.

● **On Your Own**

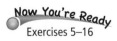
Now You're Ready
Exercises 5–16

Evaluate the expression.

1. 4^{-2} 2. $(-2)^{-5}$ 3. $6^{-8} \cdot 6^8$

4. $\dfrac{(-3)^5}{(-3)^6}$ 5. $\dfrac{1}{5^7} \cdot \dfrac{1}{5^{-4}}$ 6. $\dfrac{4^5 \cdot 4^{-3}}{4^2}$

EXAMPLE **2** **Simplifying Expressions**

a. $-5x^0 = -5(1)$ Definition of zero exponent

           $= -5$ Multiply.

b. $\dfrac{9y^{-3}}{y^5} = 9y^{-3-5}$ Quotient of Powers Property

           $= 9y^{-8}$ Simplify.

           $= \dfrac{9}{y^8}$ Definition of negative exponent

● **On Your Own**

Exercises 20–27

Simplify. Write the expression using only positive exponents.

7. $8x^{-2}$ **8.** $b^0 \cdot b^{-10}$ **9.** $\dfrac{z^6}{15z^9}$

EXAMPLE **3** **Real-Life Application**

Drop of water: 50^{-2} liter

A drop of water leaks from a faucet every second. How many liters of water leak from the faucet in 1 hour?

Convert 1 hour to seconds.

$$1 \cancel{h} \times \frac{60 \cancel{\text{min}}}{1 \cancel{h}} \times \frac{60 \text{ sec}}{1 \cancel{\text{min}}} = 3600 \text{ sec}$$

Water leaks from the faucet at a rate of 50^{-2} liter per second. Multiply the time by the rate.

$3600 \text{ sec} \cdot 50^{-2} \dfrac{\text{L}}{\text{sec}} = 3600 \cdot \dfrac{1}{50^2}$ Definition of negative exponent

                 $= 3600 \cdot \dfrac{1}{2500}$ Evaluate power.

                 $= \dfrac{3600}{2500}$ Multiply.

                 $= 1\dfrac{11}{25} = 1.44 \text{ L}$ Simplify.

∴ So, 1.44 liters of water leak from the faucet in 1 hour.

● **On Your Own**

10. **WHAT IF?** The faucet leaks water at a rate of 5^{-5} liter per second. How many liters of water leak from the faucet in 1 hour?

 **Check It Out**
Help with Homework
BigIdeasMath ✓com

✓ Vocabulary and Concept Check

1. **VOCABULARY** If a is a nonzero number, does the value of a^0 depend on the value of a? Explain.

2. **WRITING** Explain how to evaluate 10^{-3}.

3. **NUMBER SENSE** Without evaluating, order 5^0, 5^4, and 5^{-5} from least to greatest.

4. **DIFFERENT WORDS, SAME QUESTION** Which is different? Find "both" answers.

   Rewrite $\dfrac{1}{3 \cdot 3 \cdot 3}$ using a negative exponent.

   Write 3 to the negative third.

   Write $\dfrac{1}{3}$ cubed as a power.

   Write $(-3) \cdot (-3) \cdot (-3)$ as a power.

Practice and Problem Solving

Evaluate the expression.

① 5. $\dfrac{8^7}{8^7}$

6. $5^0 \cdot 5^3$

7. $(-2)^{-8} \cdot (-2)^8$

8. $9^4 \cdot 9^{-4}$

9. 6^{-2}

10. 158^0

11. $\dfrac{4^3}{4^5}$

12. $\dfrac{-3}{(-3)^2}$

13. $4 \cdot 2^{-4} + 5$

14. $3^{-3} \cdot 3^{-2}$

15. $\dfrac{1}{5^{-3}} \cdot \dfrac{1}{5^6}$

16. $\dfrac{(1.5)^2}{(1.5)^{-2} \cdot (1.5)^4}$

17. **ERROR ANALYSIS** Describe and correct the error in evaluating the expression.

   ✗ $(4)^{-3} = (-4)(-4)(-4)$
   $= -64$

18. **SAND** The mass of a grain of sand is about 10^{-3} gram. About how many grains of sand are in the bag of sand?

19. **CRITICAL THINKING** How can you write the number 1 as 2 to a power? 10 to a power?

Simplify. Write the expression using only positive exponents.

② 20. $6y^{-4}$

21. $8^{-2} \cdot a^7$

22. $\dfrac{9c^3}{c^{-4}}$

23. $\dfrac{5b^{-2}}{b^{-3}}$

24. $\dfrac{8x^3}{2x^9}$

25. $3d^{-4} \cdot 4d^4$

26. $m^{-2} \cdot n^3$

27. $\dfrac{3^{-2} \cdot k^0 \cdot w^0}{w^{-6}}$

28. **OPEN-ENDED** Write two different powers with negative exponents that have the same value.

METRIC UNITS In Exercises 29–32, use the table.

29. How many millimeters are in a decimeter?

30. How many micrometers are in a centimeter?

31. How many nanometers are in a millimeter?

32. How many micrometers are in a meter?

Unit of Length	Length (meter)
Decimeter	10^{-1}
Centimeter	10^{-2}
Millimeter	10^{-3}
Micrometer	10^{-6}
Nanometer	10^{-9}

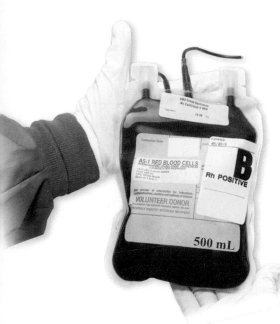

33. **BACTERIA** A species of bacteria is 10 micrometers long. A virus is 10,000 times smaller than the bacteria.

   a. Using the table above, find the length of the virus in meters.

   b. Is the answer to part (a) *less than*, *greater than*, or *equal to* one nanometer?

34. **BLOOD DONATION** Every 2 seconds, someone in the United States needs blood. A sample blood donation is shown. ($1 \text{ mm}^3 = 10^{-3} \text{ mL}$)

   a. One cubic millimeter of blood contains about 10^4 white blood cells. How many white blood cells are in the donation? Write your answer in words.

   b. One cubic millimeter of blood contains about 5×10^6 red blood cells. How many red blood cells are in the donation? Write your answer in words.

   c. Compare your answers for parts (a) and (b).

35. **PRECISION** Describe how to rewrite a power with a positive exponent so that the exponent is in the denominator. Use the definition of negative exponents to justify your reasoning.

36. **Reasoning** The rule for negative exponents states that $a^{-n} = \dfrac{1}{a^n}$. Explain why this rule does not apply when $a = 0$.

Fair Game Review What you learned in previous grades & lessons

Simplify the expression. Write your answer as a power. *(Section 16.2 and Section 16.3)*

37. $10^3 \cdot 10^6$

38. $10^2 \cdot 10$

39. $\dfrac{10^8}{10^4}$

40. **MULTIPLE CHOICE** What is the volume of a cylinder with a radius of 4 inches and a height of 5 inches? *(Section 15.1)*

   (A) 16π in.^{3}

   (B) 72π in.^{3}

   (C) 80π in.^{3}

   (D) 100π in.^{3}

You can use an **information wheel** to organize information about a topic. Here is an example of an information wheel for exponents.

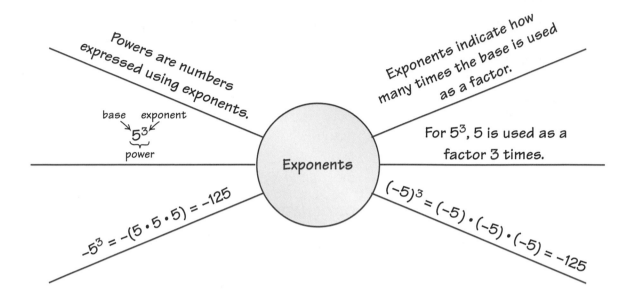

On Your Own

Make information wheels to help you study these topics.

1. Product of Powers Property

2. Quotient of Powers Property

3. zero and negative exponents

After you complete this chapter, make information wheels for the following topics.

4. writing numbers in scientific notation

5. writing numbers in standard form

6. adding and subtracting numbers in scientific notation

"I decided to color code the different flavors in my information wheel."

7. multiplying and dividing numbers in scientific notation

8. Choose three other topics you studied earlier in this course. Make an information wheel for each topic to summarize what you know about them.

Write the product using exponents. *(Section 16.1)*

1. $(-5) \cdot (-5) \cdot (-5) \cdot (-5)$

2. $7 \cdot 7 \cdot m \cdot m \cdot m$

Evaluate the expression. *(Section 16.1 and Section 16.4)*

3. 5^4

4. $(-2)^6$

5. $(-4.8)^{-9} \cdot (-4.8)^9$

6. $\dfrac{5^4}{5^7}$

Simplify the expression. Write your answer as a power. *(Section 16.2)*

7. $3^8 \cdot 3$

8. $(a^5)^3$

Simplify the expression. *(Section 16.2)*

9. $(3c)^4$

10. $\left(-\dfrac{2}{7}p\right)^2$

Simplify the expression. Write your answer as a power. *(Section 16.3)*

11. $\dfrac{8^7}{8^4}$

12. $\dfrac{6^3 \cdot 6^7}{6^2}$

13. $\dfrac{\pi^{15}}{\pi^3 \cdot \pi^9}$

14. $\dfrac{t^{13}}{t^5} \cdot \dfrac{t^8}{t^6}$

Simplify. Write the expression using only positive exponents. *(Section 16.4)*

15. $8d^{-6}$

16. $\dfrac{12x^5}{4x^7}$

17. **ORGANISM** A one-celled, aquatic organism called a dinoflagellate is 1000 micrometers long. *(Section 16.4)*

  a. One micrometer is 10^{-6} meter. What is the length of the dinoflagellate in meters?

  b. Is the length of the dinoflagellate equal to 1 millimeter or 1 kilometer? Explain.

18. **EARTHQUAKES** An earthquake of magnitude 3.0 is 10^2 times stronger than an earthquake of magnitude 1.0. An earthquake of magnitude 8.0 is 10^7 times stronger than an earthquake of magnitude 1.0. How many times stronger is an earthquake of magnitude 8.0 than an earthquake of magnitude 3.0? *(Section 16.3)*

Essential Question How can you read numbers that are written in scientific notation?

1 ACTIVITY: Very Large Numbers

Work with a partner.

- Use a calculator. Experiment with multiplying large numbers until your calculator displays an answer that is *not* in standard form.

- When the calculator at the right was used to multiply 2 billion by 3 billion, it listed the result as

    6.0E+18.

- Multiply 2 billion by 3 billion by hand. Use the result to explain what 6.0E+18 means.

- Check your explanation by calculating the products of other large numbers.

- Why didn't the calculator show the answer in standard form?

- Experiment to find the maximum number of digits your calculator displays. For instance, if you multiply 1000 by 1000 and your calculator shows 1,000,000, then it can display seven digits.

2 ACTIVITY: Very Small Numbers

Work with a partner.

- Use a calculator. Experiment with multiplying very small numbers until your calculator displays an answer that is *not* in standard form.

- When the calculator at the right was used to multiply 2 billionths by 3 billionths, it listed the result as

    6.0E−18.

- Multiply 2 billionths by 3 billionths by hand. Use the result to explain what 6.0E−18 means.

- Check your explanation by calculating the products of other very small numbers.

COMMON CORE

Scientific Notation

In this lesson, you will
- identify numbers written in scientific notation.
- write numbers in standard form.
- compare numbers in scientific notation.

Learning Standards
8.EE.3
8.EE.4

3 ACTIVITY: Powers of 10 Matching Game

Math Practice 4

Analyze Relationships

How are the pictures related? How can you order the pictures to find the correct power of 10?

Work with a partner. Match each picture with its power of 10. Explain your reasoning.

10^5 m 10^2 m 10^0 m 10^{-1} m 10^{-2} m 10^{-5} m

A.

B.

C.

D.

E.

F.

4 ACTIVITY: Choosing Appropriate Units

Work with a partner. Match each unit with its most appropriate measurement.

inches   centimeters   feet   millimeters   meters

A. Height of a door:
2×10^0

B. Height of a volcano:
1.6×10^4

C. Length of a pen:
1.4×10^2

D. Diameter of a steel ball bearing:
6.3×10^{-1}

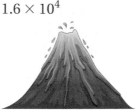

E. Circumference of a beach ball:
7.5×10^1

What Is Your Answer?

5. **IN YOUR OWN WORDS** How can you read numbers that are written in scientific notation? Why do you think this type of notation is called *scientific notation*? Why is scientific notation important?

Practice Use what you learned about reading scientific notation to complete Exercises 3–5 on page 740.

Key Vocabulary 🔊
scientific notation,
p. 738

 **Key Idea**

Scientific Notation

A number is written in **scientific notation** when it is represented as the product of a factor and a power of 10. The factor must be greater than or equal to 1 and less than 10.

The factor is greater than or equal to 1 and less than 10. ➝ 8.3×10^{-7} ⬅ The power of 10 has an integer exponent.

Study Tip

Scientific notation is used to write very small and very large numbers.

EXAMPLE ① **Identifying Numbers Written in Scientific Notation**

Tell whether the number is written in scientific notation. Explain.

a. 5.9×10^{-6}

⠿ The factor is greater than or equal to 1 and less than 10. The power of 10 has an integer exponent. So, the number is written in scientific notation.

b. 0.9×10^{8}

⠿ The factor is less than 1. So, the number is not written in scientific notation.

 **Key Idea**

Writing Numbers in Standard Form

The absolute value of the exponent indicates how many places to move the decimal point.

- If the exponent is negative, move the decimal point to the left.
- If the exponent is positive, move the decimal point to the right.

EXAMPLE ② **Writing Numbers in Standard Form**

a. Write 3.22×10^{-4} in standard form.

$$3.22 \times 10^{-4} = 0.000322$$
$$\underset{4}{}$$

Move decimal point $|-4| = 4$ places to the left.

b. Write 7.9×10^{5} in standard form.

$$7.9 \times 10^{5} = 790,000$$
$$\underset{5}{}$$

Move decimal point $|5| = 5$ places to the right.

🔊 Multi-Language Glossary at BigIdeasMath ✓com

Now You're Ready
Exercises 6–23

On Your Own

1. Is 12×10^4 written in scientific notation? Explain.

Write the number in standard form.

2. 6×10^7　　　　3. 9.9×10^{-5}　　　　4. 1.285×10^4

EXAMPLE ③ **Comparing Numbers in Scientific Notation**

An object with a lesser density than water will float. An object with a greater density than water will sink. Use each given density (in kilograms per cubic meter) to explain what happens when you place a brick and an apple in water.

Water: 1.0×10^3　　　**Brick:** 1.84×10^3　　　**Apple:** 6.41×10^2

You can compare the densities by writing each in standard form.

Water	Brick	Apple
$1.0 \times 10^3 = 1000$	$1.84 \times 10^3 = 1840$	$6.41 \times 10^2 = 641$

∴ The apple is less dense than water, so it will float. The brick is denser than water, so it will sink.

EXAMPLE ④ **Real-Life Application**

A female flea consumes about 1.4×10^{-5} liter of blood per day.

A dog has 100 female fleas. How much blood do the fleas consume per day?

$1.4 \times 10^{-5} \cdot 100 = 0.000014 \cdot 100$ 　　Write in standard form.

$= 0.0014$ 　　Multiply.

∴ The fleas consume about 0.0014 liter, or 1.4 milliliters of blood per day.

On Your Own

Now You're Ready
Exercise 27

5. **WHAT IF?** In Example 3, the density of lead is 1.14×10^4 kilograms per cubic meter. What happens when you place lead in water?

6. **WHAT IF?** In Example 4, a dog has 75 female fleas. How much blood do the fleas consume per day?

✓ **Vocabulary and Concept Check**

1. **WRITING** Describe the difference between scientific notation and standard form.

2. **WHICH ONE DOESN'T BELONG?** Which number does *not* belong with the other three? Explain.

2.8×10^{15} 4.3×10^{-30} 1.05×10^{28} 10×9.2^{-13}

 Practice and Problem Solving

Write the number shown on the calculator display in standard form.

3. `5.6E12`

4. `2.1E-10`

5. `8.73E16`

Tell whether the number is written in scientific notation. Explain.

① 6. 1.8×10^9 7. 3.45×10^{14} 8. 0.26×10^{-25}

9. 10.5×10^{12} 10. 46×10^{-17} 11. 5×10^{-19}

12. 7.814×10^{-36} 13. 0.999×10^{42} 14. 6.022×10^{23}

Write the number in standard form.

② 15. 7×10^7 16. 8×10^{-3} 17. 5×10^2

18. 2.7×10^{-4} 19. 4.4×10^{-5} 20. 2.1×10^3

21. 1.66×10^9 22. 3.85×10^{-8} 23. 9.725×10^6

24. **ERROR ANALYSIS** Describe and correct the error in writing the number in standard form.

✗ $4.1 \times 10^{-6} = 4{,}100{,}000$

25. **PLATELETS** Platelets are cell-like particles in the blood that help form blood clots.

a. How many platelets are in 3 milliliters of blood? Write your answer in standard form.

b. An adult human body contains about 5 liters of blood. How many platelets are in an adult human body?

2.7×10^8 platelets per milliliter

26. **REASONING** A googol is 1.0×10^{100}. How many zeros are in a googol?

③ 27. **STARS** The table shows the surface temperatures of five stars.

    a. Which star has the highest surface temperature?

    b. Which star has the lowest surface temperature?

Star	Betelgeuse	Bellatrix	Sun	Aldebaran	Rigel
Surface Temperature (°F)	6.2×10^3	3.8×10^4	1.1×10^4	7.2×10^3	2.2×10^4

28. **NUMBER SENSE** Describe how the value of a number written in scientific notation changes when you increase the exponent by 1.

29. **CORAL REEF** The area of the Florida Keys National Marine Sanctuary is about 9.6×10^3 square kilometers. The area of the Florida Reef Tract is about 16.2% of the area of the sanctuary. What is the area of the Florida Reef Tract in square kilometers?

30. **REASONING** A gigameter is 1.0×10^6 kilometers. How many square kilometers are in 5 square gigameters?

31. **WATER** There are about 1.4×10^9 cubic kilometers of water on Earth. About 2.5% of the water is fresh water. How much fresh water is on Earth?

32. **Critical Thinking** The table shows the speed of light through five media.

    a. In which medium does light travel the fastest?

    b. In which medium does light travel the slowest?

Medium	Speed
Air	6.7×10^8 mi/h
Glass	6.6×10^8 ft/sec
Ice	2.3×10^5 km/sec
Vacuum	3.0×10^8 m/sec
Water	2.3×10^{10} cm/sec

 Fair Game Review *What you learned in previous grades & lessons*

Write the product using exponents. *(Section 16.1)*

33. $4 \cdot 4 \cdot 4 \cdot 4 \cdot 4$

34. $3 \cdot 3 \cdot 3 \cdot y \cdot y \cdot y$

35. $(-2) \cdot (-2) \cdot (-2)$

36. **MULTIPLE CHOICE** What is the length of the hypotenuse of the right triangle? *(Section 14.3)*

    Ⓐ $\sqrt{18}$ in.   Ⓑ $\sqrt{41}$ in.

    Ⓒ 18 in.   Ⓓ 41 in.

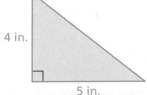

Essential Question
How can you write a number in scientific notation?

1 ACTIVITY: Finding pH Levels

Work with a partner. In chemistry, pH is a measure of the activity of dissolved hydrogen ions (H^+). Liquids with low pH values are called *acids*. Liquids with high pH values are called *bases*.

Find the pH of each liquid. Is the liquid a base, neutral, or an acid?

a. Lime juice:
$[H^+] = 0.01$

b. Egg:
$[H^+] = 0.00000001$

c. Distilled water:
$[H^+] = 0.0000001$

d. Ammonia water:
$[H^+] = 0.00000000001$

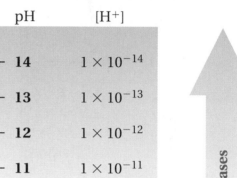

e. Tomato juice:
$[H^+] = 0.0001$

f. Hydrochloric acid:
$[H^+] = 1$

pH	$[H^+]$
14	1×10^{-14}
13	1×10^{-13}
12	1×10^{-12}
11	1×10^{-11}
10	1×10^{-10}
9	1×10^{-9}
8	1×10^{-8}
7	1×10^{-7}
6	1×10^{-6}
5	1×10^{-5}
4	1×10^{-4}
3	1×10^{-3}
2	1×10^{-2}
1	1×10^{-1}
0	1×10^{0}

Bases

Neutral

Acids

COMMON CORE

Scientific Notation

In this lesson, you will
- write large and small numbers in scientific notation.
- perform operations with numbers written in scientific notation.

Learning Standards
8.EE.3
8.EE.4

ACTIVITY: Writing Scientific Notation

Work with a partner. Match each planet with its distance from the Sun. Then write each distance in scientific notation. Do you think it is easier to match the distances when they are written in standard form or in scientific notation? Explain.

Neptune

Uranus

Saturn

Jupiter

Mars

Earth

Venus

Mercury

Sun

a. 1,800,000,000 miles

b. 67,000,000 miles

c. 890,000,000 miles

d. 93,000,000 miles

e. 140,000,000 miles

f. 2,800,000,000 miles

g. 480,000,000 miles

h. 36,000,000 miles

3 **ACTIVITY: Making a Scale Drawing**

Work with a partner. The illustration in Activity 2 is not drawn to scale. Use the instructions below to make a scale drawing of the distances in our solar system.

> • **Cut a sheet of paper into three strips of equal width. Tape the strips together to make one long piece.**
>
> • **Draw a long number line. Label the number line in hundreds of millions of miles.**
>
> • **Locate each planet's position on the number line.**

Math Practice 6

Calculate Accurately

How can you verify that you have accurately written each distance in scientific notation?

What Is Your Answer?

4. IN YOUR OWN WORDS How can you write a number in scientific notation?

Practice ➤ Use what you learned about writing scientific notation to complete Exercises 3–5 on page 746.

Check It Out
Lesson Tutorials
BigIdeasMath  com

🔑 Key Idea

Writing Numbers in Scientific Notation

Step 1: Move the decimal point so it is located to the right of the leading nonzero digit.

Step 2: Count the number of places you moved the decimal point. This indicates the exponent of the power of 10, as shown below.

> **Study Tip**
>
> When you write a number greater than or equal to 1 and less than 10 in scientific notation, use zero as the exponent.
> $6 = 6 \times 10^0$

Number Greater Than or Equal to 10

Use a positive exponent when you move the decimal point to the left.

$$8600 = 8.6 \times 10^3$$
$$3$$

Number Between 0 and 1

Use a negative exponent when you move the decimal point to the right.

$$0.0024 = 2.4 \times 10^{-3}$$
$$3$$

EXAMPLE 1 **Writing Large Numbers in Scientific Notation**

Google purchased YouTube for $1,650,000,000. Write this number in scientific notation.

Move the decimal point 9 places to the left.

$$1{,}650{,}000{,}000 = 1.65 \times 10^9$$
$$9$$

The number is greater than 10. So, the exponent is positive.

EXAMPLE 2 **Writing Small Numbers in Scientific Notation**

The 2004 Indonesian earthquake slowed the rotation of Earth, making the length of a day 0.00000268 second shorter. Write this number in scientific notation.

Move the decimal point 6 places to the right.

$$0.00000268 = 2.68 \times 10^{-6}$$
$$6$$

The number is between 0 and 1. So, the exponent is negative.

● On Your Own

Now You're Ready
Exercises 3–11

Write the number in scientific notation.

1. 50,000
2. 25,000,000
3. 683
4. 0.005
5. 0.00000033
6. 0.000506

EXAMPLE **3** **Using Scientific Notation**

An album receives an award when it
sells 10,000,000 copies.

**An album has sold 8,780,000 copies. How many more copies does
it need to sell to receive the award?**

Ⓐ 1.22×10^{-7} Ⓑ 1.22×10^{-6}

Ⓒ 1.22×10^6 Ⓓ 1.22×10^7

Use a model to solve the problem.

$$\dfrac{\text{Remaining sales}}{\text{needed for award}} = \dfrac{\text{Sales required}}{\text{for award}} - \dfrac{\text{Current sales}}{\text{total}}$$

$$= 10{,}000{,}000 - 8{,}780{,}000$$

$$= 1{,}220{,}000$$

$$= 1.22 \times 10^6$$

:•: The album must sell 1.22×10^6 more copies to receive the
award. So, the correct answer is Ⓒ.

EXAMPLE **4** **Real-Life Application**

**The table shows when the last
three geologic eras began.
Order the eras from earliest
to most recent.**

Era	Began
Paleozoic	5.42×10^8 years ago
Cenozoic	6.55×10^7 years ago
Mesozoic	2.51×10^8 years ago

Step 1: Compare the powers of 10.

Because $10^7 < 10^8$,

$6.55 \times 10^7 < 5.42 \times 10^8$ and
$6.55 \times 10^7 < 2.51 \times 10^8$.

Step 2: Compare the factors when the powers of 10 are the same.

Because $2.51 < 5.42$,
$2.51 \times 10^8 < 5.42 \times 10^8$.

Common Error

To use the method in
Example 4, the numbers
must be written in
scientific notation.

From greatest to least, the order is 5.42×10^8, 2.51×10^8, and 6.55×10^7.

:•: So, the eras in order from earliest to most recent are the Paleozoic
era, Mesozoic era, and Cenozoic era.

On Your Own

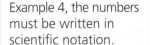

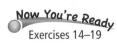
Exercises 14–19

7. **WHAT IF?** In Example 3, an album has sold 955,000 copies.
How many more copies does it need to sell to receive the
award? Write your answer in scientific notation.

8. The *Tyrannosaurus rex* lived 7.0×10^7 years ago. Consider
the eras given in Example 4. During which era did the
Tyrannosaurus rex live?

 Vocabulary and Concept Check

1. **REASONING** How do you know whether a number written in standard form will have a positive or a negative exponent when written in scientific notation?

2. **WRITING** When is it appropriate to use scientific notation instead of standard form?

 Practice and Problem Solving

Write the number in scientific notation.

3. 0.0021   **4.** 5,430,000   **5.** 321,000,000

6. 0.00000625   **7.** 0.00004   **8.** 10,700,000

9. 45,600,000,000   **10.** 0.000000000009256   **11.** 840,000

ERROR ANALYSIS Describe and correct the error in writing the number in scientific notation.

12.

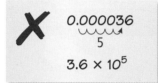

$$3.6 \times 10^5$$

13.

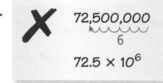

$$72.5 \times 10^6$$

Order the numbers from least to greatest.

14. 1.2×10^8, 1.19×10^8, 1.12×10^8

15. 6.8×10^{-5}, 6.09×10^{-5}, 6.78×10^{-5}

16. 5.76×10^{12}, 9.66×10^{11}, 5.7×10^{10}

17. 4.8×10^{-6}, 4.8×10^{-5}, 4.8×10^{-8}

18. 9.9×10^{-15}, 1.01×10^{-14}, 7.6×10^{-15}

19. 5.78×10^{23}, 6.88×10^{-23}, 5.82×10^{23}

20. HAIR What is the diameter of a human hair written in scientific notation?

21. EARTH What is the circumference of Earth written in scientific notation?

Diameter: 0.000099 meter

Circumference at the equator:
about 40,100,000 meters

22. CHOOSING UNITS In Exercise 21, name a unit of measurement that would be more appropriate for the circumference. Explain.

Order the numbers from least to greatest.

23. $\dfrac{68,500}{10}$, 680, 6.8×10^3

24. $\dfrac{5}{241}$, 0.02, 2.1×10^{-2}

25. 6.3%, 6.25×10^{-3}, $6\dfrac{1}{4}$, 0.625

26. 3033.4, 305%, $\dfrac{10,000}{3}$, 3.3×10^2

27. SPACE SHUTTLE The total power of a space shuttle during launch is the sum of the power from its solid rocket boosters and the power from its main engines. The power from the solid rocket boosters is 9,750,000,000 watts. What is the power from the main engines?

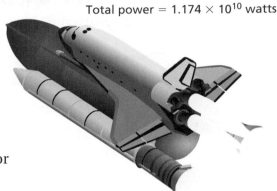

Total power = 1.174×10^{10} watts

28. CHOOSE TOOLS Explain how to use a calculator to verify your answer to Exercise 27.

Equivalent to 1 Atomic Mass Unit
8.3×10^{-24} carat
1.66×10^{-21} milligram

29. ATOMIC MASS The mass of an atom or molecule is measured in atomic mass units. Which is greater, a *carat* or a *milligram*? Explain.

30. **Reasoning** In Example 4, the Paleozoic era ended when the Mesozoic era began. The Mesozoic era ended when the Cenozoic era began. The Cenozoic era is the current era.

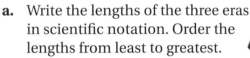

  a. Write the lengths of the three eras in scientific notation. Order the lengths from least to greatest.

  b. Make a time line to show when the three eras occurred and how long each era lasted.

  c. What do you notice about the lengths of the three eras? Use the Internet to determine whether your observation is true for *all* the geologic eras. Explain your results.

 Fair Game Review What you learned in previous grades & lessons

Classify the real number. *(Section 14.4)*

31. 15

32. $\sqrt[3]{-8}$

33. $\sqrt{73}$

34. What is the surface area of the prism? *(Section 9.1)*

  (A) 5 in.^{2}

  (B) 5.5 in.^{2}

  (C) 10 in.^{2}

  (D) 19 in.^{2}

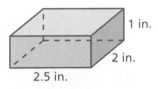

1 in.
2 in.
2.5 in.

Essential Question How can you perform operations with numbers written in scientific notation?

1 ACTIVITY: Adding Numbers in Scientific Notation

Work with a partner. Consider the numbers 2.4×10^3 and 7.1×10^3.

a. Explain how to use order of operations to find the sum of these numbers. Then find the sum.

$$2.4 \times 10^3 + 7.1 \times 10^3$$

b. The factor ▭ is common to both numbers. How can you use the Distributive Property to rewrite the sum $\left(2.4 \times 10^3\right) + \left(7.1 \times 10^3\right)$?

$$\left(2.4 \times 10^3\right) + \left(7.1 \times 10^3\right) = \boxed{} \qquad \text{Distributive Property}$$

c. Use order of operations to evaluate the expression you wrote in part (b). Compare the result with your answer in part (a).

d. **STRUCTURE** Write a rule you can use to add numbers written in scientific notation where the powers of 10 are the same. Then test your rule using the sums below.

- $\left(4.9 \times 10^5\right) + \left(1.8 \times 10^5\right) = \boxed{}$
- $\left(3.85 \times 10^4\right) + \left(5.72 \times 10^4\right) = \boxed{}$

2 ACTIVITY: Adding Numbers in Scientific Notation

Work with a partner. Consider the numbers 2.4×10^3 and 7.1×10^4.

a. Explain how to use order of operations to find the sum of these numbers. Then find the sum.

$$2.4 \times 10^3 + 7.1 \times 10^4$$

b. How is this pair of numbers different from the pairs of numbers in Activity 1?

c. Explain why you cannot immediately use the rule you wrote in Activity 1(d) to find this sum.

d. **STRUCTURE** How can you rewrite one of the numbers so that you can use the rule you wrote in Activity 1(d)? Rewrite one of the numbers. Then find the sum using your rule and compare the result with your answer in part (a).

e. **REASONING** Do these procedures work when subtracting numbers written in scientific notation? Justify your answer by evaluating the differences below.

- $\left(8.2 \times 10^5\right) - \left(4.6 \times 10^5\right) = \boxed{}$
- $\left(5.88 \times 10^5\right) - \left(1.5 \times 10^4\right) = \boxed{}$

COMMON CORE

Scientific Notation

In this lesson, you will

- add, subtract, multiply, and divide numbers written in scientific notation.

Learning Standards
8.EE.3
8.EE.4

3 ACTIVITY: Multiplying Numbers in Scientific Notation

Math Practice 3

Justify Conclusions

Which step of the procedure would be affected if the powers of 10 were different? Explain.

Work with a partner. Match each step with the correct description.

Step   **Description**

$(2.4 \times 10^3) \times (7.1 \times 10^3)$ Original expression

1. $= 2.4 \times 7.1 \times 10^3 \times 10^3$ **A.** Write in standard form.

2. $= (2.4 \times 7.1) \times (10^3 \times 10^3)$ **B.** Product of Powers Property

3. $= 17.04 \times 10^6$ **C.** Write in scientific notation.

4. $= 1.704 \times 10^1 \times 10^6$ **D.** Commutative Property of Multiplication

5. $= 1.704 \times 10^7$ **E.** Simplify.

6. $= 17,040,000$ **F.** Associative Property of Multiplication

Does this procedure work when the numbers have different powers of 10? Justify your answer by using this procedure to evaluate the products below.

- $(1.9 \times 10^2) \times (2.3 \times 10^5) =$
- $(8.4 \times 10^6) \times (5.7 \times 10^{-4}) =$

4 ACTIVITY: Using Scientific Notation to Estimate

Work with a partner. A person normally breathes about 6 liters of air per minute. The life expectancy of a person in the United States at birth is about 80 years. Use scientific notation to estimate the total amount of air a person born in the United States breathes over a lifetime.

What Is Your Answer?

5. IN YOUR OWN WORDS How can you perform operations with numbers written in scientific notation?

6. Use a calculator to evaluate the expression. Write your answer in scientific notation and in standard form.

    a. $(1.5 \times 10^4) + (6.3 \times 10^4)$ **b.** $(7.2 \times 10^5) - (2.2 \times 10^3)$

    c. $(4.1 \times 10^{-3}) \times (4.3 \times 10^{-3})$ **d.** $(4.75 \times 10^{-6}) \times (1.34 \times 10^7)$

Practice

Use what you learned about evaluating expressions involving scientific notation to complete Exercises 3–6 on page 752.

To add or subtract numbers written in scientific notation with the same power of 10, add or subtract the factors. When the numbers have different powers of 10, first rewrite the numbers so they have the same power of 10.

EXAMPLE 1 **Adding and Subtracting Numbers in Scientific Notation**

Find the sum or difference. Write your answer in scientific notation.

a. $(4.6 \times 10^3) + (8.72 \times 10^3)$

$= (4.6 + 8.72) \times 10^3$	Distributive Property
$= 13.32 \times 10^3$	Add.
$= (1.332 \times 10^1) \times 10^3$	Write 13.32 in scientific notation.
$= 1.332 \times 10^4$	Product of Powers Property

> **Study Tip**
>
> In Example 1(b), you will get the same answer when you start by rewriting 3.5×10^{-2} as 35×10^{-3}.

b. $(3.5 \times 10^{-2}) - (6.6 \times 10^{-3})$

Rewrite 6.6×10^{-3} so that it has the same power of 10 as 3.5×10^{-2}.

$6.6 \times 10^{-3} = 6.6 \times 10^{-1} \times 10^{-2}$	Rewrite 10^{-3} as $10^{-1} \times 10^{-2}$.
$= 0.66 \times 10^{-2}$	Rewrite 6.6×10^{-1} as 0.66.

Subtract the factors.

$$(3.5 \times 10^{-2}) - (0.66 \times 10^{-2})$$

$= (3.5 - 0.66) \times 10^{-2}$	Distributive Property
$= 2.84 \times 10^{-2}$	Subtract.

On Your Own

Now You're Ready
Exercises 7–14

Find the sum or difference. Write your answer in scientific notation.

1. $(8.2 \times 10^2) + (3.41 \times 10^{-1})$ **2.** $(7.8 \times 10^{-5}) - (4.5 \times 10^{-5})$

To multiply or divide numbers written in scientific notation, multiply or divide the factors and powers of 10 separately.

EXAMPLE 2 **Multiplying Numbers in Scientific Notation**

> **Study Tip**
>
> You can check your answer using standard form.
> (3×10^{-5})
> $\times (5 \times 10^{-2})$
> $= 0.00003 \times 0.05$
> $= 0.0000015$
> $= 1.5 \times 10^{-6}$

Find $(3 \times 10^{-5}) \times (5 \times 10^{-2})$. Write your answer in scientific notation.

$$(3 \times 10^{-5}) \times (5 \times 10^{-2})$$

$= 3 \times 5 \times 10^{-5} \times 10^{-2}$	Commutative Property of Multiplication
$= (3 \times 5) \times (10^{-5} \times 10^{-2})$	Associative Property of Multiplication
$= 15 \times 10^{-7}$	Simplify.
$= 1.5 \times 10^1 \times 10^{-7}$	Write 15 in scientific notation.
$= 1.5 \times 10^{-6}$	Product of Powers Property

EXAMPLE ③ Dividing Numbers in Scientific Notation

Find $\dfrac{1.5 \times 10^{-8}}{6 \times 10^7}$. Write your answer in scientific notation.

$$\dfrac{1.5 \times 10^{-8}}{6 \times 10^7} = \dfrac{1.5}{6} \times \dfrac{10^{-8}}{10^7} \qquad \text{Rewrite as a product of fractions.}$$

$$= 0.25 \times \dfrac{10^{-8}}{10^7} \qquad \text{Divide 1.5 by 6.}$$

$$= 0.25 \times 10^{-15} \qquad \text{Quotient of Powers Property}$$

$$= 2.5 \times 10^{-1} \times 10^{-15} \qquad \text{Write 0.25 in scientific notation.}$$

$$= 2.5 \times 10^{-16} \qquad \text{Product of Powers Property}$$

On Your Own

Now You're Ready
Exercises 16–23

Find the product or quotient. Write your answer in scientific notation.

3. $6 \times \left(8 \times 10^{-5}\right)$

4. $\left(7 \times 10^2\right) \times \left(3 \times 10^5\right)$

5. $\left(9.2 \times 10^{12}\right) \div 4.6$

6. $\left(1.5 \times 10^{-3}\right) \div \left(7.5 \times 10^2\right)$

EXAMPLE ④ Real-Life Application

Diameter = 1,400,000 km

How many times greater is the diameter of the Sun than the diameter of Earth?

Write the diameter of the Sun in scientific notation.

Diameter = 1.28×10^4 km

$$1{,}400{,}000 = 1.4 \times 10^6$$
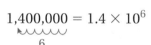

Divide the diameter of the Sun by the diameter of Earth.

$$\dfrac{1.4 \times 10^6}{1.28 \times 10^4} = \dfrac{1.4}{1.28} \times \dfrac{10^6}{10^4} \qquad \text{Rewrite as a product of fractions.}$$

$$= 1.09375 \times \dfrac{10^6}{10^4} \qquad \text{Divide 1.4 by 1.28.}$$

$$= 1.09375 \times 10^2 \qquad \text{Quotient of Powers Property}$$

$$= 109.375 \qquad \text{Write in standard form.}$$

∴ The diameter of the Sun is about 109 times greater than the diameter of Earth.

On Your Own

7. How many more kilometers is the radius of the Sun than the radius of Earth? Write your answer in standard form.

✓ Vocabulary and Concept Check

1. **WRITING** Describe how to subtract two numbers written in scientific notation with the same power of 10.

2. **NUMBER SENSE** You are multiplying two numbers written in scientific notation with different powers of 10. Do you have to rewrite the numbers so they have the same power of 10 before multiplying? Explain.

 ## Practice and Problem Solving

Evaluate the expression using two different methods. Write your answer in scientific notation.

3. $(2.74 \times 10^7) + (5.6 \times 10^7)$ 4. $(8.3 \times 10^6) + (3.4 \times 10^5)$

5. $(5.1 \times 10^5) \times (9.7 \times 10^5)$ 6. $(4.5 \times 10^4) \times (6.2 \times 10^3)$

Find the sum or difference. Write your answer in scientific notation.

 7. $(2 \times 10^5) + (3.8 \times 10^5)$ 8. $(6.33 \times 10^{-9}) - (4.5 \times 10^{-9})$

9. $(9.2 \times 10^8) - (4 \times 10^8)$ 10. $(7.2 \times 10^{-6}) + (5.44 \times 10^{-6})$

11. $(7.8 \times 10^7) - (2.45 \times 10^6)$ 12. $(5 \times 10^{-5}) + (2.46 \times 10^{-3})$

13. $(9.7 \times 10^6) + (6.7 \times 10^5)$ 14. $(2.4 \times 10^{-1}) - (5.5 \times 10^{-2})$

15. **ERROR ANALYSIS** Describe and correct the error in finding the sum of the numbers.

$$\times \quad (2.5 \times 10^9) + (5.3 \times 10^8) = (2.5 + 5.3) \times (10^9 \times 10^8)$$
$$= 7.8 \times 10^{17}$$

Find the product or quotient. Write your answer in scientific notation.

16. $5 \times (7 \times 10^7)$ 17. $(5.8 \times 10^{-6}) \div (2 \times 10^{-3})$

18. $(1.2 \times 10^{-5}) \div 4$ 19. $(5 \times 10^{-7}) \times (3 \times 10^6)$

20. $(3.6 \times 10^7) \div (7.2 \times 10^7)$ 21. $(7.2 \times 10^{-1}) \times (4 \times 10^{-7})$

22. $(6.5 \times 10^8) \times (1.4 \times 10^{-5})$ 23. $(2.8 \times 10^4) \div (2.5 \times 10^6)$

24. **MONEY** How many times greater is the thickness of a dime than the thickness of a dollar bill?

Thickness = 0.135 cm

Thickness = 1.0922×10^{-2} cm

Evaluate the expression. Write your answer in scientific notation.

25. $5{,}200{,}000 \times (8.3 \times 10^2) - (3.1 \times 10^8)$

26. $(9 \times 10^{-3}) + (2.4 \times 10^{-5}) \div 0.0012$

27. **GEOMETRY** Find the perimeter of the rectangle.

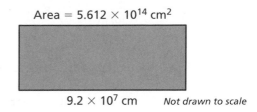

Area = 5.612×10^{14} cm²

9.2×10^7 cm   *Not drawn to scale*

28. **BLOOD SUPPLY** A human heart pumps about 7×10^{-2} liter of blood per heartbeat. The average human heart beats about 72 times per minute. How many liters of blood does a heart pump in 1 year? in 70 years? Write your answers in scientific notation. Then use estimation to justify your answers.

$H \leftarrow 0.000074$ cm

$H \leftarrow 0.000032$ cm

4.26 cm

29. **DVDS** On a DVD, information is stored on bumps that spiral around the disk. There are 73,000 ridges (with bumps) and 73,000 valleys (without bumps) across the diameter of the DVD. What is the diameter of the DVD in centimeters?

30. **PROJECT** Use the Internet or some other reference to find the populations and areas (in square miles) of India, China, Argentina, the United States, and Egypt. Round each population to the nearest million and each area to the nearest thousand square miles.

   a. Write each population and area in scientific notation.

   b. Use your answers to part (a) to find and order the population densities (people per square mile) of each country from least to greatest.

31. Albert Einstein's most famous equation is $E = mc^2$, where E is the energy of an object (in joules), m is the mass of an object (in kilograms), and c is the speed of light (in meters per second). A hydrogen atom has 15.066×10^{-11} joule of energy and a mass of 1.674×10^{-27} kilogram. What is the speed of light? Write your answer in scientific notation.

Fair Game Review *What you learned in previous grades & lessons*

Find the cube root.   *(Section 14.2)*

32. $\sqrt[3]{-729}$

33. $\sqrt[3]{\dfrac{1}{512}}$

34. $\sqrt[3]{-\dfrac{125}{343}}$

35. **MULTIPLE CHOICE** What is the volume of the cone?   *(Section 15.2)*

   Ⓐ $16\pi \text{ cm}^3$ Ⓑ $108\pi \text{ cm}^3$

   Ⓒ $48\pi \text{ cm}^3$ Ⓓ $144\pi \text{ cm}^3$

4 cm

9 cm

16.5–16.7 Quiz

Tell whether the number is written in scientific notation. Explain. *(Section 16.5)*

1. 23×10^9

2. 0.6×10^{-7}

Write the number in standard form. *(Section 16.5)*

3. 8×10^6

4. 1.6×10^{-2}

Write the number in scientific notation. *(Section 16.6)*

5. 0.00524

6. $892{,}000{,}000$

Evaluate the expression. Write your answer in scientific notation. *(Section 16.7)*

7. $(7.26 \times 10^4) + (3.4 \times 10^4)$

8. $(2.8 \times 10^{-5}) - (1.6 \times 10^{-6})$

9. $(2.4 \times 10^4) \times (3.8 \times 10^{-6})$

10. $(5.2 \times 10^{-3}) \div (1.3 \times 10^{-12})$

11. PLANETS The table shows the equatorial radii of the eight planets in our solar system. *(Section 16.5)*

    a. Which planet has the second-smallest equatorial radius?

    b. Which planet has the second-largest equatorial radius?

Planet	Equatorial Radius (km)
Mercury	2.44×10^3
Venus	6.05×10^3
Earth	6.38×10^3
Mars	3.4×10^3
Jupiter	7.15×10^4
Saturn	6.03×10^4
Uranus	2.56×10^4
Neptune	2.48×10^4

12. OORT CLOUD The Oort cloud is a spherical cloud that surrounds our solar system. It is about 2×10^5 astronomical units from the Sun. An astronomical unit is about 1.5×10^8 kilometers. How far is the Oort cloud from the Sun in kilometers? *(Section 16.6)*

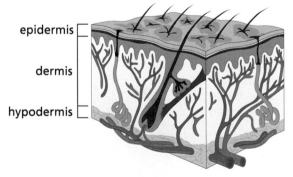

epidermis

dermis

hypodermis

13. EPIDERMIS The outer layer of skin is called the *epidermis*. On the palm of your hand, the epidermis is 0.0015 meter thick. Write this number in scientific notation. *(Section 16.6)*

14. ORBITS It takes the Sun about 2.3×10^8 years to orbit the center of the Milky Way. It takes Pluto about 2.5×10^2 years to orbit the Sun. How many times does Pluto orbit the Sun while the Sun completes one orbit around the Milky Way? Write your answer in standard form. *(Section 16.7)*

Review Key Vocabulary

power, *p. 712*   exponent, *p. 712*
base, *p. 712*   scientific notation, *p. 738*

Review Examples and Exercises

16.1 Exponents *(pp. 710–715)*

Write $(-4) \cdot (-4) \cdot (-4) \cdot y \cdot y$ using exponents.

Because -4 is used as a factor 3 times, its exponent is 3. Because y is used as a factor 2 times, its exponent is 2.

So, $(-4) \cdot (-4) \cdot (-4) \cdot y \cdot y = (-4)^3 y^2$.

Exercises

Write the product using exponents.

1. $(-9) \cdot (-9) \cdot (-9) \cdot (-9) \cdot (-9)$ **2.** $2 \cdot 2 \cdot 2 \cdot n \cdot n$

Evaluate the expression.

3. 6^3 **4.** $-\left(\dfrac{1}{2}\right)^4$ **5.** $\left|\dfrac{1}{2}(16 - 6^3)\right|$

16.2 Product of Powers Property *(pp. 716–721)*

a. $\left(-\dfrac{1}{8}\right)^7 \cdot \left(-\dfrac{1}{8}\right)^4 = \left(-\dfrac{1}{8}\right)^{7+4}$ Product of Powers Property

$= \left(-\dfrac{1}{8}\right)^{11}$ Simplify.

b. $(2.5^7)^2 = 2.5^{7 \cdot 2}$ Power of a Power Property

$= 2.5^{14}$ Simplify.

c. $(3m)^2 = 3^2 \cdot m^2$ Power of a Product Property

$= 9m^2$ Simplify.

Exercises

Simplify the expression.

6. $p^5 \cdot p^2$ **7.** $(n^{11})^2$ **8.** $(5y)^3$ **9.** $(-2k)^4$

16.3 Quotient of Powers Property (pp. 722–727)

a. $\dfrac{(-4)^9}{(-4)^6} = (-4)^{9-6}$ Quotient of Powers Property

         $= (-4)^3$ Simplify.

b. $\dfrac{x^4}{x^3} = x^{4-3}$ Quotient of Powers Property

    $= x^1$

    $= x$ Simplify.

Exercises

Simplify the expression. Write your answer as a power.

10. $\dfrac{8^8}{8^3}$

11. $\dfrac{5^2 \cdot 5^9}{5}$

12. $\dfrac{w^8}{w^7} \cdot \dfrac{w^5}{w^2}$

Simplify the expression.

13. $\dfrac{2^2 \cdot 2^5}{2^3}$

14. $\dfrac{(6c)^3}{c}$

15. $\dfrac{m^8}{m^6} \cdot \dfrac{m^{10}}{m^9}$

16.4 Zero and Negative Exponents (pp. 728–733)

a. $10^{-3} = \dfrac{1}{10^3}$ Definition of negative exponent

     $= \dfrac{1}{1000}$ Evaluate power.

b. $(-0.5)^{-5} \cdot (-0.5)^5 = (-0.5)^{-5+5}$ Product of Powers Property

           $= (-0.5)^0$ Simplify.

           $= 1$ Definition of zero exponent

Exercises

Evaluate the expression.

16. 2^{-4}

17. 95^0

18. $\dfrac{8^2}{8^4}$

19. $(-12)^{-7} \cdot (-12)^7$

20. $\dfrac{1}{7^9} \cdot \dfrac{1}{7^{-6}}$

21. $\dfrac{9^4 \cdot 9^{-2}}{9^2}$

16.5 Reading Scientific Notation (pp. 736–741)

Write (a) 5.9×10^4 and (b) 7.31×10^{-6} in standard notation.

a. $5.9 \times 10^4 = 59{,}000$

Move decimal point $|4| = 4$ places to the right.

b. $7.31 \times 10^{-6} = 0.00000731$

Move decimal point $|-6| = 6$ places to the left.

Exercises

Write the number in standard form.

22. 2×10^7

23. 3.4×10^{-2}

24. 1.5×10^{-9}

25. 5.9×10^{10}

26. 4.8×10^{-3}

27. 6.25×10^5

16.6 Writing Scientific Notation (pp. 742–747)

Write (a) 309,000,000 and (b) 0.00056 in scientific notation.

a. $309{,}000{,}000 = 3.09 \times 10^8$

The number is greater than 10. So, the exponent is positive.

b. $0.00056 = 5.6 \times 10^{-4}$

The number is between 0 and 1. So, the exponent is negative.

Exercises

Write the number in scientific notation.

28. 0.00036

29. 800,000

30. 79,200,000

16.7 Operations in Scientific Notation (pp. 748–753)

Find $(2.6 \times 10^5) + (3.1 \times 10^5)$.

$(2.6 \times 10^5) + (3.1 \times 10^5) = (2.6 + 3.1) \times 10^5$ Distributive Property

$= 5.7 \times 10^5$ Add.

Exercises

Evaluate the expression. Write your answer in scientific notation.

31. $(4.2 \times 10^8) + (5.9 \times 10^9)$

32. $(5.9 \times 10^{-4}) - (1.8 \times 10^{-4})$

33. $(7.7 \times 10^8) \times (4.9 \times 10^{-5})$

34. $(3.6 \times 10^5) \div (1.8 \times 10^9)$

Check It Out
Test Practice
BigIdeasMath.com

Write the product using exponents.

1. $(-15) \cdot (-15) \cdot (-15)$

2. $\left(\frac{1}{12}\right) \cdot \left(\frac{1}{12}\right) \cdot \left(\frac{1}{12}\right) \cdot \left(\frac{1}{12}\right) \cdot \left(\frac{1}{12}\right)$

Evaluate the expression.

3. -2^3

4. $10 + 3^3 \div 9$

Simplify the expression. Write your answer as a power.

5. $9^{10} \cdot 9$

6. $\left(6^6\right)^5$

7. $(2 \cdot 10)^7$

8. $\frac{(-3.5)^{13}}{(-3.5)^9}$

Evaluate the expression.

9. $5^{-2} \cdot 5^2$

10. $\frac{-8}{(-8)^3}$

Write the number in standard form.

11. 3×10^7

12. 9.05×10^{-3}

Evaluate the expression. Write your answer in scientific notation.

13. $\left(7.8 \times 10^7\right) + \left(9.9 \times 10^7\right)$

14. $\left(6.4 \times 10^5\right) - \left(5.4 \times 10^4\right)$

15. $\left(3.1 \times 10^6\right) \times \left(2.7 \times 10^{-2}\right)$

16. $\left(9.6 \times 10^7\right) \div \left(1.2 \times 10^{-4}\right)$

17. **CRITICAL THINKING** Is $\left(xy^2\right)^3$ the same as $\left(xy^3\right)^2$? Explain.

18. **RICE** A grain of rice weighs about 3^3 milligrams. About how many grains of rice are in one scoop?

19. **TASTE BUDS** There are about 10,000 taste buds on a human tongue. Write this number in scientific notation.

One scoop of rice weighs about 3^9 milligrams.

20. **LEAD** From 1978 to 2008, the amount of lead allowed in the air in the United States was 1.5×10^{-6} gram per cubic meter. In 2008, the amount allowed was reduced by 90%. What is the new amount of lead allowed in the air?

1. Mercury's distance from the Sun is approximately 5.79×10^7 kilometers. What is this distance in standard form? *(8.EE.4)*

   A. 5,790,000,000 km

   B. 579,000,000 km

   C. 57,900,000 km

   D. 5,790,000 km

2. The steps Jim took to answer the question are shown below. What should Jim change to correctly answer the question? *(8.G.5)*

   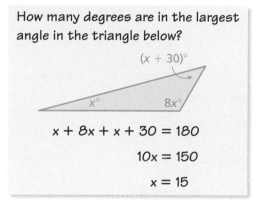

   How many degrees are in the largest angle in the triangle below?

   $$x + 8x + x + 30 = 180$$
   $$10x = 150$$
   $$x = 15$$

   F. The left side of the equation should equal $360°$ instead of $180°$.

   G. The sum of the acute angles should equal $90°$.

   H. Evaluate the smallest angle when $x = 15$.

   I. Evaluate the largest angle when $x = 15$.

3. Which expression is equivalent to the expression below? *(8.EE.1)*

   $$2^4 2^3$$

   A. 2^{12}

   B. 4^7

   C. 48

   D. 128

4. In the figure below, $\triangle ABC$ is a dilation of $\triangle DEF$.

   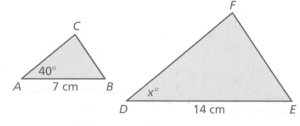

   What is the value of x? *(8.G.4)*

5. A bank account pays interest so that the amount in the account doubles every 10 years. The account started with $5,000 in 1940. Which expression represents the amount (in dollars) in the account n decades later? *(8.EE.1)*

   F. $2^n \cdot 5000$ **H.** 5000^n

   G. $5000(n + 1)$ **I.** $2^n + 5000$

6. The formula for the volume V of a pyramid is $V = \dfrac{1}{3}Bh$. Solve the formula for the height h. *(8.EE.7b)*

   A. $h = \dfrac{1}{3}VB$ **C.** $h = \dfrac{V}{3B}$

   B. $h = \dfrac{3V}{B}$ **D.** $h = V - \dfrac{1}{3}B$

7. The gross domestic product (GDP) is a way to measure how much a country produces economically in a year. The table below shows the approximate population and GDP for the United States. *(8.EE.4)*

United States 2012	
Population	312 million (312,000,000)
GDP	15.1 trillion dollars ($15,100,000,000,000)

   Part A Find the GDP per person for the United States. Show your work and explain your reasoning.

   Part B Write the population and the GDP using scientific notation.

   Part C Find the GDP per person for the United States using your answers from Part B. Write your answer in scientific notation. Show your work and explain your reasoning.

8. What is the equation of the line shown in the graph? *(8.EE.6)*

   F. $y = -\dfrac{1}{3}x + 3$ **H.** $y = -3x + 3$

   G. $y = \dfrac{1}{3}x + 1$ **I.** $y = 3x - \dfrac{1}{3}$

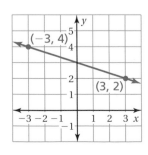

9. A cylinder and its dimensions are shown below.

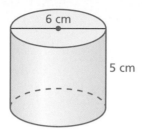

6 cm

5 cm

What is the volume of the cylinder? (Use 3.14 for π.) *(8.G.9)*

A. 47.1 cm^3

B. 94.2 cm^3

C. 141.3 cm^3

D. 565.2 cm^3

10. Find $(-2.5)^{-2}$. *(8.EE.1)*

11. Two lines have the same y-intercept. The slope of one line is 1, and the slope of the other line is -1. What can you conclude? *(8.EE.6)*

F. The lines are parallel.

G. The lines are perpendicular.

H. The lines are neither parallel nor perpendicular.

I. The situation described is impossible.

12. The volume of a sphere is $4\frac{1}{2}\pi$ cubic centimeters. What is the radius of the sphere? *(8.G.9)*

A. $\frac{4}{3}$ cm

B. $1\frac{1}{2}$ cm

C. π cm

D. $4\frac{1}{2}$ cm

Additional Topics

"Dear Sir: Here is my suggestion for a good math problem."

"A box contains a total of 30 dog and cat treats. There are 5 times more dog treats than cat treats."

"I need to learn to type so that I can write the story problems."

"How many of each type of treat are there?"

I think $D = RT$ stands for Descartes is Really Tired.

"Push faster, Descartes! According to the formula $R = D \div T$, the time needs to be 10 minutes or less to break our all-time speed record!"

Topic 1 Solving Multi-Step Equations

COMMON CORE

Solving Equations
In this lesson, you will
- use inverse operations to solve multi-step equations.
- use the Distributive Property to solve multi-step equations.

Learning Standards
8.EE.7a
8.EE.7b

 **Key Idea**

Solving Multi-Step Equations

To solve multi-step equations, use inverse operations to isolate the variable.

EXAMPLE 1 Solving a Two-Step Equation

The height (in feet) of a tree after x years is $1.5x + 15$. After how many years is the tree 24 feet tall?

$1.5x + 15 =$	24	Write an equation.	

Undo the addition. → $\underline{-15 \quad -15}$ — Subtraction Property of Equality

$1.5x = 9$ — Simplify.

Undo the multiplication. → $\dfrac{1.5x}{1.5} = \dfrac{9}{1.5}$ — Division Property of Equality

$x = 6$ — Simplify.

So, the tree is 24 feet tall after 6 years.

EXAMPLE 2 Combining Like Terms to Solve an Equation

Solve $8x - 6x - 25 = -35$.

$8x - 6x - 25 = -35$ — Write the equation.

$2x - 25 = -35$ — Combine like terms.

Undo the subtraction. → $\underline{+25 \quad +25}$ — Addition Property of Equality

$2x = -10$ — Simplify.

Undo the multiplication. → $\dfrac{2x}{2} = \dfrac{-10}{2}$ — Division Property of Equality

$x = -5$ — Simplify.

The solution is $x = -5$.

On Your Own

Now You're Ready
Exercises 6–9

Solve the equation. Check your solution.

1. $-3z + 1 = 7$ 2. $\dfrac{1}{2}x - 9 = -25$ 3. $-4n - 8n + 17 = 23$

EXAMPLE **3** | **Using the Distributive Property to Solve an Equation**

Solve $2(1 - 5x) + 4 = -8$.

$2(1 - 5x) + 4 = -8$	Write the equation.
$2(1) - 2(5x) + 4 = -8$	Distributive Property
$2 - 10x + 4 = -8$	Multiply.
$-10x + 6 = -8$	Combine like terms.
$\underline{\ -6\quad -6}$	Subtraction Property of Equality
$-10x = -14$	Simplify.
$\dfrac{-10x}{-10} = \dfrac{-14}{-10}$	Division Property of Equality
$x = 1.4$	Simplify.

Study Tip

Here is another way to solve the equation in Example 3.

$2(1 - 5x) + 4 = -8$
$2(1 - 5x) = -12$
$1 - 5x = -6$
$-5x = -7$
$x = 1.4$

EXAMPLE **4** | **Real-Life Application**

Use the table to find the number of miles x you need to run on Friday so that the mean number of miles run per day is 1.5.

Day	Miles
Monday	2
Tuesday	0
Wednesday	1.5
Thursday	0
Friday	x

Write an equation using the definition of *mean*.

sum of the data

number of values

$\dfrac{2 + 0 + 1.5 + 0 + x}{5} = 1.5$ Write the equation.

$\dfrac{3.5 + x}{5} = 1.5$ Combine like terms.

Undo the division. $\longrightarrow 5 \cdot \dfrac{3.5 + x}{5} = 5 \cdot 1.5$ Multiplication Property of Equality

$3.5 + x = 7.5$ Simplify.

Undo the addition. $\longrightarrow \underline{-3.5 \qquad\quad -3.5}$ Subtraction Property of Equality

$x = 4$ Simplify.

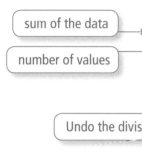

So, you need to run 4 miles on Friday.

● **On Your Own**

Now You're Ready
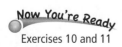
Exercises 10 and 11

Solve the equation. Check your solution.

4. $-3(x + 2) + 5x = -9$ **5.** $5 + 1.5(2d - 1) = 0.5$

6. You scored 88, 92, and 87 on three tests. Write and solve an equation to find the score you need on the fourth test so that your mean test score is 90.

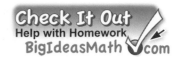

Vocabulary and Concept Check

1. **WRITING** Write the verbal statement as an equation. Then solve.

   > 2 more than 3 times a number is 17.

2. **OPEN-ENDED** Explain how to solve the equation $2(4x - 11) + 9 = 19$.

Practice and Problem Solving

CHOOSE TOOLS Find the value of the variable. Then find the angle measures of the polygon. Use a protractor to check the reasonableness of your answer.

3.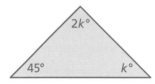

   $2k°$

   $45°$ $k°$

   Sum of angle measures: 180°

4.

   $a°$

   $2a°$ $2a°$

   $a°$

   Sum of angle measures: 360°

5.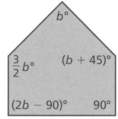

   $b°$

   $\frac{3}{2}b°$ $(b + 45)°$

   $(2b - 90)°$ $90°$

   Sum of angle measures: 540°

Solve the equation. Check your solution.

① ② **6.** $10x + 2 = 32$

7. $19 - 4c = 17$

8. $1.1x + 1.2x - 5.4 = -10$

9. $\frac{2}{3}h - \frac{1}{3}h + 11 = 8$

③ **10.** $6(5 - 8v) + 12 = -54$

11. $21(2 - x) + 12x = 44$

12. **ERROR ANALYSIS** Describe and correct the error in solving the equation.

    ✗
    $-2(7 - y) + 4 = -4$
    $-14 - 2y + 4 = -4$
    $-10 - 2y = -4$
    $-2y = 6$
    $y = -3$

13. **WATCHES** The cost C (in dollars) of making n watches is represented by $C = 15n + 85$. How many watches are made when the cost is $385?

14. **HOUSE** The height of the house is 26 feet. What is the height x of each story?

6 ft

x

x

In Exercises 15–17, write and solve an equation to answer the question.

15. **POSTCARD** The area of the postcard is 24 square inches. What is the width b of the message (in inches)?

16. **BREAKFAST** You order two servings of pancakes and a fruit cup. The cost of the fruit cup is $1.50. You leave a 15% tip. Your total bill is $11.50. How much does one serving of pancakes cost?

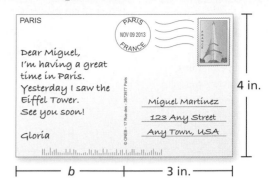

4 in.

b ─── 3 in. ───

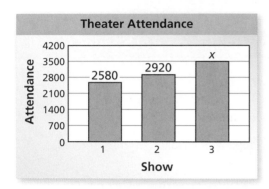

Theater Attendance

17. **THEATER** How many people must attend the third show so that the average attendance per show is 3000?

18. **DIVING** Divers in a competition are scored by an international panel of judges. The highest and the lowest scores are dropped. The total of the remaining scores is multiplied by the degree of difficulty of the dive. This product is multiplied by 0.6 to determine the final score.

   a. A diver's final score is 77.7. What is the degree of difficulty of the dive?

Judge	Russia	China	Mexico	Germany	Italy	Japan	Brazil
Score	7.5	8.0	6.5	8.5	7.0	7.5	7.0

   b. **Critical Thinking** The degree of difficulty of a dive is 4.0. The diver's final score is 97.2. Judges award half or whole points from 0 to 10. What scores could the judges have given the diver?

Fair Game Review *What you learned in previous grades & lessons*

Let $a = 3$ and $b = -2$. Copy and complete the statement using <, >, or =.
(Section 1.2, Section 1.3, and Section 1.4)

19. $-5a$ ▢ 4

20. 5 ▢ $b + 7$

21. $a - 4$ ▢ $10b + 8$

22. **MULTIPLE CHOICE** What value of x makes the equation $x + 5 = 2x$ true?
   (Skills Review Handbook)

   Ⓐ -1 Ⓑ 0 Ⓒ 3 Ⓓ 5

COMMON CORE

Solving Equations

In this lesson, you will
- solve equations with variables on both sides.
- determine whether equations have no solution or infinitely many solutions.

Learning Standards
8.EE.7a
8.EE.7b

 **Key Idea**

Solving Equations with Variables on Both Sides

To solve equations with variables on both sides, collect the variable terms on one side and the constant terms on the other side.

EXAMPLE ① **Solving an Equation with Variables on Both Sides**

Solve $15 - 2x = -7x$. Check your solution.

$$15 - 2x = -7x$$ Write the equation.

Undo the subtraction. → $\underline{+ 2x \quad + 2x}$ Addition Property of Equality

$$15 = -5x$$ Simplify.

Undo the multiplication. → $\dfrac{15}{-5} = \dfrac{-5x}{-5}$ Division Property of Equality

$$-3 = x$$ Simplify.

∴ The solution is $x = -3$.

Check

$$15 - 2x = -7x$$

$$15 - 2(-3) \overset{?}{=} -7(-3)$$

$$21 = 21 \checkmark$$

EXAMPLE ② **Using the Distributive Property to Solve an Equation**

Solve $-2(x - 5) = 6\left(2 - \dfrac{1}{2}x\right)$.

$$-2(x - 5) = 6\left(2 - \dfrac{1}{2}x\right)$$ Write the equation.

$$-2x + 10 = 12 - 3x$$ Distributive Property

Undo the subtraction. → $\underline{+ 3x \qquad\qquad + 3x}$ Addition Property of Equality

$$x + 10 = 12$$ Simplify.

Undo the addition. → $\underline{\quad - 10 \quad - 10}$ Subtraction Property of Equality

$$x = 2$$ Simplify.

∴ The solution is $x = 2$.

● **On Your Own**

 *Now You're Ready*
Exercises 6–14

Solve the equation. Check your solution.

1. $-3x = 2x + 19$ **2.** $2.5y + 6 = 4.5y - 1$ **3.** $6(4 - z) = 2z$

Some equations do not have one solution. Equations can also have no solution or infinitely many solutions.

When solving an equation that has no solution, you will obtain an equivalent equation that is not true for any value of the variable, such as $0 = 2$.

EXAMPLE **3** **Solving Equations with No Solution**

Solve $3 - 4x = -7 - 4x$.

$$3 - 4x = -7 - 4x \qquad \text{Write the equation.}$$

Undo the subtraction. $\longrightarrow$ $\underline{\quad + 4x \qquad + 4x\quad}$ Addition Property of Equality

$$3 = -7 \;\; \textbf{✗} \qquad \text{Simplify.}$$

⋮ The equation $3 = -7$ is never true. So, the equation has no solution.

When solving an equation that has infinitely many solutions, you will obtain an equivalent equation that is true for all values of the variable, such as $-5 = -5$.

EXAMPLE **4** **Solving Equations with Infinitely Many Solutions**

Solve $6x + 4 = 4\left(\dfrac{3}{2}x + 1\right)$.

$$6x + 4 = 4\left(\dfrac{3}{2}x + 1\right) \qquad \text{Write the equation.}$$

$$6x + 4 = 6x + 4 \qquad \text{Distributive Property}$$

Undo the addition. $\longrightarrow$ $\underline{\quad - 6x \qquad - 6x\quad}$ Subtraction Property of Equality

$$4 = 4 \qquad \text{Simplify.}$$

⋮ The equation $4 = 4$ is always true. So, the equation has infinitely many solutions.

● **On Your Own**

Now You're Ready
Exercises 18–29

Solve the equation.

4. $2x + 1 = 2x - 1$

5. $\dfrac{1}{2}(6t - 4) = 3t - 2$

6. $\dfrac{1}{3}(2b + 9) = \dfrac{2}{3}\left(b + \dfrac{9}{2}\right)$

7. $6(5 - 2v) = -4(3v + 1)$

Writing and Solving an Equation

The squares are identical. What is the area of each square?

Ⓐ 2   Ⓑ 4   Ⓒ 16   Ⓓ 32

The squares are identical, so the side length of each square is the same.

$$x + 2 = 2x \qquad \text{Write an equation.}$$

$$\underline{-x \qquad \quad -x} \qquad \text{Subtraction Property of Equality}$$

$$2 = x \qquad \text{Simplify.}$$

Because the side length of each square is 4, the area of each square is $4^2 = 16$.

x + 2

2x

So, the correct answer is Ⓒ.

EXAMPLE 6   Real-Life Application

A boat travels x miles per hour upstream on the Mississippi River. On the return trip, the boat travels 2 miles per hour faster. How far does the boat travel upstream?

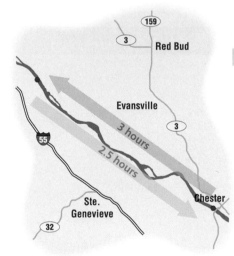

The speed of the boat on the return trip is $(x + 2)$ miles per hour.

Distance upstream = Distance of return trip

$$3x = 2.5(x + 2) \qquad \text{Write an equation.}$$

$$3x = 2.5x + 5 \qquad \text{Distributive Property}$$

$$\underline{-2.5x \quad -2.5x} \qquad \text{Subtraction Property of Equality}$$

$$0.5x = 5 \qquad \text{Simplify.}$$

$$\frac{0.5x}{0.5} = \frac{5}{0.5} \qquad \text{Division Property of Equality}$$

$$x = 10 \qquad \text{Simplify.}$$

The boat travels 10 miles per hour for 3 hours upstream. So, it travels 30 miles upstream.

On Your Own

8. **WHAT IF?** In Example 5, the side length of the purple square is $3x$. What is the area of each square?

9. A boat travels x miles per hour from one island to another island in 2.5 hours. The boat travels 5 miles per hour faster on the return trip of 2 hours. What is the distance between the islands?

 Vocabulary and Concept Check

1. **WRITING** Is $x = 3$ a solution of the equation $3x - 5 = 4x - 9$? Explain.

2. **OPEN-ENDED** Write an equation that has variables on both sides and has a solution of -3.

 Practice and Problem Solving

The value of the solid's surface area is equal to the value of the solid's volume. Find the value of x.

3.

11 in.   3 in.

4.
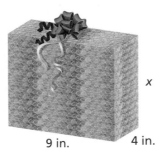
9 in.   4 in.

5.

6 in.
5 in.
x

Solve the equation. Check your solution.

 6. $m - 4 = 2m$

7. $3k - 1 = 7k + 2$

8. $6.7x = 5.2x + 12.3$

9. $-24 - \dfrac{1}{8}p = \dfrac{3}{8}p$

10. $12(2w - 3) = 6w$

11. $2(n - 3) = 4n + 1$

12. $2(4z - 1) = 3(z + 2)$

13. $0.1x = 0.2(x + 2)$

14. $\dfrac{1}{6}d + \dfrac{2}{3} = \dfrac{1}{4}(d - 2)$

15. **ERROR ANALYSIS** Describe and correct the error in solving the equation.

$$✗$$
$$3x - 4 = 2x + 1$$
$$3x - 4 - 2x = 2x + 1 - 2x$$
$$x - 4 = 1$$
$$x - 4 + 4 = 1 - 4$$
$$x = -3$$

16. **TRAIL MIX** The equation $4.05p + 14.40 = 4.50(p + 3)$ represents the number p of pounds of peanuts you need to make trail mix. How many pounds of peanuts do you need for the trail mix?

17. **CARS** Write and solve an equation to find the number of miles you must drive to have the same cost for each of the car rentals.

$15 plus $0.50 per mile

$25 plus $0.25 per mile

Solve the equation. Check your solution, if possible.

③ ④ 18. $x + 6 = x$

19. $3x - 1 = 1 - 3x$

20. $4x - 9 = 3.5x - 9$

21. $\frac{1}{2}x + \frac{1}{2}x = x + 1$

22. $3x + 15 = 3(x + 5)$

23. $\frac{1}{3}(9x + 3) = 3x + 1$

24. $5x - 7 = 4x - 1$

25. $2x + 4 = -(-7x + 6)$

26. $5.5 - x = -4.5 - x$

27. $10x - \frac{8}{3} - 4x = 6x$

28. $-3(2x - 3) = -6x + 9$

29. $6(7x + 7) = 7(6x + 6)$

30. ERROR ANALYSIS Describe and correct the error in solving the equation.

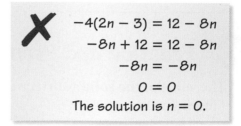

$$-4(2n - 3) = 12 - 8n$$
$$-8n + 12 = 12 - 8n$$
$$-8n = -8n$$
$$0 = 0$$
The solution is $n = 0$.

31. OPEN-ENDED Write an equation with variables on both sides that has no solution. Explain why it has no solution.

32. GEOMETRY Are there any values of x for which the areas of the figures are the same? Explain.

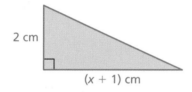

2 cm

$(x + 1)$ cm

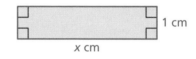

1 cm

x cm

33. SATELLITE TV Provider A charges $75 for installation and charges $39.95 per month for the basic package. Provider B offers free installation and charges $39.95 per month for the basic package. Your neighbor subscribes to Provider A the same month you subscribe to Provider B. After how many months is your neighbor's total cost the same as your total cost for satellite TV?

34. PIZZA CRUST Pepe's Pizza makes 52 pizza crusts the first week and 180 pizza crusts each subsequent week. Dianne's Delicatessen makes 26 pizza crusts the first week and 90 pizza crusts each subsequent week. In how many weeks will the total number of pizza crusts made by Pepe's Pizza equal twice the total number of pizza crusts made by Dianne's Delicatessen?

35. PRECISION Is the triangle an equilateral triangle? Explain.

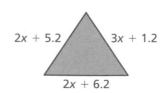

$2x + 5.2$ $3x + 1.2$

$2x + 6.2$

A polygon is *regular* if each of its sides has the same length. Find the perimeter of the regular polygon.

36.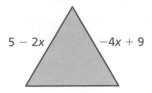

$5 - 2x$ $-4x + 9$

37.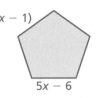

$3(x - 1)$

$5x - 6$

38.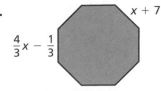

$x + 7$

$\frac{4}{3}x - \frac{1}{3}$

39. PRECISION The cost of mailing a DVD in an envelope by Express Mail® is equal to the cost of mailing a DVD in a box by Priority Mail®. What is the weight of the DVD with its packing material? Round your answer to the nearest hundredth.

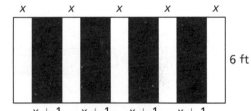

	Packing Material	Priority Mail®	Express Mail®
Box	$2.25	$2.50 per lb	$8.50 per lb
Envelope	$1.10	$2.50 per lb	$8.50 per lb

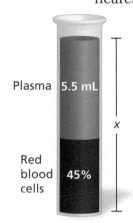

Plasma 5.5 mL

x

Red blood cells 45%

40. PROBLEM SOLVING Would you solve the equation $0.25x + 7 = \frac{1}{3}x - 8$ using fractions or decimals? Explain.

41. BLOOD SAMPLE The amount of red blood cells in a blood sample is equal to the total amount in the sample minus the amount of plasma. What is the total amount x of blood drawn?

42. NUTRITION One serving of oatmeal provides 16% of the fiber you need daily. You must get the remaining 21 grams of fiber from other sources. How many grams of fiber should you consume daily?

43. ⟨Geometry⟩ A 6-foot-wide hallway is painted as shown, using equal amounts of white and black paint.

a. How long is the hallway?

b. Can this same hallway be painted with the same pattern, but using twice as much black paint as white paint? Explain.

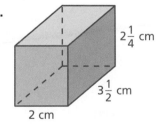

x x x x x

6 ft

$x + 1$ $x + 1$ $x + 1$ $x + 1$

 Fair Game Review *What you learned in previous grades & lessons*

Find the volume of the prism. *(Skills Review Handbook)*

44.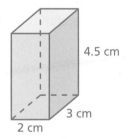

4.5 cm

3 cm

2 cm

45.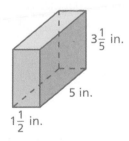

$2\frac{1}{4}$ cm

$3\frac{1}{2}$ cm

2 cm

46.

$3\frac{1}{5}$ in.

5 in.

$1\frac{1}{2}$ in.

47. MULTIPLE CHOICE A car travels 480 miles on 15 gallons of gasoline. How many miles does the car travel per gallon? *(Skills Review Handbook)*

Ⓐ 28 mi/gal Ⓑ 30 mi/gal Ⓒ 32 mi/gal Ⓓ 35 mi/gal

Topic 3 Rewriting Equations and Formulas

Check It Out
Lesson Tutorials
BigIdeasMath.com

Key Vocabulary
literal equation,
p. 774

An equation that has two or more variables is called a **literal equation**. To rewrite a literal equation, solve for one variable in terms of the other variable(s).

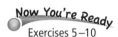

EXAMPLE **1** **Rewriting an Equation**

Solve the equation $2y + 5x = 6$ for y.

$2y + 5x = 6$	Write the equation.

Undo the addition. ⟶ $2y + 5x - 5x = 6 - 5x$ Subtraction Property of Equality

$2y = 6 - 5x$ Simplify.

Undo the multiplication. ⟶ $\dfrac{2y}{2} = \dfrac{6 - 5x}{2}$ Division Property of Equality

$y = 3 - \dfrac{5}{2}x$ Simplify.

On Your Own

Now You're Ready
Exercises 5–10

Solve the equation for y.

1. $5y - x = 10$ **2.** $4x - 4y = 1$ **3.** $12 = 6x + 3y$

EXAMPLE **2** **Rewriting a Formula**

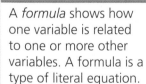

Remember

A *formula* shows how one variable is related to one or more other variables. A formula is a type of literal equation.

The formula for the perimeter P of a rectangle is $P = 2\ell + 2w$. Solve the formula for the length ℓ.

$P = 2\ell + 2w$ Write the formula.

$P - 2w = 2\ell + 2w - 2w$ Subtraction Property of Equality

$P - 2w = 2\ell$ Simplify.

$\dfrac{P - 2w}{2} = \dfrac{2\ell}{2}$ Division Property of Equality

$\dfrac{P - 2w}{2} = \ell$ Simplify.

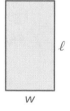

On Your Own

Now You're Ready
Exercises 14–19

Solve the formula for the red variable.

4. Area of rectangle: $A = bh$ **5.** Simple interest: $I = Prt$

6. Height of a dropped object: $h = -16t^2 + s$

Multi-Language Glossary at BigIdeasMath.com

COMMON CORE

Solving Equations
In this lesson, you will

• rewrite equations to solve for one variable in terms of the other variable(s).

Applying Standard
8.EE.7

 **Key Idea**

Temperature Conversion

A formula for converting from degrees Fahrenheit F to degrees Celsius C is

$$C = \frac{5}{9}(F - 32).$$

EXAMPLE ③ Rewriting the Temperature Formula

Solve the temperature formula for F.

$$C = \frac{5}{9}(F - 32) \qquad \text{Write the temperature formula.}$$

Use the reciprocal. ⟶ $\dfrac{9}{5} \cdot C = \dfrac{9}{5} \cdot \dfrac{5}{9}(F - 32)$ Multiplication Property of Equality

$$\frac{9}{5}C = F - 32 \qquad \text{Simplify.}$$

Undo the subtraction. ⟶ $\dfrac{9}{5}C + 32 = F - 32 + 32$ Addition Property of Equality

$$\frac{9}{5}C + 32 = F \qquad \text{Simplify.}$$

∴ The rewritten formula is $F = \dfrac{9}{5}C + 32$.

EXAMPLE ④ Real-Life Application

Sun
11,000°F

Lightning
30,000°C

Which has the greater temperature?

Convert the Celsius temperature of lightning to Fahrenheit.

$$F = \frac{9}{5}C + 32 \qquad \text{Write the rewritten formula from Example 3.}$$

$$= \frac{9}{5}(30,000) + 32 \qquad \text{Substitute 30,000 for } C.$$

$$= 54,032 \qquad \text{Simplify.}$$

∴ Because 54,032°F is greater than 11,000°F, lightning has the greater temperature.

On Your Own

7. Room temperature is considered to be 70°F. Suppose the temperature is 23°C. Is this greater than or less than room temperature?

Check It Out
Help with Homework
BigIdeasMath  com

 Vocabulary and Concept Check

1. **VOCABULARY** Is $-2x = \dfrac{3}{8}$ a literal equation? Explain.

2. **DIFFERENT WORDS, SAME QUESTION** Which is different? Find "both" answers.

   | Solve $4x - 2y = 6$ for y. | | Solve $6 = 4x - 2y$ for y. |

   | Solve $4x - 2y = 6$ for y in terms of x. | | Solve $4x - 2y = 6$ for x in terms of y. |

 Practice and Problem Solving

3. **a.** Write a formula for the area A of a triangle.

   b. Solve the formula for b.

   c. Use the new formula to find the base of the triangle.

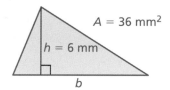

4. **a.** Write a formula for the volume V of a prism.

   b. Solve the formula for B.

   c. Use the new formula to find the area of the base of the prism.

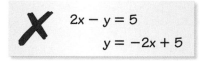

Solve the equation for y.

① 5. $\dfrac{1}{3}x + y = 4$

6. $3x + \dfrac{1}{5}y = 7$

7. $6 = 4x + 9y$

8. $3 = 7x - 2y$

9. $4.2x - 1.4y = 2.1$

10. $6y - 1.5x = 8$

11. **ERROR ANALYSIS** Describe and correct the error in rewriting the equation.

   ✗ $2x - y = 5$
   $y = -2x + 5$

12. **TEMPERATURE** The formula $K = C + 273.15$ converts temperatures from Celsius C to Kelvin K.

   a. Solve the formula for C.

   b. Convert 300 Kelvin to Celsius.

13. **INTEREST** The formula for simple interest is $I = Prt$.

   a. Solve the formula for t.

   b. Use the new formula to find the value of t in the table.

I	$75
P	$500
r	5%
t	

Solve the equation for the red variable.

14. $d = rt$

15. $e = mc^2$

16. $R - C = P$

17. $A = \frac{1}{2}h(b_1 + b_2)$

18. $B = 3\frac{V}{h}$

19. $g = \frac{1}{6}(w + 40)$

20. LOGIC Why is it useful to rewrite a formula in terms of another variable?

21. REASONING The formula $K = \frac{5}{9}(F - 32) + 273.15$ converts temperatures from Fahrenheit F to Kelvin K.

  a. Solve the formula for F.

  b. The freezing point of water is 273.15 Kelvin. What is this temperature in Fahrenheit?

  c. The temperature of dry ice is $-78.5\ °C$. Which is colder, dry ice or liquid nitrogen?

Liquid nitrogen

77.35 K

Navy Pier Ferris Wheel

C = 439.6 ft

22. FERRIS WHEEL The distance around a circle is called the *circumference*. The Navy Pier Ferris Wheel in Chicago has a circumference C that is 56% of the circumference of the first Ferris wheel built in 1893.

  a. The *radius* of a circle is the distance from the center to any point on the circle. The circumference of a circle is about 6.28 times the radius. What is the radius of the Navy Pier Ferris Wheel?

  b. What was the radius of the first Ferris wheel?

  c. The first Ferris wheel took 9 minutes to make a complete revolution. How fast was the wheel moving?

23. *Repeated Reasoning* The *radius* of a sphere is the distance from the center to any point on the sphere. The formula for the volume V of a sphere is $V = \frac{4}{3}\pi r^3$, where r is the radius and π is a constant whose value is about 3.14. Solve the formula for r^3. Use Guess, Check, and Revise to find the radius of the sphere.

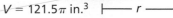

$V = 121.5\pi$ in.^{3} $\longmapsto r \longmapsto$

Fair Game Review *What you learned in previous grades & lessons*

Multiply. *(Skills Review Handbook)*

24. $5 \times \frac{3}{4}$

25. $-2 \times \frac{8}{3}$

26. $\frac{1}{4} \times \frac{3}{2} \times \frac{8}{9}$

27. $25 \times \frac{3}{5} \times \frac{1}{12}$

28. MULTIPLE CHOICE Which of the following is not equivalent to $\frac{3}{4}$?
(Skills Review Handbook)

  Ⓐ 0.75   **Ⓑ** 3 : 4   **Ⓒ** 75%   **Ⓓ** 4 : 3

Appendix A
My Big Ideas Projects

My Big Ideas Projects

The Mathematics of Jules Verne

1 Project Overview

Jules Verne (1828–1905) was a famous French science fiction writer. He wrote about space, air, and underwater travel before aircraft and submarines were commonplace, and before any means of space travel had been devised.

For example, in his 1865 novel *From the Earth to the Moon*, he wrote about three astronauts who were launched from Florida and recovered through a splash landing. The first actual moon landing wasn't until 1969.

Essential Question How does the knowledge of mathematics influence science fiction writing?

Read one of Jules Verne's science fiction novels. Then write a book report about some of the mathematics used in the novel.

Sample: A league is an old measure of distance. It is approximately equal to 4 kilometers. You can convert 20,000 leagues to miles as follows.

$$20{,}000 \text{ leagues} \cdot \frac{4 \text{ km}}{1 \text{ league}} \cdot \frac{1 \text{ mile}}{1.61 \text{ km}} \approx 50{,}000 \text{ miles}$$

2   Things to Include

- Describe the major events in the plot.

- Write a brief paragraph describing the setting of the story.

- List and identify the main characters. Explain the contribution of each character to the story.

- Explain the major conflict in the story.

- Describe at least four examples of mathematics used in the story.

- Which of Jules Verne's scientific predictions have come true since he wrote the novel?

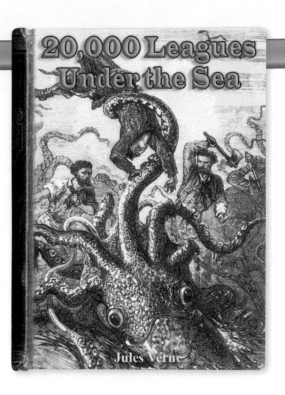

Jules Verne (1828–1905)

3   Things to Remember

- You can download one of Jules Verne's novels at *BigIdeasMath.com*.

- Add your own illustrations to your project.

- Organize your report in a folder, and think of a title for your report.

Mathematics in Ancient Greece

1 Getting Started

The ancient Greek period began around 1100 B.C. and lasted until the Roman conquest of Greece in 146 B.C.

The civilization of the ancient Greeks influenced the languages, politics, educational systems, philosophy, science, mathematics, and arts of Western Civilization. It was a primary force in the birth of the Renaissance in Europe between the 14th and 17th centuries.

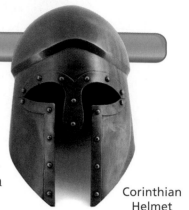

Corinthian Helmet

Essential Question How do you use mathematical knowledge that was originally discovered by the Greeks?

Sample: Ancient Greek symbols for the numbers from 1 through 10 are shown in the table.

I	II	III	IIII	Γ	ΓI	ΓII	ΓIII	ΓIIII	△
1	2	3	4	5	6	7	8	9	10

These same symbols were used to write the numbers between 11 and 39. Here are some examples.

$$\triangle\ \Gamma\ \text{III} = 18 \qquad \triangle\triangle\triangle\Gamma = 35 \qquad \triangle\triangle\text{IIII} = 24$$

Alexander the Great

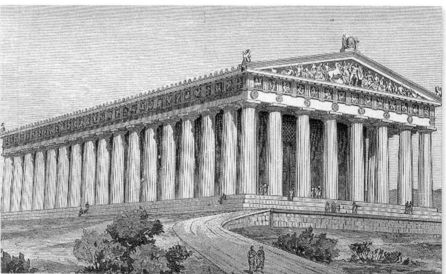

Parthenon

Things to Include

- Describe at least one contribution that each of the following people made to mathematics.

  Pythagoras (c. 570 B.C.–c. 490 B.C.)

  Aristotle (c. 384 B.C.–c. 322 B.C.)

  Euclid (c. 300 B.C.)

  Archimedes (c. 287 B.C.–c. 212 B.C.)

  Eratosthenes (c. 276 B.C.–c. 194 B.C.)

- Which of the people listed above was the teacher of Alexander the Great? What subjects did Alexander the Great study when he was in school?

- How did the ancient Greeks represent fractions?

- Describe how the ancient Greeks used mathematics. How does this compare with the ways in which mathematics is used today?

A	α	alpha	N	ν	nu
B	β	beta	Ξ	ξ	xi
Γ	γ	gamma	O	o	omicron
Δ	δ	delta	Π	π	pi
E	ε	epsilon	P	ρ	rho
Z	ζ	zeta	Σ	σ	sigma
H	η	eta	T	τ	tau
Θ	θ	theta	Υ	υ	upsilon
I	ι	iota	Φ	φ	phi
K	κ	kappa	X	χ	chi
Λ	λ	lambda	Ψ	ψ	psi
M	μ	mu	Ω	ω	omega

3 Things to Remember

- Add your own illustrations to your project.

- Try to include as many different math concepts as possible. Your goal is to include at least one concept from each of the chapters you studied this year.

- Organize your report in a folder, and think of a title for your report.

Greek Pottery

Trireme Greek Warship

A.3 Art Project

Circle Art

1 Getting Started

Circles have been used in art for thousands of years.

Essential Question How have circles influenced ancient and modern art?

Find examples of art in which circles were used. Describe how the artist might have used properties of circles to make each piece of art.

Sample: Here is a technique for making a pattern that uses circles in a stained glass window.

1. Use a compass to draw a circle.

2. Without changing the radius, draw a second circle whose center lies on the first circle.

3. Without changing the radius, draw a third circle whose center is one of the points of intersection.

4. Without changing the radius, continue to draw new circles at the points of intersection.

5. Color the design to make a pattern.

Mexican Tile

Ancient Roman Mosaic

2 Things to Include

- Describe how ancient artists drew circles.

- Describe the symbolism of circles in ancient art.

- Find examples of how circles are used to create mosaic tile patterns.

- Measure the angles that are formed by the patterns in the circle art you find. For instance, you might describe the angles formed by the netting in the Native American dreamcatcher.

- Find examples of the use of circles in modern art.

- Use circles to create your own art. Describe how you used mathematics and the properties of circles to make your art.

Native American
Dreamcatcher

3 Things to Remember

- Add your own illustrations to your project.

- Try to include as many different math concepts as possible. Your goal is to include at least one concept from each of the chapters you studied this year.

- Organize your report in a folder, and think of a title for your report.

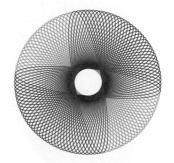

A.4 Science Project

Classifying Animals

1 Getting Started

Biologists classify animals by placing them in phylums, or groups, with similar characteristics. Latin names, such as Chordata (having a spinal cord) or Arthropoda (having jointed limbs and rigid bodies) are used to describe these groups.

Biological classification is difficult, and scientists are still developing a complete system. There are seven main ranks of life on Earth; kingdom, phylum, class, order, family, genus, and species. However, scientists usually use more than these seven ranks to classify organisms.

Essential Question How does the classification of living organisms help you understand the similarities and differences of animals?

Write a report about how animals are classified. Choose several different animals and list the phylum, class, and order of each animal.

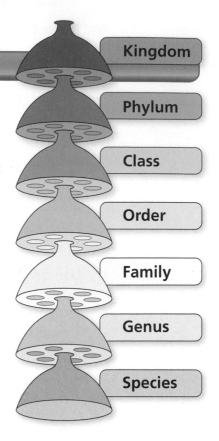

Kingdom

Phylum

Class

Order

Family

Genus

Species

Wasp
Phylum: Arthropoda
Class: Insecta
Order: Hymenoptera
   (membranous wing)

Sample: A bat is classified as an animal in the phylum Chordata, class Mammalia, and order Chiroptera. *Chiroptera* is a Greek word meaning "hand-wing."

Bat
Phylum: Chordata
Class: Mammalia
Order: Chiroptera
   (hand-wing)

Monkey
Phylum: Chordata
Class: Mammalia
Order: Primate (large brain)

Kangaroo
Phylum: Chordata
Class: Mammalia
Order: Diprotodontia
   (two front teeth)

2 Things to Include

- List the different classes of phylum Chordata. Have you seen a member of each class?

- List the different classes of phylum Arthropoda. Have you seen a member of each class?

- Show how you can use graphic organizers to help classify animals. Which types of graphic organizers seem to be most helpful? Explain your reasoning.

- Summarize the number of species in each phlyum in an organized way. Be sure to include fractions, decimals, and percents.

Parrot
Phylum: Chordata
Class: Aves
Order: Psittaciformes
       (strong, curved bill)

Frog
Phylum: Chordata
Class: Amphibia
Order: Anura (no tail)

Spider
Phylum: Arthropoda
Class: Arachnida
Order: Araneae (spider)

3 Things to Remember

- Organize your report in a folder, and think of a title for your report.

Lobster
Phylum: Arthropoda
Class: Malacostraca
Order: Decopada (ten footed)

Crocodile
Phylum: Chordata
Class: Reptilia
Order: Crocodilia
      (pebble-worm)

Cougar
Phylum: Chordata
Class: Mammalia
Order: Carnivora (meat eater)

Selected Answers

Section 1.1   Integers and Absolute Value
(pages 6 and 7)

1. 9, −1, 15

3. −6; All of the other expressions are equal to 6.

5. 6   **7.** 10   **9.** 13   **11.** 12   **13.** 8   **15.** 18

17. 45   **19.** 125   **21.** $|-4| < 7$ **23.** $|-4| > -6$ **25.** $|5| = |-5|$

27. Because $|-5| = 5$, the statement is incorrect; $|-5| > 4$

29. −8, 5   **31.** −7, −6, $|5|$, $|-6|$, 8   **33.** −17, $|-11|$, $|20|$, 21, $|-34|$

35. −4

37. a. MATE;      **b.** TEAM;

39. $n \geq 0$ **41.** The number closer to 0 is the greater integer.

43. a. Player 3   **b.** Player 2   **c.** Player 1

45. false; The absolute value of zero is zero, which is neither positive nor negative.

47. 144   **49.** 3170

Section 1.2   Adding Integers
(pages 12 and 13)

1. Change the sign of the integer.

3. positive; 20 has the greater absolute value and is positive.

5. negative; The common sign is a negative sign.

7. false; A positive integer and its absolute value are equal, not opposites.

9. −10   **11.** 7   **13.** 0   **15.** 10

17. −4   **19.** −11   **21.** −4   **23.** −34

25. −10 and −10 are not opposites; $-10 + (-10) = -20$ **27.** \$48

29. Use the Associative Property to add 13 and −13 first; −8

31. *Sample answer:* Use the Commutative Property to switch the last two terms; −12

33. *Sample answer:* Use the Commutative Property to switch the last two terms; 11

35. −27   **37.** 21   **39.** −85

41. *Sample answer:* $-26 + 1$; $-12 + (-13)$ **43.** −3

45. $d = -10$ **47.** $m = -7$

49. Find the number in each row or column that already has two numbers in it before guessing.

51. 8   **53.** 183

Section 1.3

Subtracting Integers
(pages 18 and 19)

1. You add the integer's opposite.

3. What is 3 less than -2?; -5; 5

5. C

7. B

9. 13

11. -5

13. -10

15. 3

17. 17

19. 1

21. -22

23. -20

25. $-3 - 9$

27. 6

29. 9

31. 8

33. $m = 14$

35. $c = 15$

37. 2

39. 3

41. *Sample answer:* $x = -2, y = -1$; $x = -3, y = -2$

43. sometimes; It's positive only if the first integer is greater.

45. always; It's always positive because the first integer is always greater.

47. all values of a and b

49. when a and b have the same sign and $|a| \geq |b|$ or $b = 0$

51. -45

53. 468

55. 2378

Section 1.4

Multiplying Integers
(pages 26 and 27)

1. a. They are the same. **b.** They are different.

3. negative; different signs

5. negative; different signs

7. false; The product of the first two negative integers is positive. The product of the positive result and the third negative integer is negative.

9. -21 **11.** 12 **13.** 27 **15.** 12 **17.** 0 **19.** -30

21. 78 **23.** 121 **25.** $-240,000$ **27.** 54 **29.** -700 **31.** 0

33. -1 **35.** -36 **37.** 54

39. The answer should be negative; $-10^2 = -(10 \cdot 10) = -100$

41. 32 **43.** $-7500, 37,500$ **45.** -12

47. a.

Month	Price of Skates	
June	165	$= \$165$
July	$165 + (-12)$	$= \$153$
August	$165 + 2(-12)$	$= \$141$
September	$165 + 3(-12)$	$= \$129$

b. The price drops $12 every month.

c. no; yes; In August you have $135 but the cost is $141. In September you have $153 and the cost is only $129.

49. 3

51. 14

53. D

Dividing Integers
(pages 32 and 33)

1. They have the same sign; They have different signs; The dividend is zero.

3. *Sample answer:* $-4, 2$ 5. negative   7. negative

9. -3 11. 6   13. 0   15. -6 17. 7   19. -10

21. undefined   23. 12

25. The quotient should be 0; $0 \div (-5) = 0$ 27. 15 pages

29. -8 31. 65   33. 5

35. 4   37. -400 ft/min   39. 5

41. *Sample answer:* $-20, -15, -10, -5, 0$; Start with -10, then pair -15 with -5 and -20 with 0.

43.   45. B

Rational Numbers
(pages 48 and 49)

1. no; The denominator cannot be 0.

3. rational numbers, integers   5. rational numbers, integers, whole numbers

7. repeating   9. terminating

11. 0.875   13. $-0.\overline{7}$ 15. $1.8\overline{3}$ 17. $-5.58\overline{3}$

19. The bar should be over both digits to the right of the decimal point; $-\dfrac{7}{11} = -0.\overline{63}$

21. $\dfrac{9}{20}$ 23. $-\dfrac{39}{125}$ 25. $-1\dfrac{16}{25}$ 27. $-12\dfrac{81}{200}$

29. $-2.5, -1.1, -\dfrac{4}{5}, 0.8, \dfrac{9}{5}$ 31. $-\dfrac{9}{4}, -0.75, -\dfrac{6}{10}, \dfrac{5}{3}, 2.1$

33. $-2.4, -2.25, -\dfrac{11}{5}, \dfrac{15}{10}, 1.6$ 35. spotted turtle

37. $-1.82 < -1.81$ 39. $-4\dfrac{6}{10} > -4.65$

41. $-2\dfrac{13}{16} < -2\dfrac{11}{14}$ 43. Michelle

45. no; The base of the skating pool is at -10 feet, which is deeper than $-9\dfrac{5}{6}$ feet.

47. **a.** when a is negative
    b. when a and b have the same sign, $a \neq 0 \neq b$

49. $\dfrac{7}{30}$ 51. 21.15

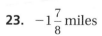

Section 2.2 — Adding Rational Numbers
(pages 54 and 55)

1. Because $|-8.46| > |5.31|$, subtract $|5.31|$ from $|-8.46|$ and the sign is negative.

3. What is the distance between -4.5 and 3.5?; 8; -1

5. $-1\frac{4}{5}$ **7.** $-\frac{5}{14}$ **9.** $-\frac{7}{12}$ **11.** 1.844

13. The decimals are not lined up correctly; Line up the decimals; -3.95

15. $-1\frac{5}{12}$ **17.** $1\frac{5}{12}$ **19.** $\frac{1}{2}$ **21.** $-9\frac{3}{4}$

23. The sum is an integer when the sum of the fractional parts of the numbers adds up to an integer.

25. less than; The water level for the three-month period compared to the normal level is $-1\frac{7}{16}$.

27. no; This is only true when a and b have the same sign.

29. Commutative Property of Addition; 7   **31.** Associative Property of Addition; $1\frac{1}{8}$

33. A

Section 2.3 — Subtracting Rational Numbers
(pages 62 and 63)

1. Instead of subtracting, add the opposite of $\frac{3}{5}$, $-\frac{3}{5}$. Then, add $\left|-\frac{4}{5}\right|$ and $\left|-\frac{3}{5}\right|$, and the sign is negative.

3. $1\frac{1}{2}$ **5.** -3.5 **7.** $-18\frac{13}{24}$ **9.** -2.6

11. 14.963 **13.** $3\frac{1}{4}$ **15.** $3\frac{1}{3}$

17. a. 410.7 feet   **b.** 136.9 feet per hour   **19.** 1.2

21. The difference is an integer when (1) the decimals have the same sign and the digits to the right of the decimal point are the same, or (2) the decimals have different signs and the sum of the decimal parts of the numbers add up to 1.

23. $-1\frac{7}{8}$ miles   **25.** Subtract the least number from the greatest number.

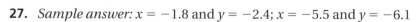

27. *Sample answer:* $x = -1.8$ and $y = -2.4$; $x = -5.5$ and $y = -6.1$

29. always; It's always positive because the first decimal is always greater.

31. 35.88 **33.** $8\frac{2}{3}$ **35.** C

Multiplying and Dividing Rational Numbers
(pages 68 and 69)

1. The same rules for signs of integers are applied to rational numbers.

3. positive   **5.** negative   **7.** $-\dfrac{4}{5}$ **9.** 0.25

11. $-\dfrac{2}{3}$ **13.** $-\dfrac{1}{100}$ **15.** $2\dfrac{5}{14}$ **17.** $3.\overline{63}$

19. -6 **21.** -2.5875 **23.** $-\dfrac{4}{9}$ **25.** 9

27. 0.025   **29.** $8\dfrac{1}{4}$

31. The answer should be negative; $-2.2 \times 3.7 = -8.14$

33. $-66°$ **35.** -19.59 **37.** -22.667 **39.** $-5\dfrac{11}{24}$

41. *Sample answer:* $-\dfrac{9}{10}, \dfrac{2}{3}$ **43.** $3\dfrac{5}{8}$ gal   **45.** -1.28 sec

47. -1.5 **49.** $4\dfrac{1}{2}$ **51.** D

Algebraic Expressions
(pages 84 and 85)

1. Terms of an expression are separated by addition. Rewrite the expression as $3y + (-4) + (-5y)$. The terms in the expression are $3y$, -4, and $-5y$.

3. no; The like terms $3x$ and $2x$ should be combined.
$$3x + 2x - 4 = (3 + 2)x - 4$$
$$= 5x - 4$$

5. Terms: $t, 8, 3t$; Like terms: t and $3t$ **7.** Terms: $2n, -n, -4, 7n$; Like terms: $2n, -n$, and $7n$

9. Terms: $1.4y, 5, -4.2, -5y^2, z$; Like terms: 5 and -4.2

11. $2x^2$ is not a like term because x is squared. The like terms are $3x$ and $9x$.

13. $11x + 2$ **15.** $-2.3v - 5$ **17.** $3 - \dfrac{1}{2}y$ **19.** $-p - 30$

21. $10.2x$; The weight carried by each hiker is 10.2 pounds.

23. yes; Both expressions simplify to $11x^2 + 3y$.

25. $(9 + 3x)$ ft^2

27. *Sample answer:*

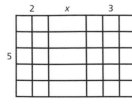

$5x + 25$

29. When you subtract the two red strips, you subtract their intersection twice. So, you need to add it back into the expression once.

Good to know.

31. 0.52 m, 0.545 m, 0.55 m, 0.6 m, 0.65 m

Section 3.2 — Adding and Subtracting Linear Expressions
(pages 90 and 91)

1. not linear; An exponent of a variable is not equal to 1.

3. not linear; An exponent of a variable is not equal to 1.

5. What is x more than $3x - 1$?; $4x - 1$; $2x - 1$

7. *Sample answer:* $(2x + 7) - (2x - 4) = 11$

9. $2b + 9$

11. $6x - 18$

13. 17

15. $m + 1$

17. $55w + 145$

19. $-3g - 4$

21. $-12y + 20$

23. $-2c$

25. The -3 was not distributed to both terms inside the parentheses.

$$(4m + 9) - 3(2m - 5) = 4m + 9 - 6m + 15$$
$$= 4m - 6m + 9 + 15$$
$$= -2m + 24$$

27. no; If the variable terms are opposites, the sum is a numerical expression.

29. $0.25x + 0.15$

31. $\left| x - 3 \right|$, or equivalently $\left| -x + 3 \right|$; 0; 6

33. $\dfrac{2}{5}$

35. D

Extension 3.2 — Factoring Expressions
(page 93)

1. $3(3 + 7)$

3. $2(4x + 1)$

5. $4(5z - 2)$

7. $4(9a + 4b)$

9. $\dfrac{1}{3}(b - 1)$

11. $2.2(x + 2)$

13. $-\dfrac{1}{2}(x - 12)$

15. $(3x - 8)$ ft

17. *Sample answer:* $2x - 1$ and x, $2x$ and $x - 1$

Section 3.3 — Solving Equations Using Addition or Subtraction *(pages 100 and 101)*

1. Subtraction Property of Equality

3. No, $m = -8$ not -2 in the first equation.

5. $a = 19$

7. $k = -20$

9. $c = 3.6$

11. $q = -\dfrac{1}{6}$

13. $g = -10$

15. $y = -2.08$

17. $q = -\dfrac{7}{18}$

19. $w = -1\dfrac{13}{24}$

21. The 8 should have been subtracted rather than added.

$$\begin{array}{r} x + 8 = 10 \\ -8 \quad -8 \\ \hline x = 2 \end{array}$$

23. $c + 10 = 3$; $c = -7$

25. $p - 6 = -14$; $p = -8$

27. $p + 2.54 = 1.38$; $-\$1.16$ million

29. $x + 8 = 12$; 4 cm

31. $x + 22.7 = 34.6$; 11.9 ft

33. Because your first jump is higher, your second jump went a farther distance than your first jump.

35. $m + 30.3 + 40.8 = 180$; 108.9°

37. -9 **39.** $6, -6$ **41.** -56

43. -9 **45.** B

Section 3.4

Solving Equations Using Multiplication or Division *(pages 106 and 107)*

1. Multiplication is the inverse operation of division, so it can undo division.

3. dividing by 5 **5.** multiplying by -8

7. $h = 5$ **9.** $n = -14$ **11.** $m = -2$ **13.** $x = -8$

15. $p = -8$ **17.** $n = 8$ **19.** $g = -16$ **21.** $f = 6\dfrac{3}{4}$

23. They should divide by -4.2.

$$-4.2x = 21$$

$$\dfrac{-4.2x}{-4.2} = \dfrac{21}{-4.2}$$

$$x = -5$$

25. $\dfrac{2}{5}x = \dfrac{3}{20}$; $x = \dfrac{3}{8}$

27. $\dfrac{x}{-1.5} = 21$; $x = -31.5$

29. $\dfrac{x}{30} = 12\dfrac{3}{5}$; 378 ft

31 and 33. Sample answers are given.

31. a. $-2x = 4.4$ **b.** $\dfrac{x}{1.1} = -2$

33. a. $4x = -5$ **b.** $\dfrac{x}{5} = -\dfrac{1}{4}$

35. $-1.26n = -10.08$; 8 days **37.** -50 ft **39.** $-5, 5$

41. -7 **43.** 12 **45.** B

Section 3.5

Solving Two-Step Equations *(pages 112 and 113)*

1. Eliminate the constants on the side with the variable. Then solve for the variable using either division or multiplication.

3. D **5.** A **7.** $b = -3$ **9.** $t = -4$

11. $g = 4.22$ **13.** $p = 3\dfrac{1}{2}$ **15.** $h = -3.5$ **17.** $y = -6.4$

19. Each side should be divided by -3, not 3.

$$-3x + 2 = -7$$
$$-3x = -9$$
$$\frac{-3x}{-3} = \frac{-9}{-3}$$
$$x = 3$$

21. $a = 1\dfrac{1}{3}$

23. $b = 13\dfrac{1}{2}$

25. $v = -\dfrac{1}{30}$

27. $2.5 + 2.25x = 9.25$; 3 games

29. $v = -5$

31. $d = -12$

33. $m = -9$

35. *Sample answer:* You travel halfway up a ladder. Then you climb down two feet and are 8 feet above the ground. How long is the ladder? $x = 20$

37. the initial fee

39. Find the number of insects remaining and then find the number of insects you caught.

Hmmm.

41. decrease the length by 10 cm; $2(25 + x) + 2(12) = 54$

43. $-6\dfrac{2}{3}$

45. 6.2

Section 4.1   Writing and Graphing Inequalities
(pages 128 and 129)

1. A closed circle would be used because -42 is a solution.

3. no; $x < 5$ is all values of x less than 5. $5 < x$ is all values of x greater than 5.

5. $x \leq -4$; all values of x less than or equal to -4

7. $w + 2.3 > 18$

9. $b - 4.2 < -7.5$

11. yes

13. no

15. yes

17.

19.

21. $p \geq 53$

23. yes

25. yes

27. a. any value that is greater than -2

b. any value that is less than or equal to -2; $b \leq -2$

c. They represent the entire set of real numbers; yes

29. $p = 11$

31. $x = -7$

Section 4.2

Solving Inequalities Using Addition or Subtraction (pages 134 and 135)

1. Yes, because of the Subtraction Property of Inequality.

3. $x \geq 11$;

5. $-4 \leq g$;

7. $-6 < y$;

9. $t \leq -2$;

11. $-\dfrac{3}{7} > b$;

13. $-2.8 < d$;

15. $\dfrac{3}{4} \geq m$;

17. $h \leq -2.4$;

19. The wrong side of the number line is shaded;

21. $7 + 7 + x < 28$; $x < 14$ ft

23. $8 + 8 + 10 + 10 + x \leq 51$; $x \leq 15$ m

25. $x - 3 \geq 5$; $x \geq 8$ ft

Hint

27. The three items must use less than 2400 watts total.

29. $x = 9$

31. $b = -22$

33. A

Section 4.3

Solving Inequalities Using Multiplication or Division (pages 143–145)

1. Multiply each side by 3.

3. *Sample answer:* $-4x < 16$

5. $x \geq -1$

7. $x \leq -35$

9. $x \leq \dfrac{3}{2}$

11. $c \leq -36$;

13. $x < -32$;

15. $k > 2$;

17. $y \leq -3$;

19. The inequality sign should not have been reversed.

$$\dfrac{x}{3} < -9$$
$$3 \cdot \dfrac{x}{3} < 3 \cdot (-9)$$
$$x < -27$$

21. $\dfrac{x}{7} < -3$; $x < -21$

23. $-2x > 30$; $x < -15$

25. **a.** $2.40x \leq 9.60$; $x \leq 4$ avocados

b. no; You must buy a whole number of avocados.

27. $n \geq -3$;

29. $h \leq -24$;

31. $y > \dfrac{14}{3}$;

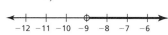

33. $m > -27$;

35. $b > 6$;

37. $-2.5x < -20$; $x > 8$ h

39. $10x \geq 120$; $x \geq 12$ cm

41. $\dfrac{x}{5} < 100$; $x < \$500$

43. *Answer should include, but is not limited to:* Use the correct number of months that the novel has been out.

45. $n \geq -12$ and $n \leq -5$;

47. $s < 14$;

49. $v = 45$

51. $m = 4$

Solving Two-Step Inequalities
(pages 150 and 151)

1. *Sample answer:* They use the same techniques, but when solving an inequality, you must be careful to reverse the inequality symbol when you multiply or divide by a negative number.

3. C

5. $y < 1$;

7. $h > \dfrac{9}{2}$;

9. $b \leq -6$;

11. They did not perform the operations in the proper order.

$$\dfrac{x}{3} + 4 < 6$$
$$\dfrac{x}{3} < 2$$
$$x < 6$$

13. $w \leq 3$;

15. $d > -9$;

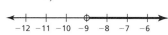

17. $c \geq -1.95$;

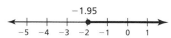

19. $x \geq 4$;

21. $-12x - 38 < -200$; $x > 13.5$ min

23. a. $9.5(70 + x) \geq 1000$; $x \geq 35\dfrac{5}{19}$, which means that at least 36 more tickets must be sold.

b. Because each ticket costs $1 more, fewer tickets will be needed for the theater to earn $1000.

25.

Flutes	7	21	28
Clarinets	4	12	16

$7:4$, $21:12$, and $28:16$

27. A

Section 5.1 — Ratios and Rates
(pages 167–169)

1. It has a denominator of 1.

3. *Sample answer:* A basketball player runs 10 feet down the court in 2 seconds.

5. $0.10 per fluid ounce

7. $72

9. 870 MB

11. $\dfrac{5}{9}$

13. $\dfrac{7}{3}$

15. $\dfrac{4}{3}$

17. 60 miles per hour

19. $2.40 per pound

21. 54 words per minute

23. 4.5 servings per package

25. 4.8 MB per minute

27. 280 square feet per hour

29. no; Although the relative number of boys and girls are the same, the two ratios are inverses.

31. $51

33. 2 cups of juice concentrate, 16 cups of water

35. **a.** rest: 72 beats per minute
    running: 150 beats per minute
    b. 234 beats

37. Try searching for "fire hydrant colors."

39. **a.** you; $\dfrac{1}{3}$ mile per hour faster
    b. $3\dfrac{1}{2}$ hours
    c. you; $1\dfrac{1}{6}$ miles

41. <

43. B

Section 5.2 — Proportions
(pages 174 and 175)

1. Both ratios are equal.

3. *Sample answer:* $\dfrac{6}{10}, \dfrac{12}{20}$

5. yes

7. no

9. yes

11. no

13. no

15. yes

17. no

19. yes

21. yes; Both can do 45 sit-ups per minute.

23. yes

25. yes

27. yes; The ratio of height to base for both triangles is $\dfrac{4}{5}$.

29. Organize the information by using a table.

31. no; The ratios are not equivalent; $\dfrac{13}{19} \neq \dfrac{14}{20} \neq \dfrac{15}{21}$ etc.

33. -13

35. -18

37. D

Extension 5.2   Graphing Proportional Relationships
(pages 176 and 177)

1. 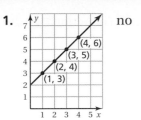 no

3. (0, 0): You earn $0 for working 0 hours.

(1, 15): You earn $15 for working 1 hour; unit rate: $\dfrac{\$15}{1\,h}$

(4, 60): You earn $60 for working 4 hours; unit rate: $\dfrac{\$60}{4\,h} = \dfrac{\$15}{1\,h}$

5. yes; 5 ft/h

7. $y = \dfrac{4}{3}$

Section 5.3   Writing Proportions
(pages 182 and 183)

1. You can use the columns or the rows of the table to write a proportion.

3. *Sample answer:* $\dfrac{x}{12} = \dfrac{5}{6}$; $x = 10$

5. $\dfrac{x}{50} = \dfrac{78}{100}$

7. $\dfrac{x}{150} = \dfrac{96}{100}$

9. $\dfrac{n \text{ winners}}{85 \text{ entries}} = \dfrac{34 \text{ winners}}{170 \text{ entries}}$

11. $\dfrac{100 \text{ meters}}{x \text{ seconds}} = \dfrac{200 \text{ meters}}{22.4 \text{ seconds}}$

13. $\dfrac{\$24}{3 \text{ shirts}} = \dfrac{c}{7 \text{ shirts}}$

15. $\dfrac{5 \text{ 7th grade swimmers}}{16 \text{ swimmers}} = \dfrac{s \text{ 7th grade swimmers}}{80 \text{ swimmers}}$

17. $y = 16$

19. $c = 24$

21. $g = 14$

23. $\dfrac{1}{200} = \dfrac{19.5}{x}$; Dimensions for the model are in the numerators and the corresponding dimensions for the actual space shuttle are in the denominators.

25. Draw a diagram of the given information.  Hint

27. $x = 9$

29. $x = 140$

Section 5.4   Solving Proportions
(pages 190 and 191)

1. mental math; Multiplication Property of Equality; Cross Products Property

3. yes; Both cross products give the equation $3x = 60$.

5. $h = 80$

7. $n = 15$

9. $y = 7\dfrac{1}{3}$

11. $k = 5.6$

13. $n = 10$

15. $d = 5.76$

17. $m = 20$

19. $d = 15$

21. $k = 5.4$

23. 108 pens

25. $x = 1.5$

27. $k = 4$

29. $\frac{2.5}{x} = \frac{1}{0.26}$; about 0.65

31. true; Both cross products give the equation $3a = 2b$.

33. 15.5 lb

35. no; The relationship is not proportional. It should take more people less time to build the swing set.

37. 4 bags

39 and 41.

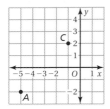

43. D

Section 5.5

Slope
(pages 196 and 197)

1. yes; Slope is the rate of change of a line.

3. 5; A ramp with a slope of 5 increases 5 units vertically for every 1 unit horizontally. A ramp with a slope of $\frac{1}{5}$ increases 1 unit vertically for every 5 units horizontally.

5. $\frac{3}{2}$

7. 1

9. $\frac{4}{5}$

11.

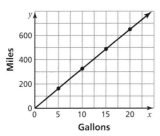

slope = 32.5;
32.5 miles per gallon

13.

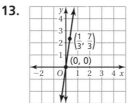

slope = 7

15.

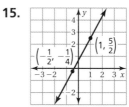

slope = $\frac{11}{6}$

17. a.

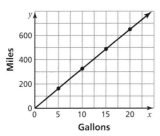

b. 2.5; Every millimeter represents 2.5 miles.

c. 120 mi

d. 90 mm

19. $y = 6$

21. $-\frac{3}{5}$

23. C

Section 5.6

Direct Variation
(pages 202 and 203)

1. $y = kx$, where k is a number and $k \neq 0$.

3. Is the graph of the relationship a line?; yes; no

5.

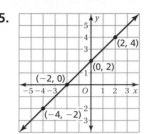

no; The line does not pass through the origin.

7. no; The line does not pass through the origin.

9. yes; The line passes through the origin; $k = \dfrac{2}{3}$

11. yes; The equation can be written as $y = kx$; $k = \dfrac{5}{2}$

13. no; The equation cannot be written as $y = kx$.

15. yes; The equation can be written as $y = kx$; $k = \dfrac{1}{2}$

17. no; The equation cannot be written as $y = kx$.

19.

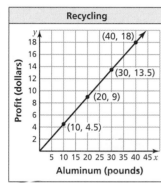

yes; $y = 0.45x$

21. $k = \dfrac{5}{3}$; $y = \dfrac{5}{3}x$

23. $y = 2.54x$

25. When $x = 0$, $y = 0$. So, the graph of a proportional relationship always passes through the origin.

27. no

29. Every graph of direct variation is a line; however, not all lines show direct variation because the line must pass through the origin.

31. 0.5625

33. 0.96

Section 6.1

Percents and Decimals
(pages 218 and 219)

1. B

3. C

5. *Sample answer:* 0.11, 0.13, 0.19

7. 0.78

9. 0.185

11. 0.33

13. 0.4763

15. 1.66

17. 0.0006

19. 74%

21. 89%

23. 99%

25. 48.7%

27. 368%

29. 3.71%

31. The decimal point was moved in the wrong direction. $0.86 = 0.86 = 86\%$

33. 34%

35. $\dfrac{9}{25} = 0.36$

37. $\dfrac{203}{1250} = 0.1624$

39. 40%

41. **a.** $16.\overline{6}\%$ or $16\dfrac{2}{3}\%$ **b.** $\dfrac{5}{6}$

43. $\dfrac{31}{100}$

45. $4\dfrac{8}{25}$

47. $-1.6n - 1$

49. $-b - \dfrac{3}{2}$

1.

Fraction	Decimal	Percent
$\frac{18}{25}$	0.72	72%
$\frac{17}{20}$	0.85	85%
$\frac{13}{50}$	0.26	26%
$\frac{31}{50}$	0.62	62%
$\frac{9}{20}$	0.45	45%

3. 0.04; 0.04 = 4%, but 40%, $\frac{2}{5}$, and 0.4 are all equal to 40%.

5. 20%   **7.** $\frac{13}{25}$ **9.** 76%

11. 0.12   **13.** 140%   **15.** 80%

17.

19.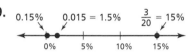

21.

23. Russia, Brazil, United States, India

25. 21%, 0.2$\overline{1}$, $\frac{11}{50}$, $\frac{2}{9}$

27. *D*

29. *C*

31. Write the numbers as percents or decimals to make the ordering easier.

33. yes

35. yes

Section 6.3

The Percent Proportion

(pages 230 and 231)

1. The percent proportion is $\dfrac{a}{w} = \dfrac{p}{100}$ where *a* is part of the whole *w*, and *p*%, or $\dfrac{p}{100}$, is the percent.

3. $\dfrac{a}{40} = \dfrac{60}{100}$; $a = 24$ **5.** 19.2   **7.** 50   **9.** about 38.5%

11. $\dfrac{12}{25} = \dfrac{p}{100}$; $p = 48$ **13.** $\dfrac{9}{w} = \dfrac{25}{100}$; $w = 36$

15. $\dfrac{a}{124} = \dfrac{75}{100}$; $a = 93$ **17.** $\dfrac{a}{40} = \dfrac{0.4}{100}$; $a = 0.16$

19. 34 represents the part, not the whole.   **21.** $6000

$$\frac{a}{w} = \frac{p}{100}$$

$$\frac{34}{w} = \frac{40}{100}$$

$$w = 85$$

23. $\dfrac{14.2}{w} = \dfrac{35.5}{100}$; $w = 40$ **25.** $\dfrac{a}{\frac{7}{8}} = \dfrac{25}{100}$; $a = \dfrac{7}{32}$ **27.** $8.40

29. a. a scale along the vertical axis

b. 6.25%; *Sample answer:* Although you do not know the actual number of votes, you can visualize each bar as a model with the horizontal lines breaking the data into equal parts. The sum of all the parts is 16. Greg has the least parts with 1, which is 100% ÷ 16 = 6.25%.

c. 31 votes

31. a. 62.5%　　**b.** 52x　　　　　**33.** −0.6　　　　　　　**35.** B

Section 6.4

The Percent Equation
(pages 236 and 237)

1. A part of the whole is equal to a percent times the whole.

3. 55 is 20% of what number?; 275; 11

5. 37.5%　　　　　　　　　**7.** 84　　　　　　　　　　**9.** 64

11. 45 = p • 60; 75%　　　　　　　**13.** 0.008 • 150; 1.2

15. 12 = 0.005 • w; 2400　　　　　　**17.** 102 = 1.2 • w; 85

19. 30 represents the part of the whole.

30 = 0.6 • w

50 = w

21. $5400　　　　　　　**23.** 26 years old　　　　**25.** 56 signers

27. If the percent is less than 100%, the percent of a number is less than the number; 50% of 80 is 40; If the percent is equal to 100%, the percent of a number will equal the number; 100% of 80 is 80; If the percent is greater than 100%, the percent of a number is greater than the number; 150% of 80 is 120.

29. Remember when writing a proportion that either the units are the same on each side of the proportion, or the numerators have the same units and the denominators have the same units.

31. 92%　　　　　　**33.** 0.88　　　　　**35.** 0.36

Section 6.5

Percents of Increase and Decrease
(pages 244 and 245)

1. If the original amount decreases, the percent of change is a percent of decrease. If the original amount increases, the percent of change is a percent of increase.

3. The new amount is now 0.　　　　　**5.** 24 L

7. 17 penalties　　　　　**9.** decrease; 66.7%　　　　**11.** increase; 225%

13. decrease; 12.5%　　　　**15.** decrease; 37.5%　　　　**17.** 12.5% decrease

19. a. about 16.7%

b. 280 people; To get the same percent error, the amount of error needs to be the same. Because your estimate was 40 people below the actual attendance, an estimate of 40 people above the actual attendance will give the same percent error.

Section 6.5 — Percents of Increase and Decrease *(continued)*
(pages 244 and 245)

21. decrease; 25%

23. decrease; 70%

25. a. about 16.95% increase

b. 161,391 people

27. 15.6 ounces; 16.4 ounces

29. less than; *Sample answer:* Let x represent the number. A 10% increase is equal to $x + 0.1x$, or $1.1x$. A 10% decrease of this new number is equal to $1.1x - 0.1(1.1x)$, or $0.99x$. Because $0.99x < x$, the result is less than the original number.

31. 10 girls

33. $39.2 = p \cdot 112$; 35%

35. $18 = 0.32 \cdot w$; 56.25

Section 6.6 — Discounts and Markups
(pages 250 and 251)

1. *Sample answer:* Multiply the original price by $100\% - 25\% = 75\%$ to find the sale price.

3. a. 6% tax on a discounted price; The discounted price is less, so the tax is less.

b. 30% markup on a $30 shirt; 30% of $30 is less than $30.

5. $35.70

7. $76.16

9. $53.33

11. $450

13. $172.40

15. 20%

17. $55

19. $175

21. "Multiply $45.85 by 0.1" and "Multiply $45.85 by 0.9, then subtract from $45.85." Both will give the sale price of $4.59. The first method is easier because it is only one step.

23. no; $31.08

25. $30

27. 180

29. C

Section 6.7 — Simple Interest
(pages 256 and 257)

1. I = simple interest, P = principal, r = annual interest rate (in decimal form), t = time (in years)

3. You have to change 6% to a decimal and 8 months to a fraction of a year.

5. a. $300   **b.** $1800

7. a. $292.50   **b.** $2092.50

9. a. $308.20   **b.** $1983.20

11. a. $1722.24   **b.** $6922.24

13. 3%

15. 4%

17. 2 yr

19. 1.5 yr

21. $1440

23. 2 yr

25. $2720

27. $6700.80

29. $8500

31. 5.25%

33. 4 yr

35. 12.5 yr; Substitute $2000 for P and I, 0.08 for r, and solve for t.

37. Year 1 = $520; Year 2 = $540.80; Year 3 = $562.43

39. $b \geq 1$;
-2 -1 0 1 2 3 4

41. A

Section 7.1

Adjacent and Vertical Angles
(pages 274 and 275)

1. two pairs; four pairs

3. $\angle ABC = 120°$, $\angle CBD = 60°$, $\angle DBE = 120°$, $\angle ABE = 60°$

5. *Sample answer:* adjacent: $\angle FGH$ and $\angle HGJ$, $\angle FGK$ and $\angle KGJ$; vertical: $\angle FGH$ and $\angle JGK$, $\angle FGK$ and $\angle JGH$

7. $\angle ACB$ and $\angle BCD$ are adjacent angles, not vertical angles.

9. vertical; 128   11. vertical; 25   13. adjacent; 20

15. 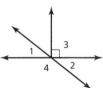   17.

19. **a.** *Sample answer:*   **b.** *Sample answer:*   **c.** *Sample answer:*

21. never   23. sometimes

25. 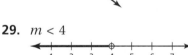   27. $n < -9$

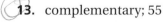

29. $m < 4$

Section 7.2

Complementary and Supplementary Angles
(pages 280 and 281)

1. The sum of the measures of two complementary angles is 90°. The sum of the measures of two supplementary angles is 180°.

3. sometimes; Either x or y may be obtuse.

5. never; Because x and y must both be less than 90° and greater than 0°.

7. complementary   9. supplementary   11. neither

13. complementary; 55   15. $\angle 1 = 130°$, $\angle 2 = 50°$, $\angle 3 = 130°$

17. 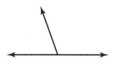   19.

Section 7.2

Complementary and Supplementary Angles
(continued) *(pages 280 and 281)*

21. *Sample answer:* 1) Draw one angle, then draw the other using a side of the first angle;
2) Draw a right angle, then draw the shared side.

23. a. 25°   **b.** 65°

25. 54°

27. $x = 10$; $y = 20$

29. $n = -\dfrac{5}{12}$

31. B

Section 7.3

Triangles
(pages 286 and 287)

1. *Angles:* When a triangle has 3 acute angles, it is an acute triangle. When a triangle has 1 obtuse angle, it is an obtuse triangle. When a triangle has 1 right angle, it is a right triangle. When a triangle has 3 congruent angles, it is an equiangular triangle.

Sides: When a triangle has no congruent sides, it is a scalene triangle. When a triangle has 2 congruent sides, it is an isosceles triangle. When a triangle has 3 congruent sides, it is an equilateral triangle.

3. *Sample answer:*

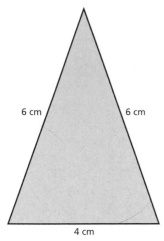

5. *Sample answer:*

7. equilateral equiangular

9. right scalene

11. obtuse scalene

13. acute isosceles

15.

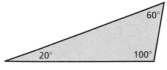

obtuse scalene triangle

17.

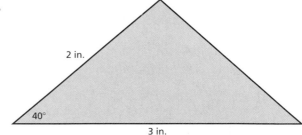

19.

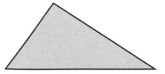

21. no; The sum of the angle measures must be 180°.

23. many; You can change the angle formed by the two given sides to create many triangles.

25. no; The sum of any two side lengths must be greater than the remaining length.

Adjacent and Vertical Angles
(pages 274 and 275)

1. two pairs; four pairs

3. $\angle ABC = 120°$, $\angle CBD = 60°$, $\angle DBE = 120°$, $\angle ABE = 60°$

5. *Sample answer:* adjacent: $\angle FGH$ and $\angle HGJ$, $\angle FGK$ and $\angle KGJ$; vertical: $\angle FGH$ and $\angle JGK$, $\angle FGK$ and $\angle JGH$

7. $\angle ACB$ and $\angle BCD$ are adjacent angles, not vertical angles.

9. vertical; 128

11. vertical; 25

13. adjacent; 20

15.

17.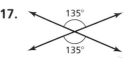

19. **a.** *Sample answer:*   **b.** *Sample answer:*   **c.** *Sample answer:*

21. never

23. sometimes

25.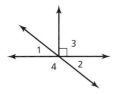

27. $n < -9$

29. $m < 4$

Complementary and Supplementary Angles
(pages 280 and 281)

1. The sum of the measures of two complementary angles is 90°. The sum of the measures of two supplementary angles is 180°.

3. sometimes; Either x or y may be obtuse.

5. never; Because x and y must both be less than 90° and greater than 0°.

7. complementary

9. supplementary

11. neither

13. complementary; 55

15. $\angle 1 = 130°$, $\angle 2 = 50°$, $\angle 3 = 130°$

17.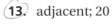

19.

Selected Answers   **A27**

21. *Sample answer:* 1) Draw one angle, then draw the other using a side of the first angle; 2) Draw a right angle, then draw the shared side.

30°

23. **a.** 25°   **b.** 65°

25. 54°

27. $x = 10; y = 20$

29. $n = -\dfrac{5}{12}$

31. B

Section 7.3

Triangles
(pages 286 and 287)

1. *Angles:* When a triangle has 3 acute angles, it is an acute triangle. When a triangle has 1 obtuse angle, it is an obtuse triangle. When a triangle has 1 right angle, it is a right triangle. When a triangle has 3 congruent angles, it is an equiangular triangle.

   Sides: When a triangle has no congruent sides, it is a scalene triangle. When a triangle has 2 congruent sides, it is an isosceles triangle. When a triangle has 3 congruent sides, it is an equilateral triangle.

3. *Sample answer:*

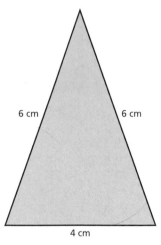
6 cm   6 cm
4 cm

5. *Sample answer:*

55°
65°   60°

7. equilateral equiangular

9. right scalene

11. obtuse scalene

13. acute isosceles

15.

60°
20°   100°

obtuse scalene triangle

17.

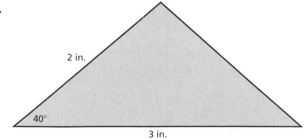
2 in.
40°
3 in.

19.

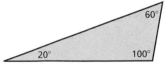

21. no; The sum of the angle measures must be 180°.

23. many; You can change the angle formed by the two given sides to create many triangles.

25. no; The sum of any two side lengths must be greater than the remaining length.

27. a. green: 65; purple: 25; red: 45

b. The angles opposite the congruent sides are congruent.

c. An isosceles triangle has at least two angles that are congruent.

29. no; The equation cannot be written as $y = kx$.　　　　**31.** B

Extension 7.3　Angle Measures of Triangles
(pages 288 and 289)

1. 91; obtuse scalene triangle

3. 90; right scalene triangle

5. 48; acute isosceles triangle

7. yes

9. no; $28\frac{2}{3}$

11. 67.5; acute isosceles triangle

13. 24; obtuse isosceles triangle

15. 35; obtuse scalene triangle

17. a. 72

b. You can change the distance between the bottoms of the two upright cards; yes; x must be greater than 60 and less than 90; If x were less than or equal to 60, the two upright cards would have to be exactly on the edges of the base card or off the base card. It is not possible to stack cards at these angles. If x were equal to 90, then the two upright cards would be vertical, which is not possible. The card structure would not be stable. In practice, the limits on x are probably closer to $70 < x < 80$.

Section 7.4　Quadrilaterals
(pages 296 and 297)

1. all of them

3. kite; It is the only type of quadrilateral listed that does not have opposite sides that are parallel and congruent.

5. trapezoid

7. kite

9. rectangle

11. 110

13. 58°

15.
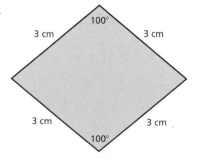

17.

19. always

21. never

23. sometimes

25.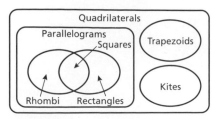

27. $\dfrac{1}{4}$

29. $\dfrac{6}{5}$

Section 7.5
Scale Drawings
(pages 303–305)

1. A scale is the ratio that compares the measurements of the drawing or model with the actual measurements. A scale factor is a scale without any units.

3. Convert one of the lengths into the same units as the other length. Then, form the scale and simplify.

5. 10 ft by 10 ft

7. 112.5%

9. 50 mi

11. 110 mi

13. 15 in.

15. 21.6 yd

17. The 5 cm should be in the numerator.

$$\frac{1\,\text{cm}}{20\,\text{m}} = \frac{5\,\text{cm}}{x\,\text{m}}$$

$$x = 100\,\text{m}$$

19. 2.4 cm; 1 cm : 10 mm

21. a. *Answer should include, but is not limited to:* Make sure words and picture match the product.

    b. Answers will vary.

23. a. 16 cm; 16 cm^2 **b.** 40 mm; 100 mm^2

25.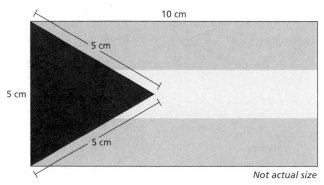

Not actual size

27. 15 ft^2

29. 3 ft^2

31. Find the size of the object that would represent the model of the Sun.

33 and 35.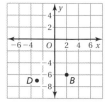

Section 8.1

Circles and Circumference
(pages 321–323)

1. The radius is one-half the diameter.

3. 2.5 cm

5. $1\frac{3}{4}$ in.

7. 4 in.

9. about 31.4 in.

11. about 56.52 in.

13. **a.** about 25 m; about 50 m

   b. about 2 times greater

15. about 7.71 ft

17. about 31.4 cm; about 62.8 cm

19. about 69.08 m; about 138.16 m

21. about 200.96 cm

23. Draw a diagram of the given information.

25. **a.** about 254.34 mm; First find the length of the minute hand. Then find $\frac{3}{4}$ of the circumference of a circle whose radius is the length of the minute hand.

   b. about 320.28 mm; Subtract $\frac{1}{12}$ of the circumference of a circle whose radius is the length of the hour hand from the circumference of a circle whose radius is the length of the minute hand.

27. 20 m

29. D

Section 8.2

Perimeters of Composite Figures
(pages 328 and 329)

1. no; The perimeter of the composite figure does not include the measure of the shared side.

3. 19.5 units

5. 25.5 units

7. 19 units

9. 56 m

11. 30 cm

13. about 26.85 in.

15. about 36.84 ft

17. First find the total distance you run.

19. *Sample answer:* By adding the triangle shown by the dashed line to the L-shaped figure, you *reduce* the perimeter.

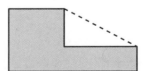

21. 279.68

23. 205

Section 8.3

Areas of Circles
(pages 336 and 337)

1. Divide the diameter by 2 to get the radius. Then use the formula $A = \pi r^2$ to find the area.

3. about 254.34 mm^2

5. about 314 in.^{2}

7. about 3.14 cm^2

9. about 2461.76 mm^2

11. about 113.04 in.^{2}

13. about 628 cm^2

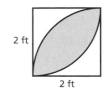

15. about 1.57 ft^2

17. What fraction of the circle is the dog's running area?

19. about 9.8125 in.^{2}; The two regions are identical, so find one-half the area of the circle.

21. about 4.56 ft^2; Find the area of the shaded regions by subtracting the areas of both unshaded regions from the area of the quarter-circle containing them. The area of each unshaded region can be found by subtracting the area of the smaller shaded region from the semicircle. The area of the smaller shaded region can be found by drawing a square about the region.

Subtract the area of a quarter-circle from the area of the square to find an unshaded area. Then subtract both unshaded areas from the square's area to find the shaded region's area.

23. 53

25. A

Section 8.4

Areas of Composite Figures
(pages 342 and 343)

1. *Sample answer:* You could add the areas of an 8-inch × 4-inch rectangle and a triangle with a base of 6 inches and a height of 6 inches. Also you could add the area of a 2-inch × 4-inch rectangle to the area of a trapezoid with a height of 6 inches, and base lengths of 4 inches and 10 inches.

3. 28.5 units^2

5. 25 units^2

7. 25 units^2

9. 132 cm^2

11. *Answer will include but is not limited to:* Tracings of a hand and foot on grid paper, estimates of the areas, and a statement of which is greater.

13. 23.5 in.^{2}

15. 24 m^2

17. Each envelope can be broken up into 5 smaller figures to find the area.

19. $y \div 6$

21. $7w$

Surface Areas of Prisms
(pages 359–361)

1. *Sample answer:* 1) Use a net. 2) Use the formula $S = 2\ell w + 2\ell h + 2wh$.

3. Find the area of the bases of the prism; 24 in.^{2}; 122 in.^{2}

5.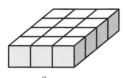

   38 in.^{2}

7. 324 m^2

9. 49.2 yd^2

11. 136 m^2

13. 294 yd^2

15. $2\frac{2}{3}$ ft^2

17. 177 in.^{2}

19. yes; Because you do not need to frost the bottom of the cake, you only need 249 square inches of frosting.

21. 68 m^2

23. $x = 4$ in.

25. The dimensions of the red prism are three times the dimensions of the blue prism. The surface area of the red prism is 9 times greater than the surface area of the blue prism.

27. **a.** 0.125 pint   **b.** 1.125 pints

    c. red and green: The ratio of the paint amounts (red to green) is 4 : 1 and the ratio of the side lengths is 2 : 1.

    green and blue: The ratio of the paint amounts (blue to green) is 9 : 1 and the ratio of the side lengths is 3 : 1.

    The ratio of the paint amounts is the square of the ratio of the side lengths.

29. 160 ft^2

31. 28 ft^2

Surface Areas of Pyramids
(pages 366 and 367)

1. no; The lateral faces of a pyramid are triangles.

3. triangular pyramid; The other three are names for the pyramid.

5. 178.3 mm^2

7. 144 ft^2

9. 170.1 yd^2

11. 1240.4 mm^2

13. 6 m

 Hint

15. 283.5 cm^2

17. Determine how long the fabric needs to be so you can cut the fabric most efficiently.

19. 124 cm^2

21. $A \approx 452.16$ units^2; $C \approx 75.36$ units

23. $A \approx 572.265$ units^2; $C \approx 84.78$ units

Section 9.3

Surface Areas of Cylinders
(pages 372 and 373)

1. $2\pi rh$

3.

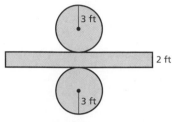

$30\pi \approx 94.2 \text{ ft}^2$

5.

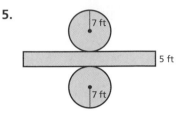

$168\pi \approx 527.5 \text{ ft}^2$

7. $156\pi \approx 489.8 \text{ ft}^2$

9. $120\pi \approx 376.8 \text{ ft}^2$

11. $28\pi \approx 87.9 \text{ m}^2$

13. $432\pi \approx 1356.48 \text{ ft}^2$

15. The surface area of the cylinder with the height of 8.5 inches is greater than the surface area of the cylinder with the height of 11 inches.

17. After removing the wedge, is there any new surface area added?

19. 10 ft^2

21. 47.5 in.^2

Section 9.4

Volumes of Prisms
(pages 380 and 381)

1. cubic units

3. The volume of an object is the amount of space it occupies. The surface area of an object is the sum of the areas of all its faces.

5. 288 cm^3

7. 210 yd^3

9. 420 mm^3

11. 645 mm^3

13. The area of the base is wrong.

$$V = \frac{1}{2}(7)(5) \cdot 10$$
$$= 175 \text{ cm}^3$$

15. 225 in.^3

17. 7200 ft^3

19. 1728 in.^3

$1 \times 1 \times 1 = 1 \text{ ft}^3$ $12 \times 12 \times 12 = 1728 \text{ in.}^3$

21. 20 cm

23. You can write the volume in cubic inches and use prime factorization to find the dimensions.

25. $90

27. $240.50

Section 9.5   Volumes of Pyramids
(pages 386 and 387)

1. The volume of a pyramid is $\frac{1}{3}$ times the area of the base times the height. The volume of a prism is the area of the base times the height.

3. 3 times

5. 20 mm^3

7. 80 in.^3

9. 252 mm^3

11. 700 mm^3

13. 156 ft^3

15. 340.4 in.^3

17. $12{,}000 \text{ in.}^3$; The volume of one paperweight is 12 cubic inches. So, 12 cubic inches of glass is needed to make one paperweight. So, it takes $12 \times 1000 = 12{,}000$ cubic inches to make 1000 paperweights.

19. *Sample answer:* 5 ft by 4 ft

21. $153°$; $63°$

23. $60°$; none

Extension 9.5   Cross Sections of Three-Dimensional Figures
(pages 388 and 389)

1. triangle

3. rectangle

5. triangle

7. The intersection is the shape of the base.

9. circle

11. circle

13. rectangle

15. The intersection occurs at the vertex of the cone.

Section 10.1   Outcomes and Events
(pages 404 and 405)

1. event; It is a collection of several outcomes.

3. 8

5. 1, 2, 3, 4, 5, 6, 7, 8, 9

7. 1, 3, 5, 7, 9

9. 1, 3

11. 3, 6, 9

13. **a.** 1 way   **b.** green

15. **a.** 1 way   **b.** yellow

17. **a.** 7 ways

     b. red, red, red, purple, purple, green, yellow

19. 7 ways

21. true

23. true

25. 30 rock CDs

27. $x = 2$

29. $w = 12$

31. C

Section 10.2 — Probability
(pages 410 and 411)

1. The probability of an event is the ratio of the number of favorable outcomes to the number of possible outcomes.

3. *Sample answer:* You will not have any homework this week.; You will fall asleep tonight.

5. either; Both spinners have the same number of chances to land on "Forward."

7. impossible

9. unlikely

11. $\dfrac{1}{10}$

13. $\dfrac{9}{10}$

15. 0

17. 20

19. **a.** $\dfrac{2}{3}$; likely   **b.** $\dfrac{1}{3}$; unlikely

   c. $\dfrac{1}{2}$; equally likely to happen or not happen

21. There are 2 combinations for each.

23. $x < 4$;

25. $w > -3$;

27. C

Section 10.3 — Experimental and Theoretical Probability
(pages 417–419)

1. Perform an experiment several times. Count how often the event occurs and divide by the number of trials.

3. There is a 50% chance you will get a favorable outcome.

5. experimental probability; The population is too large to survey every person, so a sample will be used to predict the outcome.

7. $\dfrac{12}{25}$, or 48%

9. $\dfrac{7}{25}$, or 28%

11. 0, or 0%

13. 45 tiles

15. $\dfrac{1}{3}$, or about 33.3%

17. $\dfrac{1}{2}$, or 50%

19. 1, or 100%

21. $\dfrac{25}{26}$, or about 96.2%

23. 36 songs

25. theoretical: $\dfrac{1}{5}$, or 20%;

   experimental: $\dfrac{37}{200}$, or 18.5%

   The experimental probability is close to the theoretical probability.

27. theoretical: $\dfrac{1}{5}$, or 20%;

   experimental: $\dfrac{1}{5}$, or 20%

   The probabilities are equal.

29. **a.** $\dfrac{1}{12}$; 50 times   **b.** $\dfrac{11}{50}$; 132 times

   c. A larger number of trials should result in a more accurate probability, which gives a more accurate prediction.

31. Make a list of all possible ways to get each sum.

35. 4%

37. D

33. a. As a number of trials increases, the most likely sum will change from 6 to 7.

b. As an experiment is repeated over and over, the experimental probability of an event approaches the theoretical probability of the event.

Section 10.4 Compound Events
(pages 425–427)

1. A sample space is the set of all possible outcomes of an event. Use a table or tree diagram to list all the possible outcomes.

3. You could use a tree diagram or the Fundamental Counting Principle. Either way, the total number of possible outcomes is 30.

5. 125,000

7. Sample space: Realistic Lion, Realistic Bear, Realistic Hawk, Realistic Dragon, Cartoon Lion, Cartoon Bear, Cartoon Hawk, Cartoon Dragon; 8 possible outcomes

9. 20
11. 60

13. The possible outcomes of each question should be multiplied, not added. The correct answer is $2 \times 2 \times 2 \times 2 \times 2 = 32$.

15. $\frac{1}{10}$, or 10%

17. $\frac{1}{5}$, or 20%

19. $\frac{2}{5}$, or 40%

21. $\frac{1}{18}$, or $5\frac{5}{9}$%

23. $\frac{1}{9}$, or $11\frac{1}{9}$%

25. a. $\frac{1}{9}$, or about 11.1%

b. It increases the probability that your guesses are correct to $\frac{1}{4}$, or 25%, because you are only choosing between 2 choices for each question.

27. a. $\frac{1}{1000}$, or 0.1%

b. There are 1000 possible combinations. With 5 tries, someone would guess 5 out of the 1000 possibilities. So, the probability of getting the correct combination is $\frac{5}{1000}$, or 0.5%.

29. a. The Fundamental Counting Principle is more efficient. A tree diagram would be too large.

b. 1,000,000,000 or one billion

c. *Sample answer:* Not all possible number combinations are used for Social Security Numbers (SSN). SSNs are coded into geographical, group, and serial numbers. Some SSNs are reserved for commercial use and some are forbidden for various reasons.

31. *Sample answer:* adjacent: $\angle XWY$ and $\angle ZWY$, $\angle XWY$ and $\angle XWV$; vertical: $\angle VWX$ and $\angle YWZ$, $\angle YWX$ and $\angle VWZ$

33. B

1. What is the probability of choosing a 1 and then a blue chip?; $\frac{1}{15}$; $\frac{1}{10}$

3. independent; The outcome of the first roll does not affect the outcome of the second roll.

5. $\frac{1}{8}$

7. $\frac{3}{8}$

9. $\frac{1}{42}$

11. $\frac{2}{21}$

13. The two events are dependent, so the probability of the second event is $\frac{1}{3}$.

$$P(\text{red and green}) = \frac{1}{4} \cdot \frac{1}{3} = \frac{1}{12}$$

15. $\frac{1}{6}$, or about 16.7%

17. $\frac{2}{35}$

19. $\frac{5}{162}$, or about 3.1%

21. $\frac{4}{81}$, or about 4.9%

23. $\frac{3}{4}$

25. **a.** Because the probability that both you and your best friend are chosen is $\frac{1}{132}$, you and your best friend are not in the same group. The probability that you both are chosen would be 0 because only one leader is chosen from each group.

   b. $\frac{1}{11}$ **c.** 23

27.

right scalene

29.

acute isosceles

1. **a.** *Sample answer:* Roll four number cubes. Let an odd number represent a correct answer and an even number represent an incorrect answer. Run 40 trials.

   b. Check students' work. The probability should be "close" to 6.25% (depending on the number of trials, because that is the theoretical probability).

3. *Sample answer:* Using the spreadsheet in Example 3 and using digits 1–4 as successes, the experimental probability is 16%.

1. Samples are easier to obtain.

3. Population: Residents of New Jersey
   Sample: Residents of Ocean County

5. biased; The sample is not selected at random and is not representative of the population because students in a band class play a musical instrument.

7. biased; The sample is not representative of the population because people who go to a park are more likely to think that the park needs to be remodeled.

9. no; the sample is not representative of the population because people going to the baseball stadium are more likely to support building a new baseball stadium. So, the sample is biased and the conclusion is not valid.

11. Sample A; It is representative of the population.

13. a sample; It is much easier to collect sample data in this situation.

15. a sample; It is much easier to collect sample data in this situation.

17. Not everyone has an email address, so the sample may not be representative of the entire population. *Sample answer:* When the survey question is about technology or which email service you use, the sample may be representative of the entire population.

19. Use the survey results to find the number of students in the school who plan to attend college.

21. 140

23. 3

1. **a.** Check students' work. **b.** Check students' work.

   c. Check students' work. *Sample answer:* yes; increase the number of random samples

3. **Step 1:** 7, 7.5, 8, 4.5, 8, 10, 5, 11

   Step 2:

   4.5   6   7.75   9   11

   Median hours worked each week
   3 4 5 6 7 8 9 10 11 12 13

   Step 3: *Sample answer:* The actual median number of hours probably lies within the interval 6 to 9 hours (the box). So, about 7.5 is a good estimate.

   The median of the data is 8. So, the estimate is close.

5. The more samples you have, the more accurate your inferences will be. By taking multiple random samples, you can find an interval where the actual measurement of a population may lie.

1. When comparing two populations, use the mean and the MAD when each distribution is symmetric. Use the median and the IQR when either one or both distributions are skewed.

3. **a.** garter snake: mean = 25, median = 24.5, mode = 24, range = 20, IQR = 7.5, MAD ≈ 4.33
   water snake: mean = 31.5, median = 32, mode = 32, range = 20, IQR = 10, MAD ≈ 5.08

   b. The water snakes have greater measures of center because the mean, median, and mode are greater. The water snakes also have greater measures of variation because the interquartile range and mean absolute deviation are greater.

5. **a.** Class A: median = 90, IQR = 12.5
   Class B: median = 80, IQR = 10

   The variation in the test scores is about the same, but Class A has greater test scores.

   b. The difference in the medians is 0.8 to 1 times the IQR.

7. Arrange the dot plots in Exercise 5 vertically and construct a double box-and-whisker plot in Exercise 6 to help you visualize the distributions.

9. **a.** Check students' work. Experiments should include taking many samples of a manageable size from each grade level. This will be more doable if the work of sampling is divided among the whole class, and the results are pooled together.

   b. Check students' work. The data may or may not support a conclusion.

11.

13.

1. **a.** ∠*A* and ∠*D*, ∠*B* and ∠*E*, ∠*C* and ∠*F*

   b. Side *AB* and Side *DE*, Side *BC* and Side *EF*, Side *AC* and Side *DF*

3. ∠*V* does not belong. The other three angles are congruent to each other, but not to ∠*V*.

5. congruent

7. ∠*P* and ∠*W*, ∠*Q* and ∠*V*, ∠*R* and ∠*Z*, ∠*S* and ∠*Y*, ∠*T* and ∠*X*;
   Side *PQ* and Side *WV*, Side *QR* and Side *VZ*, Side *RS* and Side *ZY*,
   Side *ST* and Side *YX*, Side *TP* and Side *XW*

9. not congruent; Corresponding side lengths are not congruent.

11. The corresponding angles are not congruent, so the two figures are not congruent.

13. What figures have you seen in this section that have at least one right angle?

15. **a.** true; Side *AB* corresponds to Side *YZ*.

    b. true; ∠*A* and ∠*X* have the same measure.

    c. false; ∠*A* corresponds to ∠*Y*.

    d. true; The measure of ∠*A* is 90°, the measure of ∠*B* is 140°, the measure of ∠*C* is 40°, and the measure of ∠*D* is 90°. So, the sum of the angle measures of *ABCD* is 90° + 140° + 40° + 90° = 360°.

17 and 19.

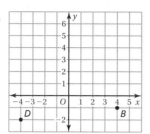

Section 11.2

Translations
(pages 476 and 477)

1. A

3. yes; Translate the letters T and O to the end.

5. no

7. yes

9. no

11. $A'(-3, 0), B'(0, -1), C'(1, -4), D'(-3, -5)$

13.

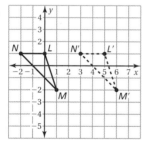

15.

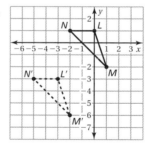

17. 2 units left and 2 units up

19. 6 units right and 3 units down

21. **a.** 5 units right and 1 unit up

    b. no; It would hit the island.

    c. 4 units up and 4 units right

23. If you are doing more than 10 moves and have not moved the knight to g5, you might want to start over.

25. no

27. yes

Reflections

(pages 482 and 483)

1. The third one because it is not a reflection.

3. Quadrant IV

5. yes

7. no

9. no

11. $M'(-2, -1), N'(0, -3), P'(2, -2)$

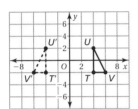

13. $D'(-2, 1), E'(0, 1), F'(0, 5), G'(-2, 5)$

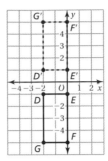

15. $T'(-4, -2), U'(-4, 2), V'(-6, -2)$

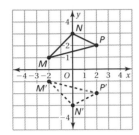

17. $J'(-2, 2), K'(-7, 4), L'(-9, -2), M'(-3, -1)$

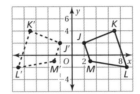

19. x-axis

21. y-axis

23. $R'(3, -4), S'(3, -1), T'(1, -4)$

25. yes; Translations and reflections produce images that are congruent to the original figure.

27. If you are driving a vehicle and want to see who is following you, where would you look?

29. obtuse

31. right

33. B

Rotations

(pages 489–491)

1. $(0, 0); (1, -3)$

3. Quadrant IV

5. Quadrant II

7. reflection

9. translation

11. yes; 90° counterclockwise

13. $A'(2, 2), B'(1, 4), C'(3, 4), D'(4, 2)$

15. $J'(0, -3), K'(0, -5), L'(-4, -3)$

17. $W'(-2, 6), X'(-2, 2), Y'(-6, 2), Z'(-6, 5)$

19. It only needs to rotate 120° to produce an identical image.

21. It only needs to rotate 180° to produce an identical image.

23. $J''(4, 4)$, $K''(3, 4)$, $L''(1, 1)$, $M''(4, 1)$

25. *Sample answer:* Rotate 180° about the origin and then rotate 90° clockwise about vertex $(-1, 0)$; Rotate 90° counterclockwise about the origin and then translate 1 unit left and 1 unit down.

Hint

27. Use Guess, Check, and Revise to solve this problem.

29. $(2, 4)$, $(4, 1)$, $(1, 1)$

31. yes

33. no

1. They are congruent.

3. Yes, because the angles are congruent and the side lengths are proportional.

5. not similar; Corresponding side lengths are not proportional.

7.

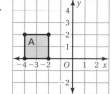 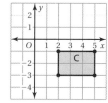 

A and B; Corresponding side lengths are proportional and corresponding angles are congruent.

9. $6\dfrac{2}{3}$ **11.** 14

13. 30 in.

15. What types of quadrilaterals can have the given angle measures?

17. 3 times

19. **a.** yes

     b. yes; It represents the fact that the sides are proportional because you can split the isosceles triangles into smaller right triangles that will be similar.

21. $\dfrac{16}{81}$

23. $\dfrac{49}{16}$

25. C

1. The ratio of the perimeters is equal to the ratio of the corresponding side lengths.

3. Because the ratio of the corresponding side lengths is $\frac{1}{2}$, the ratio of the areas is equal to $\left(\frac{1}{2}\right)^2$. To find the area, solve the proportion $\frac{30}{x} = \frac{1}{4}$ to get $x = 120$ square inches.

5. $\frac{5}{8}; \frac{25}{64}$

7. $\frac{14}{9}; \frac{196}{81}$

9. The area is 9 times larger.

11. 25.6

13. 39 in.; 93.5 in.^{2}

15. 108 yd

17. **a.** 400 times greater; The ratio of the corresponding lengths is $\frac{120 \text{ in.}}{6 \text{ in.}} = \frac{20}{1}$.
    So, the ratio of the areas is $\left(\frac{20}{1}\right)^2 = \frac{400}{1}$.

    b. 1250 ft^2

19. 15 m

21. $x = -2$

23. $n = -4$

Section 11.7 **Dilations**
(pages 511–513)

1. A dilation changes the size of a figure. The image is similar, not congruent, to the original figure.

3. The middle red figure is not a dilation of the blue figure because the height is half of the blue figure and the base is the same. The left red figure is a reduction of the blue figure and the right red figure is an enlargement of the blue figure.

5.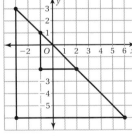

   The triangles are similar.

7. yes

9. no

11. yes

13.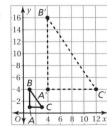

    enlargement

15.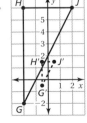

    reduction

17.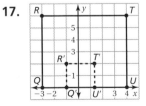

    reduction

19. Each coordinate was multiplied by 2 instead of divided by 2. The coordinates should be $A'(1, 2.5)$, $B'(1, 0)$, and $C'(2, 0)$.

21. reduction; $\dfrac{1}{4}$

23. $A''(10, 6)$, $B''(4, 6)$, $C''(4, 2)$, $D''(10, 2)$ **25.** $J''(3, -3)$, $K''(12, -9)$, $L''(3, -15)$

27. *Sample answer:* Rotate 90° counterclockwise about the origin and then dilate with respect to the origin using a scale factor of 2

29. Exercise 27: yes; Exercise 28: no; Explanations will vary based on sequences chosen in Exercises 27 and 28.

31. **a.** enlargement

     b. center of dilation

     c. $\dfrac{4}{3}$

     d. The shadow on the wall becomes larger. The scale factor will become larger.

33. The transformations are a dilation using a scale factor of 2 and then a translation of 4 units right and 3 units down; similar; A dilation produces a similar figure and a translation produces a congruent figure, so the final image is similar.

35. The transformations are a dilation using a scale factor of $\dfrac{1}{3}$ and then a reflection in the x-axis; similar; A dilation produces a similar figure and a reflection produces a congruent figure, so the final image is similar.

37. $A'(-2, 3)$, $B'(6, 3)$, $C'(12, -7)$, $D'(-2, -7)$; Methods will vary.

39. supplementary; $x = 16$

41. B

Section 12.1   Parallel Lines and Transversals
(pages 531–533)

1. *Sample answer:*

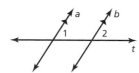

3. m and n

5. 8

7. $\angle 1 = 107°$, $\angle 2 = 73°$

9. $\angle 5 = 49°$, $\angle 6 = 131°$ **11.** 60°; Corresponding angles are congruent.

13. *Sample answer:* rotate 180° and translate down

15. $\angle 6 = 61°$; $\angle 6$ and the given angle are vertical angles.
     $\angle 5 = 119°$ and $\angle 7 = 119°$; $\angle 5$ and $\angle 7$ are supplementary to the given angle.
     $\angle 1 = 61°$; $\angle 1$ and the given angle are corresponding angles.
     $\angle 3 = 61°$; $\angle 1$ and $\angle 3$ are vertical angles.
     $\angle 2 = 119°$ and $\angle 4 = 119°$; $\angle 2$ and $\angle 4$ are supplementary to $\angle 1$.

17. $\angle 2 = 90°$; $\angle 2$ and the given angle are vertical angles.
     $\angle 1 = 90°$ and $\angle 3 = 90°$; $\angle 1$ and $\angle 3$ are supplementary to the given angle.
     $\angle 4 = 90°$; $\angle 4$ and the given angle are corresponding angles.
     $\angle 6 = 90°$; $\angle 4$ and $\angle 6$ are vertical angles.
     $\angle 5 = 90°$ and $\angle 7 = 90°$; $\angle 5$ and $\angle 7$ are supplementary to $\angle 4$.

Parallel Lines and Transversals *(continued)*
(pages 531–533)

19. 132°; *Sample answer:* ∠2 and ∠4 are alternate interior angles and ∠4 and ∠3 are supplementary.

21. 120°; *Sample answer:* ∠6 and ∠8 are alternate exterior angles.

23. 61.3°; *Sample answer:* ∠3 and ∠1 are alternate interior angles and ∠1 and ∠2 are supplementary.

25. They are all right angles because perpendicular lines form 90° angles.

27. 130

29. a. no; They look like they are spreading apart. **b.** Check students' work.

31. 13   **33.** 51   **35.** B

Angles of Triangles
(pages 538 and 539)

1. Subtract the sum of the given measures from 180°.

3. 115°, 120°, 125°   **5.** 40°, 65°, 75°   **7.** 25°, 45°, 110°

9. 48°, 59°, 73°   **11.** 45   **13.** 140°

15. The measure of the exterior angle is equal to the sum of the measures of the two nonadjacent interior angles. The sum of all three angles is not 180°;

$$(2x - 12) = x + 30$$
$$x = 42$$

The exterior angle is $(2(42) - 12)° = 72°$.

17. 126°

19. sometimes; The sum of the angle measures must equal 180°.

21. never; If a triangle had more than one vertex with an acute exterior angle, then it would have to have more than one obtuse interior angle which is impossible.

23. $x = -4$ **25.** $n = -3$

Angles of Polygons
(pages 547–549)

1. *Sample answer:*

3. What is the measure of an interior angle of a regular pentagon?; 108°; 540°

5. 1260°   **7.** 360°   **9.** 1260°

11. no; The interior angle measures given add up to 535°, but the sum of the interior angle measures of a pentagon is 540°.

13. 90°, 135°, 135°, 135°, 135°, 90°

15. 140°   **17.** 140°

19. The sum of the interior angle measures should have been divided by the number of angles, 20. 3240° ÷ 20 = 162°; The measure of each interior angle is 162°.

21. 24 sides   **23.** 75°, 93°, 85°, 107°

25. 60°; The sum of the interior angle measures of a hexagon is 720°. Because it is regular, each angle has the same measure. So, each interior angle is 720° ÷ 6 = 120° and each exterior angle is 60°.

27. 120°, 120°, 120°   **29.** interior: 135°; exterior: 45°

31. 120°

33. a. *Sample answer:*

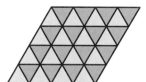

b. *Sample answer:*
square, regular hexagon

c. *Sample answer:*

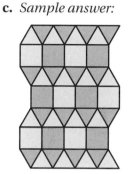

d. *Answer should include, but is not limited to:* a discussion of the interior and exterior angles of the polygons in the tessellation and how they combine to add up to 360° where the vertices meet.

35. 2   **37.** 6

Section 12.4

Using Similar Triangles
(pages 554 and 555)

1. Write a proportion that uses the missing measurement because the ratios of corresponding side lengths are equal.

3. *Sample answer:* Two of the angles are congruent, so they have the same sum. When you subtract this from 180°, you will get the same third angle.

5. Student should draw a triangle with the same angle measures as the ones given in the textbook.

If the student's triangle is larger than the one given, then the ratio of the corresponding side lengths, $\frac{\text{student's triangle length}}{\text{book's triangle length}}$, should be greater than 1. If the student's triangle is smaller than the one given, then the ratio of the corresponding side lengths, $\frac{\text{student's triangle length}}{\text{book's triangle length}}$, should be less than 1.

7. no; The triangles do not have two pairs of congruent angles.

9. yes; The triangles have the same angle measures, 81°, 51°, and 48°.

11. yes; The triangles have two pairs of congruent angles.

13. Think of the different ways that you can show that two triangles are similar.

15. 30 ft

17. maybe; They are similar when both have measures of 30°, 60°, 90° or both have measures of 45°, 45°, 90°. They are not similar when one has measures of 30°, 60°, 90° and the other has measures of 45°, 45°, 90°.

19. $y = 5x + 3$

21. $y = 8x - 4$

1. a line

3. *Sample answer:*

x	0	1
$y = 3x - 1$	-1	2

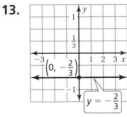

5.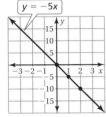

7.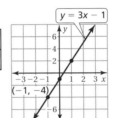

9.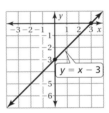

11.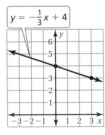

13.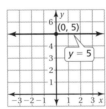

15.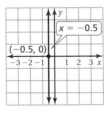

17. The equation $x = 4$ is graphed, not $y = 4$.

19. a.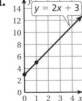

b. about $5

c. $5.25

21. $y = -\dfrac{5}{2}x + 2$

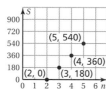

23. $y = -2x + 3$

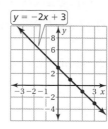

25. a. *Sample answer:*

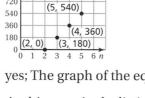

yes; The graph of the equation is a line.

b. No, $n = 3.5$ does not make sense because a polygon cannot have half a side.

27. Begin this exercise by listing all of the given information.

29. $(-6, 6)$

31. $(-4, -3)$

Section 13.2 Slope of a Line
(pages 577–579)

1. a. B and C

b. A

c. no; None of the lines are vertical.

3. The line is horizontal.

5.

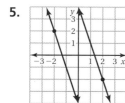

The lines are parallel.

7. $\dfrac{3}{4}$

9. $-\dfrac{3}{5}$

11. 0

13. 0

15. undefined

17. $-\dfrac{11}{6}$

19. The denominator should be $2 - 4$.
$m = -1$

21. 4

23. $-\dfrac{3}{4}$

25. $\dfrac{1}{3}$

27. $k = 11$

29. $k = -5$

31. a. $\dfrac{3}{40}$

b. The cost increases by $3 for every 40 miles you drive, or the cost increases by $0.075 for every mile you drive.

33. yes; The slopes are the same between the points.

35. When you switch the coordinates, the differences in the numerator and denominator are the opposite of the numbers when using the slope formula. You still get the same slope.

37. $b = 25$

39. $x = 7.5$

Extension 13.2 — Slopes of Parallel and Perpendicular Lines

(pages 580 and 581)

1. blue and red; They both have a slope of -3.

3. yes; Both lines are horizontal and have a slope of 0.

5. yes; Both lines are vertical and have an undefined slope.

7. blue and green; The blue line has a slope of 6. The green line has a slope of $-\frac{1}{6}$. The product of their slopes is $6 \cdot \left(-\frac{1}{6}\right) = -1$.

9. yes; The line $x = -2$ is vertical. The line $y = 8$ is horizontal. A vertical line is perpendicular to a horizontal line.

11. yes; The line $x = 0$ is vertical. The line $y = 0$ is horizontal. A vertical line is perpendicular to a horizontal line.

Section 13.3 — Graphing Proportional Relationships

(pages 586 and 587)

1. $(0, 0)$

3. no; *Sample answer:* The graph of the equation does not pass through the origin.

5. yes; $y = \frac{1}{3}x$; *Sample answer:* The rate of change in the table is constant.

7. 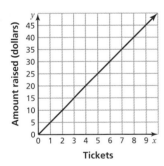   Each ticket costs $5.

9. **a.** the car; *Sample answer:* The equation for the car is $y = 25x$. Because 25 is greater than 18, the car gets better gas mileage.

   b. 56 miles

11. Consider the direct variation equation and that the graph passes through the origin.

13. a. yes; The equation is $d = 6t$, which represents a proportional relationship.

b. yes; The equation is $d = 50r$, which represents a proportional relationship.

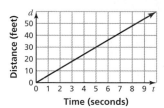

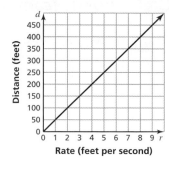

c. no; The equation is $t = \dfrac{300}{r}$, which does not represent a proportional relationship.

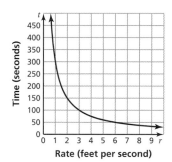

d. part c; It is called inverse variation because when the rate increases, the time decreases, and when the rate decreases, the time increases.

15.

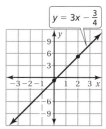

$y = 3x - \dfrac{3}{4}$

17. B

Section 13.4 — Graphing Linear Equations in Slope-Intercept Form *(pages 594 and 595)*

1. Find the x-coordinate of the point where the graph crosses the x-axis.

3. *Sample answer:* The amount of gasoline y (in gallons) left in your tank after you travel x miles is $y = -\dfrac{1}{20}x + 20$. The slope of $-\dfrac{1}{20}$ means the car uses 1 gallon of gas for every 20 miles driven. The y-intercept of 20 means there is originally 20 gallons of gas in the tank.

5. A; slope: $\dfrac{1}{3}$; y-intercept: -2

7. slope: 4; y-intercept: -5

9. slope: $-\dfrac{4}{5}$; y-intercept: -2

11. slope: $\dfrac{4}{3}$; y-intercept: -1

13. slope: -2; y-intercept: 3.5

15. slope: 1.5; y-intercept: 11

17. a.
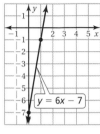

b. The x-intercept of 300 means the skydiver lands on the ground after 300 seconds. The slope of -10 means that the skydiver falls to the ground at a rate of 10 feet per second.

19.
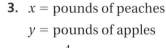

x-intercept: $\dfrac{7}{6}$

21.
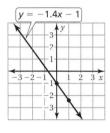

x-intercept: $-\dfrac{5}{7}$

23.
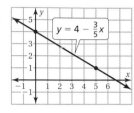

x-intercept: $\dfrac{20}{3}$

25. a. $y = 2x + 4$ and $y = 2x - 3$ are parallel because the slope of each line is 2;
$y = -3x - 2$ and $y = -3x + 5$ are parallel because the slope of each line is -3.

b. $y = 2x + 4$ and $y = -\dfrac{1}{2}x + 2$ are perpendicular because the product of their slopes is -1;

$y = 2x - 3$ and $y = -\dfrac{1}{2}x + 2$ are perpendicular because the product of their slopes is -1;

$y = -\dfrac{1}{3}x - 1$ and $y = 3x + 3$ are perpendicular because the product of their slopes is -1.

27. $y = 2x + 3$

29. $y = \dfrac{2}{3}x - 2$

31. B

Section 13.5

Graphing Linear Equations in Standard Form *(pages 600 and 601)*

1. no; The equation is in slope-intercept form.

3. $x =$ pounds of peaches
$y =$ pounds of apples
$y = -\dfrac{4}{3}x + 10$

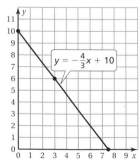

5. $y = -2x + 17$

7. $y = \dfrac{1}{2}x + 10$

9.
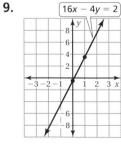

11. B

13. C

15. a.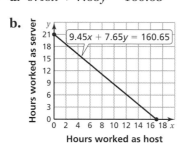

$-25x + y = 65$

b. $390

17.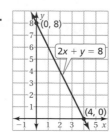

$2x + y = 8$

$(0, 8)$

$(4, 0)$

19. x-intercept: 9

y-intercept: 7

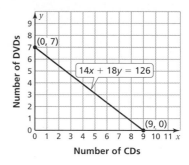

Number of DVDs

$(0, 7)$

$14x + 18y = 126$

$(9, 0)$

Number of CDs

21. a. $9.45x + 7.65y = 160.65$

b.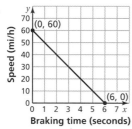

Hours worked as server

$9.45x + 7.65y = 160.65$

Hours worked as host

23. a. $y = 40x + 70$

b. x-intercept: $-\dfrac{7}{4}$; no;

You cannot have a negative time.

c.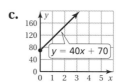

$y = 40x + 70$

25. $\dfrac{1}{2}$

Section 13.6

Writing Equations in Slope-Intercept Form
(pages 606 and 607)

1. *Sample answer:* Find the ratio of the rise to the run between the intercepts.

3. $y = 3x + 2$; $y = 3x - 10$; $y = 5$; $y = -1$

5. $y = x + 4$

7. $y = \dfrac{1}{4}x + 1$

9. $y = \dfrac{1}{3}x - 3$

11. The x-intercept was used instead of the y-intercept. $y = \dfrac{1}{2}x - 2$

13. $y = 5$

15. $y = -2$

17. a–b.

Speed (mi/h)

$(0, 60)$

$(6, 0)$

Braking time (seconds)

$(0, 60)$ represents the speed of the automobile before braking. $(6, 0)$ represents the amount of time it takes to stop. The line represents the speed y of the automobile after x seconds of braking.

c. $y = -10x + 60$

19. Be sure to check that your rate of growth will not lead to a 0-year-old tree with a negative height.

Hint

21 and 23.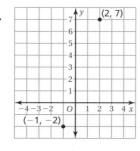

$(2, 7)$

$(-1, -2)$

Writing Equations in Point-Slope Form
(pages 612 and 613)

1. $m = -2;\ (-1, 3)$

3. $y - 0 = \frac{1}{2}(x + 2)$

5. $y + 1 = -3(x - 3)$

7. $y - 8 = \frac{3}{4}(x - 4)$

9. $y + 5 = -\frac{1}{7}(x - 7)$

11. $y + 4 = -2(x + 1)$

13. $y = 2x$

15. $y = \frac{1}{4}x$

17. $y = x + 1$

19. a. $V = -4000x + 30{,}000$

    b. $30,000

21. The rate of change is 0.25 degree per chirp.

23. a. $y = 14x - 108.5$

    b. 4 meters

25.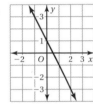

27. D

Finding Square Roots
(pages 630 and 631)

1. no; There is no integer whose square is 26.

3. $\sqrt{256}$ represents the positive square root because there is not a – or a ± in front.

5. $s = 1.3$ km

7. 3 and -3

9. 2 and -2

11. 25

13. $\frac{1}{31}$ and $-\frac{1}{31}$

15. 2.2 and -2.2

17. -19

19. The positive and negative square roots should have been given.

$$\pm\sqrt{\frac{1}{4}} = \frac{1}{2} \text{ and } -\frac{1}{2}$$

21. -116

23. 9

25. 25

27. 40

29. because a negative radius does not make sense

31. $=$

33. 9 ft

35. 8 m/sec

37. 2.5 ft

39. $y = 3x - 2$

41. $y = \frac{3}{5}x + 1$

Finding Cube Roots
(pages 636 and 637)

1. no; There is no integer that equals 25 when cubed.

3. 50 in.

5. 0.4 m

7. -5

9. 12

11. $\dfrac{7}{4}$

13. $3\dfrac{5}{8}$

15. $\dfrac{7}{12}$

17. 74

19. -276

21. 30 cm

23. $>$

25. $<$

27. $-1, 0, 1$

29. The side length of the square base is 18 inches and the height of the pyramid is 9 inches.

31. $x = 3$

33. $x = 4$

35. 289

37. 49

The Pythagorean Theorem
(pages 642 and 643)

1. The hypotenuse is the longest side and the legs are the other two sides.

3. 29 km

5. 9 in.

7. 24 cm

9. The length of the hypotenuse was substituted for the wrong variable.

$$a^2 + b^2 = c^2$$
$$7^2 + b^2 = 25^2$$
$$49 + b^2 = 625$$
$$b^2 = 576$$
$$b = 24$$

11. 16 cm

13. Use a right triangle to find the distance.

15. *Sample answer:* length $= 20$ ft, width $= 48$ ft, height $= 10$ ft;
$BC = 52$ ft, $AB = \sqrt{2804}$ ft

17. a. *Sample answer:*

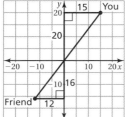

b. 45 ft

19. 6 and -6

21. 13

23. C

Approximating Square Roots
(pages 651–653)

1. A rational number can be written as the ratio of two integers. An irrational number cannot be written as the ratio of two integers.

3. all rational and irrational numbers; *Sample answer:* $-2, \dfrac{1}{8}, \sqrt{7}$

5. yes   **7.** no   **9.** whole, integer, rational

11. irrational   **13.** rational   **15.** irrational

17. 144 is a perfect square. So, $\sqrt{144}$ is rational.

19. a. If the last digit is 0, it is a whole number. Otherwise, it is a natural number.

    b. irrational number   **c.** irrational number

21. a. 26   **23. a.** -10 **25. a.** -13

    b. 26.2   **b.** -10.2 **b.** -12.9

27. $\sqrt{15}$; $\sqrt{15}$ is positive and -3.5 is negative.   **29.** $\dfrac{2}{3}$; $\dfrac{2}{3}$ is to the right of $\sqrt{\dfrac{16}{81}}$.

31. $-\sqrt{182}$; $-\sqrt{182}$ is to the right of $-\sqrt{192}$.

33. true   **35.** 8.1 ft   **37.** 8.5 ft   **39.** 20.6 in.

41. Create a table of integers whose cubes are close to the radicand. Determine which two integers the cube root is between. Then create another table of numbers between those two integers whose cubes are close to the radicand. Determine which cube is closest to the radicand; 2.4

43. *Sample answer:* $a = 82, b = 97$ **45.** 1.1   **47.** 30.1 m/sec

49. Falling objects do not fall at a linear rate. Their speed increases with each second they are falling.

Neat

51. 40 m

53. 9 cm

Repeating Decimals
(pages 654 and 655)

1. $\dfrac{1}{9}$ **3.** $-1\dfrac{2}{9}$

5. Because the solution does not change when adding/subtracting two equivalent equations; Multiply by 10 so that when you subtract the original equation, the repeating part is removed.

7. $-\dfrac{13}{30}$ **9.** $\dfrac{3}{11}$

11. Pattern: Digits that repeat are in the numerator and 99 is in the denominator; Use 9 as the integer part, 4 as the numerator, and 99 as the denominator of the fractional part.

Using the Pythagorean Theorem
(pages 660 and 661)

1. the Pythagorean Theorem and the distance formula

3. If a^2 is odd, then a is an odd number; true when a is an integer; A product of two integers is odd only when each integer is odd.

5. yes   7. no   9. yes   11. $\sqrt{52}$ 13. $\sqrt{29}$ 15. $\sqrt{85}$

17. The squared quantities under the radical should be added not subtracted; $\sqrt{136}$

19. yes   21. yes

23. no; The measures of the side lengths are $\sqrt{5000}$, $\sqrt{3700}$, and $\sqrt{8500}$ and $\left(\sqrt{5000}\right)^2 + \left(\sqrt{3700}\right)^2 \neq \left(\sqrt{8500}\right)^2$.

25. Notice that the picture is not drawn to scale. Use right triangles.

27. mean: 13; median: 12.5; mode: 12

29. mean: 58; median: 59; mode: 59

Section 15.1

Volumes of Cylinders
(pages 676 and 677)

1. How much does it take to cover the cylinder?; $170\pi \approx 534.1$ cm^2; $300\pi \approx 942.5$ cm^3

3. $486\pi \approx 1526.8$ ft^3 5. $245\pi \approx 769.7$ ft^3 7. $90\pi \approx 282.7$ mm^3

9. $252\pi \approx 791.7$ in.^{3} 11. $256\pi \approx 804.2$ cm^3 13. $\dfrac{125}{8\pi} \approx 5$ ft

15. $\sqrt{\dfrac{150{,}000}{19\pi}} \approx 50$ cm

17. Divide the volume of one round bale by the volume of one square bale.

19. $8325 - 729\pi \approx 6035$ m^3 21. yes

23. no

Section 15.2

Volumes of Cones
(pages 682 and 683)

1. The height of a cone is the perpendicular distance from the base to the vertex.

3. Divide by 3.   5. $9\pi \approx 28.3$ m^3

7. $\dfrac{2\pi}{3} \approx 2.1$ ft^3 9. $\dfrac{147\pi}{4} \approx 115.5$ yd^3 11. $\dfrac{125\pi}{6} \approx 65.4$ in.^{3}

13. The diameter was used instead of the radius; $V = \dfrac{1}{3}(\pi)(1)^2(3) = \pi$ m^3

15. 1.5 ft   17. $2\sqrt{\dfrac{10.8}{4.2\pi}} \approx 1.8$ in.   19. 24.1 min

21. $3y$ 23. $A'(-1, 1)$, $B'(-3, 4)$, $C'(-1, 4)$

25. D

1. A hemisphere is one-half of a sphere.

3. $\dfrac{500\pi}{3} \approx 523.6$ in.^{3}

5. $972\pi \approx 3053.6$ mm^3

7. $36\pi \approx 113.1$ cm^3

9. 9 mm

11. 4.5 ft

13. 2.5 in.

15. $256\pi + 128\pi = 384\pi \approx 1206.4$ ft^3

17. $r = \dfrac{3}{4}h$

19. 5400 in.^{2}; 27,000 in.^{3}

21. enlargement; 2

23. A

Section 15.4

Surface Areas and Volumes of Similar Solids
(pages 697–699)

1. Similar solids are solids of the same type that have proportional corresponding linear measures.

3. a. $\dfrac{9}{4}$; because $\left(\dfrac{3}{2}\right)^2 = \dfrac{9}{4}$

   b. $\dfrac{27}{8}$; because $\left(\dfrac{3}{2}\right)^3 = \dfrac{27}{8}$

5. no

7. no

9. $b = 18$ m; $c = 19.5$ m; $h = 9$ m

11. 1012.5 in.^{2}

13. 13,564.8 ft^3

15. 673.75 cm^2

17. a. 9483 pounds; The ratio of the height of the original statue to the height of the small statue is 8.4 : 1. So, the ratio of the weights, or volumes is $\left(\dfrac{8.4}{1}\right)^3$.

   b. 221,184 lb

19. a. yes; Because all circles are similar, the slant height and the circumference of the base of the cones are proportional.

   b. no; because the ratio of the volumes of similar solids is equal to the cube of the ratio of their corresponding linear measures

21.

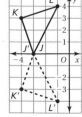

$J'(-3, 0), K'(-4, -3), L'(-1, -4)$

Exponents
(pages 714 and 715)

1. -3^4 is the negative of 3^4, so the base is 3, the exponent is 4, and its value is -81. $(-3)^4$ has a base of -3, an exponent of 4, and a value of 81.

3. 3^4

5. $\left(-\dfrac{1}{2}\right)^3$

7. $\pi^3 x^4$

9. $(6.4)^4 b^3$

11. 25

13. 1

15. $\dfrac{1}{144}$

17. The negative sign is not part of the base; $-6^2 = -(6 \cdot 6) = -36$.

19. $-\left(\dfrac{1}{4}\right)^4$

21. 29

23. 5

25. 66

27.
h	1	2	3	4	5
$2^h - 1$	1	3	7	15	31
2^{h-1}	1	2	4	8	16

$2^h - 1$; The option $2^h - 1$ pays you more money when $h > 1$.

Hint

29. Remember to add the black keys when finding how many notes you travel.

31. Associative Property of Multiplication

33. B

Product of Powers Property
(pages 720 and 721)

1. when multiplying powers with the same base

3. 3^4

5. $(-4)^{12}$

7. h^7

9. $\left(-\dfrac{5}{7}\right)^{17}$

11. 5^{12}

13. 3.8^{12}

15. The bases should not be multiplied. $5^2 \cdot 5^9 = 5^{2+9} = 5^{11}$

17. $216g^3$

19. $\dfrac{1}{25}k^2$

21. $r^{12} t^{12}$

23. no; $3^2 + 3^3 = 9 + 27 = 36$ and $3^5 = 243$

25. 496

27. 78,125

29. **a.** $16\pi \approx 50.27$ in.^{3}

   b. $192\pi \approx 603.19$ in.^{3} Squaring each of the dimensions causes the volume to be 12 times larger.

Hint

31. Use the Commutative and Associative Properties of Multiplication to group the powers.

33. 4

35. 3

37. B

Quotient of Powers Property
(pages 726 and 727)

1. To divide powers means to divide out the common factors of the numerator and denominator. To divide powers with the same base, write the power with the common base and an exponent found by subtracting the exponent in the denominator from the exponent in the numerator.

3. 6^6

5. $(-3)^3$

7. 5^6

9. $(-17)^3$

11. $(-6.4)^2$

13. b^{13}

15. You should subtract the exponents instead of dividing them. $\dfrac{6^{15}}{6^5} = 6^{15-5} = 6^{10}$

17. 2^9

19. π^8

21. k^{14}

23. $64x$

25. $125a^3b^2$

27. x^7y^6

29. You are checking to see if there is a linear relationship between memory and price, not if the change in price is constant for consecutive sizes of MP3 players.

31. 10^{13} galaxies

33. -9

35. 61

37. B

Zero and Negative Exponents
(pages 732 and 733)

1. no; Any nonzero base raised to a zero exponent is always 1.

3. $5^{-5}, 5^0, 5^4$

5. 1

7. 1

9. $\dfrac{1}{36}$

11. $\dfrac{1}{16}$

13. $5\dfrac{1}{4}$

15. $\dfrac{1}{125}$

17. The negative sign goes with the exponent, not the base. $(4)^{-3} = \dfrac{1}{4^3} = \dfrac{1}{64}$

19. $2^0; 10^0$

21. $\dfrac{a^7}{64}$

23. $5b$

25. 12

27. $\dfrac{w^6}{9}$

29. 100 mm

31. 1,000,000 nanometers

33. **a.** 10^{-9} m   **b.** equal to

35. Write the power as 1 divided by the power and use a negative exponent. Justifications will vary.

37. 10^9

39. 10^4

Reading Scientific Notation
(pages 740 and 741)

1. Scientific notation uses a factor greater than or equal to 1 but less than 10 multiplied by a power of 10. A number in standard form is written out with all the zeros and place values included.

3. 5,600,000,000,000

5. 87,300,000,000,000,000

7. yes; The factor is greater than or equal to 1 and less than 10. The power of 10 has an integer exponent.

9. no; The factor is greater than 10.

11. yes; The factor is greater than or equal to 1 and less than 10. The power of 10 has an integer exponent.

13. no; The factor is less than 1.

15. 70,000,000

17. 500

19. 0.000044

21. 1,660,000,000

23. 9,725,000

25. **a.** 810,000,000 platelets
    b. 1,350,000,000,000 platelets

27. **a.** Bellatrix
    b. Betelgeuse

29. 1555.2 km^2

31. $35,000,000 \text{ km}^3$

33. 4^5

35. $(-2)^3$

Writing Scientific Notation
(pages 746 and 747)

1. If the number is greater than or equal to 10, the exponent will be positive. If the number is less than 1 and greater than 0, the exponent will be negative.

3. 2.1×10^{-3}

5. 3.21×10^8

7. 4×10^{-5}

9. 4.56×10^{10}

11. 8.4×10^5

13. 72.5 is not less than 10. The decimal point needs to move one more place to the left.
    7.25×10^7

15. $6.09 \times 10^{-5}, 6.78 \times 10^{-5}, 6.8 \times 10^{-5}$

17. $4.8 \times 10^{-8}, 4.8 \times 10^{-6}, 4.8 \times 10^{-5}$

19. $6.88 \times 10^{-23}, 5.78 \times 10^{23}, 5.82 \times 10^{23}$

21. $4.01 \times 10^7 \text{ m}$

23. $680, 6.8 \times 10^3, \dfrac{68,500}{10}$

25. $6.25 \times 10^{-3}, 6.3\%, 0.625, 6\dfrac{1}{4}$

27. 1.99×10^9 watts

29. carat; Because 1 carat $= 1.2 \times 10^{23}$ atomic mass units and 1 milligram $= 6.02 \times 10^{20}$ atomic mass units, and $1.2 \times 10^{23} > 6.02 \times 10^{20}$.

31. natural, whole, integer, rational

33. irrational

Section 16.7 — Operations in Scientific Notation
(pages 752 and 753)

1. Use the Distributive Property to group the factors together. Then subtract the factors and write it with the power of 10. The number may need to be rewritten so that it is still in scientific notation.

3. 8.34×10^7

5. 4.947×10^{11}

7. 5.8×10^5

9. 5.2×10^8

11. 7.555×10^7

13. 1.037×10^7

15. You have to rewrite the numbers so they have the same power of 10 before adding; 3.03×10^9

17. 2.9×10^{-3}

19. 1.5×10^0

21. 2.88×10^{-7}

23. 1.12×10^{-2}

25. 4.006×10^9

27. 1.962×10^8 cm

29. First find the total length of the ridges and valleys.

31. 3×10^8 m/sec

33. $\dfrac{1}{8}$

35. C

Topic 1 — Solving Multi-Step Equations
(pages 766 and 767)

1. $2 + 3x = 17; x = 5$

3. $k = 45; 45°, 45°, 90°$

5. $b = 90; 90°, 135°, 90°, 90°, 135°$

7. $c = 0.5$

9. $h = -9$

11. $x = -\dfrac{2}{9}$

13. 20 watches

15. $4(b + 3) = 24; 3$ in.

17. $\dfrac{2580 + 2920 + x}{3} = 3000; 3500$ people

19. $<$

21. $>$

Topic 2 — Solving Equations with Variables on Both Sides
(pages 771–773)

1. no; When 3 is substituted for x, the left side simplifies to 4 and the right side simplifies to 3.

3. $x = 13.2$ in.

5. $x = 7.5$ in.

7. $k = -0.75$

9. $p = -48$

11. $n = -3.5$

13. $x = -4$

15. The 4 should have been added to the right side.

$$3x - 4 = 2x + 1$$
$$3x - 2x - 4 = 2x + 1 - 2x$$
$$x - 4 = 1$$
$$x - 4 + 4 = 1 + 4$$
$$x = 5$$

17. $15 + 0.5m = 25 + 0.25m; 40$ mi

19. $x = \dfrac{1}{3}$

21. no solution

23. infinitely many solutions

25. $x = 2$

27. no solution

29. infinitely many solutions

31. *Sample answer:* $8x + 2 = 8x$; The number $8x$ cannot be equal to 2 more than itself.

33. It's never the same. Your neighbor's total cost will always be $75 more than your total cost.

35. no; $2x + 5.2$ can never equal $2x + 6.2$.

37. 7.5 units

39. Remember that the box is with priority mail and the envelope is with express mail.

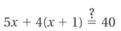

41. 10 mL

43. **a.** 40 ft

b. no;

$$2(\text{white area}) = \text{black area}$$
$$2[5(6x)] = 4[6(x + 1)]$$
$$60x = 24x + 24$$
$$36x = 24$$
$$x = \frac{2}{3}$$

$$5x + 4(x + 1) \overset{?}{=} 40$$
$$\text{Length of hallway is } 5\left(\frac{2}{3}\right) + 4\left(\frac{2}{3} + 1\right) \overset{?}{=} 40$$
$$10 \neq 40$$

45. $15\frac{3}{4} \text{ cm}^3$

47. C

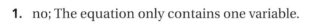

Topic 3

Rewriting Equations and Formulas
(pages 776 and 777)

1. no; The equation only contains one variable.

3. **a.** $A = \frac{1}{2}bh$ **b.** $b = \frac{2A}{h}$ **c.** $b = 12$ mm

5. $y = 4 - \frac{1}{3}x$

7. $y = \frac{2}{3} - \frac{4}{9}x$

9. $y = 3x - 1.5$

11. The y should have a negative sign in front of it.
$$2x - y = 5$$
$$-y = -2x + 5$$
$$y = 2x - 5$$

13. **a.** $t = \frac{I}{Pr}$

b. $t = 3$ yr

15. $m = \frac{e}{c^2}$

17. $h = \frac{2A}{b_1 + b_2}$

19. $w = 6g - 40$

21. **a.** $F = 32 + \frac{9}{5}(K - 273.15)$

b. 32°F

c. liquid nitrogen

23. $r^3 = \frac{3V}{4\pi}; r = 4.5$ in.

25. $-5\frac{1}{3}$

27. $1\frac{1}{4}$

Key Vocabulary Index

Mathematical terms are best understood when you see them used and defined *in context*. This index lists where you will find key vocabulary. A full glossary is available in your Record and Practice Journal and at *BigIdeasMath.com*.

Student Index

This student-friendly index will help you find vocabulary, key ideas, and concepts. It is easily accessible and designed to be a reference for you whether you are looking for a definition, real-life application, or help with avoiding common errors.

A

Absolute value, 2–7
  defined, 4
  error analysis, 6
  real-life application, 5
Addition
  of expressions
    linear, 86–91
    modeling, 86–87
  of integers, 8–15
    with different signs, 8–11
    error analysis, 12
    with the same sign, 8–11
  Property
    of Equality, 98
    of Inequality, 132
    real-life application, 99
  of rational numbers, 50–55
    error analysis, 54
    real-life application, 53
    writing, 54, 55, 65
  to solve inequalities, 130–135
Addition Property of Equality, 98
  real-life application, 99
Addition Property of Inequality, 132
Additive inverse
  defined, 10
  writing, 12
Additive Inverse Property, 10
Adjacent angles
  constructions, 270–275
  defined, 272
Algebra
  equations
    equivalent, 98
    graphing linear, 566–567
    literal, 774
    modeling, 96–97, 102, 108–109
    multi-step, 764–767
    rewriting, 774–777
    solving, 96–113
    two-step, 108–113
    with variables on both sides, 768–773
    writing, 100, 106, 112

expressions
  linear, 86–91
  modeling, 86–87, 91
  simplifying, 80–85
  writing, 84, 90
formulas, *See* Formulas
linear equations
  graphing, 566–567
  slope of a line, 572–581
  slope intercept form, 590–607
  standard form, 596–601
properties, *See* Properties
Algebra tiles
  equations, 96–97, 102, 108–109
  expressions, 86–87
Angle(s)
  adjacent
    constructions, 270–275
    defined, 272
  alternate exterior, 530
  alternate interior, 530
  classifying, 277–278
    triangles by, 284
  complementary
    constructions, 276–281
    defined, 278
  congruent
    defined, 272
    reading, 284
  corresponding, 528–529
    defined, 468
    error analysis, 531
  exterior
    defined, 529
    error analysis, 539
  interior, defined, 529
  measures
    of a quadrilateral, 295
    of a triangle, 288–289
  naming, 272
    error analysis, 274
  of polygons, 542–549
    defined, 536
    error analysis, 547, 548
    reading, 544
    real-life application, 545
    similar, 550–555

of rotation, 486
sums
  for a quadrilateral, 295
  for a triangle, 288–289
supplementary
  constructions, 276–281
  defined, 278
of triangles, 534–539
  exterior, 536
  interior, 536
  real-life application, 537
  similar, 552
vertical
  constructions, 270–275
  defined, 272
Angle of rotation, defined, 486
Area, *See also* Surface area
  of a circle, 332–337
    formula, 334
  of a composite figure, 338–343
Area of similar figures, 500–505
  formula, 502
  writing, 504

B

Base, defined, 712
Biased sample(s), defined, 442

C

Center, defined, 318
Center of dilation, defined, 508
Center of rotation, defined, 486
Choose Tools, *Throughout. For example, see:*
  angles, 271
  circles and circumference, 321
  graphing linear equations, 567
  indirect measurement, 551
  probability, 406, 426
  scientific notation, 747
  slope, 578
Circle(s)
  area of, 332–337
    formula, 334
  center of, 318
  circumference and, 316–323
    defined, 319

Student Index

Student Index

real-life application, 31
  with the same sign, 28–30
  writing, 32
multiplying, 22–27
  with different signs, 22–24
  error analysis, 26
  modeling, 27
  real-life application, 25
  with the same sign, 22–24
  writing, 26
subtracting, 14–19
  error analysis, 18
  real-life application, 17
  writing, 18

Interior angle(s)
alternate, 530
defined, 529, 536
of triangles, 534–539
  real-life application, 537

Interior angles of a polygon,
defined, 536

Interest
defined, 254
principal, 254
simple, 252–257
  defined, 254
  error analysis, 256
writing, 256

Irrational number(s), defined, 648

K

Kite, defined, 294

L

Lateral surface area, defined, 358
Leg(s), defined, 640
Like terms
combining to solve equations, 764
defined, 82
error analysis, 84
writing, 84
Line(s)
  graphing
    horizontal, 568
    vertical, 568
  parallel, 526–533
    defined, 528
    error analysis, 531
    project, 532
    slope of, 580
    symbol, 528

perpendicular
  defined, 528
  slope of, 581
  symbol, 528
of reflection, 56
slope of, 572–581
transversals, 526–533
x-intercept of, 592
y-intercept of, 592

Line of reflection, defined, 480
Linear equation(s), *See also*
    Equations, Proportional
    relationships
defined, 568
graphing, 566–571
  error analysis, 570
  horizontal lines, 568
  real-life applications, 569, 599
  in slope-intercept form, 590–595
  in standard form, 596–601
  vertical lines, 568
point-slope form
  defined, 610
  real-life application, 611
  writing, 612
  writing in, 608–613
slope of a line, 572–581
  defined, 572, 574
  error analysis, 578
  formula, 572, 574
  reading, 574
slope-intercept form
  defined, 592
  error analysis, 594, 606
  real-life applications, 593, 605
  writing in, 593–607
  x-intercept, 592
  y-intercept, 592
solution of, 568
standard form, 596–601
  defined, 598
  error analysis, 600
  modeling, 601
  real-life application, 599
  writing, 600

Linear expression(s)
adding, 86–91
  modeling, 86–87, 91
  writing, 90
factoring, 92–93
subtracting, 86–91
  error analysis, 91

modeling, 87
real-life application, 89
writing, 90
Literal equation(s), defined, 774
Logic, *Throughout. For example,*
    see:
absolute value, 3
angles, 281
  interior, 534
  measures, 532
circles, 323
circumference, 323
constructing a triangle, 287
cube roots, 637
inequalities
  solving, 145
  writing, 129
linear equations
  graphing, 566, 601
  in slope-intercept form, 591
linear expression, 91
percent equations, 237
prisms, 381
probability
  dependent events, 434
  experimental, 417
  independent events, 434
  theoretical, 417
rewriting equations, 777
samples, 445
transformations
  similar figures, 499

M

Markup(s), 246–251
defined, 248
writing, 250
Meaning of a Word
adjacent, 270
dilate, 506
opposite, 10
proportional, 170
rate, 162
rational, 44
reflection, 478
rotate, 484
translate, 472
transverse, 526
Mental Math, *Throughout. For*
    example, see:
integers
  adding, 13
  subtracting, 19

error analysis, 642
modeling, 638
project, 643
real-life applications, 638, 659
using, 656–661
distance formula, 658
error analysis, 660
writing, 660

Q

Quadrilateral(s), 292–297
classifying, 294
constructions, 292–297
reading, 294
defined, 292
kite, 294
parallelogram, 294, 341
rectangle, 294, 341
rhombus, 294
square, 294
sum of angle measures, 295
trapezoid, 294
Quotient of Powers Property,
722–727
defined, 724
error analysis, 726
real-life application, 725
writing, 726

R

Radical sign, defined, 628
Radicand, defined, 628
Radius, defined, 318
Rate(s)
defined, 164
ratios and, 162–169
research, 169
unit rate
defined, 164
writing, 167
Ratio(s), *See also* Proportions,
Rates
complex fraction, 165
defined, 164
proportions and, 170–189
cross products, 173
Cross Products Property, 173
error analysis, 182, 190
proportional, 172
real-life application, 189
solving, 186–191
writing, 178–183

rates and, 162–169
complex fractions, 165
writing, 167
scale and, 300
similar figures
areas of, 502
perimeters of, 502
slope, 192–197
error analysis, 196
Rational number(s), *See also*
Fractions, Decimals
adding, 50–55
error analysis, 54
real-life application, 53
writing, 54, 55, 65
defined, 44, 46, 648
dividing, 64–69
error analysis, 68
real-life application, 67
writing, 65, 68
multiplying, 64–69
error analysis, 68
writing, 65, 68
ordering, 47
repeating decimals
defined, 46
error analysis, 48
writing, 46
subtracting, 58–63
error analysis, 62
real-life application, 61
writing, 62
terminating decimals
defined, 46
writing, 46
writing as decimals, 46
Reading
congruent angles, 284
congruent sides, 284
experiments, 402
images, 474
inequalities
solving, 133
symbols of, 126
outcomes, 402
polygons, 544
proportional relationship, 172
quadrilaterals, 294
slope, 574
symbol
congruent, 466
prime, 474
similar, 496

Real number(s), 648–653
classifying, 648
defined, 648
error analysis, 651
Real-Life Applications, *Throughout.*
For example, see:
absolute value, 5
angles of triangles, 537
cube roots, 635
decimals, 217, 222, 223
direct variation, 201
distance formula, 659
equations
multi-step, 765
rewriting, 775
with variables on both sides,
770
exponents
evaluating expressions, 713
negative, 731
Quotient of Powers Property,
725
expressions
linear, 89
simplifying, 83
fractions, 217, 222, 223
integers
dividing, 31
multiplying, 25
subtracting, 17
interior angles of a polygon, 545
linear equations
graphing, 569
in point-slope form, 611
in slope-intercept form, 593,
605
in standard form, 599
writing, 605
percent, 217, 222, 223
equations, 235
proportions, 229
proportions, 189, 229
Pythagorean Theorem, 638, 659
rational numbers
adding, 53
dividing, 67
subtracting, 61
scientific notation
operations in, 751
reading numbers in, 739
writing numbers in, 745
similar figures, 497
solving equations, 99, 105, 111

A76 Student Index

Photo Credits

Chapter 13

564 ©iStockphoto.com/Alistair Cotton; **569** NASA; **570** ©iStockphoto.com/David Morgan; **571** *top right* NASA; *center left* ©iStockphoto.com/jsemeniuk; **578** ©iStockphoto.com/Amanda Rohde; **579** Julian Rovagnati/Shutterstock.com; **583** RyFlip/Shutterstock.com; **586** Luke Wein/Shutterstock.com; **589** AVAVA/Shutterstock.com; **594** ©iStockphoto.com/Dreamframer; **595** *top right* Jerry Horbert/Shutterstock.com; *center left* ©iStockphoto.com/Chris Schmidt; **597** ©iStockphoto.com/biffspandex; **600** ©iStockphoto.com/Stephen Pothier; **601** *top left* Gina Smith/Shutterstock.com; *center left* Dewayne Flowers/Shutterstock.com; **605** Herrenknecht AG; **606** ©iStockphoto.com/Adam Mattel; **607** *top left* ©iStockphoto.com/Gene Chutka; *center right* ©iStockphoto.com/marcellus2070, ©iStockphoto.com/beetle8; **611** ©iStockphoto.com/Connie Maher; **612** ©iStockphoto.com/Jacom Stephens; **613** *top right* ©iStockphoto.com/Petr Podzemny; *bottom left* ©iStockphoto.com/adrian beesley; **614** Richard Goldberg/Shutterstock.com; **620** Thomas M Perkins/Shutterstock.com

Chapter 14

624 ©iStockphoto/Michael Flippo, ©iStockphoto.com/Ann Marie Kurtz; **629** Perfectblue97; **630** ©iStockphoto.com/Benjamin Lazare; **631** *top right* ©iStockphoto.com/iShootPhotos, LLC; *center left* ©iStockphoto.com/Jill Chen, Oleksiy Mark/Shutterstock.com; **636** Gary Whitton/Shutterstock.com; **637** Michael Stokes/Shutterstock.com; **638** ©Oxford Science Archive/Heritage Images/Imagestate; **642** ©iStockphoto.com/Melissa Carroll; **645** *center left* ©iStockphoto.com/Yvan Dubé; *bottom right* Snvv/Shutterstock.com; **646** ©iStockphoto.com/Kais Tolmats; **650** *top left* ©iStockphoto.com/Don Bayley; *center left* ©iStockphoto.com/iLexx; **653** ©iStockphoto.com/Marcio Silva; **657** Monkey Business Images/Shutterstock.com; **665** LoopAll/Shutterstock.com; **666** CD Lanzen/Shutterstock.com

Chapter 15

670 ©iStockphoto.com/ALEAIMAGE, ©iStockphoto.com/Ann Marie Kurtz; **672** ©iStockphoto.com/Jill Chen; **678** ©iStockphoto.com/camilla wisbauer; **677** *Exercises 13 and 14* ©iStockphoto.com/Prill Mediendesigns & Fotografie; *Exercise 15* ©iStockphoto.com/subjug; *center left* ©iStockphoto.com/Matthew Dixon; *center right* ©iStockphoto.com/nilgun bostanci; **683** ©iStockphoto.com/Stefano Tiraboschi; **689** Donald Joski/Shutterstock.com; **690** ©iStockphoto.com/Yury Kosourov; **691** Carlos Caetano/Shutterstock.com; **698** Courtesy of Green Light Collectibles; **699** *top right* ©iStockphoto.com/wrangel; *center left* ©iStockphoto.com/ivanastar; *bottom left* ©iStockphoto.com/Daniel Cardiff; **700** Eric Isselée/Shutterstock.com; **704** ©iStockphoto.com/Daniel Loiselle

Chapter 16

708 Varina and Jay Patel/Shutterstock.com, ©iStockphoto.com/Ann Marie Kurtz; **710** ©iStockphoto.com/Franck Boston; **711** *Activity 3a* ©iStockphoto.com/Manfred Konrad; *Activity 3b* NASA/JPL-Caltech/R.Hurt (SSC); *Activity 3c and d* NASA; *bottom right* Stevyn Colgan; **713** ©iStockphoto.com/Philippa Banks; **714** ©iStockphoto.com/clotilde hulin; **715** ©iStockphoto.com/Boris Yankov; **720** ©iStockphoto.com/VIKTORIIA KULISH; **721** *top right* ©iStockphoto.com/Paul Tessier; *center left* ©iStockphoto.com/subjug, ©iStockphoto.com/Valerie Loiseleux, ©iStockphoto.com/Linda Steward; **726** ©iStockphoto.com/Petrovich9; **727** *top right* Dash/Shutterstock.com; *center left* NASA/JPL-Caltech/L.Cieza (UT Austin); **731** ©iStockphoto.com/Aliaksandr Autayeu; **732** EugeneF/Shutterstock.com; **733** ©iStockphoto.com/Nancy Louie; **735** ©iStockphoto.com/Dan Moore; **736** ©iStockphoto.com/Kais Tolmats; **737** *Activity 3a and d* Tom C Amon/Shutterstock.com; *Activity 3b* Olga Gabay/Shutterstock.com; *Activity 3c* NASA/MODIS Rapid Response/Jeff Schmaltz; *Activity 3f* HuHu/Shutterstock.com; *Activity 4a* PILart/Shutterstock.com; *Activity 4b* Matthew Cole/Shutterstock.com; *Activity 4c* Yanas/Shutterstock.com; *Activity 4e* unkreativ/Shutterstock.com; **739** *top left* ©iStockphoto.com/Mark Stay; *top center* ©iStockphoto.com/Frank Wright; *top right* ©iStockphoto.com/Evgeniy Ivanov; *bottom left* ©iStockphoto.com/Oliver Sun Kim; **740** ©iStockphoto.com/Christian Jasiuk; **741** Microgen/Shutterstock.com; **742** *Activity 1a* ©iStockphoto.com/Susan Trigg; *Activity 1b* ©iStockphoto.com/subjug; *Activity 1c* ©iStockphoto.com/camilla wisbauer; *Activity 1d* ©iStockphoto.com/Joe Belanger; *Activity 1e* ©iStockphoto.com/thumb; *Activity 1f* ©iStockphoto.com/David Freund; **743** NASA; **744** *center* Google and YouTube logos are registered trademarks of Google Inc., used with permission.; **745** *top left* Elaine Barker/Shutterstock.com; *center right* ©iStockphoto.com/breckeni; **746** *bottom left* ©iStockphoto.com/Max Delson Martins Santos; *bottom right* ©iStockphoto.com/Jan Rysavy; **747** *top right* BORTEL Pavel/Shutterstock.com; *center right* ©iStockphoto.com/breckeni; **751** *center left* Sebastian Kaulitzki/Shutterstock.com; *center right* ©iStockphoto.com/Jan Rysavy; **753** ©iStockphoto.com/Boris Yankov; **754** mmutlu/Shutterstock.com; **758** *bottom right* ©iStockphoto.com/Eric Holsinger; *bottom left* TranceDrumer/Shutterstock.com

Additional Topics

763 ©iStockphoto.com/ALEAIMAGE, ©iStockphoto.com/Ann Marie Kurtz; **764** ©iStockphoto.com/Harley McCabe; **765** ©iStockphoto.com/Jacom Stephens; **766** ©iStockphoto.com/Harry Hu; **767** ©iStockphoto.com/Ralf Hettler, Vibrant Image Studio/Shutterstock.com; **771** ©iStockphoto.com/Andrey Krasnov; **772** Shawn Hempel/Shutterstock.com; **777** *top right* ©iStockphoto.com/Alan Crawford; *center left* ©iStockphoto.com/Julio Yeste; *bottom right* ©iStockphoto.com/Mark Stay

Appendix A

A0 *background* ©iStockphoto.com/Björn Kindler; *top right* ©iStockphoto.com/MichaelMattner; *top left* AKaiser/Shutterstock.com; **A1** *top right* ©iStockphoto.com/daver2002ua; *bottom left* ©iStockphoto.com/Eric Isselée; *bottom right* ©iStockphoto.com/Victor Paez; **A2** sgame/Shutterstock.com; **A3** *top right* AKaiser/Shutterstock.com; *bottom left* Photononstop/SuperStock, ©iStockphoto.com/Adam Radosavljevic; *bottom right* ©iStockphoto.com/MichaelMattner; **A4** ©iStockphoto.com/Duncan Walker; **A5** *top right* ©iStockphoto.com/DNY59; *bottom left* ©iStockphoto.com/Boris Katsman; *bottom right* © User:MatthiasKabel/Wikimedia Commons / CC-BY-SA-3.0 / GFDL; **A6** *top right* ©iStockphoto.com/Keith Webber Jr.; *bottom left* ©iStockphoto.com/Eliza Snow; *bottom right* ©iStockphoto.com/David H. Seymour; **A7** *top right* ©iStockphoto.com/daver2002ua; *bottom left* ©iStockphoto.com/Victor Paez; **A8** *wasp* ©iStockphoto.com/arlindo71; *bat* ©iStockphoto.com/Alexei Zaycev; *monkey and kangaroo* ©iStockphoto.com/Eric Isselée; **A9** *macaw* ©iStockphoto.com/Eric Isselée; *frog* ©iStockphoto.com/Brandon Alms; *spider* ©iStockphoto.com/arlindo71; *puma* ©iStockphoto.com/Eric Isselée; *lobster* ©iStockphoto.com/Joan Vicent Cantó Roig; *crocodile* ©iStockphoto.com/Mehmet Salih Guler

Cartoon illustrations Tyler Stout

Common Core State Standards

Kindergarten

Counting and Cardinality	– Count to 100 by Ones and Tens; Compare Numbers
Operations and Algebraic Thinking	– Understand and Model Addition and Subtraction
Number and Operations in Base Ten	– Work with Numbers 11–19 to Gain Foundations for Place Value
Measurement and Data	– Describe and Compare Measurable Attributes; Classify Objects into Categories
Geometry	– Identify and Describe Shapes

Grade 1

Operations and Algebraic Thinking	– Represent and Solve Addition and Subtraction Problems
Number and Operations in Base Ten	– Understand Place Value for Two-Digit Numbers; Use Place Value and Properties to Add and Subtract
Measurement and Data	– Measure Lengths Indirectly; Write and Tell Time; Represent and Interpret Data
Geometry	– Draw Shapes; Partition Circles and Rectangles into Two and Four Equal Shares

Grade 2

Operations and Algebraic Thinking	– Solve One- and Two-Step Problems Involving Addition and Subtraction; Build a Foundation for Multiplication
Number and Operations in Base Ten	– Understand Place Value for Three-Digit Numbers; Use Place Value and Properties to Add and Subtract
Measurement and Data	– Measure and Estimate Lengths in Standard Units; Work with Time and Money
Geometry	– Draw and Identify Shapes; Partition Circles and Rectangles into Two, Three, and Four Equal Shares

Grade 3

Operations and Algebraic Thinking	– Represent and Solve Problems Involving Multiplication and Division; Solve Two-Step Problems Involving Four Operations
Number and Operations in Base Ten	– Round Whole Numbers; Add, Subtract, and Multiply Multi-Digit Whole Numbers
Number and Operations—Fractions	– Understand Fractions as Numbers
Measurement and Data	– Solve Time, Liquid Volume, and Mass Problems; Understand Perimeter and Area
Geometry	– Reason with Shapes and Their Attributes

Grade 4

Operations and Algebraic Thinking	– Use the Four Operations with Whole Numbers to Solve Problems; Understand Factors and Multiples
Number and Operations in Base Ten	– Generalize Place Value Understanding; Perform Multi-Digit Arithmetic
Number and Operations—Fractions	– Build Fractions from Unit Fractions; Understand Decimal Notation for Fractions
Measurement and Data	– Convert Measurements; Understand and Measure Angles
Geometry	– Draw and Identify Lines and Angles; Classify Shapes

Grade 5

Operations and Algebraic Thinking	– Write and Interpret Numerical Expressions
Number and Operations in Base Ten	– Perform Operations with Multi-Digit Numbers and Decimals to Hundredths
Number and Operations—Fractions	– Add, Subtract, Multiply, and Divide Fractions
Measurement and Data	– Convert Measurements within a Measurement System; Understand Volume
Geometry	– Graph Points in the First Quadrant of the Coordinate Plane; Classify Two-Dimensional Figures

Mathematics Reference Sheet

Conversions

U.S. Customary
1 foot = 12 inches
1 yard = 3 feet
1 mile = 5280 feet
1 acre $\approx$ 43,560 square feet
1 cup = 8 fluid ounces
1 pint = 2 cups
1 quart = 2 pints
1 gallon = 4 quarts
1 gallon = 231 cubic inches
1 pound = 16 ounces
1 ton = 2000 pounds
1 cubic foot $\approx$ 7.5 gallons

U.S. Customary to Metric
1 inch = 2.54 centimeters
1 foot $\approx$ 0.3 meter
1 mile $\approx$ 1.61 kilometers
1 quart $\approx$ 0.95 liter
1 gallon $\approx$ 3.79 liters
1 cup $\approx$ 237 milliliters
1 pound $\approx$ 0.45 kilogram
1 ounce $\approx$ 28.3 grams
1 gallon $\approx$ 3785 cubic centimeters

Time
1 minute = 60 seconds
1 hour = 60 minutes
1 hour = 3600 seconds
1 year = 52 weeks

Temperature
$$C = \frac{5}{9}(F - 32)$$

$$F = \frac{9}{5}C + 32$$

Metric
1 centimeter = 10 millimeters
1 meter = 100 centimeters
1 kilometer = 1000 meters
1 liter = 1000 milliliters
1 kiloliter = 1000 liters
1 milliliter = 1 cubic centimeter
1 liter = 1000 cubic centimeters
1 cubic millimeter = 0.001 milliliter
1 gram = 1000 milligrams
1 kilogram = 1000 grams

Metric to U.S. Customary
1 centimeter $\approx$ 0.39 inch
1 meter $\approx$ 3.28 feet
1 kilometer $\approx$ 0.62 mile
1 liter $\approx$ 1.06 quarts
1 liter $\approx$ 0.26 gallon
1 kilogram $\approx$ 2.2 pounds
1 gram $\approx$ 0.035 ounce
1 cubic meter $\approx$ 264 gallons

Number Properties

Commutative Properties of Addition and Multiplication
$$a + b = b + a$$
$$a \cdot b = b \cdot a$$

Associative Properties of Addition and Multiplication
$$(a + b) + c = a + (b + c)$$
$$(a \cdot b) \cdot c = a \cdot (b \cdot c)$$

Addition Property of Zero
$$a + 0 = a$$

Multiplication Properties of Zero and One
$$a \cdot 0 = 0$$
$$a \cdot 1 = a$$

Distributive Property:
$$a(b + c) = ab + ac$$
$$a(b - c) = ab - ac$$

Properties of Equality

Addition Property of Equality
If $a = b$, then $a + c = b + c$.

Subtraction Property of Equality
If $a = b$, then $a - c = b - c$.

Multiplication Property of Equality
If $a = b$, then $a \cdot c = b \cdot c$.

Multiplicative Inverse Property
$$n \cdot \frac{1}{n} = \frac{1}{n} \cdot n = 1, n \neq 0$$

Division Property of Equality
If $a = b$, then $a \div c = b \div c, c \neq 0$.

Squaring both sides of an equation
If $a = b$, then $a^2 = b^2$.

Cubing both sides of an equation
If $a = b$, then $a^3 = b^3$.

Properties of Inequality

Addition Property of Inequality
If $a > b$, then $a + c > b + c$.

Subtraction Property of Inequality
If $a > b$, then $a - c > b - c$.

Multiplication Property of Inequality
If $a > b$ and c is positive, then $a \cdot c > b \cdot c$.
If $a > b$ and c is negative, then $a \cdot c < b \cdot c$.

Division Property of Inequality
If $a > b$ and c is positive, then $a \div c > b \div c$.
If $a > b$ and c is negative, then $a \div c < b \div c$.

Properties of Exponents

Product of Powers Property: $a^m \cdot a^n = a^{m+n}$

Quotient of Powers Property: $\dfrac{a^m}{a^n} = a^{m-n}, a \neq 0$

Power of a Power Property: $(a^m)^n = a^{mn}$

Power of a Product Property: $(ab)^m = a^m b^m$

Zero Exponents: $a^0 = 1, a \neq 0$

Negative Exponents: $a^{-n} = \dfrac{1}{a^n}, a \neq 0$

Slope

$m = \dfrac{\text{rise}}{\text{run}}$

$= \dfrac{\text{change in } y}{\text{change in } x}$

$= \dfrac{y_2 - y_1}{x_2 - x_1}$

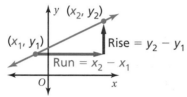

Equations of Lines

Slope-intercept form
$y = mx + b$

Standard form
$ax + by = c, a, b \neq 0$

Point-slope form
$y - y_1 = m(x - x_1)$

Angles of Polygons

Interior Angle Measures of a Triangle

$x + y + z = 180$

Interior Angle Measures of a Polygon

The sum S of the interior angle measures of a polygon with n sides is $S = (n - 2) \cdot 180°$.

Exterior Angle Measures of a Polygon

$w + x + y + z = 360$

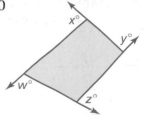

Formulas in Geometry

Prism

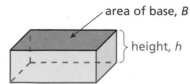

$S = $ areas of bases
$\quad$ + areas of
$\quad$ lateral faces
$V = Bh$

Pyramid

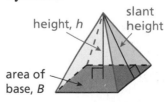

$S = $ area of base
$\quad$ + areas of
$\quad$ lateral faces
$V = \dfrac{1}{3}Bh$

Circle

$C = \pi d$ or $C = 2\pi r$
$A = \pi r^2$

Cylinder

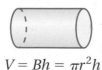

$V = Bh = \pi r^2 h$

Cone

$V = \dfrac{1}{3}Bh = \dfrac{1}{3}\pi r^2 h$

Sphere

$V = \dfrac{4}{3}\pi r^3$